This unique volume provides a comprehensive and up-to-date examination of all aspects of the biology of the Old World monkey genus, *Theropithecus*, which evolved alongside our human ancestors. The authors explore the fossil history and evolution of the genus, its biogeography, comparative evolutionary biology and anatomy, and the behaviour and socioecology of the living and extinct representatives of the genus. This highly multidisciplinary approach provides a rare and insightful account of the evolutionary biology of this fascinating and once highly successful group of primates.

Theropithecus will be of interest to researchers in the fields of primatology, anthropology, palaeontology, and mammalian behaviour, physiology, and anatomy.

T0211561

Theropithecus

Theropithecus

The Rise and Fall of a Primate Genus

Edited by

NINA G. JABLONSKI

CAMBRIDGE
UNIVERSITY PRESS

CAMBRIDGE UNIVERSITY PRESS
Cambridge, New York, Melbourne, Madrid, Cape Town, Singapore, São Paulo

Cambridge University Press
The Edinburgh Building, Cambridge CB2 2RU, UK

Published in the United States of America by Cambridge University Press, New York

www.cambridge.org
Information on this title: www.cambridge.org/9780521411059

First published 1993
This digitally printed first paperback version 2005

A catalogue record for this publication is available from the British Library

Library of Congress Cataloguing in Publication data

Theropithecus : the rise and fall of a primate genus / edited by Nina
G. Jablonski.
p. cm.
ISBN 0 521 41105 X (hc)
1. Theropithecus — Congresses. 2. Theropithecus — Evolution —
Congresses. I. Jablonski, Nina G.
OL737.P93T53 1993
599.8′2–dc20 92–10889 CIP

ISBN-13 978-0-521-41105-9 hardback
ISBN-10 0-521-41105-X hardback

ISBN-13 978-0-521-01849-4 paperback
ISBN-10 0-521-01849-8 paperback

To my mother, the memory of my father, and Erika

Contents

Contributors

A. DAVID BEYNON Department of Oral Biology, University of Newcastle-Upon-Tyne, Newcastle-Upon-Tyne NE2 4BW, England, UK.

DAVID DEAN The Graduate School, City University of New York, New York, New York, USA.

ERIC DELSON Department of Vertebrate Paleontology, American Museum of Natural History, New York, New York 10024 USA.

ROBIN I. M. DUNBAR Department of Anthropology, University College London, Gower Street, London WS1E 6BT, England, UK.

GERALD G. ECK Department of Anthropology, DH–05, University of Washington, Seattle, Washington 98195 USA.

ROBERT A. FOLEY Department of Biological Anthropology, University of Cambridge, Downing Street, Cambridge CB2 3DZ, England, UK.

ROBERT HOFFSTETTER Institut de Paléontologie, Muséum national d'Histoire Naturelle, 8, rue Buffon, 75005 Paris, France.

TOSHITAKA IWAMOTO Department of Biology, Miyazaki University, Miyazaki-shi, 889–21, Japan.

NINA G. JABLONSKI Department of Anatomy and Human Biology, and Centre for Human Biology, University of Western Australia, Nedlands, Western Australia, 6009 Australia.

HARTMUT B. KRENTZ Department of Anthropology, University of Victoria, P.O. Box 1700, Victoria, BC, V8S 4L8, Canada.

MEAVE G. LEAKEY Department of Palaeontology, National Museums of Kenya, P.O. Box 40658, Nairobi, Kenya.

PHYLLIS C. LEE Department of Biological Anthropology, University of Cambridge, Downing Street, Cambridge CB2 3DZ, England, UK.

ROBERT D. MARTIN Anthropologisches Institut und Museum, Universität Zürich-Irchel, Winterthurerstrasse 190, CH-8057, Zürich, Switzerland.

MARTIN PICKFORD Institut de Paléontologie, Muséum national d'Histoire Naturelle, 8, rue Buffon, 75005 Paris, France.

DARIS R. SWINDLER 1212 8th Avenue North, Edmonds, Washington 98020, USA.

MARK F. TEAFORD Department of Cell Biology and Anatomy, The Johns Hopkins University School of Medicine, 725 North Wolfe Street, Baltimore, Maryland 21205, USA.

Most of the participants in the symposium, 'Theropithecus as a case-study in primate evolutionary biology', 9–12 April 1990, King's College, University of Cambridge. From left to right, Eric Delson, Cliff Jolly, Meave Leakey, Marta Lahr, Rob Foley, Harriet Eeley, Prosper Ndessokia, Chris Dean, Bob Martin, Daris Swindler, Chris Wood, Gerry Eck, Dave Beynon, Hartmut Krentz, Robin Dunbar, Toshitaka Iwamoto, Bernard Wood, Nina Jablonski, and Phyllis Lee.

Preface

This book is the result of a serendipitous meeting between Rob Foley and myself in the palaeontology collections of the National Museums of Kenya in July, 1987. Rob was introduced to me one afternoon while I was measuring skulls of *Theropithecus oswaldi*. He too was visiting to look at fossil *Theropithecus* and we soon struck up a conversation about our mutual fossil friends. Over the course of that afternoon and the next few days we discussed many aspects of the evolution and ecology of fossil and living *Theropithecus*. We quickly realized that the last twenty-five years had witnessed a dramatic increase in our knowledge of biology of the genus, ranging from behavioural and ecological information on the extant species, *Theropithecus gelada*, to information drawn from molecular biology, karyology, and palaeontology bearing on the evolutionary relationships of the genus. What could we piece together of the history and palaeoecology of the genus from the relatively large samples of fossil materials of the various extinct species available for study? What could this body of information tell us about the evolution of *Theropithecus* itself, and what general lessons about speciation, rates of evolution, extinction, and other phenomena might this information hold for all students of primate and human evolution? Realizing that many of the people instrumental in bringing this evidence to light were still very much alive and taking an active interest in *Theropithecus* biology, we decided that a meeting of these individuals would be timely and appropriate. Thus, the idea for the symposium '*Theropithecus* as a case-study in primate evolutionary biology' was born.

Over the course of the next two years, the details of the symposium were organized and funding was sought from the Wenner-Gren Foundation for Anthropological Research. Thanks to a generous grant from Wenner-Gren, 15 scientists from eight countries were invited to attend and contribute papers for discussion. The symposium was convened at King's College, University of Cambridge, on 9 April 1990. A total of 23 participants, including 15 contributors of papers and 8 discussants, took part in the three full days of discussions. On 12 April, the enthused, if exhausted, symposiasts took their leave of Cambridge and preparation of this volume began.

This volume reflects fairly well our current knowledge of the evolutionary biology of *Theropithecus*, including the controversies and problems that remain unresolved. Most of the chapters were based on papers initially discussed at the

Cambridge symposium, but a few papers and contributors were added after the symposium to make for as comprehensive a volume as possible. All of the chapters underwent thorough and rigorous peer review, and most underwent extensive revision in the light of discussions at the symposium and the comments of outside reviewers.

There are gaps in our knowledge still, but I hope that the contents of this volume will entice old and new students to use their insight and imagination to help fill them.

Nina Jablonski,
Perth, Australia

Acknowledgements

The symposium, 'Theropithecus as a case-study in primate evolutionary biology' could not have taken place without the generous financial support of the Wenner-Gren Foundation for Anthropological Research. For their interest in the problems of primate evolution and for their money, I am very grateful.

The meeting itself would not have proceeded smoothly, if at all, without the skills, patience, and plain hard work of Rob Foley, the co-organizer of the symposium. Rob designed and coordinated arrangements for the symposium at King's and, from pâté to projectors, did a masterful job. Most importantly, Rob was a personal inspiration to me, as he had a way of prying me out from between molar cusps to talk about problems of animals, environments and evolution. He is to be thanked also for providing the title of this volume.

At the symposium, Harriet Eeley and Marta Lahr provided able and enthusiastic technical assistance, and helped to make sure that participants did not get lost in the maze of corridors at King's.

Discussions at the symposium benefited greatly from the many original fossil specimens and high quality casts of fossils that were available for participants to view, compare, photograph, and, in some cases, measure. For providing specimens from the British Museum (Natural History), including several original fossils from Kanjera, I am extremely grateful to Peter Andrews. I thank Meave Leakey for bringing specimens from the National Museums of Kenya, including exciting new specimens from West Turkana. I am also grateful to Prosper Ndessokia of the Department of Antiquities of Tanzania for bringing with him the original fossil cranium of a well-preserved female *Theropithecus oswaldi* from Peninj. Eric Delson is hereby thanked for bringing with him several original fossils from South Africa and for arranging for the only known specimen of *Theropithecus delsoni* to be brought from India to the USA and finally to Cambridge for the symposium.

I am very grateful also to the following individuals not associated with the symposium who helped to review various chapters: Marianne Bouvier, Tony Collins, Len Freedman, Bill Kimbel, Lawrence Martin, Eric Meikle, Brian Shea, and Andrew Whiten.

The Centre for Human Biology of The University of Western Australia is warmly thanked for making a donation to defray the cost of publishing this book.

I was first introduced to the world of *Theropithecus* by Gerry Eck, and much of my work on the genus has been inspired by the questions he has asked over the years. I am grateful to him for his steady guidance, healthy scepticism, and friendship. Finally, I wish to thank my husband, George Chaplin, for his constant support, good humour and inspiration.

1 Introduction

NINA G. JABLONSKI

Many would argue that the only non-hominid species that stir the hearts of palaeoanthropologists are those that have the potential to shed light on one or more aspects of human evolution. Species of *Theropithecus* enjoy this distinction and in some of the chapters of this book the parallels between hominid and theropith evolution are explored in detail. Perhaps more importantly, though, in this book we celebrate *Theropithecus* as an unsung hero of primate evolution in its own right. The study of the evolutionary biology of *Theropithecus* species has forced us to confront a series of difficult and controversial questions ranging from the nature of species and speciation events through to the nature of events precipitating extinction.

This book is divided into four major sections and two appendices. The first two sections deal with specific and general aspects of the evolutionary history of *Theropithecus*, the third with aspects of the anatomy of the genus, and the fourth with the behaviour and ecology of extinct and extant *Theropithecus* species. The first appendix contains a partial catalogue of fossil specimens of *Theropithecus* and the second appendix comprises a report on the conservation of the sole extant species of the genus, *T. gelada*.

With two exceptions, all the papers that were originally presented at the 1990 Cambridge, UK, conference, 'Theropithecus as a Case-Study in Primate Evolutionary Biology' have been developed into chapters for this book. One of the two exceptions was Cliff Jolly's contribution in which the biology of the once species-rich *Theropithecus* was examined through the lens of the biology of the now species-rich *Papio* and the entirety of the Tribe Papionini. Jolly's discussion set the tone for much of the subsequent forum on theropith evolution because it stressed the importance of examining papionin evolutionary history in the light of modern papionin population biology, mostly as revealed by field studies of the genus *Papio*. Before continuing with an outline of the contents of this book, it is worth dwelling for a moment on Jolly's observations.

Papio, Theropithecus and the biology of the Papionini

'How many species of *Theropithecus* existed at any given time in the past?'

1

'What were their signal attributes?' 'How long did they survive?' 'When and why did lineage splitting occur?' Have these questions remained unanswered because of inadequacies of the fossil record, or our abilities to analyse it, or have they persisted because they are simply the wrong questions? Jolly's remarks at Cambridge and in a subsequent conference presentation (Jolly, 1991) suggested that the latter was a more pressing problem than the former. It seems that in the study of *Theropithecus* and *Papio*, we have been obsessed with the issue of species definitions, to the detriment of our understanding of the animals' biology.

Drawing upon his observations of *Papio* populations, Jolly pointed out that the Papionini are unusual among mammals in being able to hybridize for prolonged periods of time while maintaining the integrity of distinct core populations that are visually distinct from one another. Wherever there can be a hybrid zone in nature, there is. Given these conditions, how many species can be said to exist? The potential number of species of *Papio* that can be defined depends entirely on the class of criteria one decides to apply. Using the classic guide-line of the biological species concept, only one species, *P. hamadryas*, would be recognized, because modern *Papio* is composed of a 'string' of forms with intervening hybrid zones and the ability to hybridize in any combination in captivity: 'if they meet, they mate'. Using ecological criteria, we could define two species: a desert group, *P. hamadryas*, and a savanna group comprising all the rest. Applying strictly anatomical criteria might result in two species

(small *P. anubis* and all others) if one examined dental features; five are possible (*P. hamadryas*, *P. papio*, *P. anubis*, *P. cynocephalus*, and *P. ursinus*) if one considered pelage and facial colouration; six, seven or more 'species' are possible depending on the criteria chosen. By applying strict 'zygo-criteria' based on the position of naturally occurring hybrid zones, three species might be defined: *P. hamadryas*, *P. anubis*, and another comprising all other forms. The list goes on: 'what you choose to see is what you get'.

Having successfully made his point with respect to potential numbers of species, Jolly argued that *Papio* is best viewed as 'one large smear down central Africa, with side branches'. The interbreeding and species patterns of today can be traced to the past and specifically to the effects of refugia and the workings of the founder flush principle and cumulative lineage effect. He made the important point that in populations of *Papio*, genetic variation is partitioned unevenly, according to matrilines. Anatomical distinctions are not necessarily products of natural selection, but are just as likely to be the products of small population effects, having little or no adaptive significance. Jolly emphasized that the Papionini are particularly susceptible to this type of evolutionary change because of matriline fissioning.

For the study of extant and extinct papionins, Jolly asserted that the lessons are clear: recognize the 'tyranny of the biological species concept' and do not worry about species definitions, but about the *kinds* of animals living at various times:

energy presently expended in arguing about how many species there 'really are' would be better devoted to gathering and describing the (more) objective data pertaining to patterns of phenotypic variation, patterns of gene-flow, and the external influences that historically gave rise to these patterns.

(Jolly, 1991:100).

Applying this framework to our understanding of the evolution of *Theropithecus*, Jolly distinguished the widespread Plio-Pleistocene species, *T. oswaldi*, as the 'central species' of the genus, and this met with widespread agreement. Several distinct regional and temporal variants of *T. oswaldi* no doubt existed, in much the same way as variants of *Papio* do today. Recognition of *T. darti* as a distinct species ancestral to *T. oswaldi* was considered to be more a matter of professional taste than biological reality. *Theropithecus darti* and *T. oswaldi* were seen to fulfil the criteria of true chronospecies because they represented a temporal sequence of the same *kind* of animal. Apart from *T. darti*, Jolly recognized two other offshoots. One of these was *T. brumpti*, the extinct forest dweller of the Omo and Turkana Basins in East Africa, which was viewed as an 'oddity thrown up by the cumulative lineage effect'. The other was the only extant theropith, *T. gelada*, which he considered as the highland survivor of a small number of founders who had succeeded in breaking through the montane forest barrier by founder flush.

In a later presentation (Jolly, 1991), Jolly further developed his ideas on species recognition in the Papionini by suggesting that we explicitly recognize the qualitative differences between extinct and extant species. Palaeospecies, he contended, were *phenospecies*, because we could only recognize them by the most restricted of morphological criteria (generally the remains of bones and teeth only). Living species were *zygospecies* because their integrity could be assessed by natural reproductive criteria as well as by morphological characteristics.

Jolly's remarks provide a sound basis for development of a balanced biological perspective on the evolution of *Theropithecus* and it is hoped that the reader will keep his words in mind as they consider subsequent chapters and see the details of a rich biological story unfold.

The evolutionary history of *Theropithecus*

In the section entitled 'Fossil evidence and phylogeny', four chapters (2, 3, 5, and 6) present descriptions and critical discussions of previously undescribed or incompletely described collections of fossil *Theropithecus*. An important supplement to these chapters and others is the partial catalogue of fossil remains of *Theropithecus* from all sites presented in Appendix I.

While the authors in their respective chapters bring forth different opinions about the systematics and taxonomy of the genus, all agree that within *Theropithecus* three lineages can be recognized: one comprising *T. darti* and *T. oswaldi*, one comprising *T. brumpti* and its putative forebears, and one comprising the sole extant species *T. gelada*, which has no known fossil record.

Our knowledge of the early evolution of the *T. oswaldi* lineage is expanded significantly by Eck's description of the earliest known fossils of *T. darti* from the Hadar formation in Ethiopia in chapter 2. These fossils have been recovered from sediments that have a radiometrically determined age of 3.4–2.9 Ma. Of particular importance in this chapter are the descriptions of well-preserved specimens of male crania and mandibles, which complement the previously published description of a female skull from Makapansgat (Maier, 1972), South Africa. From his comparison of the Hadar and Makapansgat remains, Eck concludes that the *Theropithecus* fossils from the two sites shared sufficient features distinct from other later populations to warrant their placement together in *T. darti*. The larger and more derived morphology of the Makapansgat remains suggests to him, however, that these fossils were younger in age than the Hadar specimens. Critical to our understanding of the early evolution of the *T. oswaldi* lineage is Eck's finding that the environment at Hadar during the span of *T. darti* was relatively dry and open woodland and grassland, very similar to the environments inhabited later by *T. oswaldi*.

In the Omo and Turkana Basins to the south of Hadar, the majority of *Theropithecus* history, from 3.5 Ma onward, is recorded more or less continuously. In chapter 3, Leakey describes the evolution of both the *T. oswaldi* and *T. brumpti* lineages in the Turkana Basin in a detailed account which complements the previously published reports of Eck & Jablonski (1987) and Eck (1987) on *Theropithecus* from the Lower Omo Basin.

Leakey describes new, early specimens from West Turkana belonging to the *T. brumpti* lineage and also takes the dramatic step of sinking into a single species all the known representatives of the *T. brumpti* lineage, on account of the continuous record of morphological change in the early Turkana Basin fossils. Here she departs from both Delson (chapter 4) and Eck & Jablonski (1984) in sinking *T. baringensis* and *T. quadratirostris* into *T. brumpti*. Specimens previously assigned to *T. baringensis* are accommodated into the more primitive of Leakey's two new subspecies, *T. brumpti baringensis*, while material formerly referred to *T. quadratirostris* of 'intermediate evolutionary status' is placed in *T. brumpti* subspecies indeterminate. *Theropithecus brumpti brumpti* is then reserved for the more highly derived forms from later horizons. Leakey develops a similar arrangement for the specimens of *Theropithecus* from the Turkana Basin attributable to the *T. oswaldi* lineage. Noting the morphological continuum in body size and tooth dimensions from early to later forms, she recognizes a single species, *T. oswaldi*, with three subspecies, *T. oswaldi darti*, *T. oswaldi oswaldi* and *T. oswaldi leakeyi* that replace each other chronologically. Leakey contrasts the habitat preferences and feeding adaptations of *T. brumpti* and *T. oswaldi* and concludes that the former species preferred more forested environments and probably enjoyed a diet of mixed vegetation whereas the latter was associated with more open habitats and was almost certainly a habitual consumer of grasses.

The taxonomic status of controversial

fossils attributed to the *T. brumpti* lineage is the subject of Delson and Dean's contribution (chapter 4). In this account, the authors review the evidence for and against placement of *Papio baringensis* and *P. quadratirostris* into *Theropithecus*, an arrangement originally advocated by Eck & Jablonski (1984). The key point made by Delson and Dean is that *P. baringensis* and *P. quadratirostris* share with *Papio* but not with *Theropithecus* a configuration of hafting of the facial skeleton on the neurocranium known as klinorhynchy. They conclude that, on balance, *P. baringensis* probably is an early and primitive member of the *T. brumpti* lineage and should be named accordingly, but that *P. quadratirostris* is best accommodated as a species in *Papio (Dinopithecus)*. This interpretation of the placement of the *quadratirostris* species differs from that of Leakey (chapter 3) and Eck & Jablonski (1984), and it seems unlikely that a consensus on the status of this species will be achieved until a larger collection of fossils can be unearthed and examined.

Delson's interpretation of the taxonomy of the genus is summarized in chapter 5, which is also devoted to important descriptions of previously undescribed or incompletely described fossils of *T. oswaldi* from Africa and India. Of signal importance is Delson's account of the first (and only) *Theropithecus* specimen described from outside Africa. This specimen was originally described as *T. delsoni* Gupta & Sahni (1981) and Delson demonstrates that its provenance from the Mirzapur area of India is definite and unambiguous. The specimen clearly represents a large, derived theropith and Delson tentatively places it in *T. oswaldi delsoni* on the basis of its close similarities to later derived African *T. oswaldi*. Delson's taxonomic scheme for the genus differs from that of Leakey (chapter 3) and Eck & Jablonski (1984) in its recognition of the *T. brumpti* lineage as the sister to the *T. gelada* + *T. oswaldi* sublineages. This is reflected by his description of the new subgenus *T. (Omopithecus)* for the *T. brumpti* lineage. With respect to African *T. oswaldi*, Delson's scheme departs from Leakey's only in its preservation of *T. darti* as a species within the *T. oswaldi* lineage.

The descriptions of new *Theropithecus* fossils are completed in chapter 6, where Delson and Hoffstetter describe for the first time the remains of the genus from Ternifine, Algeria. The Ternifine remains represent a later, derived theropith referred to *T. oswaldi leakeyi* that may date to about 700 Ka. The authors make the important point in their discussion that although populations of *T. oswaldi* increased in size through the Pleistocene, the rate of size increase varied significantly through time.

In the last chapter of the section, Jablonski (chapter 7) provides a summary of the phylogeny of *Theropithecus* based on a review of molecular, chromosomal, morphological, and palaeontological evidence. In addition to summarizing the various arguments bearing on the phyletic relationships within *Theropithecus*, she examines the evidence bearing on the position of *Theropithecus* within the Cercopithecinae. The position of *Theropithecus* among the Old World monkeys has been a subject of debate largely because the large number of anatomical distinctions of the genus were traditionally interpreted as reflecting the genus' long, independent evolutionary history. Opinions were divided as to whether *Theropithecus* was mostly closely related to *Papio*, *Macaca*, or to some other form. The weight of neontological and palaeontological evidence now, however, supports the view that among the Papionini, *Theropithecus* most recently shared a common ancestor with *Papio*, between about 7 and 3.5 Ma.

Despite a lack of consensus on the most appropriate taxonomic scheme for the genus, there is agreement among all students that the three lineages of *Theropithecus* – however one wishes formally to designate them – were established early in the history of the genus. The *T. oswaldi* lineage was by far the most expansive in time and space, enjoying a stable adaptation to dry, open, lowland environments. The great evolutionary success of this lineage appears to have been due in large measure to the animals' successful exploitation of a terrestrial, grass-eating niche. The *T. brumpti* and *T. gelada* lineages, in contrast to the *T. oswaldi* lineage, were far more restricted in time and space. This is particularly true of *T. gelada*, which today enjoys a narrow distribution in the cold, dry highlands of central Ethiopia. Although the gelada is often singled out as a mere relict of a once widespread and successful genus, its adaptation as a relatively small-bodied, montane grass-eater has helped to shed light on the factors leading to extinction of its larger lowland congeners (discussed more fully below and in chapters 18 and

19). *Theropithecus brumpti* and its fore-bears present the most extensive suite of departures from the basic *Theropithecus* plan. *Theropithecus brumpti* is the only known species of its genus to have been a forest dweller, and aspects of its peculiar and unusual feeding and locomotor adaptations are the subject of extensive discussions in chapters 11, 12, and 14.

The reasons underlying the evolutionary patterns seen in *Theropithecus* through time are the subjects of chapters 8 and 9. In both essays, climatic factors are strongly implicated as major forces shaping the course of theropith evolution. In chapter 8, Pickford develops the hypothesis that changes in the distribution of extinct theropith species were the direct responses to local and global climatic changes. He ascribes some of the changes in species distribution to changes in climate in eastern Africa brought about by the rise of the 'Roof of Africa' during the Plio-Pleistocene. Others appear to have been related to global-scale climatic changes, which resulted in latitudinal shifts in the boundaries between the Ethiopian and Palaearctic biogeographic realms. The radiations of *Papio*, *Homo*, and several herbivores – in particular bovids and suids – are also considered by Pickford to have contributed to the shrinkage of *Theropithecus* territory.

Foley (chapter 9) takes a different tack by comparing the radiations of the several 'African Terrestrial Primates' (ATP), namely the genera of Hominidae and the various taxa of baboons including *Theropithecus*, in order to draw inferences about evolutionary process and pattern in Plio-Pleistocene Africa. In his wide-ranging essay, Foley explores questions related to speciation and diversity, taxonomic longevity, competition and coevolution, the effect of climate, and the direct interactions between the hominids and the baboons. Foley makes the important point that, with respect to the biogeography of the ATP, East Africa is clearly the focus for most evolutionary novelty and diversity, while other areas of Africa show some characteristics of refugia. He points out that baboons and hominids show different kinds of successional patterns of evolution, with genera replacing each other during the Pleistocene with greater (hominids) and lesser (baboons) amounts of overlap between genera. In contrast to Pickford's biogeographical interpretation, Foley sees a relationship between the *frequency* of climatic oscillations and species diversity of the ATP in the Plio- Pleistocene. Among the many topics requiring further refinement and exploration in the future will be the nature of the relationship between climatic changes and species diversity.

Anatomy and adaptation in *Theropithecus*

The anatomical peculiarities of *Theropithecus*, and in particular of the theropith skull, have been discussed at length by many students. The discussions contained in Part III 'Anatomy of the fossil and living species of *Theropithecus*' clearly reflect the maturity and sophistication of modern functional anatomical techniques and the growing integration of anatomical and ecological studies. Nonetheless, the palaeontologist's tradi-

tional preoccupation with skulls and in particular with jaws and teeth is reflected in the proportion of anatomical essays in this section (four out of five) devoted to descriptive and functional accounts of the skull and masticatory apparatus in living and fossil *Theropithecus*.

Martin (chapter 10) critically examines several of the generalizations about canine size, sexual dimorphism and brain size that have become part of the anatomical litany of *Theropithecus*. Through a series of scaling (allometric) analyses, Martin demonstrates that, contrary to certain statements in the literature, there is no evidence of reduction in overall canine size in extant *Theropithecus* and that further studies are required to determine whether there has been actual reduction in relative size of the canines in fossil *Theropithecus*, as has been claimed by many students. In other scaling analyses, Martin shows that extant *Theropithecus* exhibits a somewhat greater degree of sexual dimorphism than expected from female body size and that large-bodied fossil forms probably showed even more pronounced sexual dimorphism in body size. The often cited claim of relatively small brain size in extant *Theropithecus* is supported by Martin's analyses of the scaling of the brain in Old World monkeys. His results show that the relative brain size of extant *Theropithecus* is intermediate between that of colobines and other cercopithecines. Noting the relationship between specific dietary habits such as leaf eating, relatively small brain size and low basal metabolic rates, Martin takes this opportunity to develop further his hypothesis that low maternal metabolic levels constrain foetal brain development and hence limit completed adult brain size. He also ventures on the basis of preliminary results that relative brain sizes in Theropithecus oswaldi were at least as small as in modern geladas.

Comprehensive treatments of the highly specialized masticatory apparatuses of theropith species are the subject of chapters 11 and 12. These chapters in many respects pick up where Leakey left off in her exploration of the dietary and niche differences between T. brumpti and T. oswaldi in the Turkana Basin. In chapter 11, Jablonski employs simple biomechanical and scaling analyses to trace the evolution of the masticatory apparatus in the three lineages of Theropithecus. She shows that members of the T. brumpti and T. oswaldi lineages have specialized in divergent directions. In T. brumpti and its forebears, evolution of a longer muzzle was correlated with the anterior and lateral expansion of the origin of the masseter muscle. The masticatory apparatus of T. brumpti, like that of species of Papio and Mandrillus and other cercopithecines, has been designed to meet the requirements of a large gape. In Papio and Mandrillus, however, a wide gape serves mostly to facilitate male canine displays and there is no evidence that a wide gape is associated with food processing in these animals. In contrast, the masticatory apparatus of T. brumpti was engineered to meet the requirements of a wide gape during canine displays and feeding. Jablonski suggests that the jaws of T. brumpti could generate high occlusal pressures between the molar teeth when the jaws were relatively widely separated, and that it is possible that the diet of T. brumpti included large or thick-husked items such as roots, tubers, or large fruits that were severed and processed between the molars. Theropithecus oswaldi, in contrast, appears to have gradually lost much of its ability to gape widely as it increased its efficiency as a grass chewing machine. The anatomy of the temporomandibular joint and the teeth of this species suggest that – especially in the later, largest-sized forms – wide gape displays were sacrificed to the demands of heavy molar chewing of large quantities of low quality, high fibre vegetation. This modification may have had interesting implications for predator defence and maintenance of social order in T. oswaldi groups.

Leakey's and Jablonski's inferences about dietary differences between T. oswaldi and T. brumpti are supported by the results of Teaford's analysis of dental microwear in the genus (chapter 12). Using T. gelada as the basis for comparison, Teaford demonstrates that the molar microwear patterns of T. oswaldi are more similar to those of the extant gelada than they are to those of T. brumpti. This would suggest that the diet of T. oswaldi was similar to that of T. gelada, consisting largely of grasses, an interpretation also supported by recent isotopic evidence (Lee-Thorp, van der Merwe & Brain, 1989). Theropithecus brumpti, on the other hand, shows a distinctly different pattern, which Teaford sees as reflecting a diet that may have included more soft fruit.

The story of Theropithecus teeth is completed by a comprehensive chapter on dental development by Swindler and

Beynon (chapter 13). Their investigation revealed that at birth the teeth of *T. gelada* were comparable in development to those of *Papio* and *Macaca*. The sequence of permanent tooth eruption in *T. gelada*, as well as in *T. oswaldi* and *T. brumpti*, was found to be similar to that of other cercopithecids. Examination of the pattern of enamel formation in the molars of *T. gelada* also revealed the interesting fact that the species appears to be unique in the absence of hypoplastic bands or accentuated striae. The absence of accentuated growth lines in the molars of *T. gelada* suggests that the animals grow up in a stable environment, lacking environmental or dietary stresses or illnesses during the period of dental development. This finding is supported by ample ecological evidence that indicates that the geladas live in a constant environment in which seasonal changes of diet are subtle and threats from predators and other external agents are few and far between.

The final chapter of this section (chapter 14) is devoted to a long-overdue, detailed examination of the functional anatomy of the postcranial skeleton in the species of *Theropithecus* by Krentz. His study is based on extensive examinations of all available fossil theropith postcrania that have been unambiguously given a specific assignment, and comparisons of these remains with the postcranial skeletons of geladas and other representative cercopithecids. His conclusions indicate a basic dichotomy in the genus between *T. brumpti* and a 'non-*brumpti* group' comprising *T. darti*, *T. oswaldi*, and *T. gelada*, with *T. brumpti* sharing many traits with arboreal quadrupeds such as *Colobus*. Krentz's enumeration of a series of postcranial features in *T. brumpti* that reflect an arboreal habitus provides unequivocal support for the hypothesis that the adaptation of the species as a forest-dwelling arborealist was unique for the genus. In general, however, Krentz characterizes the postcrania of *Theropithecus* by a suite of features that include adaptations toward terrestrial locomotion, possession of a hand adapted to 'manual grazing', specializations of the femur and hindlimb for long periods of upright sitting and pronounced sexual dimorphism. The fact that these features are present in the oldest known members of the genus suggests to him that the postural and locomotor adaptations that distinguish *Theropithecus* today were elements of the original adaptation of the genus.

Behaviour and ecology of *Theropithecus*

Theropithecus gelada was among the first non-human primate species to be studied in its natural environment. Since the landmark study by Crook & Aldrich-Blake (1968), a series of long-term ecological and behavioural studies of *T. gelada* have made significant contributions to our basic understanding of primate social organization and socioecology. The results of the major long-term studies of *T. gelada* in the wild are summarized in three comprehensive chapters by Dunbar (chapter 15) and Iwamoto (chapters 16 and 17).

Dunbar's review of the social organization of *T. gelada* describes the species' social system as multi-level, comprising at least three increasingly inclusive group-

ings: the coalition, the reproductive unit
and the band. The basic social group is
identified as the one-male reproductive
unit that consists of a single reproductive
male and up to 12 reproductive females
plus their dependent offspring. The cohesion of the reproductive units is maintained mostly by the relationships among
the reproductive females, with the males
being relatively peripheral. Females
remain in their natal units throughout
their lives, whereas males leave their natal
units as subadults to join an all-male
group. Some two to four years later males
return to reproductive units to acquire
their own females, and Dunbar identifies
two options that males can pursue toward
this end. In connection with the demography of gelada populations, Dunbar
notes that while geladas have low
reproductive rates, their very low
mortality rates make for population
growth rates that are among the highest
for any primate species. Such population
growth rates can apparently be successfully sustained because of the species'
highly efficient exploitation of a rich and
stable environment.

Iwamoto's extensive examination of the
ecology of *T. gelada* (chapter 16) begins
with a clear description of the species'
habitat in the highlands of central Ethiopia. In this environment geladas achieve
higher population and biomass densities
than other sympatric primates and ungulates because of their highly efficient food
processing techniques and anti-predator
strategies that include group living and the
use of cliffs as refuges. He documents fully
the graminivorous habits of the gelada and
makes the important point that geladas
spend more time feeding than any other
herbivorous primate. Iwamoto's discussion of digestive processes and energetics
in *T. gelada* (chapter 17) follows naturally
from his discussion of basic aspects of the
species' ecology. Several students have
speculated that the gelada's successful
exploitation of grasses hinges on their
capacity to break down cellulose in the
gut. Iwamoto tested the hypothesis of
hindgut fermentation in *T. gelada* by
examining the ability of captive animals to
digest crude fibre and by evaluating the
extent of bulk feeding by animals in the
wild during the dry season. The results of
his experiment on captive animals suggested that the geladas have a microbial
flora in the hindgut that decomposes fibre.
When the geladas are faced with lower
food quality in the wild during the dry
season, they significantly increased their
food intake by prolonging their feeding
time and by changing their staple food
from grass to the herb *Trifolium*. A natural
outgrowth of Iwamoto's work would be a
direct evaluation of Martin's hypothesis of
a relatively low basal metabolic rate in
geladas (chapter 10).

The reasons for the extinction of the
Plio-Pleistocene and Late Pleistocene
theropiths have been discussed at length
by many students. Up to now, the effects
of climatic change have been implicated as
being the most important factors to have
precipitated the extinction of the
T. brumpti and *T. oswaldi* lineages, but
the nature of these effects has never been
fully investigated. In the case of
T. oswaldi, hominid predation and competition from relatively larger-brained
Papio baboons have also been said to have

contributed to their demise. In the last two chapters of this book, the causes of the extinction of *T. oswaldi* lineage are explored in detail through two innovative modelling approaches.

In chapter 18, Dunbar makes use of a systems model of the socioecology of *T. oswaldi* to predict maximum group sizes for populations of extinct theropiths under different climatic conditions. He shows that the theropiths of the Plio-Pleistocene were not large enough to be able to exploit low quality vegetation at low altitudes and that they must have had an even more restricted distribution than the extant gelada. Dunbar indicates that later Pleistocene theropiths were even more narrowly adapted because they could have only survived in the places they did if the ambient temperatures were somewhat cooler than those of today or if the quality of the graze was three times higher than that eaten by geladas. This suggests that they were limited to the immediate vicinity of permanent water sources and thus highly vulnerable to extinction.

The precarious nature of the adaptation of giant Late Pleistocene *T. oswaldi* is explored from a different perspective by Lee and Foley (chapter 19). Modelling the energetic requirements of *T. oswaldi* along with correlated life history parameters, they demonstrate that the extinction of the giant theropiths resulted from limited nutritional intake, which led to a reduced reproductive rate. Lee and Foley suggest that the animals were time-limited with respect to exploitation of their resource base of low quality forage. Their inability to meet energetic require-

ments reduced reproductive rates to the point where populations became vulnerable to minor environmental perturbations. Both sets of authors emphasize that an understanding of extinction relies on a clear understanding of how climatic changes are mediated through behaviour. In this connection Dunbar makes the important point that, prior to their extinction, the giant theropiths simply could not meet both the basic social and nutritional requirements for survival. In particular, the need to spend increasing amounts of time in food harvesting reduced social cohesion in Late Pleistocene theropith groups by significantly curtailing grooming time. Thus it appears that climatic changes triggered a cascade of behavioural changes in the giant theropiths that made their populations more vulnerable to extinction by decreasing their rate of reproduction and their degree of social cohesiveness.

Having now more clearly defined the factors likely to have contributed to the extinction of theropith species, we can better assess the prospects for survival of the extant species, *T. gelada*. Fortunately, the conservation status of the gelada now seems very good, and in the final section of this book (Appendix II), Dunbar reviews this topic and presents an up-to-date map of the species' distribution.

References

CROOK, J.H. & ALDRICH-BLAKE, P. (1968). Ecological and behavioural contrasts between sympatric ground dwelling primates in Ethiopia. *Folia Primatologica*, 8, 192–227.

ECK, G.G. (1987). *Theropithecus oswaldi* from

the Shungura Formation, Lower Omo Basin, southwestern Ethiopia. In *Les Faunes Plio-Pléistocènes de la Basse Vallée de l'Omo (Éthiopie)*, Tome 3, Cercopithecidae de la Formation de Shungura, ed. Y. Coppens & F.C. Howell, pp. 123–39. Cahiers de Paléontologie, Travaux de Paléontologie Est-Africaine. Paris: Editions du Centre National de la Recherche Scientifique.

ECK, G.G. & JABLONSKI, N.G. (1984). A reassessment of the taxonomic status and phyletic relationships of *Papio baringensis* and *Papio quadratirostris* (Primates: Cercopithecidae). *American Journal of Physical Anthropology*, 65, 109–34.

ECK, G.G. & JABLONSKI, N.G. (1987). The skull of *Theropithecus brumpti* compared with those of other species of the genus *Theropithecus*. In *Les Faunes Plio-Pléistocènes de la Basse Vallée de l'Omo (Éthiopie)*, Tome 3, Cercopithecidae de la Formation de Shungura, ed. Y. Coppens &

F.C. Howell, pp. 11–122. Cahiers de Paléontologie, Travaux de Paléontologie Est-Africaine. Paris: Editions du Centre National de la Recherche Scientifique.

GUPTA, V.J. & SAHNI, A. (1981). *Theropithecus delsoni*, a new cercopithecine species from the Upper Siwaliks of India. *Bulletin of the Indian Geological Association*, 14, 16–71.

JOLLY, C. (1991). Species definitions and variations in extant *Papio*. *American Journal of Physical Anthropology*, Supplement 12, p. 100.

LEE-THORP, J.A., VAN DER MERWE, N.J. & BRAIN, C.K. (1989). Isotopic evidence for dietary differences between two extinct baboon species from Swartkrans. *Journal of Human Evolution*, 18, 183–90.

MAIER, W. (1972). The first complete skull of *Simopithecus darti* from Makapansgat, South Africa, and its systematic position. *Journal of Human Evolution*, 1, 395–405.

PART I

Fossil evidence and phylogeny

2 *Theropithecus darti* from the Hadar Formation, Ethiopia

GERALD G. ECK

Summary

1. The well-preserved sample of specimens of *Theropithecus darti* (Broom & Jensen, 1946) from the Hadar Formation of northeastern Ethiopia makes significant contributions to our understanding of the early evolution of the *T. oswaldi* lineage and to the initial radiation of the genus *Theropithecus*.

2. The Hadar specimens are the first of the species to be associated with radiometrically dated sediments. The age of these sediments indicates that the specimens are between 3.4 and 2.9 Ma old and are, thus, the oldest of the species yet described.

3. The well-preserved adult male cranial and mandibular specimens allow, for the first time, description of the adult male skull of the species. Characteristics of Hadar males, as well as females, confirm the close relationship of *T. darti* to *T. oswaldi* proposed by others based on specimens from Makapansgat in South Africa. The Hadar specimens share with those from Makapansgat a small suite of primitive characters that are used to differentiate the two species and, in conjunction with the derived characters shared with *T. oswaldi*, indicate an ancestor-descendant relationship between them. The larger size of the specimens from Makapansgat and the occurrence of more derived dental and mandibular characters suggest that they are younger in age than those from Hadar, as does evidence from suid biostratigraphy.

4. The occurrence of *T. darti* in the middle Pliocene of East Africa and the late Pliocene of South Africa indicates that the species was probably widely distributed in sub-Saharan Africa during later Pliocene times. Simple statistical analyses of the association between specimens of *T. darti* and those of other mammals at Hadar suggest that *T. darti* preferred relatively dry and open woodlands and grasslands. The wide geographical distribution and preference for dry open habitats typical of the later *T. oswaldi* (Andrews, 1916), thus, seem to be established

early in the evolution of the lineage.

5. The specimens of *T. darti* from Hadar show that the *T. oswaldi* lineage was well established by about 3.4 Ma, as was the *T. brumpti* lineage. The many apomorphic characters seen in the skull of *T. gelada* suggest that it too has had a long history of separate evolution. Taken together, the available evidence indicates that the radiation of genus *Theropithecus* occurred during earlier Pliocene times.

Introduction

By Pleistocene times, the genus *Theropithecus* had apparently established itself as the most common large monkey in the lowland woodland and grassland habitats of sub-Saharan Africa. Although of geographically limited extent, specimens of *T. brumpti* (Arambourg, 1947) contribute a major component of the mammalian fauna from the Shungura Formation derived from sediments between 2.0 and 2.9 Ma and are also found in deposits of similar age in the Koobi Fora and Nachukui formations. Showing a far larger distribution and often equally impressive numbers, specimens of *T. oswaldi* are regularly found in most fossil-bearing sites of Early Pleistocene age from Ethiopia in the north to the Republic of South Africa in the south. The origin of the genus and the early divergence of its several lineages is, by comparison, poorly documented. Sites containing sediments older than 3 Ma are not common and those known often produce small samples of monkeys or highly fragmented specimens.

Specimens of *Theropithecus* older than 3.0 Ma have been recovered from the Hadar, Usno, and Shungura formations of Ethiopia, the Koobi Fora, Nachukui, Lothagam, and Chemeron formations of Kenya. The more complete of these from all but one site appear to inform us about the evolution of *T. brumpti* and its ancestors, leaving only the Hadar Formation as a source of information about the ancestry and early evolution of *T. oswaldi*, the most common and widespread Pleistocene species of the genus.

The Hadar sample of *T. darti*, recovered during the 1973 to 1976 field seasons of the International Afar Research Expedition, is important for a number of reasons. First, it is the largest and best preserved of its age that has so far been found in Africa. Secondly, the well-preserved specimens allow unambiguous statements about their morphological and taxonomic status and relationships to later forms found elsewhere on the continent. Thirdly, the specimens, for the first time, clearly link an East African cercopithecoid taxon with one of those from the South African site of Makapansgat. And fourthly, the association of the monkey specimens with a large sample of other mammalian species, as well as with sedimentary structures, provides the basis for analyzing the habitat preferences of this early form of *Theropithecus*.

The following report presents (1) a brief summary of the geological characteristics of the Hadar Formation, (2) descriptions of the best-preserved specimens of *T. darti* and discussions of their taxonomic and phyletic relationships with specimens found elsewhere in Africa, (3) comments on the habitat preferences of the Hadar taxon, and (4) a catalogue of all specimens assigned to the species from the Hadar Formation.

Geological summary

The Hadar region is located at approximately 11° 10′ north and 40° 35′ east in the west central Afar Triangle, about 300 km northeast of Addis Ababa, Ethiopia. It encompasses approximately 100 square km of middle Pliocene sediments that are eroded into spectacular badlands on either side of the Awash river.

The stratigraphy of the Hadar Formation, as presently known, comprises some 280 m of sediments, which have been divided into four members, three of which are further divided into submembers (for more detailed discussion see Aronson & Taieb, 1981; Johanson, Taieb & Coppens, 1982; and Walter & Aronson, 1982). The Basal Member (BM) is the oldest of these and is composed, in the main, of upwards of 100 m of massive detrital clays that were deposited in a shallow basin. The overlying Sidi Hakoma Member (SH) is composed of 60 to 150 m of sands, silts, and clays that were laid down in flood plain, delta plain and margin, and shallow lacustrine depositional environments. This member is divided into four submembers labeled SH-1 through SH-4, from bottom to top. In the eastern part of the exposures, the middle portion of SH-4 submember contains the Kadada Moumou Basalt. The Denen Dora Member (DD) is the next in the sequence and comprises about 30 m of sands, silts, and clays that were deposited under shallow lacustrine,

swamp, and flood plain conditions. The DD Member is subdivided into three submembers: DD-1, DD-2, and DD-3. The Kada Hadar Member (KH), last in the Pliocene sequence, is unconformably capped by a cobble gravel of middle Pleistocene age. The member consists primarily of silty clays with intercalated sands and conglomerates that are approximately 90 m thick. Sedimentation occurred in lacustrine and flood plain environments. The KH Member is subdivided into four submembers (KH-1 through KH-4); the BKT tuff complex occurs in the upper part of the KH-4 submember.

The age of the top of the Hadar Formation appears reasonably well established. A suite of 14 conventional K/Ar determinations indicate an age of 2.88 ± 0.08 Ma for the upper unit of the BKT-2 tuff. Fission track ages of 20 zircon samples from this tuff have a mean of 2.7 ± 0.2 Ma, lending support to the K/Ar age estimate. Of the two, the age estimate produced by the K/Ar technique is thought to be the most reliable for technical reasons (Walter & Aronson, 1982). As noted above, the BKT-2 tuff occurs in the upper part of the KH-4 submember. It, thus, overlies the majority of sediments that have produced vertebrate fossils and all of those producing primates.

The age of the bottom of the formation remains a matter of dispute. Walter & Aronson (1982) (see also Aronson, Walter & Taieb, 1983) argue that new radiometric analyses of the Kadada Moumou Basalt, thought to be located in the upper part of the Sidi Hakoma Member, indicate that it has an age of 3.60 ± 0.15 Ma. They suggest that the new age for the basalt, taken in conjunction with the magnetostratigraphy of the lower Hadar Formation, implies an age 'somewhat older than 3.6 Myr' for the base of the Sidi Hakoma Member.

In contrast, Brown (1982, 1983) argues that the chemical composition of the Sidi Hakoma Tuff, which by definition forms the base of the Sidi Hakoma Member, is so similar to that of the Tulu Bor Tuff of the Koobi Fora Formation that they represent the same volcanic eruption. Although the Tulu Bor Tuff and its correlates in the Turkana Basin have not been dated by radiometric means, in the Koobi Fora Formation, the tuff directly underlies the Toroto Tuff, which dates to 3.32 ± 0.02 Ma (McDougall, 1985; Brown et al., 1985). Brown (1983) further argues that the normal polarity of the Sidi Hakoma-Tulu Bor Tuffs requires that they should be either younger than 3.41 Ma or older than 3.82 Ma and that stratigraphic considerations in both areas suggest the younger age most likely. Thus, the conflicting age of 3.6 Ma of the Kadada Moumou Basalt results from errors in dating induced by chemical alteration of the basalt or its stratigraphic misplacement within the Formation. Recent discovery of volcanic ash layers, chemically identical to tuffs (including the Tulu Bor Tuff) of the Koobi Fora Formation, in drilling cores from the Gulf of Aden persuade me of the widespread distribution of these tephra and an age slightly less than 3.4 Ma for the Tulu Bor Tuff (Sarna-Wojcicki et al., 1985). Thus, in this report, the age of the base of the Sidi Hakoma Member and the maximum age for all cercopithecids from the Hadar Formation will be taken as just less than 3.4 Ma.

Analysis of sedimentary and faunal evidence indicates that the Hadar sediments were deposited in a basin that was occupied by a lake fed by rivers and streams flowing off the highlands to the west. The water level of the lake fluctuated substantially through time, causing local changes in depositional environments and habitat types (Gray, 1980; Johanson et al., 1982). In the upper portion of the Basal Member and the SH-1 and SH-2 submembers, a complex of marshy lake edge and bush/woodland habitats is suggested by the association of clays and silts with a fauna dominated by Aepyceros, Ugandax, and Kolpochoerus and sands with a fauna dominated by Notochoerus, Nyanza-choerus, Tragelaphus, and elephantids. The varied lithologies of the SH-3 submember have produced a fauna with abundant alcelaphines, suggestive of extensive grasslands, although other faunal elements indicated the presence of some woodland as well. The silts and sands of the SH-4 submember and the DD Member produce a fauna in which Kobus and Hipparion are dominant and in which both Tragelaphus and alcelaphines are present in smaller but about equal numbers, suggesting a complex mosaic of woodland, bush, and grassland habitats. In the KH-1 submember, a fauna dominated by alcelaphines and equids is associated with varied sediment types that also contain Tragelaphus, indicating the presence of extensive grasslands as well as smaller amounts of bush and woodland. Finally, the silty clays of the KH-2 through KH-4 submembers have produced only a small fauna, but the available evidence suggests a possible return to the marshy lake edge, bush, and woodland habitats seen near the bottom of the formation. Aronson & Taieb (1981) point out that about 10 per cent of the sediments were deposited in lakes or marshes and about 20 per cent in river channels; the remaining 70 per cent appear to be overbank and/or interfluvial colluvial deposits. The nature of the rivers and/or streams producing the channel deposits remains unclear, for the deposits have characteristics produced by both meandering and braided streams. The paleontological importance of the riverine deposits is clear, however, for they produced 80 per cent of the vertebrate fauna recovered from the formation. In general, it appears that the habitats extant during the deposition of the Hadar Formation were composed of varying combinations of marsh, woodland, bush, and grassland with proportions that were basically controlled by changes in hydrologic regime. These changes produced a fluctuating mosaic of plant communities similar to those proposed for many other East African sites of Plio-Pleistocene age.

Stratigraphic context of *Theropithecus darti*

Eighty-seven cranial and associated post-cranial specimens of *T. darti*, derived from 61 individuals, have been recovered in 49 of the 315 collection localities established at Hadar. In many cases, it is difficult to establish the exact submember provenience of these specimens because all were found on the surface. The situation is further complicated by the fact that the erosional features at Hadar often slope steeply and can contain significant por-

tions of the stratigraphic column. Localized surface accumulations of mammalian specimens can, thus, derive from more than one submember, but, as Gray (1980) has shown, most localities contain specimens from only one or two. Allowing for these uncertainties, specimens of the species have been collected from all submembers of the three members with the exception of the uppermost two – KH-2 and KH-3. Gray's best estimate for the submember provenience of each specimen of *T. darti* can be found in Appendix 2.1 (of this chapter). The distribution of specimens among the submembers is not, however, even. Excluding the seven specimens with uncertainties in stratigraphic provenience greater than two submembers, the 17 specimens deriving from SH-1 and SH-2 comprise 31 per cent of the remaining 54 specimens and the 25 from DD-2 and DD-3 comprise 46 per cent. Of the remaining 12 specimens, three may also derive from SH-2 or SH-3, three from DD-1 or DD-2, and two from DD-3 or KH-1, leaving only one certainly from each of SH-3 and SH-4 and only two certainly from KH-1. It remains unclear whether or not DD-1 has produced specimens of the species. There appear, thus, to be two peaks in the frequency of *T. darti* in the Hadar Formation, one in SH-1, -2 and one in DD-2-3 (Fig. 2.1).

Theropithecus darti appears to have been the dominant large monkey at Hadar during the deposition of the SH-1 to KH-1 submembers. Of the 67 cercopithecid individuals represented by cranial, mandibular, and dental fragments identified at the generic level, 61 (91 per cent) belong to this species. The other seven jaw

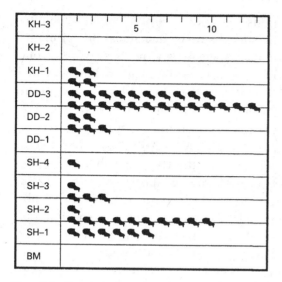

Fig. 2.1. Stratigraphic distribution of cranial, dental, and mandibular specimens of *Theropithecus darti*. Each cranium represents one individual; those on a boundary line derive from the unit above or below.

fragments and teeth, which were most probably recovered from the SH-1 and SH-2 submembers, all apparently belong to the colobine genus *Rhinocolobus* (E. Delson, pers. comm.). The only other cercopithecid taxon known from the Hadar Formation is a species of *Parapapio* that is represented by three fragmentary skulls associated with postcranial elements, all recovered from a single locality in the KH-3 submember, a unit that has produced only one other cercopithecoid specimen.

Comparative materials

Comparisons between specimens of *T. darti* from Hadar and Makapansgat are made first, if appropriate specimens from the later site are available, because of obvious similarities in size and morphology. Next, comparisons are made with specimens of *T. oswaldi* in order to elucidate the close relationship between these two species generally agreed upon by most authorities. Finally, comparisons are made with specimens of *T. brumpti* and *T. gelada* in order to clarify relationships among the various lineages of the genus.

In text and tables, specimen numbers of *T. darti* beginning with 'AL' derive from the Hadar Formation, while those beginning with 'M' or 'MP' were recovered from the Makapansgat Formation. The sample of *T. oswaldi* comprises specimens from both East and South Africa. Specimen numbers with prefixes 'F' or 'M' are from Kanjera, 'KNM-OG' from the Olorgesailie Formation, 'Old' from Olduvai Gorge, and 'SK' from the Swartkrans Formation. All specimens of *T. brumpti* were found in the Shungura Formation and have numbers beginning with 'L' or 'Omo'. The small sample of specimens of *T. gelada* were all shot in the wild and are presently housed in the Department of Anthropology, University of California, Berkeley ('Berkeley' and 'W'), the Cleveland Museum of Natural History, Cleveland ('CMNH'), and the United States National Museum, Washington ('USNM'). All measurements are presented in millimetres and were taken by me on original specimens.

Additional details about the geologic age of the above sites, the stratigraphic and/or geographic provenience of the specimens, and the citations to the literature on both of these topics can be found in Eck & Jablonski (1987).

Morphological description

Male cranium

Until the recovery of the Hadar material, the male cranium of *T. darti* was essentially unknown. The only other male cranial specimens of the species are two poorly preserved and fragmentary maxillae (MP237 and M3083, the latter briefly described by Maier, 1970) found in the South African Makapansgat Formation. The five additional specimens from Hadar now make possible a general description of the male cranium. The following description is based principally on the two adult specimens [AL134-5a (Fig. 2.13) and AL205-1a (Figs. 2.2 to 2.5)], supplemented by the two juvenile specimens [AL187-10 (Figs. 2.6 and 2.14) and AL310-19]. Specimen AL208-10b is poorly preserved and provides little morphological information. As the four well-preserved specimens are, most likely, from the DD-2, DD-3, or KH-1 submembers, they are roughly synchronic, each being approximately 3.1 Ma old.

Upon brief inspection, the specimens from Hadar appear remarkably similar in morphology to those of male *T. oswaldi* from Kanjera and Beds I and lower II of Olduvai Gorge. They differ most obviously in size; the Hadar specimens are considerably smaller (see further discussion below). As the male cranium (M32102) of *T. oswaldi* from Kanjera is the most complete early specimen described, it will serve as the basis of comparison for the Hadar male sample (for detailed treatments of M32102 see Eck & Jablonski, 1987; Jolly, 1972; and L. Leakey, 1943).

Fig. 2.2. AL205–1a, adult male cranium, lateral view, approximately 75% natural size.

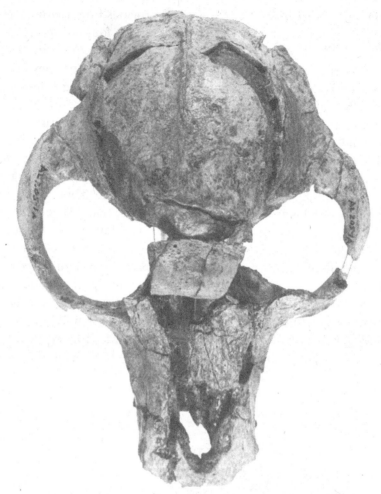

Fig. 2.3. AL205–1a, adult male cranium, superior view, approximately 75% natural size.

Muzzle Development of the maxillary ridges appears to be as variable in males of *T. darti* as it is in males of *T. oswaldi*. Ridges are absent in AL134-5a and M32102. Conversely, specimen AL205-1a possesses low, rounded ridges that arise from the bulges of the canine alveoli and are oriented at an angle of about 25° to the occlusal plane; in this respect, AL205-1a is most similar to 067/5608 from Olduvai Gorge.

The maxillary fossae in both of the Hadar specimens are very shallow and even more poorly defined than those of M32102.

The morphology of the muzzle dorsum appears to differ slightly between the Hadar specimens. Even though AL205-1a is broken in this region, its dorsum seems to form a simple, shallow, superiorly convex curve in the coronal section above the maxillary ridges and is, thus, similar in

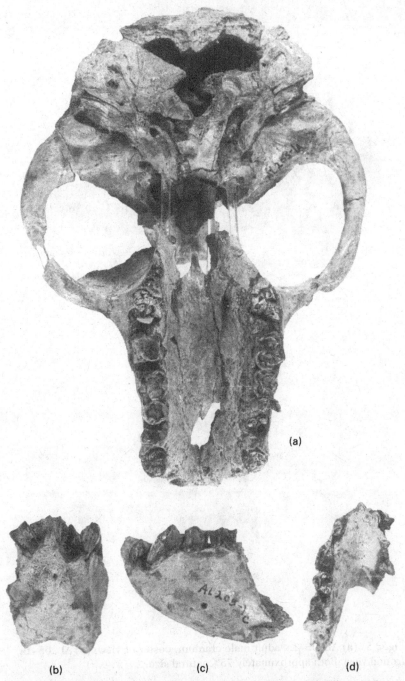

Fig. 2.4. (a) AL205–1a, adult male cranium, occlusal view; (b) AL205–1c, adult male mandible fragment, frontal view; (c) AL205–1c, lateral view; (d) AL205–1c, occlusal view. All approximately 75% natural size.

Fig. 2.5. (a) AL205–1a, adult male cranium, posterior view; (b) AL205–1a, frontal view. Both approximately 75% natural size.

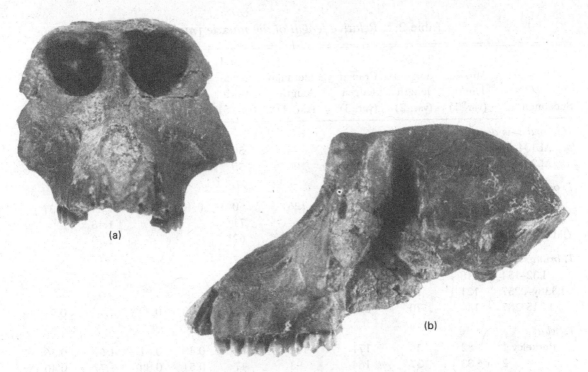

(a)

(b)

Fig. 2.6. (a) AL321–12, adult female cranium, frontal view; (b) AL321–12, lateral view. Both approximately 75% natural size.

morphology to M32102 and Old 69 S133. The dorsum of AL134-5a, in contrast, has two very shallow depressions that run in an anterior–posterior direction on either side of the midline; the nasal bones are raised substantially above these depressions, giving the dorsum a subdued, concavo-convexo-concave curvature that is unlike that of AL205-1a or M32102. Whether or not males of *T. oswaldi* vary in a similar fashion is unknown, for most specimens are badly damaged in this region.

The margins of the piriform aperture are poorly preserved in the two adult specimens from Hadar and well preserved only in the juvenile specimen, AL310-19. In this specimen and probably in the other two as well, the aperture is oval in shape as it is in M32102 and other males of *T. oswaldi*, but perhaps relatively more narrow. The inferior and lateral margins of the aperture, like those of *T. oswaldi*, are sharply defined by ridges. Only specimen AL205-1a preserves the region of the interpremaxillary suture; anterior nasal tubercles are absent as they are in specimens of *T. oswaldi*. The shallow groove on the midline seen in the Kanjeran specimen is absent in AL205-1a. The surface formed by the margins of the piriform aperture appears to be planar in the Hadar specimens and is oriented at about 45° angle to the occlusal plane, again showing marked similarity to M32102.

The morphology of the muzzle dorsum

Table 2.1. *Relative length of the muzzle (males)*

Specimen	Muzzle length (var. 1)	Muzzle dorsum length (var. 2)	Cranial length (var. 3)	Midcranial length (var. 4)	Dental arcade width (var. 5)	M.L. / C.L.	M.L. / M.c.L.	M.L. / D.A.W.	M.D.L. / M.c.L.
Theropithecus darti									
AL134–5a	—	40*	—	—	50*				
AL205–1a	83	41*	177	106	55	0.47	0.78	1.51	0.39
T. oswaldi									
M32102	92	47	208	126	60	0.44	0.73	1.53	0.37
Old 69 S133	105	52	—	—	71*			1.48	
Old 69 S196†	87	38*	—	—	62*			1.40	
T. brumpti									
L32–154	118	63*	—	—	60*			1.97	
L338y–2257	121	71	—	—	60			2.02	
L345–287	120	70	226	121	61	0.53	0.99	1.97	0.58
T. gelada									
Berkeley 1	82	36	171	101	49	0.48	0.81	1.67	0.36
2	83	37*	164	93	47	0.51	0.89	1.77	0.40
5	83	35	173	104	48	0.48	0.80	1.73	0.34
W239	77	31	163	97	46	0.47	0.79	1.67	0.32

*Estimated owing to damage.
†See Eck and Jablonski (1984).

of the Hadar specimens is, thus, very similar to that of M32102 and other males of *T. oswaldi*. Minor differences may occur in the relatively narrower piriform aperture and in the raised nasal bones and more complexly curved dorsum of at least some of the Hadar specimens. That these differences represent more than individual variation common to both species, however, will only be resolved when larger samples are known.

The Hadar males also have muzzles with relative lengths similar to those of males of *T. oswaldi*. As can be seen in Table 2.1, the ratios of muzzle length to cranial length, to midcranial length, and to dental arcade width, calculated for AL205-1a, are all very similar to those of males of *T. oswaldi* and differ consistently from those of other species of the genus. Males of both *T. darti* and *T. oswaldi* have relatively short muzzles.

Similarly, the muzzle dorsum of the Hadar males is short relative to the midcranial length as it is in males of *T. oswaldi* (Table 2.1). It should be noted here that the muzzle dorsum of *T. gelada* is also short, but that of *T. brumpti* is long, as it is in *Papio hamadryas* (Table 2.1; Eck & Jablonski, 1987).

Table 2.2. *Relative width of the maxillary dental arcade (males)*

Specimen	Dental arcade width (var. 5)	Molar row length (var. 6)	Dental arcade width at C^Us (var. 7)	Dental arcade width at I^2s (var. 8)	M.R.W. / D.A.W.	D.A.W.C. / D.A.W.	D.A.W.I^2 / D.A.W.
Theropithecus darti							
AL134–5a	50*	36*	43*	—	0.72	0.86	
AL205–1a	55	39*	47	36	0.71	0.85	0.65
T. oswaldi							
M32102	60	49	51	28	0.82	0.85	0.47
Old 68 S9	66	58	53*	30	0.88	0.80	0.45
Old 69 S133	71*	56	59	27	0.79	0.83	0.38
T. brumpti							
L345–287	61	48	59	36	0.79	0.97	0.59
L17–45	64*	45	54*	34	0.70	0.84	0.53
L32–154	60*	50	62*	36	0.83	1.03	0.60
L338y–2257	60	45	60	30	0.75	1.00	0.50
T. gelada†							
ranges	44–53	34–41	35–45	20–26			
means, n = 15	46.7	36.7	39.1	23.7	0.79	0.84	0.51

*Estimated owing to damage.
†From Eck (1977).

The maxillary dental arcade in the Hadar males is generally 'U-shaped' like those of males of *T. oswaldi* and other species of *Theropithecus*. Furthermore, the incisive arc of AL205-1a has a shallow curve and projects little beyond the canines as it does in males of *T. oswaldi* and *T. brumpti* (Eck & Jablonski, 1987). It, thus, differs from the more sharply curved and projecting arc of *T. gelada*. The ratios of molar row length and dental arcade width at the canines to dental arcade width in the two Hadar specimens are generally similar to those in males of *T. oswaldi* (Table 2.2). The very large ratio of dental arcade width at I^2 to dental arcade width, seen in AL205-1a (Table 2.2), results, at least in part, from a large diastema between the alveoli of the central incisors.

The palatal morphology of specimens AL134-5a and AL205-1a is very similar to that of M32102. The walls of the palate are subparallel in the anterior–posterior direction and tend to converge slightly before they bend sharply onto the nearly flat roof of the palate. The roof becomes more elevated as it runs from the incisive region to the region of the M^3s. The relative size, shape, and location of the incisive and greater palatine foramina are similar to those of *T. oswaldi* and unremarkable for a cercopithecine. Finally, various ratios of palatal length to other cranial measure-

Table 2.3. *Relative length of the palate (males)*

Specimen	Cranial length (var.2)	Dental arcade width (var. 5)	Palatal length (var. 9)	Inferior posterior cranial length (var. 10)	P.L. / D.A.W.	P.L. / C.L.	P.L. / I.P.C.L.
Theropithecus darti							
AL205–1a	177	55	77	100	1.40	0.44	0.77
T. oswaldi							
M32102	208	60	92	116	1.53	0.44	0.79
T. brumpti							
L345–287	226	61	109	115	1.79	0.48	0.95
L17–45	—	64*	107	—	1.67		
L32–154	—	60*	108	—	1.80		
L338y–2257	—	60	108	—	1.80		
T. gelada†							
Berkeley 1	171	49	76	96	1.55	0.44	0.79
2	164	47	78	87	1.66	0.48	0.90
5	173	48	77	96	1.60	0.45	0.80
W239	163	46	71	91	1.54	0.44	0.78
USNM 27039	177	49	81	97	1.65	0.40	0.84

*Estimated owing to damage.

ments, in specimen AL205- 1a, are similar to those of M32102 and *T. gelada*, but different from those of *T. brumpti* (Table 2.3).

In summary, the relatively short, saddle-shaped, inflated muzzle, lacking strongly developed maxillary ridges and fossae, but possessing an oval piriform aperture with margins that define a planar surface, seen in the Hadar males, is very similar to those of males of *T. oswaldi*, especially M32102 from Kanjera. The Hadar specimens appear to differ, however, from those of *T. oswaldi* in two taxonomically important ways. First, the Hadar muzzles are considerably smaller than those of *T. oswaldi*, a topic to which I will return. Secondly, although the spacing of the incisors in AL205-1a is obviously aberrant, given the large central diastema seen in the specimen, I think that some of the width of the incisive alveoli is related to the 'unreduced' size of the incisors postulated for the species by Maier (1970) and Eck & Jablonski (1987). (See below for further discussion.)

Zygomatic arch The inferior portion of the maxillary root of the zygomatic arch arises above the M^3s in both AL134-5a and AL205-1a. The anterior surfaces of the zygomatic bones curve gradually and smoothly superoposteriorly, producing an infraorbital surface very similar to that of M32102. The absence of deeply excavated maxillary fossae, and the smoothly curved transition from the walls of the muzzle to the anterior surfaces of the zygomatic bones, give this region the inflated appearance typical of males of *T. oswaldi*. The anterior portions of the inferior margin of the zygomatic arches are strongly buttressed and the temporal surfaces of the zygomatic bones are dominated by deeply excavated fossae, seen also in M32102 and other specimens of *T. oswaldi*.

In specimens AL205-1a and M32102, the anterior terminations of the attachment areas for the masseter muscles lie below and behind the level of the infraorbital foramina. When viewed from above, the shape of the arches in AL205-1a differs from that of the Kanjeran specimen in two ways. First, in the Hadar specimen, the anterior transition in the course of each arch is broadly curved, in contrast to the much sharper curvature observed in M32102. Secondly, the arches in AL205-1a appear more laterally bowed than they do in M32102. This impression is substantiated by the ratio of bizygomatic breadth to cranial length of 0.73 in the former and 0.64 in the latter (bizygomatic breadths (Variable 11) are 130 mm and 134 mm, respectively; see Table 2.3 for cranial lengths). When viewed laterally, the inferior margins of the arches of AL205-1a have slight, superiorly convex curvatures, while they are essentially straight in M32102. Like those of M32102, the arches of AL205-1a are lightly built, have ovoid cross sections at the midpoint in the length of the zygomaticotemporal sutures, and ascend slightly in relation to the occlusal plane as they run posteriorly. Finally, the superior surfaces of the arches

of AL205-1a form broad open troughs as they join the neurocranium, as they do in other male specimens of the genus.

The zygomatic arches of AL205-1a are, thus, very similar to those of M32102. They differ only in the amount of lateral bowing and in the sharpness of the curvature of the transition from their facial to lateral aspects.

Orbital region The orbital regions of both AL134-5a and AL205-1a are poorly preserved; the former retains small segments of both inferior orbital margins and the latter a small portion of the supraorbital torus above the interorbital region. Although most of the internal orbital structure is also absent from the juvenile male specimen AL187-10, its supraorbital torus is well preserved and only the lateral portions just above the zygomaticofrontal sutures are missing. When AL187-10 is viewed superiorly, the anterior margin of the torus forms a simple, anteriorly convex curve. When viewed frontally, the superior margin of the torus is generally convex superiorly although there is a shallow, narrow depression above the interorbital region; this depression is also observed in AL205-1a. The only other notable features of this region in AL187-10 are the deeply excavated supraorbital notches, which are laterally defined by relatively large spines. Overall, the supraorbital tori of AL187-10 and, probably, AL205-1a are very similar to that of M32102.

Cranial vault The anterior portion of the cranial vault of AL205-1a is poorly preserved; much of the frontal bone is either missing or damaged, as are smaller portions of the parietal bones. The vault of AL187-10, in contrast, is well preserved, having sustained only minor damage to the frontal bone. In the latter specimen, the temporal lines run medially for short distances as well-developed crests and, after curving sharply posteriorly, they run as nearly straight lines to join just behind bregma, losing their definition as they approach the point of union. The area enclosed by the temporal lines and the supraorbital torus forms a shallow supraorbital sulcus that is most deeply excavated just behind the torus at the midline. A low sagittal crest, no more than 1 or 2 mm high, rises just above lambda from the posterior extension of the united temporal lines. The sagittal crest runs posteriorly to near the tip of the occipital protuberance, where the temporal lines curve sharply laterally to join the nuchal crest. The nuchal crest is roughly circular in form and is lightly built, being no more than a few millimetres wide near inion, widening to about 5 mm near its lateral terminations behind the external auditory meati.

The anterior portions of the temporal lines are obscured by breakage in AL205-1a, although enough of the region is preserved to indicate that the specimen possessed a supraorbital sulcus very similar in morphology to that of AL187-10. It is also apparent that the temporal lines joined just anterior to bregma to form a very low sagittal crest. As the sagittal crest runs posteriorly, it gains its greatest height of approximately 3 mm about midway between bregma and inion. In AL205-1a, the nuchal crest is circular in form. From a

width of about 3 mm near inion, it broadens to about 7 mm near its terminations behind the external auditory meati. An extension of the plane defined by the nuchal crest would intersect the occlusal plane at an angle of approximately 150°.

The minor differences between AL187-10 and AL205-1a in the size and distribution of crests are most likely related to the relative immaturity of the former specimen. Furthermore, the courses of the temporal lines, the form of the supraorbital sulcus, the form of the sagittal and nuchal crests, and the orientation of the plane of the nuchal crest seen in the Hadar specimens are very similar to those of the Kanjeran male. The most striking difference between AL205-1a and M32102 is seen in the size of the sagittal and nuchal crests, for the sagittal crest of M32102 rises to a height of about 10 mm and the nuchal crest broadens from about 10 mm near inion to about 12 mm near its lateral terminations.

The vaults of the two Hadar specimens are ovoid in shape when viewed superiorly and reach their greatest widths just above the external auditory meati. In both, the coronal suture, when viewed from above, appears composed of two straight segments that converge on the midline at an angle of about 90°. Bregma is located relatively far forward. When viewed posteriorly, the vault, especially of the adult specimen, appears broad and low. The walls of the vault above the external auditory meati are oriented at about 65° to a line through the meati and curve gradually to form the vault roof. In all these features, as well as in the ratios of postorbital constriction and midparietal width

Table 2.4. *Length, width, height of the cranial vault (males)*

Specimen	Vault length (var. 12)	Vault width (var. 13)	Postorbital constriction (var. 14)	Vault height (var. 15)	Midparietal width (var. 16)	V.W. / V.L.	P.o.C. / V.W.	V.H. / V.L.	M.p.W. / V.W.
Theropithecus darti									
AL205–1a	108	91	46*	64	55	0.84	0.51	0.59	0.60
AL187–10	104	77	43	59	56	0.74	0.56	0.57	0.73
T. oswaldi									
M32102	125	99	48	—	62	0.79	0.48		0.63
Old 69 S89	138	106*	54	—	76	0.77	0.51		0.72
T. brumpti									
L345–287	119*	91	43	72	60	0.76	0.47	0.61	0.66
T. gelada									
Berkeley 1	107	84	46	69	70	0.79	0.55	0.64	0.83
2	100	76	42	68	65	0.76	0.55	0.68	0.86
5	112	83	46	71	68	0.74	0.55	0.63	0.82
W239	103	82	42	70	68	0.80	0.51	0.68	0.83
USNM 27039	108	84	42	73		0.78	0.51	0.68	

*Estimated owing to damage.

to cranial width (Table 2.4), the Hadar specimens are very similar to M32102. They are, however, smaller in absolute size.

Cranial base The cranial bases of both AL187-10 and AL205-1a are generally well preserved, although there is damage to the occipital plane, posterior margin of the foramen magnum, pterygoid plates and sphenoid of the former and the petrous temporal bones and pterygoid plates of the latter.

The occipital plane in both Hadar specimens is essentially flat and is oriented at an angle of about 145° to the occlusal plane. The most obvious features of this region consist of two, symmetric, moderately deep depressions that are located on either side of the midline and separated by a well-developed crest. The depressions serve as the areas of insertion for the rectus capitis posterior minor muscles. The mastoid regions are elevated slightly above the general level of the occipital plane and have slightly posteroinferiorly concave surfaces. The mastoid surfaces are dominated by broad, shallow, digastric grooves. The mastoid processes of both Hadar specimens are relatively thin, triangular in shape, and only moderately projecting. Although the margins of the foramen magnum in each of the Hadar specimens is damaged, the foramina appear to have been nearly circular in form.

The posterior portions of the cranial bases of male specimens of *T. oswaldi* are generally poorly preserved. Those features present in M32102, Old 69 S89 from Olduvai, and KNM-OG-1420 from Olorgesailie are, however, morphologically similar to those of the Hadar specimens. These features include the general configuration and orientation of the occipital plane, the form of the digastric groove, and the shape and degree of development of the mastoid process. The surfaces of the mastoid regions in specimens M32102 and Old 69 S89 are slightly convex posteroinferiorly, in contrast to the slightly concave surfaces in the Hadar specimens. The surface of the mastoid region in KNM-OG-1420 is nearly flat, however, and thus intermediate in form between AL187-10 and AL205-1a and the other two specimens of *T. oswaldi*.

The tympanic temporal bones are well preserved in both Hadar specimens and point posterolaterally at an angle of about 110° to the sagittal plane, an orientation similar to that in specimens M32102 and Old 69 S89. Both postglenoid processes in AL205-1a and the right process in AL187-10 are well preserved. Those of AL205-1a are only moderately tall (10 mm) and rectangular in shape, while that of AL187-10 is taller (11 mm) and triangular in shape. The rectangular shape in the former specimen is similar to that of the processes of three specimens of *T. oswaldi*, but they appear relatively shorter. The processes of Old 69 S89, M32102, and KNM-OG-1420 are about twice as tall (21 mm, 20 mm, and 25 mm, respectively) as those of AL205-1a, but their crania, although larger, were almost certainly not twice as large as the Hadar specimen. The postglenoid process of AL187-10 is most

similar to that of a female specimen of *T. oswaldi* (M14936) from Kanjera, but again is relatively shorter. In both of the Hadar males, the postglenoid processes are separated from the tympanic temporal bones by distinct grooves, while in all male specimens of *T. oswaldi* they are tightly pressed against the tympanics. The glenoid regions of the two Hadar specimens differ. The articular eminences are relatively high and the articular fossae deep in AL187-10, features shared with M32102, while the eminences are lower and the fossae more shallow in AL205-1a. Finally, when viewed inferiorly, the temporal fossae of the Hadar specimens are very similar to those of M32102. Most importantly, the posterior margins of the fossae are oriented at an angle of about 120° to the sagittal plane and the posterolateral margins are smoothly and broadly curved.

Thus, the cranial bases of males of *T. darti* appear to differ most from those of *T. oswaldi* in having relatively shorter postglenoid processes that are separated by distinct grooves from the tympanic temporal bones. The males of the two species are otherwise very similar in features of this region.

Relationship of the face and neurocranium The face of AL205-1a articulates with the neurocranium only through the right zygomatic arch. Although I think that my reconstruction of the specimen is reasonably accurate, quantification of the relationship between these two regions would imply greater precision than the condition of the specimen permits. The reconstruction does make clear, however, that the cranial base and glenoid fossa were elevated above the occlusal plane, a characteristic of all species of the genus *Theropithecus* (see Szalay & Delson (1979) and Eck & Jablonski (1987) for a review of this relationship and the literature pertaining to it).

Size of the male cranium Upon visual inspection, specimen AL205-1a appears considerably smaller than M32102. This impression is confirmed by the fact that, in all but one case (dental arcade width at I^2, Variable 8), a number of corresponding measurements of AL205-1a and M32102 yield lower values for the former specimen (Tables 2.1 through 2.4). The percentage difference in the values of the two specimens varies considerably, however, depending on the variable chosen, from 80 per cent for the molar row length (Variable 6) to 97 per cent for bizygomatic breadth (Variable 11). Thus, it appears that no single variable adequately expresses the differences in size between the two specimens. Using all of the variables discussed above, with the exceptions of Variable 8 (probably aberrantly large) and 15 (not available for *T. oswaldi*), AL205-1a is, on average, 12 per cent smaller than M32102 (Table 2.5).

Female cranium

The strongest morphological links between specimens of *T. darti* from Hadar and from the Makapansgat Formation are based on comparison of the nearly complete adult female cranium [AL321-12 (Figs. 2.7 to 2.9)] and the partial juvenile female skull [AL185-5 (Figs. 2.10 and

Table 2.5. *Percentage size difference of specimens of* Hadar *crania (males)*

Variable	AL205–1a	M32102	AL205–1a / M32102
2	177	208	0.85
3	68	80	0.85
4	114	137	0.83
5	55	60	0.92
6	39*	49	0.80
7	47	51	0.92
9	77	92	0.84
10	100	116	0.86
11	130	134	0.97
12	108	125	0.86
13	91	99	0.92
14	46*	48	0.96
16	55	62	0.89
Average			0.88

*Estimated owing to damage.

2.11)] from the former site with the well-preserved adult female cranium (MP222) and the nearly complete juvenile skull (M3073) from the latter. The specimens from Makapansgat have been described by Maier (1970, 1972) and Freedman (1976). Descriptions of the Hadar female crania are based on comparison with those from Makapansgat; summary statements about differences from, and similarities to, females of *T. oswaldi* are included.

Muzzle Maxillary ridges are absent from the adult female specimen (AL321-12) from Hadar, but low rounded ones are present in the juvenile specimen (AL185-5). Development of maxillary ridges

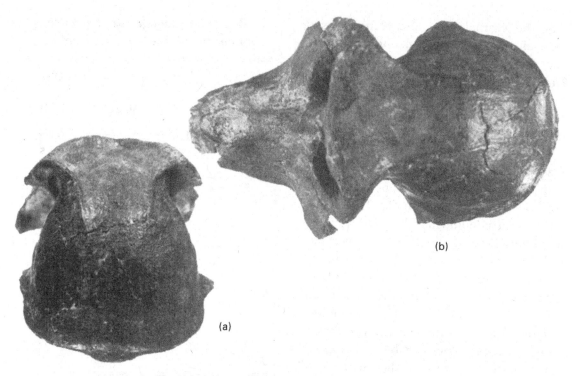

Fig. 2.7. (a) AL321–12, adult female cranium, posterior view; (b) AL321–12, superior view. Both approximately 75% natural size.

appears as variable in females as in males from the site. The maxillary ridges in both specimens from Makapansgat, noted by Freedman (1976) and Maier (1970, 1972), are more strongly developed than those of the juvenile from Hadar. The variability in the ridges of the two Hadar females is similar to that of females of *T. oswaldi*, in which they are either weakly developed or absent.

Specimen AL321-12 lacks maxillary fossae, while those of AL185-5 are very shallow and poorly defined. The shallow fossae, noted by Freedman and Maier, in both of the Makapansgat females, are more deeply excavated and sharply defined than those of AL185-5. The maxillary fossae of

females of *T. oswaldi* are similar to those of the Makapansgat specimens and, thus, more deeply excavated and sharply defined than those in AL185-5.

The muzzle dorsum of AL321-12 has the concavo-convexo-concave morphology seen in the adult male (AL134-5); distinct depressions appear on either side of the midline and the nasal bones are raised substantially above the level of the depressions. The curvature of the region is stronger in the female, making the elevation of the nasal bones more obvious than in the male. Although much of the dorsum is missing from specimen AL185-5, it is apparent that depressions were absent; in all likelihood, the dorsum had a simple,

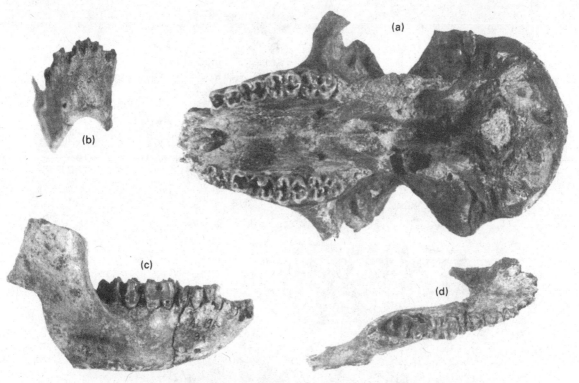

Fig. 2.8. (a) AL321–12, adult female cranium, occlusal view; (b) AL196–3a, adult female mandible fragment, frontal view; (c) AL196–3a, lateral view; (d) AL196–3a, occlusal view. All approximately 75% natural size.

superiorly convex curvature similar to that of the adult male AL205-1a. Both females from Makapansgat have the raised nasal bones and paired depressions seen in AL321-12. In contrast, most females of *T. oswaldi* lack raised nasal regions and thus have simply curved, convex dorsa. Two specimens (SK561 and SK563 from Swartkrans) may, however, have had raised nasals similar to those of the specimens from Makapansgat and AL321-12, but damage to both precludes certainty in this regard.

The margins of the piriform aperture are well preserved in the Hadar adult female. They are sharply defined by ridges in their lateral and inferior portions and form a relatively narrow, oval aperture. Near the interpremaxillary suture, the ridges form a broad, shallow groove that runs onto the incisive alveoli; there are no anterior nasal tubercles. The margins are planar and oriented with the occlusal plane at an angle of about 40°. In all characteristics of the piriform aperture, AL321-12 is very similar to the female specimens from Makapansgat and those of *T. oswaldi*. The angle of orientation of the aperture is less steep, however, in the Hadar specimen.

When viewed superiorly, the anterior margin of the premaxillae in specimen AL321-12 projects more anteriorly than it

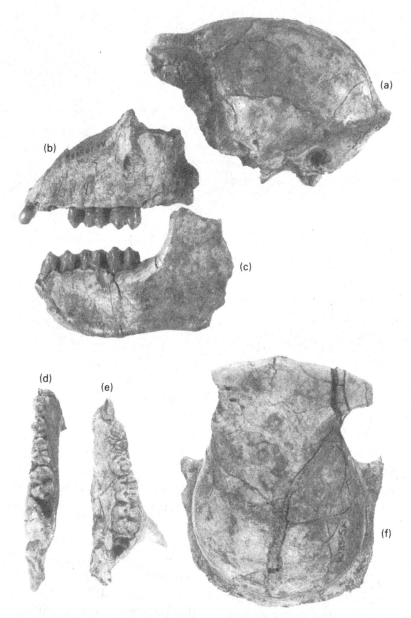

Fig. 2.9. (a) AL185–5a, juvenile female neurocranium, lateral view; (b) AL185–5b, juvenile female left facial fragment, lateral view; (c) AL185–5c, juvenile female mandible fragment, lateral view; (d) AL185–5c, occlusal view; (e) AL185–5b, occlusal view; (f) AL185–5a, superior view. All approximately 75% natural size.

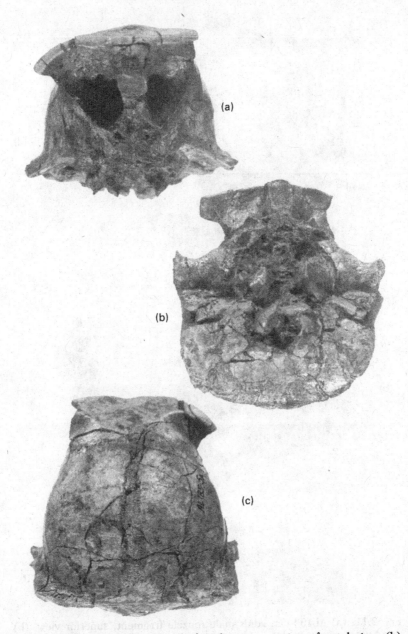

Fig. 2.10. (a) AL185–5a, juvenile female neurocranium, frontal view; (b) AL185–5a, basal view; (c) AL185–5a, posterior view. All approximately 75% natural size.

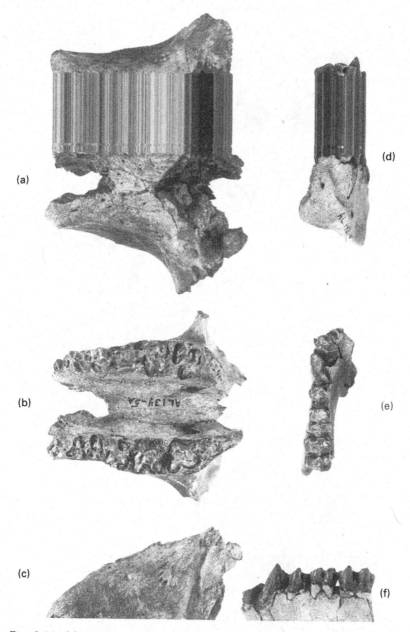

Fig. 2.11. (a) AL134–5a, adult male muzzle fragment, superior view; (b) AL134–5a, occlusal view; (c) AL134–5a, lateral view; (d) AL142–19, juvenile male mandible fragment, frontal view; (e) AL142–19, occlusal view; (f) AL142–19, lateral view. All approximately 75% natural size.

does in the male AL205-1a, although the projection is still only moderate. The nasal processes of the premaxillae are moderately robust in both of the Hadar females, being about 3 mm wide at the midpoint in the length of the piriform aperture. Overall, the morphology of the premaxillae of the Hadar adult female is

Table 2.6. *Relative length of the muzzle (females)*

Specimen	Muzzle length (var. 1)	Muzzle dorsum length (var. 2)	Cranial length (var. 3)	Midcranial length (var. 4)	Dental arcade width (var. 5)	M.L. / C.L.	M.L. / M.c.L.	M.L. / D.A.W.	M.D.L. / M.c.L.
Theropithecus darti									
AL321–12	53	33	161	116	50	0.33	0.46	1.06	0.28
MP222	65	42*	171	114	54	0.38	0.57	1.20	0.37
T. oswaldi									
M14936	60	38*	174	125	55*	0.34	0.48	1.09	0.30
T. gelada									
Berkeley 6	53		140	94	41	0.38	0.56	1.29	
CMNH B1209	51		139	94	41	0.37	0.54	1.24	

*Estimated owing to damage.

very similar to that of the adult female from Makapansgat and those of *T. oswaldi*.

While the males of *T. darti*, like those *T. oswaldi*, appear to have muzzles that are relatively shorter than those of *T. brumpti* and *T. gelada* (Table 2.1), a similar pattern is not seen among the females. The ratios of muzzle length to other cranial variables of AL321-12 are very similar to those of M14936, an adult female cranium of *T. oswaldi* from Kanjera. Those of MP222 are most similar to the ratios obtained from two female specimens of *T. gelada* (Table 2.6).

Likewise, the length of the muzzle dorsum relative to the midcranial length in the Hadar specimen is very similar to that of the female specimen of *T. oswaldi*, while that of the specimen from Makapansgat appears relatively longer (Table 2.6).

The shape of the maxillary dental arcade is generally similar among females (as among males) of all species of *Theropithecus*. The female arcade differs from the male by having slightly anteriorly convergent, rather than parallel, cheek-tooth rows and by having more strongly curved incisive regions. The adult female specimens of *T. darti* are most similar to those of *T. oswaldi* and *T. brumpti*, for, in all of the fossil specimens, the incisive region is less strongly curved than it is in females of *T. gelada*. As can be seen in Table 2.7, however, the relative width of the dental arcade in the incisive region is larger in the females of *T. darti* than it is in a female of *T. oswaldi* (SK561) and, thus, more like that of females of *T. gelada*. This implies less incisor reduction in females of *T. darti* than in *T. oswaldi*.

In most specimens of *Theropithecus*, the occlusal surfaces of the premolars and molars have a 'reversed curve of Spee'. Although this topic is discussed in more detail in the sections on the mandible, it should be noted here that the maxillary

Table 2.7. *Relative width of the maxillary dental arcade (females)*

Specimen	Dental arcade width (var. 5)	Molar row length (var. 6)	Dental arcade width at C^Us (var. 7)	Dental arcade width at I^2s (var. 8)	M.R.W. / D.A.W.	D.A.W.C. / D.A.W.	D.A.W.I^2 / D.A.W.
Theropithecus darti							
AL321–12	50	37	34	22	0.74	0.68	0.44
MP222	54	46	33	22	0.85	0.61	0.41
T. oswaldi							
SK561	59	46	34	19	0.78	0.58	0.32
T. gelada[†]							
ranges	40–46	30–35	24–30	17–22			
means, n = 15	41.9	33.0	27.5	19.7	0.79	0.66	0.47

[†]From Eck (1977).

occlusal surfaces in specimen AL321-12 appear straight rather than curved.

The morphology of the palate of AL321-12 is very similar to that of the Hadar males and to that of MP222; both females are similar in morphology to females of *T. oswaldi*.

Zygomatic arch In both Hadar females these are poorly preserved; the best preserved is the anterior portion of the left arch of AL321-12. In this specimen as in the males, the arch arises above M^3. The facial surface of the zygomatic bone is smoothly and convexly curved antero-superiorly, and the anterior termination of the area of attachment of the masseter muscle is posteroinferior and lateral to the infraorbital foramina. The anterior portion of the inferior margin of the arch is strongly buttressed and the temporal surface of the zygomatic bone consists of a broad, deeply excavated fossa. Just above this fossa, a strongly developed ridge runs inferomedially from a point that lies just below the zygomaticotemporal suture. The lateral margin of the frontal process of the zygomatic bone runs in a postero-superior–anteroinferior direction at an angle of about 110° to the occlusal plane. In all of these features, AL321-12 and MP222 are very similar, and both are very similar to females of *T. oswaldi*. The zygomatic bone of AL321-12 differs most from MP222 in having a more robust buttress on the inferior margin and a more strongly developed ridge on the temporal surface. Unfortunately, the left arch of AL321-12 is broken at the point where it begins to bend posteriorly and, thus, it is impossible to tell if the arch was as sharply curved in this region as it is in MP222 and other specimens of *T. darti* and *T. oswaldi*, or if it was as broadly curved as it is in the male specimen AL205-1a.

Orbital region The orbital margins are well preserved in AL321-12. In this specimen,

Table 2.8. *Height and width of the orbits (females)*

Specimens	Orbital height (var. 17)	Orbital width (var. 18)	O.H. / O.W.
Theropithecus darti			
AL321–12	28	26	1.08
MP222	22	29	0.76
M3073	26	26	1.00
T. oswaldi			
M14936	24	29	0.83
M32104	24	28	0.86
T. brumpti			
L32–155	27	29	0.93
T. gelada			
Berkeley 6	18	23	0.78
CMNH B1209	20	23	0.87

the inferior portions of the margins are strongly curved, producing orbits that are egg-shaped and of relatively great height (Table 2.8). The supraorbital notches are shallow and terminate laterally as low rounded tubercles. The well-preserved notches in the juvenile female (AL185-5) are more shallow and the tubercles less pronounced than those in the adult. The shape of the orbits in MP222 differs considerably from that of the Hadar adult female; as noted by Freedman (1976), the orbits of the Makapansgat adult are relatively low and broad (Table 2.8) and, in addition, have the more oblong shape common to many other specimens of the genus. In contrast, the left orbit of the Makapansgat juvenile (M3073) has the relatively high, egg-shaped morphology of

the Hadar specimen (the outline of the right orbit is obscured by breakage). It is of interest that similar variation is seen in the shape of the orbits among the specimens of *T. brumpti* from the Omo (Eck & Jablonski, 1987). The morphology of the supraorbital notch in both of the Makapansgat specimens is very similar to that of the Hadar juvenile female.

The supraorbital tori of the two Hadar females are very similar in morphology to that of the male AL187-10. When viewed superiorly, they are smoothly and convexly curved anteriorly. When viewed frontally, their most notable feature is the shallow and narrow depression that occurs above glabella. In both of these features, the Hadar females are very similar to those from Makapansgat and those of *T. oswaldi*.

Cranial vault The surface of the cranial vault is well preserved in both AL321-12 and AL185-5. In the adult specimen, the temporal lines begin as well developed crests that converge at an angle of about 45° to the sagittal plane and are smoothly concave posterolaterally. As the lines approach the level of bregma, they are transformed into low rounded ridges. The area between the temporal lines and the superior margin of the supraorbital torus forms a shallow supraorbital sulcus. The lines reach their point of closest approach (20 mm apart) about one- third of the distance between bregma and inion. As the lines run posteriorly from this point they are nearly straight and slightly divergent. Just above the level of inion, they curve sharply laterally and join the nuchal crest about midway between inion and the external auditory meati. When viewed from above, the margin of the nuchal crest is circular in form and the crest itself is of modest size, being no more than a few millimetres wide at its points of greatest development; these points lie about midway between inion and the external auditory meati. The plane of the nuchal crest is oriented at about 150° to the occlusal plane.

In the juvenile specimen AL185-5, the morphology of the supraorbital sulcus and the shape and angle of orientation of the margin of the nuchal crest are similar to those of the adult specimen. The course of the temporal lines and development of the nuchal crest differ, however. In the juvenile, the anterior portions of the temporal lines converge rapidly and join just behind bregma to form a very low, rounded ridge that runs to a point just above inion. At this point, the lines diverge as sharp curves and join the nuchal crest just lateral to the occipital protuberance. The nuchal crest is about 2 mm wide below inion and widens to approximately 6 mm about two-thirds of the distance between inion and the external auditory meati. As expected, the development of the sagittal and, perhaps, the nuchal crests appears to be less in the Hadar females than in the males. The difference between the two female specimens in the courses of the temporal lines most likely reflects individual variation, for similar variability is seen in the morphology of the lines in males of *T. gelada* (see Eck & Jablonski, 1987).

As described by Freedman (1976), the morphology of the supraorbital sulcus, the courses of the temporal lines, and the development, shape, and angle of orientation of the nuchal crest, in the adult female from Makapansgat, are very similar to those noted in the juvenile from Hadar. The Makapansgat specimen has, however, a low sagittal crest rather than a low rounded ridge. In contrast, the courses of the temporal lines of the Makapansgat juvenile are similar to those of the Hadar adult, for they are separated throughout by at least 12 mm (Maier, 1970, 1972). The supraorbital sulcus of the Makapansgat juvenile is similar to those of the Hadar specimens; the nuchal crest is so damaged as to preclude comparison.

The configuration of the temporal lines and nuchal crests in females of *T. darti* is generally similar to that of the females of *T. oswaldi*. The development of both the sagittal and nuchal crests is, however, greater in the latter than in the former species.

The shape of the cranial vault in the Hadar females is similar to that described for the males with one notable exception. As can be seen in Table 2.9, the ratio of midparietal width to cranial width in both Hadar females is large when compared to that of the males or to that of both sexes of *T. oswaldi* (see also Table 2.4). In fact, the degree of midparietal expansion in the Hadar females and the Makapansgat juvenile appears similar to that found in both sexes of *T. gelada*. As preserved, the adult female from Makapansgat would appear to have the unexpanded midparietal morphology common to the Hadar males and other fossil specimens of the genus. Close inspection reveals, however,

Table 2.9. *Length, width, height of the cranial vault (females)*

Specimen	Vault length (var. 12)	Vault width (var. 13)	Postorbital constriction (var. 14)	Vault height (var. 15)	Midparietal width (var. 16)	V.W. / V.L.	P.o.C. / V.W.	V.H. / V.L.	M.p.W. / V.W.
Theropithecus darti									
AL321–12	102*	76	45	63	61	0.75	0.59	0.62	0.80
AL185–5	101	74	46	—	60	0.73	0.62		0.81
MP222	109	76	44	63*	—	0.70	0.58	0.58	
M3073	105*	84*	51	68	70	0.80	0.61	0.65	0.83
T. oswaldi									
M14936	118	89*	50	72	63	0.75	0.56	0.61	0.71
SK561	116	94	45	67	70	0.81	0.48	0.58	0.74
Old BK II E	129	110	56	74	74	0.85	0.51	0.57	0.67
T. gelada									
Berkeley 6	98	75	42	59	64	0.77	0.56	0.60	0.85
CMNH B1209	97	71	41	64	62	0.73	0.58	0.66	0.87

*Estimated owing to damage.

that inward compression of the parietals during fossilization may have reduced the midparietal width of this specimen.

Cranial base The cranial base of AL321-12 is well preserved, having suffered significant damage only to the right postglenoid process and to the pterygoid plates. The base of AL185-5 is less well preserved although the occipital plane, mastoid, and glenoid regions are in relatively good condition.

In both specimens, the occipital plane is nearly flat. The attachment areas of the rectus capitis posterior minor muscles are well marked in the adult female, and are morphologically indistinguishable from those in the males. The mastoid regions of the females, like those of the males, are raised slightly above the level of the occipital plane. In the juvenile, the mastoid surface is essentially flat, while in the adult female, it is slightly inflated and is posteroinferiorly convexly curved. The mastoid surfaces of both females differ from those of the males, in which they are posteroinferiorly concave. Both of the female specimens have broad, shallow digastric grooves. The mastoid processes are broad and rounded; they project only slightly in the adult and little if at all in the juvenile. They, thus, differ from those of the males, in which the processes are thin, triangular, and moderately projecting. The foramen magnum of the adult female is circular in form.

The tympanic temporal bones in both of the female specimens are similar to those of the males, and point posterolaterally at an angle of about 110° to the sagittal plane.

The postglenoid process of AL321-12 is relatively tall (8 mm) and rectangular in shape and, thus, similar to those of the adult male AL205-1a; it is separated from the tympanic temporal bone by a distinct groove, which is less broad and open than that of the male. The relationship of the postglenoid process to the tympanic temporal bone is obscured by damage in the juvenile female specimen. The articular eminences and fossae in both female specimens are, like those of the adult male, relatively low and shallow, respectively. In their preserved portions, the temporal fossae of both Hadar females are similar in morphology, when viewed inferiorly, to those of the adult male. The posterior margins of the fossae in the females are oriented at about a 120° angle to the sagittal plane.

The cranial bases of the Makapansgat female specimens have been described by Freedman (1976) and Maier (1970, 1972). Based upon their descriptions and my own observations, the morphology of the Makapansgat specimens appears very similar to that of the Hadar adult, although the region of neither is as well preserved as it is in those from Hadar. The occipital planes are nearly flat, mastoid regions slightly inflated, digastric grooves broad and shallow. The tympanic temporal bones point posterolaterally at about 110° to the sagittal plane and are separated from moderately tall postglenoid processes (9 mm in MP222) by distinct grooves. In addition, the articular eminences and fossae of the Makapansgat adult are low and shallow, respectively, as they are in the Hadar specimens, and the posterior margins of the temporal fossae are similarly oriented to the sagittal plane. The mastoid processes of the Makapansgat adult are most similar in morphology to those of the Hadar juvenile, while those of the Makapansgat juvenile are most similar to those of the Hadar adult.

The morphology of the cranial base of females of *T. oswaldi* is generally similar to that of females of *T. darti*. Two regions warrant mention, however, because they exhibit interesting variation or differences. First, the mastoid surfaces of females of *T. oswaldi*, like those of *T. darti*, vary in configuration from flat to slightly posteroinferiorly convex. In addition, the mastoid processes of most females of *T. oswaldi* are broadly rounded and slightly projecting as they are in females of *T. darti*, while in one (M14936) they are triangular, moderately projecting, and similar to those of the Hadar males. Secondly, although complete postglenoid processes are present on only two female specimens of *T. oswaldi* (M14936 and Old BK II E), in both they are very tall (17 mm and 19 mm, respectively). The difference in height of the processes between females of the two species is, thus, similar to that between the males. The processes of M14936 are triangular in shape, as they are in the juvenile male from Hadar, while those of Old BK II E are rectangular and, thus, similar in shape to those of the adult male and female from Hadar and males of *T. oswaldi*. The bases of the postglenoid processes in all females of *T. oswaldi* are tightly pressed against the tympanic temporal bones rather than separated from them by fissures or grooves, as they are in females of *T. darti*.

Table 2.10. *Relationship between the glenoid region and the cheektooth row in* Hadar *specimens*

Specimen	Raw values			Standardized values†		
	(Var. 19)	(Var. 20)	(Var. 21)	(Var. 19)	(Var. 20)	(Var. 21)
Theropithecus darti						
AL321–12	59	120	56	0.251	0.511	0.238
MP222	55	123	65	0.226	0.506	0.267
T. oswaldi						
M14936	60	121	61	0.247	0.500	0.252
SK561	60*	130*	68	0.233	0.504	0.264

*Estimated owing to damage.
†See Eck (1977) for method of calculation.

Relationship of the face and neurocranium The relationship of the face to the neurocranium is well preserved in AL321-12. The cranial base and glenoid fossae of the specimen are located some distance above the occlusal plane (Table 2.10, Figure 2.12), a condition typical of the genus *Theropithecus*.

Size of the female cranium Visual inspection of the Hadar females indicates that they are slightly smaller than their counterparts from Makapansgat, although the difference in size is not nearly as great as that between the Hadar adult male and the one from Kanjera. Perusal of the values presented in Tables 2.6, 2.7, and 2.9 indicates no clear pattern of size difference between AL321-12 and MP222. In some dimensions, the Hadar female is smaller, and in others larger, than the one from Makapansgat. Calculation of the average percentage value of the variables presented in the above tables suggests, however,

that the Hadar adult is about four per cent smaller than the Makapansgat adult (Table 2.11). A similar calculation (excluding Variables 7, 8, 9, and 10) indicates that the Hadar female is about 10 per cent smaller than the nearly complete female cranium of *T. oswaldi* from Kanjera (M14936).

Male Mandible

The description of the male mandible of *T. darti* from Hadar is based on the eight best preserved specimens, none of which is complete. Two of the specimens were found in the Sidi Hakoma Member, five in the Denen Dora Member, and one in the Kada Hadar Member. The age-range of the sample, thus, nearly spans that of the complete stratigraphic column at Hadar. As no consistent morphological differences can be discerned between the geologically youngest and oldest of these, they will be treated together. The description of the

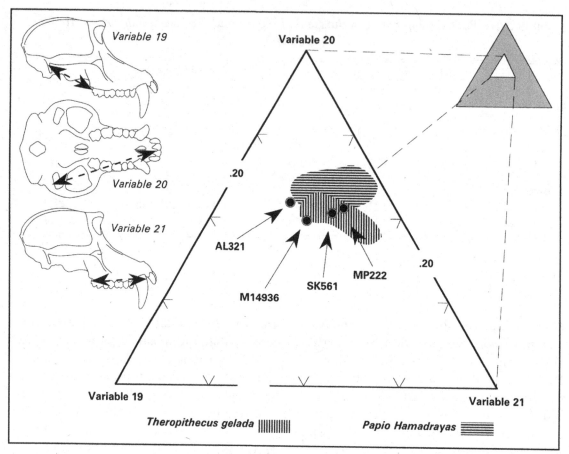

Fig. 2.12. Relationship of the occlusal plane to the temporal glenoid fossa. An increase in variable 19 in relation to variable 20 (leftward and downward) indicates elevation of the glenoid fossa.

Hadar specimens is based on comparison with three adult male mandibular fragments (MP1, formerly M201; MP44, formerly M621 or M626; and M3074) and one nearly complete juvenile mandible (M3071) from Makapansgat, which are described by Freedman (1957) and Maier (1970).

Mandibular body The anterior portion of the mandibular body is well preserved in five specimens (AL58-23, AL142-19 (Fig. 2.13), AL153-14a, AL205-1c, and AL329-

1). All of these exhibit very shallow mandibular corpus fossae, which terminate posteriorly below M_2. Similar fossae appear to have been present in the other three specimens (AL144-1, AL163-1, and AL174-10), although their damaged condition precludes certainty. All of the male mandibles from Makapansgat also have mandibular corpus fossae (*contra* Maier, 1970). The fossae in the juvenile are very shallow (their depths cannot be measured because of damage); the fossa in specimen MP1 is slightly deeper than those of the

Table 2.11. *Percentage size difference of specimens of* Hadar *crania (females)*

Variable	AL321–21	MP222	M14936	AL321–12 / MP222	AL321–12 / M14936
2	161	171	174	0.94	0.93
3	53	65	60	0.82	0.88
4	116	114	125	1.02	0.93
5	50	54	55*	0.93	0.91
6	37	46	40	0.80	0.93
7	34	33	—	1.03	
8	22	22	—	1.00	
9	67	68	—	0.99	
10	96	103	—	0.93	
12	102*	109	118	0.94	0.86
13	76	76	89*	1.00	0.85
14	45	44	50	1.02	0.90
15	63	63*	72	1.00	0.88
16	61	—	63		0.97
Average				0.96	0.90

*Estimated owing to damage.

Hadar specimens, and that of MP44 is substantially deeper than those in any of the other specimens of the species (Table 2.12). The absolute depths of the mandibular corpus fossae and the depths relative to the thickness of the ramus, in the Hadar specimens, are similar to those from Makapansgat (with the exception of MP44) and to two specimens of *T. oswaldi* from Olduvai Gorge (Old 067/5603 and Old 063/3366). The specimens from Hadar, as well as most specimens from Makapansgat, indicated that shallow mandibular corpus fossae are typical of male *T. darti*, and that the deep fossa of MP44 is exceptional (see also Dechow & Singer, 1984; Freedman, 1957; Maier, 1970). The suggestion by Jolly (1972) that

deep fossae might be characteristic of the species, indicating primitive status, is not borne out by the specimens from Hadar or the more recently recovered specimens from Makapansgat.

Portions of at least one mental ridge are preserved in all of the male mandibular specimens from Hadar, except AL144-4 and AL174- 10. Where preserved, the posterior portion of the ridge arises as a low, rounded eminence below M_1. As the ridge runs anteriorly, it is smoothly and convexly curved anteroinferiorly. In the three specimens in which most of the symphyseal region is preserved (AL142-19, AL163-11, and AL205-1c), the left and right ridges converge and nearly join on the incisive alveoli. The anterior portion of

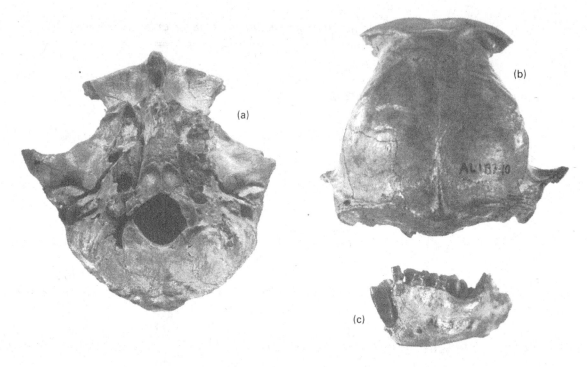

Fig. 2.13. (a) AL187–10, young adult male neurocranium, basal view; (b) AL187–10, posterior view; (c) AL284–2, juvenile mandible fragment, view of unerupted incisor. All approximately 75% natural size.

the ridges, between the incisive alveoli and the mesial root of the P_3, is slightly rugose and more sharply defined and ridge-like than the posterior portion. The angle of convergence of the two ridges is moderately wide.

Only two male mandibular specimens from Makapansgat (MP1 and M3074) have partially preserved mental ridges. The ridges in these specimens are generally similar to those of the Hadar males in their smoothly convex ·curvature and moderately wide angle of convergence. The Makapansgat specimens differ, however, from those of Hadar in two

important ways. First, the anterior portions of their ridges are rounded, not sharply defined and rugose as they are in the three Hadar specimens. Secondly, as noted by Freedman (1957) and Maier (1970), the anterior portions of the ridges in the Makapansgat specimens are more perpendicularly oriented in relation to the occlusal plane than they are in the Hadar specimens. This orientation produces a distinct chin-like, mental protuberance, which is not seen in the Hadar specimens, but is seen in male specimens of *T. oswaldi*.

Most of the inferior margin of the

Table 2.12. *Depth of the mandibular corpus fossa (males)*

Specimen	Depth (var. 22)	Thickness (var. 23)	D./T.
Theropithecus darti			
AL58–23	1.2	11.0	0.11
AL142–19	0.7	9.0	0.08
AL153–14a	0.6	12.1	0.05
AL205–1c	1.6	10.8	0.15
AL329–1	0.7	12.2	0.06
MP1	4.0	16.3	0.25
MP44	6.3	9.0	0.70
T. oswaldi			
Old 067/5603	2.7	15.6	0.17
Old 063/3366	2.8	15.8	0.18
T. brumpti			
L161–24	16.1	6.6	2.44
L199–5	8.0	7.5	1.07
L292–9	13.5	5.9	2.29
L576–8	11.1	8.0	1.39
L856–1	8.1	8.0	1.01
Omo 18–'69–495	12.9	4.2	3.07
Omo 40–'68–1405	9.9	7.0	1.41
Omo 75N–'71–C1	4.5	7.5	0.60
T. gelada			
Berkeley 1	5.4	4.4	1.23
2	5.6	4.6	1.22
5	4.1	6.7	0.61
W239	3.0	7.1	0.42
USNM 27039	5.2	6.0	0.87

mandibular body is preserved in three of the Hadar males (AL58-23, AL153-14a, and AL329-1); smaller anterior and posterior portions of the margins are present in specimens AL142-19 and AL144-1, respectively. The inferior margin of the juvenile specimen (AL329-1) is slightly anteriorly divergent, that is, it descends in relation to the occlusal plane as it runs from the level of the M_3 to that of the M_1. In the other specimens, all of which are adults, the margin has (AL58-23 and AL153-14a), or probably had (AL142-19 and AL144-1), a subparallel orientation to the occlusal plane.

The inferior margin is reasonably well preserved in three mandibular specimens from Makapansgat (MP1, MP44, and M3071). In the first two, which are adults, the margins are anteriorly divergent (as noted by Freedman, 1957), while in M3071, a juvenile, the margin has a subparallel orientation. If the ratio of the depth of the mandibular body below the M_1 to the depth below M_3 is used as a measure of the relationship of the inferior margin to the occlusal plane, then the relationship varies among adult males of *T. darti* from Hadar and Makapansgat as it does among males of the other congeneric species (Table 2.13). Jolly (1972) proposed that differences in the orientation of the inferior margin might be used to distinguish mandibular specimens of *T. darti* and *T. oswaldi*. The variation within, and the amount of overlap among, the species of the genus in this characteristic suggests that it will prove of little use in classification.

The incisive plane is well preserved in only one male mandibular specimen from Hadar, although it is partially preserved in three others. In all of the specimens, the plane terminates posteriorly at about the level of anterior P_4. The incisive plane appears relatively narrow and steeply sloped (deep) in specimens AL142-19 and AL163-11, intermediate in morphology in AL205-1c, and broad and nearly flat (shallow) in AL58-23. Similar variation in morphology is seen in three specimens from

Table 2.13. *Depth of the mandibular body below the M_1 and M_3 (males)*

Specimen	Depth below M_1 (var. 24)	Depth below M_3 (var. 25)	$D.M_1 / D.M_3$
Theropithecus darti			
AL58–23	35	30	1.17
AL142–19	27	—	
AL144–1	—	32	
AL153–14a	33	29	1.14
MP44	42	34	1.24
T. oswaldi			
Old 067/5603	44	37	1.19
Old 063/3366	45	39	1.15
T. brumpti			
L32–158	45	42	1.07
L161–24	57	49	1.16
L199–5	41	36	1.14
L292–9	52	41	1.27
L345–4	44	32	1.38
L576–8	55	45	1.22
L856–1	41	42	0.98
Omo 40–'68–1405	48	42	1.14
Omo 75N–'71–C1	36	31	1.16
T. gelada[†]			
ranges	32–40	25–31	1.20–1.44
means, n = 14	36	28	1.29

[†]From Eck (1977).

Makapansgat. As noted by Freedman (1957), the plane in MP44 appears narrow and steeply sloped, while that of MP1 is broader and more shallow. The incisive plane in M3071 is broad and very shallow. Although the incisive plane in specimens of *T. oswaldi* has often been described as broad and shallow (Andrews, 1916; Jolly, 1972; Leakey & Whitworth, 1958), its morphology varies in this species as well (Eck & Jablonski, 1987). The nature and extent of morphological variation of the incisive plane is, in fact, very similar among males of *T. darti* and *T. oswaldi*. Furthermore, Eck & Jablonski (1987) have shown that the incisive plane is also quite variable in males of both *T. brumpti* and *T. gelada*. The similar ranges of morphology of the incisive plane in males of *T. darti* and *T. oswaldi* and its variation in males of all species of the genus argues against its use in differentiating the mandibles of males of *T. darti* and *T. oswaldi* (*contra* Jolly, 1972; see also Dechow & Singer, 1984).

None of the males from Hadar have complete, mandibular dental arcades. When viewed superiorly, the preserved portions appear generally similar to the arcades of males of *Theropithecus*. The cheektooth rows are straight and slightly anteriorly convergent. The incisive alveoli are partially preserved in only three male specimens from Hadar (AL142-19, AL163-11, and AL205- 1c). In these, the anterior margin of the incisive alveoli appears to have formed a shallow arc connecting the canine regions. The arc is most similar in form to that of *T. oswaldi* and *T. brumpti*, being less strongly curved than it is in *T. gelada* (Eck & Jablonski, 1987). The width of the incisive alveoli (Variable 28) can be measured on AL163-11 and AL205-1c and estimated on AL142-19, giving values of 19 mm, 11 mm, and 11 mm, respectively. Of these specimens, only AL163-11 has a complete molar row, which is 44 mm long (Variable 30). The ratio of the width of the incisive alveoli to the length of the molar row in this specimen has a value of 0.43, which is considerably larger than that of the other specimens of the genus (see Eck & Jablonski, 1987; and Table 2.28, for comparative values). Given that the molars of AL163-11 are heavily worn and that the values of the widths of the incisive alveoli in the other two specimens are much smaller, it is likely that the relative width of the incisive alveoli in the Hadar males are generally similar to those of males of the other species of *Theropithecus*. Males of *T. darti* and *T. oswaldi* appear not to differ in the relative widths of their mandibular incisive alveoli, while they do in their maxillary alveoli.

Although mandibular dental arcades of Hadar males are quite similar to those of males of other species of the genus, they appear to differ from them in one important way. As discussed by Jablonski (1981) and Eck & Jablonski (1987), the mandibular dental arcades of most adult specimens of *Theropithecus* (including MP44 from Makapansgat) have 'reversed curves of Spee', that is, the occlusal surfaces of the P_4s through M_3s form an arc that is distinctly convex superiorly in an anterior–posterior direction. In most other catarrhines, the occlusal surfaces form either a straight line or a slightly concave (superiorly) arc. Only three adult male specimens from Hadar have intact cheektooth rows. Of these, two (AL58-23 and AL153-14a) have straight occlusal surfaces; only that of AL163-11 is slightly convexly curved. Thus, although the number of specimens is small, the sample of *T. darti* from Hadar may differ from all others of the genus, except *T. baringensis*, by having a less consistently reversed curve of Spee (see further discussion in the section on the female mandible).

Mandibular ramus Four male mandibular specimens from Hadar (AL58-23, AL144-1, AL153-14a, and AL174-10) preserve small portions of the anteroinferior ramus; the gonial region is also intact in AL153-14a. The anteroinferior apex of the triangular depression on the lateral surface of the ramus is partially preserved in the latter three specimens. In specimen AL144-1 and AL153-14a, the apex is deeply excavated and the posteroinferior margin of the depression forms a strongly developed ridge. The apex is less deeply excavated and the posteroinferior margin is more rounded in AL174-10. In all three specimens, the angle between the posteroinferior and anterior margins of the depression is moderately wide. The Hadar specimens are similar to M3071 from Makapansgat in all of these features and differ from most male specimens of *T. oswaldi* (in which the anteroinferior apex is very shallow, the posteroinferior margin is low and rounded, and the angle between the anterior and posteroinferior margins is more acute).

The most anteroinferior area of insertion of the temporalis tendon on the lateral aspect of the mandible is preserved in three of the Hadar male specimens. In AL58-23 and AL144-1, it forms a very low, rounded, roughened eminence, while in AL153-14a it forms a low, roughened ridge that is anteriorly convexly curved. This region is not preserved in the adult male specimens from Makapansgat and the area of the insertion is unmarked in the juvenile (M3071). The morphology of the insertion area in AL58-23 and AL144-1 is most similar to that of two adult male specimens of *T. oswaldi* (Old 067/5603 and Old 580,57 SC II), while the area in

AL153-14a is most similar to that in a number of male specimens of *T. brumpti*, a species in which the morphology of the area is highly variable (Eck & Jablonski, 1987).

Although the superior portion of the ramus is missing from all of the Hadar male specimens, the more inferior portions preserved in AL153-14a and AL174-10 suggest tall, upright rami similar to those consistently seen in other specimens of the genus.

Size of the male mandible The fragmentary state of preservation of the male mandibles from Hadar makes comparison of their size with specimens from Makapansgat and those of *T. oswaldi* difficult. Visually they appear, however, somewhat smaller than those from Makapansgat and substantially smaller than specimens of *T. oswaldi* from Kanjera, Beds I and II at Olduvai Gorge, and Swartkrans.

Female mandible

The description of the female mandible of *T. darti* from Hadar is based on the nine best preserved specimens, all of which are fragmentary. They were recovered from all three members of the Hadar Formation and, thus, span a substantial range of time, but as in the case of the male mandibles, there are no consistent morphological differences between the geologically oldest and youngest. The Hadar female mandibles are compared with female specimens MP56/57/95 (formerly M633, M634, and M2981; fragments probably of the same individual) and M3073; the

former is an adult and the latter a juvenile from Makapansgat, described by Freedman (1957) and Maier (1970, 1972).

Mandibular body The anterior, lateral surface of the mandibular body is well preserved in all but one (AL310-15) of the nine female mandibular specimens from Hadar. A very shallow mandibular corpus

Table 2.14. *Depth of the mandibular corpus fossa (females)*

Specimen	Depth (var. 22)	Thickness (var. 23)	D./T.
Theropithecus darti			
AL126–30	0.4	10.4	0.04
AL129–8	0.5	8.7	0.06
AL185–5c	0.6	9.2	0.07
AL186–17	0.8	11.6	0.07
AL196–3a	0.3	12.4	0.02
AL217–1	0.2	10.2	0.02
AL269–3	0.4	10.5	0.04
AL270–1	0.5	10.5	0.05
MP 56/57/95	0.6	12.1	0.05
T. oswaldi			
M11539	1.0	12.4	0.08
M14938	1.7	12.2	0.14
Old 067/2771	1.2	13.0	0.09
Old 068/6516	0.0	22.3	0.00
KNM–OG–2	1.8	15.6	0.12
SK411	0.5	15.6	0.03
T. brumpti			
L49–1	3.7	8.2	0.45
L199–3	6.5*	4.5	1.44
L305–6	6.8	5.6	1.21
L137–4	4.0	6.2	0.65
T. gelada			
Berkeley 3	5.1	3.5	1.45
6	3.1	4.6	0.67

*Estimated owing to damage.

fossa, which terminates posteriorly at the level of M$_2$, is present in these eight specimens. The morphology of the fossae in the Hadar females is very similar to that in the males, but on average those of the females appear to be slightly more shallow. The morphology and depth of the mandibular corpus fossae in the Hadar females are also very similar to those of the Makapansgat females and to those generally present in females of *T. oswaldi* (Table 2.14).

Mental ridges are well preserved on five of the female mandibles (AL126-30, AL129-8, Fig. 2.14; AL186-17, Fig. 2.15;

AL196-3a, Fig. 2.9; and AL269-3, Fig. 2.15); the region is completely absent only on AL310-15. The morphology of the ridges of the females is very similar to that described for the males; in fact, there appears to be no sexual dimorphism in these structures within the Hadar sample. The ridges are also generally similar to those of the adult female from Makapansgat and those of females of *T. oswaldi*. Only in the anterior portion of the ridges do the Hadar females differ, for their anterior ridges are sharply defined and slightly rugose, whereas those of the

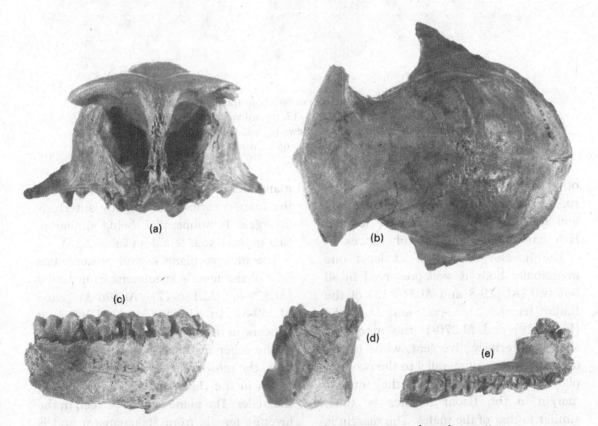

Fig. 2.14. (a) AL187–10, young adult male neurocranium, frontal view; (b) AL187–10, superior view; (c) AL129–8, adult female mandible fragment, lateral view; (d) AL129–8, frontal view; (e) AL129–8, occlusal view. All approximately 75% natural size.

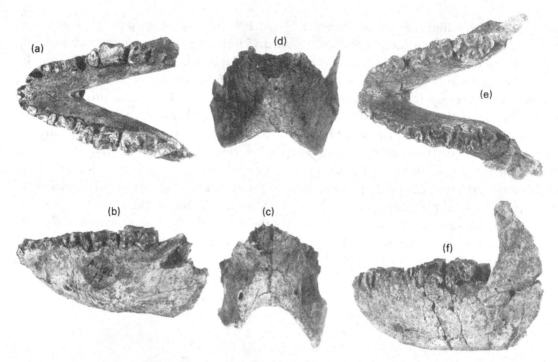

Fig. 2.15. (a) AL186–17, adult female mandible fragment, occlusal view;
(b) AL186–17, lateral view; (c) AL186–17, frontal view; (d) AL269–3, adult
female mandible fragment, frontal view; (e) Al269–3, occlusal view; (f)
AL269–3, lateral view. All approximately 65% natural size.

other females are low, rounded, and non-rugose. Like females from Makapansgat and those of *T. oswaldi*, Hadar females lack mandibular mental protuberances.

The inferior margin of at least one mandibular body is well preserved in all but two (AL129-8 and AL310-15) of the Hadar females. In specimen AL185-5c (Fig. 2.10) and AL270-1, the margin is slightly anteriorly divergent, while in the other five it lies subparallel to the occlusal plane. The orientation of the inferior margin in the Hadar females is, thus, similar to that of the males. The margin is poorly preserved and its orientation obscure in the females from Makapansgat. In a number of well-preserved female mandibles of *T. oswaldi*, the orientation of the margin varies from slightly anteriorly divergent to subparallel, being similar to that in the Hadar females (Table 2.15).

The incisive plane is well preserved in four of the female specimens from Hadar (AL129-8, AL186-17, AL196-3a, and AL269-3). In AL129-8 and AL269-3, it appears narrow and steeply sloped, while in the other two, it is broad and shallow; thus, the morphology of the incisive plane varies in the Hadar females as it does in the males. The plane cannot be seen in the juvenile female from Makapansgat and is poorly preserved in the adult, where, as noted by Jolly (1972), it appears to have been narrow and steeply sloped. The

Table 2.15. *Depth of the mandibular body below the M_1 and M_3 (females)*

Specimen	Depth below M_1 (var. 24)	Depth below M_3 (var. 25)	$D.M_1$ / $D.M_3$
Theropithecus darti			
AL126–30	26	25	1.04
AL186–17	27	27	1.00
AL196–3a	27	26	1.04
AL217–1	30	27	1.11
AL269–3	33	28	1.18
AL270–1	27	24	1.23
AL310–15	—	25	
T. oswaldi			
M11539	35	29	1.21
M14937	38*	37*	1.03
M14938	38	33	1.15
Old 067/2771	35	28	1.25
Old 068/6516	43	44	0.98
KNM–OG–2	35	32	1.09
SK575	29	28	1.04
T. brumpti			
L49–1	33	26	1.27
L199–3	38	32	1.19
L305–6	34	26	1.13
L317–4	34	28	1.21
T. gelada[†]			
ranges	26–32	23–28	0.96–1.30
means, n = 15	29	25	1.15

*Estimated owing to damage.
†From Eck (1977).

Table 2.16. *Depth and width of the incisive plane (females)*

Specimen	Depth (var. 26)	Width (var. 27)	D./W.
Theropithecus darti			
AL129–8	7	16*	0.44
AL186–17	4	18	0.22
AL196–3a	5	21*	0.24
AL269–3	7	20	0.35
T. oswaldi			
M11539	6	20*	0.30
Old 067/2771	8	20*	0.40
Old 068/6516	8	34	0.24
SK411	6	23*	0.26
T. brumpti			
L49–1	9	21*	0.43
L199–3	8	25*	0.32
L305–6	11	20*	0.55
T. gelada			
Berkeley 3	9	6	1.50
	6	8	6	1.33

*Estimated owing to damage.

incisive plane also varies among female specimens of *T. oswaldi* and the range of variation of the ratio of the depth of the incisive plane to its width is similar to that seen in females from Hadar (Table 2.16).

Complete mandibular dental arcades are present in two of the Hadar females (AL186-17 and AL269-3), although their anterior teeth are not in good condition. The cheektooth rows appear more strongly anteriorly convergent and the incisive alveoli more sharply curved in these two specimens and two others (AL129-8 and AL196-3a, from which most of the left mandibular bodies are missing) than they are in the males. The differences between females and males from Hadar are similar to those between the sexes in other species of the genus. As can be seen in Table 2.17, the relative widths of the mandibular dental arcade at the I_2s appear, on average, to be slightly larger in the Hadar females than they are in females of *T. oswaldi*. Relative widths of the

Table 2.17. *Relative widths of the mandibular incisive and canine alveoli (females)*

Specimen	Width of incisive alveoli (var. 28)	Width of canine alveoli (var. 29)	Molar row length (var. 30)	W.I.A. / M.R.L.	W.C.A. / M.R.L.
Theropithecus darti					
AL129–8	12	19	36	0.33	0.53
AL186–17	14	19	36	0.39	0.53
AL196–3a	14	21	40	0.35	0.53
AL269–3	—	18	38		0.47
T. oswaldi					
M11539	16	25	45	0.36	0.56
Old 067/2771	13	21	50	0.26	0.42
Old 068/6516	17*	38	58	0.29	0.66
SK411	14	20	46	0.30	0.43
T. brumpti					
L49–1	17	25	45	0.38	0.56
L119–3	15	23	47	0.32	0.49
L305–6	17	23	48	0.35	0.48
T. gelada[†]					
means, n = 15	13	18	34	0.38	0.53

*Estimated owing to damage.
[†]From Eck (1977)

mandibular dental arcade at the C_Ls are, however, similar. The occlusal surfaces of the P_4s through M_3s are reasonably well preserved on at least one side in all of the Hadar adult females. The surfaces form a slightly superiorly convex curve in two specimens (AL186-17 and AL270-1) and a distinctly, superiorly concave curve in one (AL217-1). The surfaces lie on a straight line in all of the others. Thus, Hadar females, like the males, have a variable configuration in the curve of Spee, whereas it is consistently 'reversed' (superiorly convex) in other adult females of the genus, including the adult female from Makapansgat.

Mandibular ramus Small portions of the anteroinferior mandibular ramus are present in three female specimens from Hadar (AL185-5c, AL196-3a, and AL269-3). The anteroinferior apex of the triangular depression is deeply excavated in specimen AL196-3a, while it is moderately deep in the other two. The posteroinferior margin of the depression is rounded in AL185-5c and AL196-3a, whereas it appears to have formed a more sharply defined ridge in AL269-3. In all three specimens, the angle between the anterior and posteroinferior margins is wide. The morphology of the triangular depression of the Hadar females is, thus, similar to that of the males. The morphology of the depression in the adult female from Makapansgat is similar to that of the Hadar females, while that of the juvenile female differs. In the latter specimen, the depression is very shallow and has a low, rounded posteroinferior margin and acute anteroinferior apex, being similar to the depressions in many female specimens of *T. oswaldi*.

The most anteroinferior attachment area for the temporalis tendon is present in the three Hadar female specimens listed above and in three others (AL126-30, AL217-1, and AL310-15). In the juvenile specimen (AL185-5c), the attachment area is unmarked. It forms an anteriorly convex ridge in AL126-30, and a low, roughened eminence in AL196-3a, AL217-1, and AL310-15. Although the region is slightly damaged on the left and missing from the right ramus of AL269-3, the left attachment area appears to have been marked by a moderately pronounced tubercle. The morphology of the insertion area of the temporalis tendon is, thus, as variable in the Hadar females as it is in females of *T. oswaldi* and *T. brumpti* (see Eck & Jablonski, 1987, for further discussion).

The poorly preserved rami of AL185-5c, AL196-3a, and AL269-3, like those of the males from Hadar, suggest the tall, upright configuration typical of the genus.

Size of the female mandible Documen-tation of the size of the female mandibles from Hadar is difficult. It is my impression, however, that they are slightly smaller than those from Makapansgat, which are in turn slightly smaller than those of females of *T. oswaldi* from Kanjera, Beds I and II at Olduvai Gorge, and Swartkrans. The differences in size between these three sets of females appear to be smaller than the differences between the equivalent sets of males. It should also be noted that the four mandibular specimens from the Sidi Hakoma Member seem very slightly smaller than the five from the Denen Dora and Kada Hadar Members.

Dentition

Incisors There is general agreement that the incisors of *T. gelada* are smaller in relation to the rest of the dentition than are those of the other large African cercopithecines (see Delson, 1975; Eck, 1977; Jolly, 1970). Furthermore, it is agreed that the relative sizes of the incisors of *T. oswaldi* have undergone substantial additional reduction when compared to those of the living species of the genus (Freedman, 1957; Jolly, 1972; Leakey & Leakey, 1973; Szalay & Delson, 1979). Although the incisors of *T. brumpti* are poorly preserved, evidence suggests that their relative sizes are similar to those of *T. gelada* (Eck & Jablonski, 1987). The incisors of *T. darti* from Makapansgat are also poorly preserved and their relative sizes not precisely defined, but Maier (1972) has stated that the relative sizes of the Makapansgat incisors are not as small as those of *T. oswaldi* and are similar to those of the living species, a statement with which I agree.

The sample of permanent incisors of *T. darti* from Hadar is also small, totalling nine teeth. Four of these teeth are I²s; two are associated with maxillary fragments (AL185-5b, left; and AL186-18, right); two are isolated (AL188-19, left; and AL200-19, right). The mandibular specimen AL284-2 (Fig. 2.6) has an unerupted left I_1, which has been excavated from its crypt. All four of the lower incisors are preserved in specimen AL329-1; all are moderately worn.

The basal lengths, basal widths, and labial heights of the incisors differ little between the samples of *T. darti*, *T. oswaldi*, and *T. gelada*; their incisors are all similar in absolute size. When these variables are, however, related to the basal length of the upper or lower M1s, the values of the ratios for the Hadar specimens are substantially larger than those of *T. oswaldi* and very similar to those of *T. gelada* (Tables 2.18, 2.19, and 2.20). The incisors of *T. darti* from Hadar and Makapansgat appear, thus, to be only moderately reduced, similar in relative size to those of *T. gelada* and substantially larger in relative size than those of *T. oswaldi*. Evidence for relatively large incisors in *T. darti* is augmented by evidence from the maxillary and mandibular dental arcades (see discussion above).

Canines Although Jolly (1970) has stated that the canines of *Theropithecus* are reduced when compared to those of *Papio*, the basis for his statement is unclear. As shown by Washburn & Ciochon (1974)

and Eck (1977), the relative lengths and heights of the crowns of the C^Us and C_Ls of both males and females of *T. gelada* are very similar to those of the savannah baboon. In addition, the relative widths of the crowns of the C^Us and C_Ls of females and C_Ls of males are similar to those of *Papio*. The only consistent differences between the canines of the two genera are seen in the slightly thinner and more sharply curved C^Us of males of *T. gelada* (Eck, 1977). Although the sample of canines of *T. brumpti* is small and not well preserved, the available evidence suggests that the morphology and relative size of the canines of both males and females of this species are very similar to those of the living species (Eck & Jablonski, 1987). Similarly, the small and poorly preserved sample of canines of *T. darti* from Makapansgat suggest strong similarity in morphology and relative size to those of *T. gelada* (Freedman, 1960; Maier, 1970, 1972; Eck & Jablonski, 1987). If one considers only relative crown height, the canines of males and females of *T. oswaldi* appear reduced (*sensu* Jolly, 1972) when compared to those of the other congeneric species. The relative widths of the crowns of the C^Us and C_Ls in males and the C^Us in females of *T. oswaldi* are, however, larger than those of the other species. The canines of *T. oswaldi* are relatively short, robust teeth (Eck & Jablonski, 1987).

Permanent canines are preserved in only nine of the Hadar specimens and, in most, they are not in good condition. Four male C^Us have been recovered. Two of these are isolated specimens (AL148-105, left; and AL200-16, right); both are slightly worn; the former lacks the apex of its

Table 2.18. *Dimensions of I^2*

Specimen	Basal length I^2 (var. 31)	Basal width I^2 (var. 32)	Labial Height I^2 (var. 33)	Basal length M^1 (var. 52)	B.W. I^2 / B.L. M^1	L.H. I^2 / B.L. M^1
Theropithecus darti						
AL185–5b	5.1	5.6	—	8.7	0.64	
AL186–16b	4.4	6.7	13.3	10.0	0.67	1.33
AL188–19	4.9	—	—	—		
AL200–18	4.1	5.4	—	—		
T. oswaldi						
M18720	4.2	6.3	—	11.2	0.56	
M18748	4.1	5.5	—	—		
M18749	3.7	5.3	—	—		
M18750	4.1	5.4	—	—		
M32104	4.0	5.6	—	10.1	0.55	
SK561	4.0	5.1	—	11.2	0.46	
Old S58	4.6	6.5	—	—		
Old 69 S133	4.9	6.5	—	14.0	0.46	
Olorgesailie						
n	39	39	6	32		
ranges	3.6–5.5	5.1–8.7	9.4–13.3	11.7–15.2		
means	4.69	6.69	11.13	13.22	0.51	0.84
T. gelada[†]						
n	34	34	4	32		
ranges	3.3–4.9	4.8–7.6	11.3–14.3	7.8–10.6		
means	3.98	5.84	12.93	8.96	0.65	1.44

[†]From Eck (1977).

crown, while the dento–enamel junction of the latter is missing. Both C^us are present in specimen AL310-19, but they are only beginning to erupt and the apices of their crowns are damaged. Although the crown heights of none of these specimens can be measured, their preserved portions do not suggest the short, squat C^us typical of males of *T. oswaldi*. The length and width of the crown can be measured only on specimen AL148-105 (Table 2.21). The value of the ratio of width to length of this specimen is 0.67, a value intermediate between those seen in males of *T. oswaldi* and those seen in males of *T. gelada*, *T. brumpti*, and a single specimen of *T. darti* from Makapansgat (compare with values presented in Eck & Jablonski, 1987, Table 35). None of the female specimens from Hadar have upper canines.

Table 2.19. *Dimensions of I_1*

Specimen	Basal length I_1 (var. 34)	Basal width I_1 (var. 35)	Labial Height I_1 (var. 36)	Basal length M_1 (var. 64)	B.W. I_1 / B.L. M_1	L.H. I_1 / B.L. M_1
Theropithecus darti						
AL284–2	4.2	6.7	14.2	10.7	0.63	1.33
AL329–1	3.3	7.2	—	9.9	0.73	
T. oswaldi						
M18747	3.3	6.0	—	—		
F3398	3.2	5.1	—	10.5	0.49	
Old 068/6640	3.7	5.8	—	—		
Olorgesailie						
n	36	36	7	38		
ranges	3.1–4.1	5.3–7.2	11.0–12.9	11.8–14.8		
means	3.52	6.37	11.76	13.27	0.48	0.89
T. gelada†						
n	30	30	1	27		
ranges	2.4–3.5	5.0–7.0		7.4–9.6		
means	3.00	5.95	11.3	8.49	0.70	1.33

†From Eck (1977).

Lower canines are present in five Hadar specimens, three males (AL142-19, AL253-2, and AL329-1) and two females (AL126-30 and AL129-8). None of the crown heights can be measured, but again the most complete specimens (AL253-1 and AL329-1) do not suggest the strong reduction in height seen in the lower canines of males of *T. oswaldi*. Width/length ratios can be calculated for all three male specimens. The values for AL142-19 and AL329-1 are very similar to those of males of *T. gelada* and *T. brumpti*, whereas the value for AL253-2 is similar to those of *T. oswaldi* (compare values in Table 2.21 with those in Eck &

Jablonski, 1987, Table 36). The value of the width/length ratio for AL129-8 is similar to that for other females of the genus.

The comparative morphology of the canines of the Hadar specimens is thus unclear. At least in the males, it appears that the canines are not as short and robust as they are in males of *T. oswaldi*, but they may not be as tall and thin as they are in *T. gelada* and *T. brumpti* (see also Dechow & Singer, 1984).

Premolars As shown by Jolly (1972), Dechow & Singer (1984), and Eck & Jablonski (1987), the P_3s of males of

Table 2.20. *Dimensions of* I_2

Specimen	Basal length I_2 (var. 37)	Basal width I_2 (var. 38)	Labial Height I_2 (var. 39)	Basal length M_1 (var. 64)	B.W. I_2 / B.L. M_1	L.H. I_2 / B.L. M_1
Theropithecus darti						
AL329–1	3.2	6.5	—	9.9	0.66	
T. oswaldi						
M18744	3.1	5.5	—	—		
F3398	2.8	5.6	—	10.5	0.53	
Olorgesailie						
n	37	37	1	38		
ranges	3.1–4.0	5.2–7.7		11.8–14.8		
means	3.40	6.32	10.0	13.27	0.48	0.75
T. gelada[†]						
n	34	34	4	27		
ranges	2.2–3.0	4.7–7.1	9.7–11.2	7.4–9.6		
means	2.65	5.65	10.65	8.49	0.66	1.25

[†]From Eck (1977).

Table 2.21. *Dimensions of* C^U *and* C_L *of Hadar specimens*

Specimen	C^U length (var. 34)	C^U width (var. 29)	C_L length (var. 30)	C_L width (var. 52)	C^UW / C^UL	C_LW / C_LW
Males						
AL148–105	15.6	10.4	—	—	0.67	
AL142–19	—	—	11.9	6.4		0.54
AL253–2	—	—	11.8	7.6		0.64
AL329–1	—	—	12.5	7.0		0.56
Females						
AL126–30	—	—	—	4.1		
AL129–8	—	—	6.9	3.8		0.56

T. oswaldi are relatively short teeth when compared to those of *T. gelada*, *T. brumpti*, and *T. darti* from Makapansgat. In addition, the relative lengths of the P_3s of males of *T. oswaldi* change through time, for those of middle Pleistocene age from Olorgesailie are relatively shorter than those of early Pleistocene age from Kanjera, Bed I at Olduvai Gorge, and Swartkrans. The case for reduction of the relative length of the P_3s in females of the species is less clear. While these teeth appear to be relatively shorter in females from Kanjera, Olduvai Gorge, Swartkrans, and Olorgesailie than in females of *T. gelada*, the difference is not as great as it is between the males of the two species (Eck & Jablonski, 1987). Furthermore, the relative length of female P_3s of latest Pliocene *T. oswaldi* from the Shungura Formation is highly variable; some of the specimens from this site are as short as those from Olorgesailie, while others are as long or longer than those of extant females (Eck, 1987).

Well-preserved P_3s are present in six male and four female specimens from Hadar. The absolute sizes of the P_3s in the Hadar sample are similar to those in samples of other species of the genus, being, on average, slightly larger than those of *T. gelada*, slightly smaller than those of *T. brumpti*, and most similar to those of *T. darti* from Makapansgat and *T. oswaldi* (compare values of length and width presented in Table 2.21 with those presented in Table 39 of Eck & Jablonski, 1987). In only three of the male specimens from Hadar are the P_3s associated with M_3s. The ratio of the length of the P_3 to the length of the M_3 for two of these (AL163-11 and

Table 2.22. *Dimensions of P_3 of Hadar specimens*

Specimen	P_3 length (var. 44)	P_3 width (var. 45)	M_3 maximum length (var. 73)	L. P_3 / M.L. M_3
Males				
AL142–19	18.7	6.0	—	
AL153–14a	17.3	6.4	19.0	0.91
AL163–11	20.0*	6.4	20.7	0.97
AL205–1	19.4	6.7	—	
AL208–10	18.7	6.3	—	
AL329–1	18.7	6.4	18.6	1.01
Females				
AL126–30	10.3*	—	17.1	0.60
AL129–8	10.4	5.0	14.9	0.70
AL185–5c	12.3	5.6	—	
AL202–3	12.6	5.5	—	

*Estimated owing to damage.

AL329-1) has values most similar to those of males of *T. gelada*, *T. brumpti*, and *T. darti* from Makapansgat. The value for the third specimen (AL153-14a) falls within the range of values for males of *T. oswaldi* (Table 2.22). Two of the female specimens from Hadar (AL126-30 and AL129-8) have associated P_3s and M_3s; the values for their ratios are most similar to the mean ratio of a sample of females of *T. gelada* and larger than those for most females of *T. oswaldi* (see Eck, 1987, and Eck & Jablonski, 1987, for further discussion of these ratios). Comparison with females of *T. brumpti* and *T. darti* from Makapansgat cannot be made because of damage to all specimens.

Thus, the P_3s of the males and females from Hadar may not be as reduced in relative length as are those of *T. oswaldi*. The small sample sizes, the low value of the ratio of AL153-14a, and the wide range in

Table 2.23. *Dimensions of* P_4 *of* Hadar specimens

Specimen	P_4 length (var. 46)	P_4 width (var. 46)
Males		
AL58–23	8.7	7.7
AL142–19	8.0	7.0
AL153–14a	8.1	7.9
AL163–11	8.7	9.0
AL205–1c	8.4*	7.9
AL329–1	9.6	7.9
Females		
AL126–30	7.2	—
AL129–8	7.7	—
AL185–5c	7.9	6.8
AL310–15	8.3	6.9
Unknown		
AL153–18	9.1	7.2

*Estimated owing to damage.

Table 2.24. *Dimensions of* P^3 *and* P^4 *of* Hadar *specimens*

Specimen	P^3 length (var. 48)	P^3 width (var. 49)	P^4 length (var. 50)	P^4 width (var. 51)
Males				
AL134–5a	—	—	—	8.4
AL205–1b	7.4*	8.8	7.8	9.2
AL208–10b	7.4	8.4	—	—
AL310–19	7.5	8.6	8.4	9.5
Females				
AL185–5b	—	—	7.3	8.2
AL321–12	5.7	7.1	6.2	7.9
Unknown				
AL52–1	—	—	7.0	—
AL200–17	7.5	8.9	—	—

*Estimated owing to damage.

variation of the ratio in the females from the Shungura Formation suggest, however, that this generalization should be treated with caution.

The other premolars of the species of *Theropithecus* are generally similar in morphology. It should be noted, however, that on average these teeth in the Hadar sample appear very slightly smaller than those of *T. darti* from Makapansgat, smaller than those of *T. oswaldi* of latest Pliocene and early Pleistocene age, and substantially smaller than those of Middle Pleistocene *T. oswaldi* (compare the values presented in Table 2.23 and 2.24 with those published in Eck, 1987; Freedman, 1957; Jolly, 1972; Leakey & Leakey, 1973; Maier, 1970).

Molars There is general agreement that the molars of the various species of *Theropithecus* share a suite of characteristics that differ from those of other cercopithecines (see Szalay & Delson, 1979, for a review). The molars of the Hadar specimens certainly exhibit this suite of characteristics. The most obvious differences between the Hadar molars, those of *T. darti* from Makapansgat, and those of *T. oswaldi* involve size. Comparison of values presented in the references cited above with those presented in Tables 2.25 and 2.26 indicates a pattern of molar size similar to that seen among the premolars. Although the numbers of premolars and upper molars from Hadar are small as they are from most sites and, thus, only suggestive of size differences, the numbers of M_2s and M_3s are considerably larger and provide a more convincing basis for comparison. For

Table 2.25. *Dimensions of M^1, M^2, M^3, and $M^?$ of* Theropithecus *specimens*

Specimen	M^1 basal length (var. 52)	M^1 maximum width (var. 53)	M^1 mesial width (var. 54)	M^1 distal width (var. 55)	M^2 basal length (var. 56)	M^2 maximum width (var. 57)	M^2 mesial width (var. 58)	M^2 distal width (var. 59)
Males								
AL134–5	8.7	—	—	—	10.2	—	—	—
AL205–1	9.8	9.8	9.7	—	11.0	—	—	—
AL208–10b	10.8	11.4	10.0	9.8	11.2	12.5	11.9	10.5
Females								
AL185–5	8.7	10.5	9.1	8.1	11.1	12.4	11.2	10.6
AL321–12	8.5	10.4	9.3	8.7	10.8	13.0	11.5	10.0
Unknown								
AL52–1	8.5	10.7	9.4	9.1	—	—	—	—
AL116–23	—	—	—	—	11.5	13.9	11.7	11.1
AL186–16	10.0	11.6	10.2	9.4	—	—	—	—
AL263–1	—	11.3	8.0	8.1	—	—	—	—

Specimen	M^3 basal length (var. 60)	M^3 maximum width (var. 61)	M^3 mesial width (var. 62)	M^3 distal width (var. 63)		Specimen	$M^?$ basal length	$M^?$ maximum width	$M^?$ mesial width	$M^?$ distal width
Males						*Unknown*				
AL208–10b	13.0	14.5	—	11.2		AL200–12	11.8	13.8	12.9	12.7
Females						AL225–5	10.1	13.0	—	10.3
AL321–12	11.0	13.6	11.3	9.7		AL300–6	8.1	10.5	9.7	9.2
Unknown										
AL185–16	11.4	13.4	11.4	9.8*						

*Estimated owing to damage.

example, comparison of the ranges of variation and means of the basal lengths of the M_2s or M_3s from Hadar with those from Makapansgat and a number of samples of *T. oswaldi* suggests a relatively consistent increase in the values as their age decreases (Table 2.27). Several points should be made about the information presented in Table 2.27. First, the basal length (Variables 68 and 72) was chosen as the best estimator of molar size because, of all molar dimensions, it is usually the least affected by interstitial wear and/or damage to the molar crown. Secondly, the dimensions of males and females were not analyzed separately because the sex of

Table 2.26. *Dimensions of M_1, M_2, and M_3 of Hadar specimens*

Specimen	M_1 basal length (var. 64)	M_1 maximum length (var. 65)	M_1 mesial width (var. 66)	M_1 distal width (var. 67)	M_2 basal length (var. 68)	M_2 maximum length (var. 69)	M_2 mesial width (var. 70)	M_2 distal width (var. 71)	M_3 basal length (var. 72)	M_3 maximum length (var. 73)	M_3 mesial width (var. 74)	M_3 distal width (var. 75)
Males												
AL58–23	9.1	11.0	7.9	8.3	11.8	13.3	10.5	10.3	18.0	18.3	12.2	10.5
AL142–19	9.1	10.2	8.0	8.4	11.7	13.1	—	—	—	—	—	—
AL144–1	—	—	—	—	12.3	13.1	11.0	10.8	16.3	17.0	11.3	9.4
AL148–107	9.7	11.1	8.5	8.9	12.1	13.1	10.5	9.3	17.8	19.0	11.7	11.0
AL152–14a	9.1	10.5	8.7	8.7	11.9	13.0	10.3	9.8	20.2	20.7	12.9	10.1
AL163–11	9.8	—	—	10.0	12.0	13.6	12.2	10.5*	20.4	20.9	14.0	12.9*
AL174–10	—	—	—	—	15.1	16.2	—	12.7	—	—	—	—
AL205–1c	9.3	—	—	—	—	—	—	—	—	—	—	—
AL208–10a	9.0	10.0	8.6	8.7	11.9	12.0	—	10.5	16.7	—	11.0	11.0
AL329–1	9.9	11.5	8.4	8.7	13.8	14.1	11.6	10.3	17.3	12.7	12.7	10.8
Females												
AL126–30	9.1	9.6	8.4	8.4	11.6	11.8	10.7	10.0	16.0	17.1	11.6*	10.0
AL129–8	8.2	9.0	7.8	8.4*	10.3	11.9	9.3	9.0	13.4	14.9	9.3	8.2
AL185–5c	9.2	10.4	—	9.0	12.3	13.0	10.2	9.3	—	17.1*	—	9.5
AL186–17	—	—	—	—	10.6	—	—	—	16.0	17.1	10.8	10.4
AL196–3a	—	—	—	—	11.2	13.0	9.8	9.7	17.1	—	—	—
AL204–4	12.2	12.7	10.2	—	—	—	—	—	—	—	—	—
AL217–1	—	—	—	—	—	—	—	—	15.5	—	—	9.7
AL269–3	—	—	—	—	11.0	—	—	—	16.7	17.1	12.4	11.0
AL270–1	—	—	—	—	—	—	—	—	16.0*	—	—	10.5
AL310–15	9.7	—	8.6	—	11.8	13.2	—	—	17.4	18.2	10.8	9.7
AL345–1	8.9	10.4	7.9	—	—	—	—	—	—	—	—	—
Unknown												
AL55–43	10.0	11.7	8.3	8.7	—	11.4	9.3	9.1	—	—	—	—
AL56–17	—	—	—	—	10.5	—	—	—	16.7	17.2	11.1*	9.3*
AL137–11	—	—	8.4*	—	12.4	—	—	—	—	—	—	—
AL137–12	9.0	10.9	8.3	8.2	13.4	14.2	11.4	10.4	—	—	—	—
AL153–18	10.0	—	—	—	10.1	12.9	9.0	8.6	15.1	—	—	—
AL158–91	—	—	—	—	10.5	11.5	9.2	8.7	15.1	15.9	9.5	8.7
AL161–23	—	—	—	—	—	—	—	—	—	—	—	—
AL183–45	8.8	10.7	8.2	8.3	—	—	—	—	—	—	—	—
AL193–1	10.7	—	—	—	—	—	—	—	16.1	—	—	10.9
AL199–4	—	—	—	—	12.0	13.8	10.6	10.5	—	—	—	—
AL200–14	—	—	—	—	—	—	—	—	17.2	—	12.7	—
AL208–6	—	—	—	—	—	—	—	—	—	—	—	11.1
AL284–2	10.8	12.8	8.4	8.7	—	—	—	—	17.8	—	12.1	—
AL304–1	—	—	—	—	—	—	—	—	17.2	18.1	10.7	10.7
AL317–2	—	—	—	—	—	—	—	—	16.1	17.5	—	9.8

*Estimated owing to damage.

Table 2.27. *The size (basal length) of the M_2s and M_3s of* Theropithecus darti *and* T. oswaldi *as a function of geological age*

Site	M_2			M_3		
	n	range	mean	n	range	mean
Middle Pleistocene						
Hopefield	2	16.8–17.8	17.3	4	24.0–27.4	25.6
Olorgesailie	42	15.1–19.5	17.3	21	22.8–27.1	25.3
Ternifine	1		16.5	3	22.4–26.2	23.8
Olduvai Gorge						
(Upper Bed II & Bed III)	3	15.7–18.5	17.3	9	22.5–26.4	24.5
Plio-Pleistocene						
Swartkrans	9	12.9–16.5	14.3	4	19.4–25.0	22.0
Olduvai Gorge						
(Bed I and Lower Bed II)	8	13.9–16.8	15.5	7	19.3–22.5	21.2
Kanjera	3	13.9–15.1	14.5	5	19.7–21.2	20.5
Shungura	16	12.9–17.0	14.5	15	17.9–24.2	20.6
Pliocene						
Makapansgat	6	12.0–16.3	13.9	6	16.7–20.5	18.8
Hadar	22	10.1–15.1	11.8	22	13.4–20.4	16.9

many of the molars from most of the sites cannot be accurately determined. Thirdly, the sites are listed in rough chronologic order, from youngest to oldest (see Eck & Jablonski, 1987, for further discussion).

As discussed by Jolly (1972), the molars of *T. oswaldi* become larger through time. This trend is even more apparent if the Kanjeran sample is of Plio-Pleistocene age, as argued by Eck & Jablonski (1987), rather than of Middle Pleistocene age. Jolly also indicates that the molars of *T. oswaldi* are generally larger than those of *T. darti* from Makapansgat, a view that is substantiated by the values in Table 2.27. Furthermore, it is now apparent that

the molars of *T. darti* also increase in size through time, for the Hadar specimens are generally smaller than those from Makapansgat. In addition, as can be seen in Table 2.28, the lower molars from the Sidi Hakoma Member appear, on average, to be slightly smaller than those from the Denen Dora and Kada Hadar Members of the Hadar Formation. Thus, it appears that the molars of the *T. darti–T. oswaldi* lineage became larger throughout its 3 Ma age-range. Unfortunately, the small sample sizes from, and the poorly known absolute ages of, many of the sites preclude reliable estimation of the rate of this change.

Table 2.28. *Size (basal length) of the lower molars of* Theropithecus darti *from the Hadar Formation*

	M_1	M_2	M_3
Denen Dora and Kadada Hadar Members			
n	10	13	13
range	9.1–10.8	10.1–15.1	15.1–20.4
mean	9.7	11.9	17.5
Sidi Hakoma Member			
n	9	9	9
range	8.2–12.2	10.3–12.4	13.4–17.2
mean	9.3	11.7	16.0

Conclusions

Taxonomy and classification

The description of the relatively large and well-preserved sample of *Theropithecus* from Hadar and its comparison with that from Makapansgat and various samples of *T. oswaldi* lead to three taxonomically important conclusions. First, the Hadar sample is most similar in morphology to that from Makapansgat and should be included in the species *T. darti*. Secondly, although there are many morphological similarities in the skulls of *T. darti* and *T. oswaldi*, there are enough differences between them to warrant the continued recognition of both species. Thirdly, when the morphology of a character differs between *T. darti* and *T. oswaldi*, that of *T. darti* is often most similar to the morphology of *T. brumpti* and *T. gelada*, probably indicating primitive status for these character states. Each of these points is discussed in greater detail below.

Even with the addition of the Hadar specimens, the sample of *T. darti* remains small, and the absence, especially of an adult male cranium from Makapansgat, makes generalization about morphological differences between, and similarities among, the two samples precarious. Comparison of specimens from the two sites (allowing for reasonable amounts of individual and sexual variation) indicates, however, that most of the characteristics of the skull are very similar in the two samples. The Hadar skulls may differ, though, from those from Makapansgat in four important ways. First, the male mandibular specimens from Hadar lack mental protuberances, structures that are present in those few male specimens from Makapansgat in which the anterior symphyseal surface is preserved and in all male specimens *T. oswaldi*. Absence of mental protuberances from Hadar females is of less interest because these structures are also absent from females from Makapansgat and those of *T. oswaldi*. Secondly, the anterior portions of the mental ridges, as they approach the incisive alveoli, form well defined, slightly rugose structures in male and female mandibles from Hadar. In contrast, specimens of both sexes from Makapansgat, like specimens of *T. oswaldi*, have low, rounded, poorly defined anterior ridges. Thirdly, the mandibular curve of Spee varies from slightly concave to slightly convex (superiorly) in the adults from Hadar, while it appears to be consistently convex in the adults from Makapansgat, as it is in other specimens of the genus (except those of *T. baringensis*). Fourthly, although the differences are small and the ranges of variation of several measurements overlap substantially, the specimens from Hadar appear to be slightly smaller than those from Makapansgat, and these in turn appear slightly smaller than specimens of *T. oswaldi* of latest Pliocene and earliest Pleistocene age. Thus, although cranial and mandibular specimens of *T. darti* from Hadar and Makapansgat are very similar in the vast majority of their characteristics, when they differ, the specimens from Makapansgat are similar in these characteristics to specimens of *T. oswaldi*.

The morphology of a number of characteristics suggest possible, but poorly documented, differences between the skulls of *T. darti* and *T. oswaldi*. First, in the four specimens of *T. darti* (two from Hadar and two from Makapansgat) in which the region is well preserved, the muzzle dorsum has the complex, concavo-convexo-concave curvature that produces distinctly raised nasal bones. In most if not all specimens of *T. oswaldi*, the dorsum has a simple, convex curvature. The morphology of the nasal bones may not differ consistently between the two species, however, because two additional specimens from Hadar, in which the dorsum is partially preserved, appear to have had the simple, convex curvature typical of *T. oswaldi*. Secondly, the piriform aperture has a relatively narrow, oval form in the four specimens from Hadar and Makapansgat in which its margins are well preserved. In contrast, the aperture in most specimens of *T. oswaldi* is more broadly oval. At least one specimen of this species (a female from Kanjera) has, however, a relatively narrow aperture

similar to those of the Hadar and Makapansgat specimens. Thirdly, the broadly bowed, smoothly curved zygomatic arches of the Hadar adult male differ from those of the few males of *T. oswaldi* in which they are preserved. Conversely, the form of the zygomatic arches in the two females from Makapansgat is very similar to that of females of *T. oswaldi*. Fourthly, the anteroinferior apex of the triangular depression of the mandibular ramus appears, in males of *T. darti*, to be deeply excavated and to have a relatively wide angle between the posteroinferior and anterior margins. In most males of *T. oswaldi*, the apex is more shallow and acutely angled. These differences are less clear cut among females of the two species, however, for the apex is shallow and acute in one female from Makapansgat, while it is deeper and broader in a number of females of *T. oswaldi*. The patterns of differences in these four regions, although as yet unclear, may become important in differentiating the two species as more specimens become available.

Another set of characteristics appear to consistently differ between *T. darti* and *T. oswaldi*, although again larger samples of specimens are needed to confirm the following generalizations. First, the widths of the incisive alveoli and several variables of the incisors, themselves, indicate that the incisors of *T. darti* are relatively much larger than those of *T. oswaldi*. Secondly, the morphology of both the canines and P_3s, especially of males, indicates that the canine mechanism of *T. darti* is not as fully 'reduced' in relative size as it is in *T. oswaldi*. Thirdly, the postglenoid processes of *T. darti* are consistently separated from the tympanic temporal bones by distinct fissures or grooves, while in *T. oswaldi* they are tightly pressed against the tympanic temporal bones. Fourthly, the postglenoid processes of *T. darti* are relatively shorter than those of *T. oswaldi*. Fifthly, the sagittal and nuchal crests of the former species are less pronounced than those of the latter. Sixthly, although the differences in size between the specimens of *T. darti* from Makapansgat and those of *T. oswaldi* of Plio-Pleistocene age is not great, the difference between the specimens from the Sidi Hakoma Member of the Hadar Formation and those of *T. oswaldi* from sites of late middle Pleistocene age is quite remarkable. Early *T. darti* was about the same size as the gelada, while late *T. oswaldi* was about as large as the gorilla.

I think it of special importance that the relative sizes of the incisors, the canine mechanism and the postglenoid process, and the relationship of the postglenoid process to the tympanic temporal bone of *T. darti* are closer to those of *T. brumpti* and *T. gelada* than to those of *T. oswaldi*. This suggests primitive status for *T. darti* in relation to *T. oswaldi*. In addition, although the primitive body size of *Theropithecus* is not known, it is likely to have been small (see Delson, 1975); thus, *T. darti* is probably primitive in this respect as well.

My concept of *T. darti* as a small, primitive species, which is directly ancestral to *T. oswaldi*, is generally similar to that of Jolly (1972). It differs, however, from his concept in two important aspects. First, he proposes that, among a number of charac-

teristics, the height of the female muzzle, the morphology of the incisive plane, the depth of the mandibular corpus fossa, and the orientation of the inferior margin of the mandible differ, or might differ, between the two species. The evidence now suggests that these four characteristics are either too variable within, or too similar among, the species to be of diagnostic significance. In addition, Jolly (1972) states that the relative size of the incisors of *T. darti* are the same as those of *T. oswaldi*, while I agree with Maier (1972) that they are not. In fact, the difference in relative size of the incisors is one of the most important differences between the two species. Further, I suggest that the differences in the relative heights of the postglenoid processes, the relationships of the postglenoid process to the tympanic temporal bone, and the degree of development of the sagittal and nuchal crests may prove to be of classificatory importance. Secondly, Jolly (1972) includes all of the specimens of *Theropithecus* from Swartkrans and Beds I and Lower II at Olduvai Gorge in *T. darti*. I think, to the contrary, that the overall size of the specimens, the relative sizes of the incisors and canine mechanisms, and the relationship of the postglenoid process to the tympanic temporal bone seen in the specimens from Swartkrans indicate that they should be placed in *T. oswaldi*. Although the sample of *Theropithecus* from Beds I and Lower II at Olduvai Gorge is small and fragmentary, the overall size of the specimens and the degree of reduction of the canine mechanism seen in some of them suggests that they too belong to this species. It should be noted here that Leakey & Leakey (1973) conclude that all of the specimens of *Theropithecus* from Bed II belong to *T. oswaldi*, while one specimen (Old 63 3050) might belong to a different species. Although I agree with their first conclusion, I think that the juvenile age and fragmentary state of Old 63 3050 makes their second difficult to substantiate.

Maier (1970, 1972) also views *T. darti* as primitive, but on grounds quite different from mine. His discussion centres on the comparative morphology of the supraorbital torus of *T. darti*, *Parapapio*, and *Papio*. He concludes that the tori of *T. darti* and *Parapapio* are similar and primitive, while that of *Papio* is derived. Maier does not make explicit his views on the relationship between *T. darti* and *T. oswaldi*, but implies a close affinity by his statements of similarities among specimens of *T. darti* from Makapansgat and *T. oswaldi* from Swartkrans.

Freedman's (1976) concept of the two species is the most different from mine, for he proposes that all specimens of *Theropithecus* from Makapansgat, Swartkrans, and Hopefield should be placed in *T. darti*, while those from sites in East and North Africa should be placed in *T. oswaldi*. The similarities among the specimens from Hadar and Makapansgat, the similarities among those from Swartkrans, Hopefield, and the sites in East and North Africa other than Hadar, and the differences between these two groups renders his proposition untenable.

Dechow (1981), Dechow & Singer (1984), Leakey & Whitworth (1958), and Singer (1962) suggest that the specimens of *T. darti* from Makapansgat differ so little

from those of *T. oswaldi* that the continued recognition of *T. darti* is, or might be, unwarranted. My opinion is, of course, to the contrary. The new material from Hadar tends to confirm the existence of a number of differences between the two species described by Freedman (1957), Jolly (1972), and Maier (1970, 1972) and adds a number of new ones as well.

Theropithecus darti, especially as represented by the sample from Hadar, appears primitive in its relatively large incisors, 'unreduced' canine mechanism, small size, and a number of other characteristics of the skull. The large number of characteristics, especially of the cranium, that *T. darti* shares with *T. oswaldi* and that neither share with *T. brumpti* or *T. gelada* indicates, however, that *T. darti* was directly ancestral to *T. oswaldi*. The size of the specimens and the morphology of the anterior symphyseal region of the male mandible from Makapansgat suggests that the transition from the former to the latter species had begun by Makapansgat times. Unfortunately, the small size of the sample of well-preserved adult specimens from this site, its uncertain age, and the general absence of specimens of the lineage from approximately 2.3 to 2.5 Ma ago precludes determination of the nature, and especially, the rate of this transition.

Habitat preferences

The large fluctuations in numbers of specimens of *T. darti* from one stratigraphic submember to the next, noted in the Introduction, suggest major changes through time in factors determining rates

Table 2.29. *Relationship of the numbers of specimens of* Theropithecus darti *to those of the family Bovidae*

Stratigraphic levels	Number of bovids	Bovids relative to *T. darti*[†]	Number of *T. darti*
KH–3	4	0.19	0
KH–2	8	0.37	0
KH–1	7	0.37	**2**
DD–3/KH–1	31	1.45	**2**
DD–3	83	3.88	10
DD–2/DD–3	477	22.27	*13*
DD–2	114	6.70	*2*
DD–1/DD–2	32	1.49	**3**
DD–1	21	0.98	0
SH–4/DD–1	5	0.23	0
SH–4	46	2.15	*1*
SH–3/SH–4	7	0.33	0
SH–3	24	1.12	1
SH–2/SH–3	29	1.35	**3**
SH–2	24	1.12	1
SH–1/SH–2	147	6.86	**10**
SH–1	11	0.51	**6**
SHT/SH–1	8	0.37	0
SHT	0	0.00	0
BM/SHT	57	2.66	0
TOTAL	1135	54.40	54

[†]See text for method of calculation.

of preservation of specimens. Given the complex sets of environments of deposition evinced by the sediments of the Hadar Formation and their alternations through time, one might conclude that much of this variation in numbers results from geological factors affecting preservation rather than from biological or ecological characteristics of the species itself. Such a conclusion is supported by a frequency analysis of bovid specimens from the formation, for the numbers of bovid cranial specimens recovered from each submember vary concordantly with those of *T. darti* (Table 2.29). Peaks in frequencies

of bovid specimens and those of *T. darti* are seen in the lower Sidi Hakoma Member and the middle to upper Denen Dora Member. Numbers of both taxa are generally much lower in other submembers of the formation. I would think that, if ecological factors controlling population sizes, densities, and distributions were primarily responsible for this variation, then variation in numbers of bovid specimens would be much lower than that of *T. darti*, given the, surely, much greater adaptive breadth of the family Bovidae. Further analysis of the numbers suggests, however, that, although geological factors may have had predominant effects on preservation rates, ecological factors also played a part and that it may be possible to make statements about the habitat preferences of *T. darti* from their geological and faunal contexts at Hadar.

Difficulties arise when comparing numbers of bovids and monkeys because of differences in their magnitudes; there are 21 times more bovid than monkey specimens available for analysis. Division of the bovid numbers by this factor produces numbers that can be viewed as composing a base line against which to compare the numbers of *T. darti* (see column 3, Table 2.29). Doing this comparison suggests that some submembers have an 'excess' of monkeys compared to bovids (bold numbers in column 4, Table 2.29) and that some have a 'dearth' (*italicized* numbers), while many (plain numbers) have about as many monkeys as one would expect if factors determining relative numbers of monkeys and bovids were equal. Although I admit that this pattern may be merely the result of chance and the vagaries of sampling, I

Table 2.30. *Frequencies of specimens of bovid tribes*

Samples	Reduncini % (n)	Bovini % (n)	Tragelaphini % (n)	Aepycerotini % (n)	Alcelaphini % (n)	Total % (n)
KH–1 + DD–3/KH–1 + DD–3	18 (21)	10 (12)	11 (13)	20 (24)	41 (48)	100 (118)
DD–2/DD–3 + DD–2	49 (303)	8 (49)	13 (83)	17 (104)	13 (76)	100 (615)
DD–1/DD–2	43 (13)	8 (1)	20 (6)	10 (3)	23 (7)	99 (30)
SH–4	24 (8)	15 (5)	24 (8)	21 (7)	18 (6)	102 (34)
SH–2/SH–3	4 (1)	22 (5)	9 (2)	35 (8)	30 (7)	100 (23)
SH–1/SH–2 + SH–1	4 (5)	13 (17)	20 (27)	48 (64)	16 (21)	101 (134)
BM/SHT	4 (2)	26 (13)	24 (12)	28 (14)	18 (9)	100 (50)

think there is more to it, because the bovid tribal composition of samples from sub-members with an 'excess' of monkeys differs in interesting ways from those with a 'dearth'. Grouping together samples of bovids from contiguous submembers with either an excess or a dearth of monkeys, in order to maintain large sample sizes, produces three sampling units with relatively large numbers – SH-1 + SH-1/SH-2 with an excess of monkeys, DD-2 + DD-2/DD-3 with too few, and DD-3 + DD-3/KH-1 + KH-1 with an excess. Four other samples with much smaller sizes can also be analyzed – BM/SHT, SH-2/SH-3, SH-4, and DD-1/DD-2 (Table 2.30).

Perusal of numbers in Table 2.30 suggest that the large samples with an excess of *T. darti* have small percentages of reduncine bovids and large percentages of *Aepyceros* or alcelaphines. The single large sample with lower than expected numbers of *T. darti*, in contrast, is dominated by the reduncines and has low percentages of *Aepyceros* and alcelaphines. The pattern of positive association between *T. darti*, *Aepyceros*, and the alcelaphines and negative association of *T. darti* with the reduncines seems strengthened by the fact that the large samples alternate in time, suggesting that the pattern is not principally determined by faunal change through time, and that two of the small samples (SH-2/SH-3 and SH-4) show similar patterns. The pattern is, admittedly, weakened because in two of the small samples (BM/SHT and DD-1/DD-2) it is reversed. The pattern of associations suggests, perhaps, that *T. darti*, like *Aepyceros* and the alcelaphines, preferred drier woodlands and grasslands rather than the moist swamps and woodlands preferred by the reduncines.

Age-range and geographic distribution

Given the concept of *T. darti* used here, the known paleontological record of the species is restricted to the Hadar Formation of Ethiopia and the Makapansgat Formation of the Republic of South Africa. The age-range of the species is not well defined, for the age at the bottom of the Hadar Formation is a matter of dispute and determination of the age of the Makapansgat Formation remains problematic. Radiometric estimates indicate, however, that the maximum age of specimens of the species is about 3.4 Ma. Suid biostratigraphy suggests that the youngest specimens may be about 2.4 Ma old (Harris & White, 1979). It seems that *T. darti* occupied both East and South Africa around 3.0 Ma ago.

Phyletic implications

The ancestral relationship of *T. darti* to *T. oswaldi*, discussed above, now seems reasonably well documented and noncontroversial. The *T. oswaldi*-lineage (*T. darti* + *T. oswaldi*) was well established by 3.3 Ma ago, having by this time developed most of its synapomorphic cranial characteristics. Thereafter, it made, by comparison, only minor morphological adjustments prior to its extinction in the late Middle Pleistocene. Although *T. brumpti* obviously differs from *T. oswaldi* in many fundamental ways (Eck & Jablonski, 1987), the species composition of its ancestral lineage remains unsettled. Eck & Jablonski (1984) propose on morphological grounds that *T. baringensis* and *T. quadratirostris* constitute this ancestry and that the three species comprise the *T. brumpti*-lineage (*T. baringensis* + *T. quadratirostris* + *T. brumpti*). They claim that the lineage arises about 4.0 Ma ago and becomes extinct around 2.3 Ma. This view is challenged by Delson (1982, pers. comm.), who thinks that *T. quadratirostris* belongs to *Dinopithecus*, a close ally if not synonym of *Papio*, and that the age of *T. baringensis* is closer to 3.0 Ma than to 4.0 Ma. Even if Delson proves to be correct, however, the ancestry of *T. brumpti* must be of substantial antiquity because typical specimens of the species approach 3.0 Ma in age (Eck & Jablonski, 1987; Harris, Brown & Leakey, 1988).

Theropithecus gelada also shows many apomorphic characters, including the shape of the margins and surface of the piriform aperture, the presence of nasal tubercles, the association of a long muzzle with a short muzzle dorsum, the shape of the supraorbital torus, and, perhaps, the shape of the neurocranium, among others. Like members of the proposed *T. brumpti*-lineage, however, it remains primitive in many aspects of its dentition, in the relationship of the postglenoid process to the tympanic temporal bone, in its deep mandibular corpus fossae, and, perhaps, small body size. This mix of autapomorphic and sympleisomorphic characters, with a general absence of synapomorphic characters at the species level, suggests that Jolly (1972) was right when he proposed that the *T. gelada*-lineage has been long separated from the others of the genus. The short muzzle dorsum of *T. gelada* may indicate that the species is more closely related to the *T. oswaldi*-lineage than to that of *T. brumpti*. Its long muzzle indicates, however, that if this is true, then *T. gelada* diverged from the *T. oswaldi*-lineage prior to the rise of *T. darti*, which has both the short muzzle and muzzle dorsum typical of the lineage. The relationships among the lineages suggest, then, that all three have probably had separate evolutionary histories since early Middle Pliocene times.

Acknowledgements

I thank the government of the Democratic Republic of Ethiopia for the loan of the Hadar specimens during the period of their study. I thank D.C. Johanson for permission to study the specimens and the Cleveland Museum of Natural History and the Institute of Human Origins for the use of facilities to do so. The University of Washington, the Cleveland Museum of Natural History, and the Institute of Human Origins provided financial support during different phases of the study and write-up, for which I am grateful. I appreciate the thoughtful discussions of various aspects of the evolution of *Theropithecus* that I have had with colleagues, especially E. Delson, M.G. Leakey, and N.G. Jablonski. I thank N. Gooch and N.E. Kahn for help in the production of the plates. Finally, I wish to thank N.G. Jablonski and R.A. Foley for my invitation to the symposium and all attendants for the stimulating presentations and discussion they provided.

References

ANDREWS, C.W. (1916). Notes on a new baboon (*Simopithecus oswaldi*, gen. et sp. nov.) from the (?) Pliocene of British East Africa. *The Annals and Magazine of Natural History*, 18, 410–19.

ARAMBOURG, C. (1947). *Mission scientifique de l'Omo 1932–1933, 1(111), Geologie-Anthropologie*. Paris: Museum National d'Histoire Naturelle.

ARONSON, J.L. & TAIEB, M. (1981). Geology and Paleogeography of the Hadar Hominid Site, Ethiopia. *Hominid Sites: Their Geologic Settings*, ed. G. Rapp, Jr. & C.F. Vondra, AAAS Selected Symposium, 63, 165–95. American Association for the Advancement of Science, Washington.

ARONSON, J.L., WALTER, R.C. & TAIEB, M. (1983). Correlation of Tulu Bor Tuff at Koobi Fora with the Sidi Hakoma Tuff at Hadar. *Nature*, 306, 209–10.

BROOM, R. & JENSEN, J.S. (1946). A new fossil baboon from the caves at Potgietersrust. *Annals of the Transvaal Museum*, 20, 337–40.

BROWN, F.H. (1982). Tulu Bor Tuff at Koobi Fora correlated with the Sidi Hakoma Tuff at Hadar. *Nature*, 300, 631–3.

BROWN, F.H. (1983). Correlation of Tulu Bor Tuff at Koobi Fora with the Sidi Hakoma Tuff at Hadar. *Nature*, 306, 209–10.

BROWN, F.H., MCDOUGALL, I., DAVIES, T. & MAIER, R. (1985). An integrated Plio-Pleistocene Chronology for the Turkana Basin. *Ancestors: The Hard Evidence* ed. E. Delson, pp. 82–90. New York: Alan R. Liss, Inc.

DECHOW, P.C. (1981). Evaluation of primitive morphological features of Makapansgat *Theropithecus*. *American Journal of Physical Anthropology*, 54, 213 (Abstract).

DECHOW, P.C. & SINGER, R. (1984). Additional fossil *Theropithecus* from Hopefield, South Africa: a comparison with other African sites and a reevaluation of its taxonomic status. *American Journal of Physical Anthropology*, 63, 405–35.

DELSON, E. (1975). Evolutionary history of the Cercopithecidae. Approaches to Primate Paleobiology. *Contributions to Primatology*, 5 (Ed. by F.S. Szalay), pp. 167–217.

DELSON, E. (1982). *Dinopithecus*, the 'giant baboon' of Plio-Pleistocene Africa. *American Journal of Physical Anthropology*, 57, 179.

ECK, G.G. (1977). *Morphometric Variability in the Dentitions and Teeth of Theropithecus and Papio*. Ph.D. Dissertation, University of California, Berkeley.

ECK, G.G. (1987). *Theropithecus oswaldi* from the Shungura Formation, lower Omo valley, southwestern Ethiopia. In *Les Faunes Plio-Pléistocènes de la Basse Vallée de l'Omo (Éthiopie)*, Tome 3, Cercopithecidae de la Formation de Shungura, ed. Y. Coppens & F.C. Howell, pp. 123–40. Cahiers de Paléontologie, Travaux de Paléontologie Est-Africaine, Paris: Editions du Centre National de la Recherche Scientifique.

ECK, G.G. & JABLONSKI, N.G. (1984). A reassessment of the taxonomic status and phyletic relationships of *Papio baringensis* and *Papio quadratirostris* (Primates: Cercopithecidae). *American Journal of Physical Anthropology*, 65, 109–34.

ECK, G.G. & JABLONSKI, N.G. (1987). The skull of *Theropithecus brumpti* compared with those of other species of the genus *Theropithecus*. In *Les Faunes Plio-Pléistocènes de la Basse Vallée de l'Omo (Éthiopie)*, Cercopithecidae de la Formation de Shungura, ed. Y. Coppens & F.C. Howell, Tome 3, pp. 11–122. Cahiers de Paléontologie. Paris: Editions du Centre National de la Recherche Scientifique.

FREEDMAN, L. (1957). The fossil Cercopithecoidea of South Africa. *Annals of the Transvaal Museum*, 23, 121–262.

FREEDMAN, L. (1960). Some new cercopithecoid specimens from Makapansgat, South Africa. *Palaeontologia Africana*, 7, 7–45.

FREEDMAN, L. (1976). South African fossil Cercopithecoidea: a re-assessment including a description of new material from Makapansgat, Sterkfontein and Taung. *Journal of Human Evolution*, 5, 297–315.

GRAY, B.T. (1980). *Environmental reconstruction of the Hadar Formation (Afar, Ethiopia)*. Ph.D. Dissertation, Case Western Reserve University.

HARRIS, J.M. & WHITE, T.D. (1979). Evolution of the Plio-Pleistocene African Suidae. *Transactions of the American Philosophical Society*, 69, 1–128.

HARRIS, J.M., BROWN, F.H. & LEAKEY, M.G. (1988). Stratigraphy and paleontology of Pliocene and Pleistocene localities west of Lake Turkana, Kenya. *Contributions in Science*, (399), 1–125. Natural History Museum of Los Angeles County, Los Angeles.

JABLONSKI, N.G. (1981). *Functional Analysis of the Masticatory Apparatus of the Gelada Baboon*, Theropithecus gelada *(Primates: Cercopithecidae)*. Ph.D. Dissertation, University of Washington, Seattle.

JOHANSON, D.C., TAIEB, M. & COPPENS, Y. (1982). Pliocene hominids from the Hadar Formation, Ethiopia (1973–1977): stratigraphic, chronologic, and paleoenvironmental contexts, with notes on hominid morphology and systematics. *American Journal of Physical Anthropology*, 57, 373–402.

JOLLY, C.J. (1970). The seed eaters: a new model of hominid differentiation based on a baboon analogy. *Man*, 5, 5–26.

JOLLY, C.J. (1972). The classification and natural history of *Theropithecus (Simopithecus)* (Andrews, 1916), baboons of the African Plio-Pleistocene. *Bulletin of the British Museum (Natural History), Geology*, 22, 1–123.

LEAKEY, L.S.B. (1943). Notes on *Simopithecus oswaldi* Andrews from the type site. *Journal of the East African Natural History Society*, 17, 39–44.

LEAKEY, L.S.B. & WHITWORTH, T. (1958). Notes on the genus *Simopithecus*, with a description of a new species from Olduvai. *Coryndon Memorial Museum Occasional Papers*, (6), 3–14.

LEAKEY, M.G. & LEAKEY, R.E.F. (1973). Further evidence of *Simopithecus*, (Mammalia, Primates) from Olduvai and Olorgesailie. In *Fossil Vertebrates of Africa*, vol. 3, ed. L.S.B. Leakey, R.J.G. Savage & S.C. Coryndon, pp. 101–20. New York: Academic Press.

MAIER, W. (1970). New fossil Cercopithecoidea

from the lower Pleistocene cave deposits of the Makapansgat Limeworks, South Africa. *Palaeontologia Africana*, 13, 69–107.

MAIER, W. (1972). The first complete skull of *Simopithecus darti* from Makapansgat, South Africa, and its systematic position. *Journal of Human Evolution*, 1, 395–405.

MCDOUGALL, I. (1985). K-Ar and ^{40}Ar/^{39}Ar dating of the hominid bearing Plio-Pleistocene sequence at Koobi Fora, Lake Turkana, northern Kenya. *Geological Society of America, Bulletin*, 96, 159–75.

SARNA-WOJCICKI, A.M., MEYER, C.E., ROTH, P.H. & BROWN, F.H. (1985). Ages of tuff beds at East African early hominid sites and sediments in the Gulf of Aden. *Nature*, 313, 306–8.

SINGER, R. (1962). *Simopithecus* from Hopefield, South Africa. *Bibliotheca Primatologica*, 1, 43–70.

SZALAY, F.S. & DELSON, E. (1979). *Evolutionary History of the Primates*. New York: Academic Press.

WALTER, R.C. & ARONSON, J.L. (1982). Revisions of K/Ar ages for the Hadar hominid site, Ethiopia. *Nature*, 296, 122–7.

WASHBURN, S.L. & CIOCHON, R.L. (1974). Canine teeth: notes on controversies in the study of human evolution. *American Anthropologist*, 76, 765–84.

Appendix 2.1 Catalogue of Specimens

The locality and specimen number, stratigraphic provenience, sex, age, and a brief description of the state of preservation of each of the Hadar specimens assigned to *Theropithecus darti* are presented below. Submembers enclosed in parentheses are thought less likely correct than those not so enclosed.

AL52-1; SH-3; sex unknown; adult (M^1 heavily worn). A right maxillary fragment preserved from anterior P^3 to posterior M^1; does not extend above the tooth roots; P^3 and P^4 are present but fragmentary; M^1 is complete but heavily worn.

AL55-43; DD-3 or KH-1; sex unknown; juvenile (M_1 slightly worn). A left mandibular fragment preserved from anterior dP_4 to posterior M_1; the margin is preserved below these teeth; dP_4 and M_1 are well preserved.

AL56-17; DD-2 or DD-3; sex unknown, young adult (M_2 moderately worn). A complete, isolated left M_2.

AL58-23; SH-4, (DD-1), or DD-2; male; young adult (M_3 very slightly worn). A left mandibular body lacking the incisive alveoli, gonial region and ascending ramus; P_4 through M_3 are well preserved.

AL116-23; DD-2 or DD-3; sex unknown; juvenile (M^2 erupted but unworn). A right maxillary fragment preserved from anterior M^1 to posterior M^2; does not extend above the tooth roots; M^1 is present but fragmentary; M^2 is complete.

AL126-30; SH-1 or SH-2; female; adult (M_3 moderately worn). A right mandibular body lacking the gonial region and ascending ramus; C_L, P_3, and P_4 are slightly damaged; M_1 through M_3 are complete.

AL129-8; SH-1 or SH-2; female; adult (M_3 slightly worn). A mandibular fragment lacking the left body and ascending ramus behind P_3, the right gonial region and ascending ramus, and the entire margin; left C_L and P_3 and right C_L through M_3 are present; right C_L is slightly damaged.

AL132-26a,b,c,e,f,g; SH-1 or SH-2; sex unknown; juvenile (dP^4 erupted but unworn): A series of associated cranial fragments.

-26a is a partial endocranial cast with much of the surface of the vault and portions of the orbital region.

-26b is a small, right maxillary fragment with dP^4.

-26c is a left maxillary fragment preserved behind dP^3 with part of the orbital rim and the zygomatic arch, and the dP^3 and dP^4.

-26e is an isolated right dP^3 crown.

-26f is a right dC^U.

-26g is a left (?) dI^1 crown.

AL133-54; DD-2 or DD-3; sex unknown, juvenile (M_2 slightly worn). A complete, isolated, right M_2.

AL134-5a,b,c; (Sh-4, DD-1), DD-2 or DD-3; adult (M^2 very heavily worn). A muzzle with associated postcranial fragments.

-5a is a nearly complete and undistorted muzzle lacking the superior margin of the piriform aperture, the incisive alveoli, the interorbital region, and the left maxillary tubercle; portions of the inferior orbital surfaces and the roots of both zygomatic arches are preserved; right M^2 and left P^3 through M^2 are present, but all are heavily worn and fragmentary.

-5b is a left distal tibia with a slightly damaged articular region.

-5c is a left proximal femur lacking the head and neck.

AL137-11; SH-1 or SH-2; sex unknown; adult (M_3 moderately worn). A fragment of a right mandibular body preserved from anterior M_2 to posterior M_3; does not extend below the tooth roots; fragmentary M_2 and slightly damaged M_3 are present.

AL137-12; SH-1 or SH-2; sex unknown; juvenile (M_1 moderately worn). A fragmentary, isolated, left M_1. Not the same individual as AL137-11 because AL137-12 is considerably less worn than is the M_2 of AL137-11.

AL142-19; SH-1 or SH-2; male; juvenile (M_2 slightly worn). A left mandibular body preserved from the symphysis to posterior M_2; the margin is intact in this region; the crown of the C_L is badly damaged; P_3 and

P_4 are well preserved; M_1 and M_2 are slightly damaged.

AL144-1; SH-2 or SH-3; probably male; adult (M_3 moderately worn). A fragment of a left mandibular body preserved from anterior M_2 to some distance behind M_3; the margin is intact in this region; M_2 and M_3 are present.

AL148-105; SH-1 or SH-2; male; juvenile (C^U slightly worn). A crown of an isolate, left C^U lacking the tip.

AL148-107; SH-1 or SH-2; male; juvenile (M_2 very slightly worn). A fragmentary mandible that preserves a small portion of the anterior right body and a nearly complete left hemimandible lacking the coronoid process and condyle; the margin is intact throughout; right and left C_Ls are present but unerupted; left P_3, M_1, and M_2 are well preserved. Not the same individual as AL148-105 because the P_3 does not have a wear facet corresponding to that on AL148-105.

AL153-14a,b; DD-2 or DD-3; male; adult (M_3 slightly worn). Two associated mandibular fragments.

-14a is a right hemimandible lacking the incisive and canine alveoli, the symphyseal region, the coronoid process and the condyle; the margin is intact; P_3 through M_3 are present.

-14b is a left mandibular body preserved from anterior P_4 to some distance behind M_3; the margin is absent; P_4 and M_1 are preserved; M_2 is fragmentary.

AL153-18; DD-2 or DD-3; sex unknown; juvenile (M_1 slightly worn). A fragment of a right mandibular body preserved from dP_3 to posterior M_2; dP_3 is fragmentary; dP_4 and M_1 are complete; P_4 and M_2 are unerupted.

AL156-28; DD-3 or KH-1; sex unknown, age unknown. A badly eroded, right upper molar.

AL158-91; DD-2 or DD-3; sex unknown; juvenile (M_2 slightly worn). A small fragment of a right mandibular body preserved from anterior M_1 to posterior M_2; the margin is absent; M_2 is complete.

AL161-23; DD-2 or DD-3; sex unknown; adult (M_2 slightly worn). A small fragment of a left mandibular body preserved from anterior M_2 to posterior M_3; the margin is absent; M_2 and M_3 are present.

AL163-11; DD-2 or DD-3; male, adult (M_3 moderately worn). A mandibular body preserved from left I_2 through posterior right M_3; the entire margin is absent; right P_3 through M_3 are present; P_3, M_1, and M_2 are slightly damaged.

AL174-10; DD-1 or DD-2; probably male; adult (M_3 slightly worn). A fragment of a right mandible preserved behind P_4, but lacking the margin, coronoid process, and condyle; fragmentary M_2 and slightly damaged M_3 are present.

AL183-45; DD-2 or (DD-3); sex unknown; old adult (M_3 heavily worn). A small fragment of a right mandibular body; the margin and much of the lateral surface are absent; M_3 is present, but fragmentary.

AL185-5a,b,c,d,e; DD-3; female; juvenile (M^2s very slightly worn). A partial skull associated with postcranial elements.

-5a is a well preserved neurocranium; the sphenoid, basiocciput, zygomatic process of the temporals, and left supraorbital torus are slightly damaged.

-5b is a left premaxilla, maxilla, and damaged zygomatic bone; I^2 and P^4 through M^2 are present; M^3 is unerupted.

-5c is a left mandible preserved behind C_L; the posterior margin of the ascending ramus, condyle, and coronoid process are absent; P_3 through unerupted M_3 are present.

-5d is a left proximal femoral fragment lacking the head.

-5e is, perhaps, a fragment of a tibial shaft.

AL185-16; DD-3; sex unknown, old adult (M^3 heavily worn). A slightly damaged, isolated, right M^3.

AL185-22a,b,c,d,e,f,g,h; DD-3; sex unknown; juvenile (M_2 slightly worn). A small mandibular fragment associated with fragmentary postcranial elements.

-22a is the greater trochanter and neck of a right femur.

-22b is a small fragment of a right proximal radius lacking the head.

-22c is a fragment of a right proximal ulna broken through the lunar notch.

-22d is the olecranon process and the proximal portion of the lunar notch of AL185-22c.

-22e is a right proximal humeral fragment with a damaged greater tuberosity.

-22f is a fragment of a right distal humerus preserving mainly the capitulum.

-22g is a right distal tibial fragment.

-22h is a very small left mandibular fragment; a fragmentary M_2 is present.

AL186-16a,b,c; DD-3; sex unknown; juvenile (M^1s very slightly worn). A maxillary fragment and associated, isolated teeth.

-16a is a small eroded fragment of a left maxilla; dP^3, dP^4, and M^1 are well preserved; P^3 and a fragment of M^2 are present but unerupted.

-16b is an unworn I^2.

-16c is a complete, right M^1.

AL186-17; DD-3; female; old adult (M_3 heavily worn). A mandible lacking the gonial regions and ascending rami; the right margin is damaged below M_2 and M_3; the left M_1 through M_3 and right M_2 and M_3 are present, but all are slightly damaged.

AL187-10; DD-2 or DD-3; male; young adult. A well-preserved neurocranium lacking the zygomatic processes of the temporal bones; the petrous temporal and sphenoid bones are slightly damaged. Male status is based on the presence of a small sagittal crest that is not seen in females from the site. Age is based on the unobliterated state of many of the cranial sutures.

AL188-19; DD-2 or DD-3; sex unknown; age unknown. An isolated, left I^2 lacking the lingual surface.

AL193-1; SH-1 or SH-2; sex unknown; juvenile (M_1 slightly worn). A fragment of a right mandibular body preserved from C_L to behind M_2; only the M_1 is present.

AL196-3a,b,c,d; DD-3; female; adult (M_3 slightly worn). A fragmentary mandible associated with postcranial elements.

-3a is a mandible lacking the left side behind P_4, the right gonial region, condyle, and coronoid process; the right P_4 and M_1 are present but fragmentary; the

right M_2 and M_3 are slightly damaged.

-3b is a left proximal tibial fragment.

-3c is a left distal humeral fragment.

-3d is a fragment of a left proximal ulna.

AL199-4; SH-1, sex unknown; adult (M_3 slightly worn). A small fragment of a right mandibular body from the region around M_3; the margin is intact; a fragmentary M_3 is present.

AL200-12; SH-1; sex unknown; age unknown. An isolated, unworn, right upper molar lacking the lingual and distal roots.

AL200-14; SH-1; sex unknown; juvenile (M_2 very slightly worn). A very small fragment of left mandibular body lacking the margin; the M_2 is present.

AL200-16; SH-1; male; juvenile (C^U very slightly worn, but not fully developed). An isolated, right C^U crown that is not fully formed at the dentoenamel junction.

AL200-17; SH-1; sex unknown; age unknown. An isolated, slightly worn, left P^3 lacking the mesial and distal roots.

AL200-18; SH-1; sex unknown; age unknown. An isolated, slightly worn, right I^2. (The size, preservation, and wear status of specimens AL200-12, AL200-14, and AL200-18 indicate that they may derive from the same individual. The size of the wear facet on AL200-16 indicates a slightly older individual than, and the preservation appears quite different from, the above specimens. The wear status of AL200-17 indicates an older individual than all of the others.)

AL202-3; SH-1 or SH-3; female; juvenile (M_1 slightly worn). A badly shattered and exploded right mandibular fragment; P_3 and M_3 are complete, but unerupted; the M_1 is slightly damaged; the dP_4 is badly damaged.

AL204-4; SH-1, SH-2, or SH-3; female; juvenile (M_1 moderately worn). A shattered and slightly exploded fragment of a left mandibular body preserved from the symphysis to posterior M_1; P_3 and P_4 are badly damaged; M_1 is slightly damaged.

AL205-1a,b,c; KH-1; male; adult (M^2s heavily worn). A cranium and associated mandibular fragment.

-1a,b constitute a nearly complete cranium; the dorsum of the muzzle, sphenoid and orbital regions, frontal and parietal bones, the occipital plane, and left zygomatic arch are damaged; the left P^3 through M^2 and right P^4 through M^2 are present but damaged.

-1c is a mandibular symphysis and left body preserved anterior to M_2; the P_3 is complete; P_4 and M_1 are fragmentary.

AL208-6; SH-2 or SH-3; sex unknown; adult (M_3 moderately worn). An isolated slightly damaged, right M_3.

AL208-10a,b; SH-2 or SH-3; male; adult (M_3s moderately worn). An associated mandible and maxillary fragment.

-10a is a badly crushed mandible lacking the left ascending ramus and right gonial region and ascending ramus; the left and right P_3s through M_3s are present, but all are slightly damaged.

-10b is a crushed left maxillary fragment preserved from anterior P^3 to posterior M^3; does not extend above the molar roots; P^4 through M^2 are complete; M^3 is fragmentary.

AL217-1; SH-1, SH-2, or SH-3; probably female; old adult (M_3 heavily worn). A right mandibular body preserved from anterior P_3 to some distance behind M_3; the margin is intact; P_4 through M_3 are present, but damaged.

AL225-5; SH-1, SH-2, or SH-3; sex unknown; age unknown. An isolated crown of a right upper molar, slightly worn and damaged.

AL253-2; SH-1 or SH-2; male; juvenile (C_L very slightly worn). An isolated, left C_L lacking the crown apex and root tip.

AL263-1; SH-4, DD-1, or DD-2; sex unknown; juvenile (dP^4 slightly worn). A left maxillary fragment preserving the region above the dPs, the lateral orbital margin, and the root of the zygomatic arch; the dP^3 is fragmentary; the dP^4 is complete; the germ of the M^1 is not fully formed.

AL269-3; DD-3; female; adult (M_3s moderately worn). A mandible lacking the left gonial margin and condyle, and the right gonial region and ascending ramus; the left M_2

and M_3 and the right P_4 through M_2 are all heavily damaged; the right M_3 is only slightly damaged.

AL270-1; SH-1, SH-2, or SH-3; female; adult (M_3 moderately worn). A right mandibular body preserved from anterior P_3 to posterior M_3; the P_4 is fragmentary; the M_1 through M_3 are slightly damaged.

AL284-2; DD-1 or DD-2; sex unknown; juvenile (M_1 very slightly worn). A left mandibular body lacking the gonial region and ascending ramus; the I_1 is unerupted; the dC_L through M_1 are well preserved.

AL288-45; KH-1; sex unknown, juvenile (dP_4 is heavily worn). An isolated crown of a left dP_4.

AL300-6; DD-1 or DD-2; sex unknown; age unknown. An isolated crown of a moderately worn M^1 or M^2.

AL304-1; DD-2 or DD-3; sex unknown; adult (M_3 slightly worn). An isolated left M_3 with damaged distal roots.

AL310-15; DD-3 or (KH-1); probably female; adult (M_3 slightly worn). A fragment of a left mandibular body preserved from anterior P_4 to posterior M_3; the margin is intact only below M_3; P_4 and M_3 are well preserved; M_1 and M_2 are slightly damaged.

AL310-19; DD-3 or (KH-1); male; juvenile (C^Us just erupting). A fragment of a muzzle preserving the piriform aperture, dorsum, right interorbital region, and part of the posterior palate; the left C^U through P^4 and right C^U and P^3 are present; the tips of both C^Us are broken.

AL317-2; DD-3 or (KH-1); sex unknown; adult (M_3 slightly worn). A very small fragment of a right mandibular body; complete M_3 is present.

AL321-12; DD-2 or DD-3; female; adult (M^3s slightly worn). A nearly complete cranium lacking the zygomatic arches; the left canine alveolus, right supraorbital torus, right postglenoid process, and pterygoid plates are damaged; the left P^3 through M^3 and right P^4 through M^3 are well preserved.

AL327-2; SH-2; sex unknown; juvenile (dP_3 erupted, but unworn). A mandibular body preserved from left I_2 through posterior right dP_4; the left I_1 and I_2, the right I_1 through dP_3, and the germ of the right dP_4 are present.

AL329-1; DD-2; male; juvenile (M_2 slightly worn). A mandibular body preserved from posterior right P_4 to just behind left M_3; it is crushed and exploded especially in the symphyseal region; the left I_1 through M_3 (unerupted) and right I_1 through P_4 are present; the tips of the C_{LS} are broken.

AL345-1; SH-4; female; juvenile (M_1 moderately worn). A small fragment of a left mandibular body preserved from anterior P_3 to posterior M_1; the margin is absent; a slightly damaged M_1 is present.

Appendix 2.2 Measured variables

The numbers and names of the measured variables used in this report follow. Where landmarks are not obvious from the variable's name, further definition is provided. In general, lengths are measured in an anterior–posterior (or mesial–distal) direction, widths in a lateral–lateral or medial–lateral direction, and heights in a superior–inferior direction.

1. Muzzle length.
    The distance between incision and the left or right arc of the infraorbital foramina.
2. Muzzle dorsum length.
    The distance between orbitale and the posterior margin of the piriform aperture at the midline.
3. Cranial length.
    The distance between incision and inion.
4. Midcranial length.
    The distance between orbitale and the tip of the occipital protuberance.
5. Width of the maxillary dental arcade.
    The distance between points on the buccal walls of the mesial lophs of the left and right M^2s.
6. Length of the upper molar row.

7. Width of the C^U alveoli.
    The distance between points on the lateral margins of the left and right C^U alveoli that lie near the midpoint in the length of the canines.
8. Width of the upper incisive alveoli.
    The distance between points on the distal margins of the alveoli of the left and right I^2s.
9. Length of the palate.
    The distance between incision and a point on the anterior margin of the nasal choanae.
10. Length of the inferior, posterior cranium.
    The distance between a point on the anterior margin of the nasal choanae and inion.
11. Bizygomatic breadth.
12. Length of the vault.
    The distance between glabella and inion.
13. Maximum width of the vault.
    The distance between points on the left and right suprameatal crests just above the external auditory meati.
14. Minimum postorbital constriction.
15. Height of the vault.
    The distance between the point of union of the temporal lines and a point on the anterior margin of the foramen magnum.
16. Width of the vault in the midparietal region.
    The distance between points on the left and right parietals that lie above the external auditory meati and midway between them and the sagittal crest or suture.
17. Height of the orbit.
18. Width of the orbit.
19. Distance between a point at the base of a postglenoid process and point at the base of the distal wall of the M^3 on the same side.
20. Distance between a point at the base of the postglenoid process and incision.
21. Length of the upper cheektooth row.
    The distance between a point at the base of the mesial wall of the C^U and a point at the base of the distal wall of the M^3 on the same side.
22. Maximum depth of the mandibular fossa.
    The distance between a line that runs from the P^4 or M^1 alveolus to the mental ridge

(perpendicular to the occlusal plane) and the point of maximum excavation of the fossa.

23. Minimum thickness of the mandibular body.
   The distance between a point at the maximum depth of excavation of the mandibular fossa and a point on the medial surface of the body located so that the measurement is taken perpendicularly to the sagittal and occlusal planes. If a mandibular fossa is absent, the lateral point lies midway between the P_4 alveolus and the mental ridge.
24. Depth of the mandibular body below M_1.
25. Depth of the mandibular body below M_3.
26. Depth of the incisive plane.
   The distance between a line that runs between the medial margins of the alveoli of the P_4s and a point on the anterosuperior margin of the genioglossal fossa.
27. Width of the incisive plane.
   The distance between points on the anteromedial margins of the alveoli of the P_4s.
28. Width of the lower incisive alveoli.
   The distance between points on the distal margins of the alveoli of the left and right I_2s.
29. Width of the lower canine alveoli.
   The distance between points on the lateral margins of the alveoli of the left and right C_Ls.
30. Length of the lower molar row.
31. Basal length of the I^2.
   The distance between a point on the dentoenamel junction on the distal margin and a point the dentoenamel junction on the mesial margin of the tooth.
32. Basal width of the I^2.
   The distance between a point on the dentoenamel junction of the labial side of the tooth and a point on the dentoenamel junction of the lingual side of the tooth.
33. Labial height of the I^2.
   The distance between the apex of the crown and a point on the dentoenamel junction on the labial margin of the tooth.
34. Basal length of the I_1.
   The distance between a point on the dentoenamel junction on the distal margin and

a point the dentoenamel junction on the mesial margin of the tooth.
35. Basal width of the I_1.
   The distance between a point on the dentoenamel junction of the labial side of the tooth and a point on the dentoenamel junction of the lingual side of the tooth.
36. Labial height of the I_1.
   The distance between a point in the middle of the incisive edge and a point on the dentoenamel junction on the labial margin of the tooth.
37. Basal length of the I_2.
   Same as Variable 34.
38. Basal width of the I_2.
   Same as Variable 35.
39. Labial height of the I_2.
   Same as Variable 36.
40. Maximum length of the C^U.
   The distance between a point on the dentoenamel junction on the distal margin and a point the dentoenamel junction on the most mesiolabial margin of the tooth.
41. Maximum width of the C^U.
   The distance between a point on the dentoenamel junction of the labial side of the tooth and a point on the dentoenamel junction of the lingual side of the tooth.
42. Maximum length of the C_L.
   The distance between a point on the dentoenamel junction of the distolingual margin and a point on the dentoenamel junction of the mesiolabial margin of the tooth.
43. Maximum width of the C_L.
   The distance between points on the dentoenamel junction of the tooth measured perpendicularly to the maximum length.
44. Maximum length of the P_3.
   The distance between a point on the most distal projection of the heel of the tooth and a point on the most anterior projection of the crown on the mesial root.
45. Maximum width of the P_3.
46. Maximum length of the P_4.
47. Maximum width of the P_4.
48. Maximum length of the P^3.
49. Maximum width of the P^3.
50. Maximum length of the P^4.
51. Maximum width of the P^4.

52. Basal length of the M^1.
    The distance between a point at the center of the dentoenamel junction on the mesial wall and a point at the center of the dentoenamel junction on the distal wall of the tooth.
53. Maximum length of M^1.
54. Mesial width of M^1.
55. Distal width of M^1.
56. Basal length of the M^2.
    Same as Variable 52.
57. Maximum length of M^2.
58. Mesial width of M^2.
59. Distal width of M^2.
60. Basal length of the M^3.
    Same as Variable 52.
61. Maximum length of M^3.

62. Mesial width of M^3.
63. Distal width of M^3.
64. Basal length of the M_1.
    Same as Variable 52.
65. Maximum length of M_1.
66. Mesial width of M_1.
67. Distal width of M_1.
68. Basal length of the M_2.
    Same as Variable 52.
69. Maximum length of M_2.
70. Mesial width of M_2.
71. Distal width of M_2.
72. Basal length of the M_3.
    Same as Variable 52.
73. Maximum length of M_3.
74. Mesial width of M_3.
75. Distal width of M_3.

3 Evolution of *Theropithecus* in the Turkana Basin

MEAVE G. LEAKEY

Summary

1. The Turkana Basin provides a unique and relatively continuous record of part of the evolution of two extinct lineages of *Theropithecus*. Representatives of the *T. brumpti* lineage are found in deposits ranging in age from approximately 3.5 to 2.0 Ma. *T. oswaldi* replaces *T. brumpti* after 2 Ma as the common cercopithecoid in the later part of the Koobi Fora Formation. The two species are found in contemporary deposits only in the Shungura Formation between Units E-3 and G-12 (approximately 2.4 to 2.0 Ma).

2. Compared with later material, the early *T. brumpti* specimens are smaller and have less complex molar teeth, they lack a reversed curve of Spee and expanded flaring zygomatic arches, and they have less developed maxillary and mandibular corpus fossae, mental ridges and mental rugosity. All these features are more progressively developed in the later specimens from the lower Omo Valley. The *T. brumpti* lineage includes specimens previously referred to *T. baringensis* and *T. quadratirostris*. Two subspecies of *T. brumpti* are now recognized: *T. b. baringensis* for plesiomorphic material from earlier horizons and *T. b. brumpti* for apomorphic material from later horizons. Specimens of intermediate evolutionary status are referred to *T. brumpti* subspecies indeterminate.

3. Through the upper portion of the Koobi Fora Formation, *T. oswaldi* shows a progressive increase in body size with time and a corresponding reduction in the relative size of the anterior dentition and increase in the size of the posterior cheek teeth. Only one species, *T. oswaldi* is recognized in this lineage which is represented by three subspecies, *T. o. darti* for the early stage, *T. o. oswaldi* for the better known intermediate stage and *T. o. leakeyi* for the late progressive stage.

4. The differing masticatory adaptations displayed by the two *Theropithecus* lineages relate to constraints imposed by the length of the muzzle while maintaining the required

degree of lateral movement. Microscopic parallel scratches on the enamel surface of the molars aligned at right angles to the long axis of the jaw attest to the existence of significant lateral movement. These differing adaptive complexes may be interpreted to relate to differing diets but may represent different adaptations for similar diets.

5. Both lineages have molars that are morphologically similar to those of *T. gelada*. *T. brumpti* molars retain a shearing function longer than those of *T. oswaldi* which, particularly in the later subspecies, rapidly wear to a flat occlusal plane. The *T. brumpti* molars may have been used for cutting and shearing in a mixed graminivorous and folivorous diet and the *T. oswaldi* molars used in grinding a more exclusively graminivorous diet. Evolution of the high crowned complex molars and the reversed curve of Spee must have occurred in parallel in the two lineages subsequent to their separation.

6. To maintain efficiency, lateral excursion of the mandible in the region of the progressively enlarged molars was necessary and must have led to increased movement at the mandibular condyles particularly in *T. o. leakeyi*. The delayed eruption of the molars, their oblique angle on eruption and the reversed curve of Spee together provide a mechanism which increases the longevity of the molars.

7. *Theropithecus* probably derived from a papionin ancestor between 3.5 and 4.0 Ma. Papionins were relatively common at Pliocene accumulations such as Laetoli (3.65 to 4.0 Ma), Lothagam (5.0 to 5.5 Ma) and the Lonyumun Member of the Koobi Fora Formation (almost 4.0 Ma). The earliest securely dated *Theropithecus* are specimens of *T. brumpti* from the Usno Formation of the lower Omo Valley and the Lokochot Member of the Koobi Fora Formation. The earliest evidence for *T. oswaldi* is from Hadar and Makapansgat.

8. *T. brumpti* preferred the more forested environments prevailing in the Turkana Basin in the predominantly fluviatile regime prior to 2 Ma. After 2 Ma, *T. oswaldi* replaced *T. brumpti* and is associated with more open woodland and grassland prevailing in the succeeding predominantly lacustrine regime. The large colobines common in the upper Burgi Member of the Koobi Fora Formation and lower members of the Shungura Formation mostly disappear after deposition of the KBS tuff. Their persistence in the Koobi Fora Formation subsequent to the disappearance of *T. brumpti* suggests that they were less closely tied to forest habitats.

9. In contrast to the widely distributed *T. oswaldi*, *T. brumpti* has only been recognized in the Turkana and adjacent Baringo basins at sites dated between 3.5 and 2.0 Ma. This may relate to the paucity of sites of this age providing suitable habitats for sampling. With the exception of Hadar and Makapansgat, both of

which have yielded *T. o. darti* and may therefore be sampling different environmental conditions, few other African sites of this age are known. One mandible from Makapansgat shows morphological features typical of *T. b. baringensis* and may indicate a wider distribution for this genus.

10. The increased body size of *T. oswaldi* probably resulted from environmental factors. Loss of widespread tree cover may have led to increased body size as a deterrent to predators. Dental adaptations which increased masticatory efficiency may have been allometric. *T. oswaldi*, like other large graminivorous herbivores of the African Pleistocene savannas, succumbed as a result of extreme climatic fluctuations which were taking place in the Late Pleistocene.

Introduction

Theropithecus I. Geoffroy, 1843 is the best represented and most comprehensively studied of the Plio-Pleistocene fossil cercopithecoids (Jolly, 1972; Leakey & Leakey, 1973; Dechow & Singer, 1984; Eck & Jablonski, 1987; Eck, 1987). The genus appears to derive from a papionin ancestor at about 4 Ma and the subsequent evolution of two distinct *Theropithecus* lineages is documented at sites ranging in age from over 3 Ma to approximately 400,000 years. Both Plio-Pleistocene lineages became extinct, the evolutionary history of the only extant species of this genus remains unknown.

The genus *Theropithecus* is charac-

terized by molars, unique among the Cercopithecoidea, that are high crowned like those of the Colobinae. But whereas the major cusps of colobine molars are linked by high transverse lophs to form a series of sharp chopping transverse ridges, the occlusal surface of *Theropithecus* molars is reduced to a plane surface at an early stage. Trituration is enhanced by additional clefts and fossae, especially mesial and distal to the major cusps and on either side of a prominent longitudinal ridge separating the median buccal cleft and talonid basin on the lower molars and the median lingual cleft and trigon basin on the uppers. The grinding occlusal surface thus presents an intricate pattern of enamel ridges similar to adaptive complexes found in the molars of grazing mammals such as equids, bovids, suids, elephantids, and microtines (Jolly, 1972). The feeding habits of the modern gelada baboon, *T. gelada*, have been extensively studied (Crooke & Aldrich-Blake, 1968; Dunbar & Dunbar, 1974; Dunbar, 1977; Iwamoto, 1979). Grasses average 90–95 per cent of the annual food intake supplemented by a seasonally variable proportion of leaves, seeds and rhizomes (Dunbar, 1977). The striking similarity between the molar morphology of *T. gelada* and extinct *Theropithecus* species suggests a similar graminivorous diet. A recent carbon stable isotope study of *T. oswaldi* from Swartkrans (Lee-Thorp & van der Merwe, 1989) indicated that *T. oswaldi* at Swartkrans fed mainly on C4 grasses.

The two fossil lineages of *Theropithecus* show contrasting adaptive features of the masticatory apparatus and cranial morphology. The *T. brumpti* (Arambourg, 1947) lineage is characterized by a long muzzle and the evolutionary development of expanded flaring anterior zygomatic arches. The *T. oswaldi* (Andrews, 1916) lineage is characterized by a shorter muzzle, high hafting of the braincase to the face and the reduction of the canines and lower third premolars, increased size of the posterior cheek teeth and increasing body size. Both lineages are well represented in the Turkana Basin where Pliocene and Pleistocene fossiliferous deposits dating from over 4 Ma to less than 1 Ma are extensive and widespread. The geology of the Usno and Shungura formations in the lower Omo Valley (de Heinzelin, 1983), the Koobi Fora Formation east of the lake (Brown *et al.*, 1985) and the Nachukui Formation to the west (Harris, Brown & Leakey, 1988) has now been well studied. The analysis of tephra from these three regions has resulted in reliable correlations (Feibel, Brown & McDougall, 1989) and has permitted the reconstruction of their geological history, palaeoenvironments and faunal assemblages (Harris, 1983; Brown & Feibel, 1987; Shipman & Harris, 1988; Harris, *et al.*, 1988; Feibel, 1988).

The Turkana Basin succession documents evolutionary change in the two extinct lineages of *Theropithecus*. Representatives of the *T. brumpti* lineage are found in deposits ranging in age from approximately 3.5 Ma to 2.0 Ma. *Theropithecus brumpti* was contemporaneous with *T. oswaldi* in the Shungura Formation of the Omo Valley between Units E-3 to G-12, (approximately 2.4 to 2.0 Ma) (Eck, 1987), an

interval poorly represented in other areas of the Basin. Only *T. oswaldi* persisted after 2 Ma. The development of very different cranial morphologies in the two lineages relates to differing adaptive strategies (Eck & Jablonski, 1987). Brief discussions of the evolution and taxonomy of these two lineages, as represented in the Lake Turkana Basin, is presented below as a background to an evaluation of their role in the Pliocene and Pleistocene assemblages of which they formed a part.

Theropithecus brumpti lineage

The oldest possible representative of this lineage was recovered from the Usno Formation at the White Sands locality. This relatively complete cranium was originally described as *Papio quadratirostris* (Iwamoto) 1982 but was later referred to *Theropithecus* (Eck & Jablonski, 1984). This attribution is discussed further by Delson and Dean in chapter 4. The specimen probably derives from Unit U-8 or U-9 (Eck & Jablonski, 1987) which, following correlations given by Feibel *et al.* (1989), is equivalent to the Lokochot Member of the Koobi Fora Formation (between 3.36 and 3.5 Ma). *Theropithecus brumpti* is best represented in the Shungura Formation where it has been recovered from Unit B-10 (2.95 Ma) to Unit G-13 (c. 2 Ma). Eck & Jablonski (1987) have provided detailed descriptions of these specimens, most of which were found in Member C. Other specimens of *T. brumpti* have been recovered from the lower, middle and upper Lomekwi Members of the Nachukui Formation (3.36–2.5 Ma) (Harris *et al.*, 1988; Feibel *et al.*, 1989) and from the

Tulu Bor and Lokochot Members of the Koobi Fora Formation (*c.* 3.5–2.68 Ma). These specimens and that from the Usno Formation thus extend the range of this lineage beyond that documented in the Shungura Formation.

While most of the East African specimens of the *T. brumpti* lineage have been recovered from the Turkana Basin, two are known from site JM 90/91 farther south in the Chemeron Formation near Lake Baringo. KNM-BC 3 (Figs 3.1–3.3), a cranium and mandible, originally described as *Papio baringensis* Leakey, 1969, was later, with the White Sands specimen, recognized as early *Theropithecus* (Eck & Jablonski, 1984). KNM-BC 1647 (Figs 3.2–3.3), a mandible, proximal femur, and distal radius, was found at the same site. Eck & Jablonski concluded that *Theropithecus baringensis*, *T. quadratirostris* and *T. brumpti* form a phyletic lineage: *T. baringensis* (4 Ma)–*T. quadratirostris* (3.3 or 3.4 Ma)–*T. brumpti* (2.0–2.8 Ma). Recent dating of site JM 90/91 near the base of the Chemeron Formation has shown it to be 3.2 Ma (A. Deino, pers. comm.). *T. baringensis* is thus somewhat younger than Eck and Jablonski had estimated, postdating *T. quadratirostris*.

Recent correlative advances in the Turkana Basin (Feibel *et al.*, 1989) now permit the ordering of this material as follows (Fig. 3.4).

Units U-8 to U-9 of the Usno Formation which have yielded the single specimen of *T. quadratirostris* correlate with the Lokochot Member of the Koobi Fora Formation.

Site JM 90/91 of the Chemeron Formation, which has yielded *T. baringensis*, correlates with the Tulu Bor Member of the Koobi Fora Formation, the lower Lomekwi Member of the Nachukui Formation and Member B of the Shungura Formation. Members C and D of the Shungura Formation, which have yielded *T. brumpti*, correlate with the upper Lomekwi Member of the Nachukui Formation.

Members E and F of the Shungura Formation, which have yielded *T. brumpti*, correlate with the Lokalalei and lower part of the Kalochoro Members of the Nachukui Formation; neither have yielded *T. brumpti*. Correlative horizons are not represented in the Koobi Fora Formation.

Upper Member G of the Shungura Formation, which has yielded *T. brumpti*, correlates with the upper part of the Kalochoro Member of the Nachukui Formation and the upper Burgi Member of the Koobi Fora Formation; neither have yielded *T. brumpti*.

Theropithecus quadratirostris The cranium of *T. quadratirostris* from the Usno Formation differs from that of *T. baringensis* in a few features which led Iwamoto (1982) to name a new species for this single specimen. The main distinguishing characters are the 'partially expanded anterior zygomatic bone, a more oblique orientation of the lateral margin of the frontal process and (perhaps) the more obtuse angle between the plane of the nuchal margin and the occlusal plane' (Eck & Jablonski, 1984). Unfortunately there are no mandibles or well-preserved dentitions of *T. quadratirostris* from the Usno Formation for comparison with the mandibular specimens, fragmentary max-

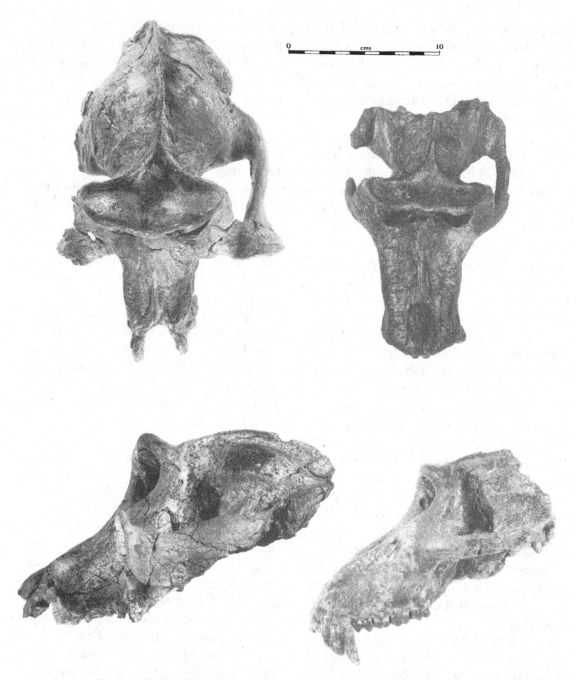

Fig. 3.1. Skulls of *Theropithecus brumpti*. Left: *T. brumpti brumpti* (KNM-ER 16828); right: *T. brumpti baringensis* (KNM-BC 3). Above, superior and below, left lateral views.

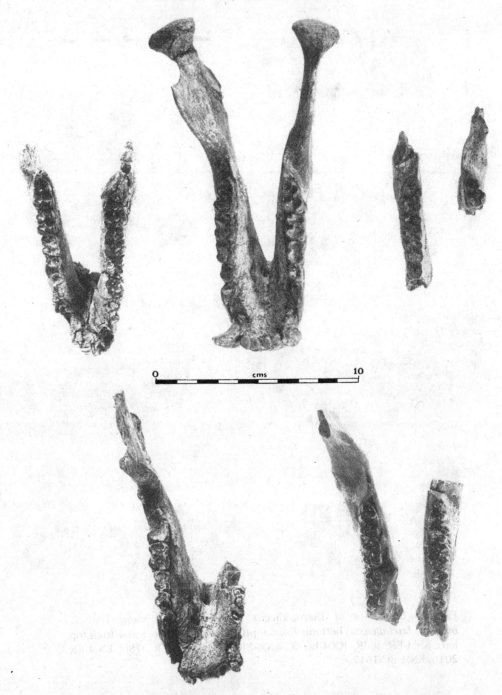

Fig. 3.2. Mandibles of *Theropithecus brumpti*, occlusal view. Top: *T. brumpti baringensis*; bottom: *T. brumpti brumpti*. From left to right: top; KNM-ER 3038, KNM-BC 3, KNM-BC 1647, bottom; KNM-ER 2015, KNM-ER 3780.

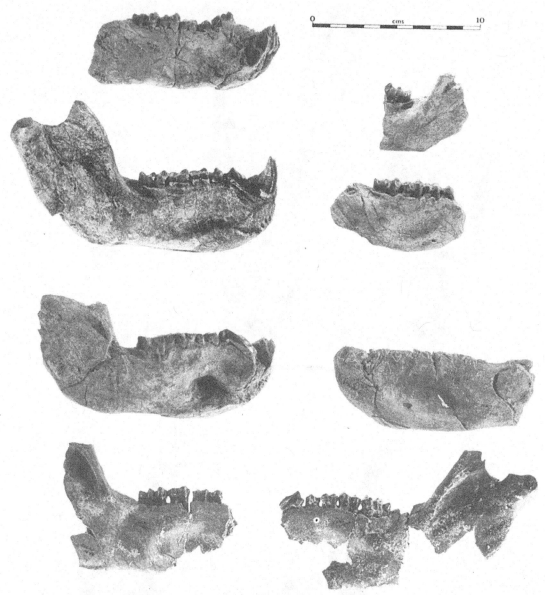

Fig. 3.3. Mandibles of *Theropithecus brumpti*, lateral view. Top: *T. brumpti baringensis*; bottom: *T. brumpti brumpti*. Anticlockwise from top left: KNM-ER 3038, KNM-BC 3, KNM-ER 2015, KNM-ER 3780, KNM-ER 2016, KNM-BC 1647.

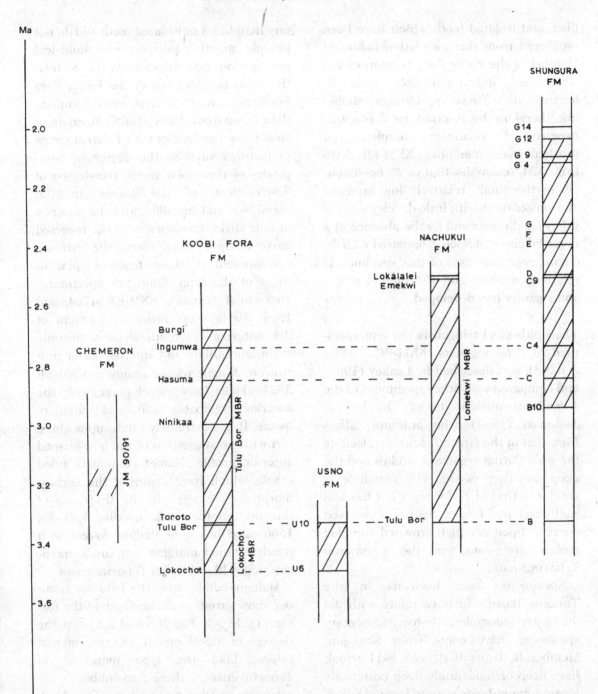

Fig. 3.4. Correlation diagram for strata in the Shungura, Usno, Koobi Fora, Nachukui and Chemeron Formations from which *Theropithecus brumpti* has been recovered.

illae, and isolated teeth which have been recovered from the correlative Lokochot Member of the Koobi Fora Formation and which are indistinguishable from *T. baringensis*. These specimens should not therefore be referred to *T. quadratirostris*. A relatively complete but crushed male mandible, KNM-ER 3038 (Fig. 3.3), resembles that of *T. baringensis* in the small, relatively low crowned and simple teeth with little development of valleys and crests and by the absence of a reversed curve of Spee. Compared with *T. baringensis* the fossa of this specimen is relatively shallow and the mental ridges and rugosity less developed.

Theropithecus baringensis The type specimen of *T. baringensis*, KNM-BC 3 (Figs 3.1–3.3), was described by Leakey (1969) and compared with other specimens of the *T. brumpti* phyletic lineage by Eck & Jablonski (1984). The cranium differs from that of the larger *T. brumpti*, lacking the wide flaring zygomatic arches and the deep maxillary fossae. The mandible is similar to that of *T. brumpti* but has less well developed fossae. Neither a reversed curve of Spee nor high crowned complex molars are found in the Chemeron *T. baringensis*.

Specimens from horizons in the Turkana Basin which correlate with JM 90/91 are incomplete. Three fragmentary specimens have come from Shungura Member B, Units B-10 and B-11 which have deep or moderately deep posteriorly located mandibular corpus fossae (Eck & Jablonski, 1987). The 14 specimens from the lower Lomekwi Member of the Nachukui Formation comprise fragmen-

tary mandibles or isolated teeth and do not provide much significant morphological information. Specimens from the correlative Tulu Bor Member of the Koobi Fora Formation are, in general, more complete than those from the Nachukui Formation. Most show similarities with *T. baringensis* in features such as the degree of complexity of the molar teeth, the degree of development of the fossae in the mandibles and maxillae, and the absence or only slight development of the reversed curve of Spee. None shows the extreme development of these features seen in some of the later Shungura specimens. Two small females, KNM-ER 4704 and 1564, display very little development of the anterior zygomatic arch with only slight lateral flare and anterosuperior protrusion. A male partial cranium, KNM-ER 3025, lacks any development of the anterior zygomatic arch, and maxillary fossae. It has extremely thick supraorbital tori with no suggestion of the bow-shaped anterior profile. Numerous matrix filled cracks which have distorted the surface morphology exaggerate the thickness of the tori. Another male maxilla, KNM-ER 1566, also has only shallow fossae with poorly defined margins and small maxillary ridges like those of *T. baringensis*.

Male mandibles from the Tulu Bor Member show variation in the depth of the fossae (Table 3.1, Figs 3.3 and 3.5) and the degree of development of the mental ridges. Like the type mandible of *T. baringensis*, these mandibles are relatively smaller, and have smaller cheek teeth compared with those of *T. brumpti*, (Table 3.2, Figs 3.6 and 3.7). In addition the triangular depression on the

Table 3.1. *Depth of the mandibular corpus fossa in* Theropithecus brumpti

Member	fossa depth
Shungura Members E–G	
mean	6.3
n	2
max	8.1
min	4.5
Shungura Members C–D, and upper Lomekwi	
mean	11.43
n	6
max	16.10
min	8.00
Tulu Bor, lower Lomekwi, and Chemeron	
mean	8.53
n	11
max	11.50
min	4.20
Lokochot Member	
mean	3.00
n	2
max	3.50
min	2.50

Note: Measurements of specimens from the Shungura Formation were taken from Eck & Jablonski (1987).

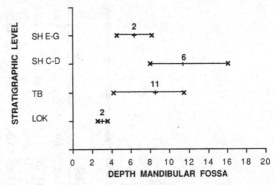

Fig. 3.5. Depth of the mandibular corpus fossa of *Theropithecus brumpti* at different intervals in time. The mean of the measurements is indicated with the number of individuals in the sample and the length of the bar indicates the range of measurements. SH E-G = Shungura Formation Members E-G; SH C-D = Shungura Formation Members C-D; TB = Tulu Bor Member Koobi Fora Formation, lower Lomekwi Member Nachukui Formation and site JM90/91 Chemeron Formation; LOK = Lokochot Member Koobi Fora Formation.

superolateral surface of the ramus, which provides an attachment for the fleshy fibres of the deep masseter muscle and tendon, is generally smaller, less deeply excavated and less clearly defined. The reversed curve of Spee, characteristic of later representatives of *T. brumpti*, is absent or poorly developed. The mental ridges tend to be rugose and well developed in these mandibles though less

prominent than in the later Shungura material.

Theropithecus brumpti The distinctive cranium of the Shungura Formation *T. brumpti* has been described in detail by Eck & Jablonski (1987). The anterior zygomatic arch is wide and flaring and the mid-section robustly built. The supraorbital tori are distinctly 'bow-shaped' along the superior margin and have a deep depression above the interorbital region. The long muzzle has strongly developed maxillary ridges and deeply excavated maxillary fossae. The mandibular body has a long, deeply excavated fossa and the mental ridges are strongly developed and highly rugose. There is a deeply excavated triangular depression on the superolateral surface of the mandibular rami. These

Table 3.2. Measurement of the lower dentition of T. brumpti

A. Males and females – total sample

		I1		I2		P4		M1			M2			M3		
		L	B	L	B	L	B	L	AB	PB	L	AB	PB	L	AB	PB
Shungura Members E–G	Mean					9.40	8.00	11.77	9.00	9.35	15.10	11.93	11.65	20.84	13.16	12.00
	n					1	1	3	2	2	3	4	4	5	5	5
	Max							12.30	9.40	9.60	15.50	13.50	13.30	22.20	15.50	13.70
	Min							11.10	8.60	9.10	14.80	11.50	10.30	19.20	11.40	11.00
Shungura Members C, D and upper Lomekwi	Mean			7.70	6.00	9.88	7.95	12.95	9.68	9.90	16.06	12.71	11.98	22.30	14.83	12.62
	n			1	1	6	6	6	5	8	7	8	7	11	10	11
	Max					10.80	8.20	14.00	10.30	11.30	17.50	16.30	13.10	24.50	15.10	14.40
	Min					9.40	7.50	12.10	9.10	9.20	15.00	11.20	10.80	16.50	11.70	11.30
Tulu Bor, lower Lomekwi and Chemeron	Mean	5.00	6.70	5.40	6.70	8.15	7.25	11.68	9.41	9.21	13.73	10.85	10.40	18.63	11.96	10.48
	n	1	1	1	1	5	5	9	8	5	10	9	7	16	14	11
	Max					9.10	7.50	13.70	12.80	10.20	15.80	11.80	11.00	22.00	15.00	12.45
	Min					7.45	6.90	10.00	8.20	8.60	12.60	9.80	9.50	16.50	10.10	8.60
Lokochot						8.10	7.50	10.00		9.30	12.50	11.10	9.70	15.80	11.60	9.30

B. Males

		C		P3		P4		M1			M2			M3		
		L	B	L	B	L	B	L	AB	PB	L	AB	PB	L	AB	PB
Shungura Members E–G	Mean	7.00	12.00	23.25	6.60	9.40	8.00	12.30	9.40	9.60	15.50	12.75	12.65	22.05	14.70	13.15
	n	1	1	2	2	1	1	1	1	1	1	2	2	2	2	2
	Max			26.00	7.10							13.50	13.30	22.20	15.50	13.70
	Min			20.5	6.10							12.00	12.00	21.90	13.90	12.60

Shungura Members

	P3		C		P4		M1			M2			M3		
	L	B	L	B	L	B	L	AB	PB	L	AB	PB	L	AB	PB
C, D and upper Lomekwi																
Mean	25.67	6.70	10.60	12.07	10.17	8.00	12.60	9.40	9.90	16.35	12.53	12.05	22.39	13.61	12.76
n	3	2	3	3	3	3	3	3	5	4	4	4	8	8	8
Max	30.10	7.30	15.50	15.30	10.80	8.20	12.90	9.90	11.30	17.50	13.40	13.10	24.50	15.10	14.40
Min	22.50	6.10	8.90	8.00	9.40	7.80	12.10	9.10	9.20	15.60	11.00	10.80	20.50	12.00	11.30
Tulu Bor, lower Lomekwi, and Chemeron																
Mean	22.93	6.72			8.55	7.17	10.97	8.65	8.75	13.98	10.62	10.07	18.24	11.38	10.18
n	3	4			2	2	4	3	3	5	4	4	5	4	4
Max	24.00	7.10			9.10	7.30	11.70	8.95	8.96	15.80	11.00	10.30	20.80	12.00	10.90
Min	21.70	6.18			8.00	7.05	10.20	8.20	8.60	13.00	9.80	9.50	16.80	10.80	9.30
Lokochot 117	25.20				8.10	7.50	10.00	9.50	9.30	12.50	11.10	9.70	15.80	11.60	9.30

C. Females

	P3		C		P4		M1			M2			M3		
	L	B	L	B	L	B	L	AB	PB	L	AB	PB	L	AB	PB
Shungura Members E–G																
Mean	11.30	6.90	5.70	9.05	9.60	7.90	11.50	8.60	9.10	14.90	11.10	10.65	20.03	12.13	11.23
n	2	2	2	2	3	3	2	1	1	2	2	2	3	3	3
Max	14.60	7.10	6.10	9.50	9.80	8.20	11.90			15.00	11.50	11.00	20.90	12.90	11.50
Min	8.00	6.70	5.30	8.60	9.50	7.50	11.10			14.80	10.70	10.30	19.20	11.40	11.00
Shungura Members C, D and upper Lomekwi																
Mean			5.00	8.50			12.95	9.70	9.65	15.67	11.77	11.56	18.60	12.80	11.55
n			1	1			2	1	2	3	2	2	2	2	2
Max							13.50	9.80		16.00	12.60	12.10	20.70	13.90	11.70
Min							12.40	9.50		15.00	11.20	11.03	16.50	11.70	11.40
Tulu Bor, lower Lomekwi, and Chemeron																
Mean					7.45	7.50				12.70	10.80	10.50	18.55	10.70	10.00
n					1	1				1	1	1	2	1	1
Max													20.60		
Min													16.50		

Note: Measurements of specimens from the Shungura Formation taken from Eck & Jablonski (1987).

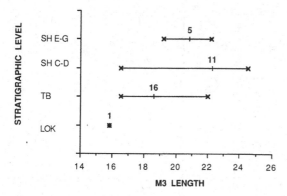

Fig. 3.6. Mesiodistal length of the lower M3 of *Theropithecus brumpti* at different intervals in time. See Fig. 3.5 for abbreviations.

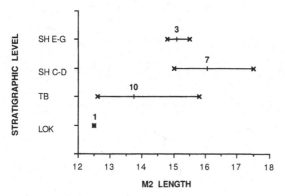

Fig. 3.7. Mesiodistal length of the lower M2 of *Theropithecus brumpti* at different intervals in time. See Fig. 3.5 for abbreviations.

features are well developed on most of the male specimens recovered from Members C and D of the Shungura Formation and to a lesser degree on the female specimens.

Eck & Jablonski (1987) were unable to observe any directional change in the size or morphology of the Shungura specimens, because the majority are from Units C-6 through D-3 and thus differ in geologic age by no more than a few thousand years (Eck & Jablonski, 1987). These specimens

are of similar age to fragmentary mandibles, maxillae, cranial, and post-cranial elements from the upper Lomekwi Member of the Nachukui Formation. One relatively complete cranium, KNM-WT 16828 (Fig. 3.1), from the upper Lomekwi Member differs from Shungura specimens of similar age by the shape and orientation of the anterior zygomatic arches which are broad and flaring but have an anteriorly directed inferior surface and an anterior border that protrudes superiorly almost to the level of the inferior margin of the orbits. This contrasts with the condition in the Shungura specimens where the inferior surface is generally antero-inferiorly directed and the most superior point is well below the level of the orbits. This difference can probably be accommodated within morphological variability displayed by the Shungura specimens. Eck & Jablonski note that the orientation of the external auditory meatus and posterior margin of the temporal fossa is perpendicular to the sagittal plane in *T. brumpti*, an observation based on very few specimens. Although the external auditory meati of WT 16828 are missing, the remaining surrounding bone suggests their orientation was oblique. Again this was probably a morphologically variable feature.

Of the 135 specimens listed by Eck & Jablonski (1987), only 11 fragmentary pieces are from the uppermost horizons in Shungura Member G. No specimens of *T. brumpti* have been recovered from the correlative upper Burgi Member of the Koobi Fora Formation nor from the Kalochoro Member of the Nachukui Formation.

Evolutionary trends

Theropithecus brumpti material from the Shungura, Usno, Nachukui, Chemeron, and Koobi Fora formations displays significant morphological changes through time. The early specimens lack the expanded zygomatic arches but have some development of the maxillary and mandibular fossae, the mental ridges and mental rugosity. The teeth are relatively simple and not unlike those of other papionids although the incisors are proportionately reduced. The development of the reversed curve of Spee appears linked with the increase in dental complexity, changes in the anterior zygomatic arches and in the shape of the supraorbital tori. Unfortunately the development of dental complexity shown by these specimens is difficult to demonstrate metrically because of the different degrees of wear shown by the teeth. Only the few unworn teeth provide meaningful measurements. The molar teeth became progressively higher crowned, with deeper and more pronounced infoldings of the enamel and deeper valleys and clefts. This can be observed but quantification is elusive. There is some indication that some of these evolutionary trends may be reversed in Shungura Members E–G. Specimens from this time interval have smaller average molar measurements (Table 3.2, Figs 3.6 and 3.7) and shallower mandibular corpus fossae (Table 3.1, Fig. 3.5). One of the latest mandibular specimens recovered, Omo 75N- 71-Cl from Unit G-12, has a very shallow mandibular fossa. The male cranium, L388y-2257 from Unit E-3, lacks the exaggerated development of

the zygomatic arch typical of specimens from the earlier Members C and D. To summarize, the evolution of *T. brumpti* in the Turkana Basin between 3.4 and 2.0 Ma is characterized by the evolutionary development of: (1) high complex molar crowns, (2) a reversed curve of Spee, (3) wide flaring zygomatic arches, (4) bow-shaped supraorbital tori, (5) deep fossae in the maxilla and mandible, (6) strong, rugose mental ridges, and (7) increasing body size.

A high degree of sexual dimorphism in body size, canine and lower P3 size is demonstrated by species of this lineage. Dental measurements indicate a similar degree of sexual dimorphism to that of *T. gelada* (Table 3.3). Female specimens from the upper Lomekwi Member of the Nachukui Formation and correlative members of the Shungura Formation have similar sized molars that are smaller than those of *T. brumpti* males. Small female crania and mandibles of *T. baringensis* from the Tulu Bor Member of the Koobi Fora Formation reflect the small body size of these earlier female specimens.

High variability is also due to the relatively long period of time over which the specimens are sampled. Characters that change with time such as the fossa in the mandible, the development of a reversed curve of Spee, and the development of the anterior zygomatic process are individually variable. The precision to which *T. brumpti* specimens can be assigned to a particular dated horizon is often no more precise than plus or minus several thousand years. 110,000 years, the estimated duration of Shungura Units C-4 to C-9, from which 21 of the 31 best speci-

Table 3.3. *Indices to show the degree of sexual dimorphism displayed in the dentition of extant and extinct species of* Theropithecus

	C		P3		M2		M3	
	L	B	L	B	L	AB	L	AB
Theropithecus oswaldi								
Koobi Fora and Olduvai Bed I								
mean male/mean female	1.79	1.84	1.58	1.16	1.07	1.04	1.04	1.02
max male/min female	2.28	2.27	2.08	1.56	1.43	1.32	1.40	1.37
Others								
mean male/mean female	1.59	1.62	1.46	1.14	1.04	1.17		
max male/min female	1.97	2.04	1.74	1.18	1.11			
Olorgesailie								
mean male/mean female	1.69	1.56	1.33	1.15				
max male/min female	2.12	1.87	1.68	1.40				
Theropithecus brumpti								
Shungura Members E, G								
mean male/mean female					1.04	1.15	1.10	1.21
max male/min female					1.05	1.26	1.16	1.20
Shungura Members C, D and upper Lomekwi								
mean male/mean female			2.27	.97	1.04	1.06	1.20	1.06
max male/min female			2.06	1.09	1.17	1.06	1.48	1.29
Tulu Bor, lower Lomekwi and Chemeron								
mean male/mean female	1.62	1.71		1.24	1.10	.98	.98	1.06
max male/min female	1.78	1.82		1.31	1.24	1.02	1.12	1.12
Theropithecus gelada								
mean male/mean female	1.88	1.65	1.74	1.13	1.04	1.06	1.04	1.04
max male/min female	3.71	2.18	2.44	1.37	1.21	1.28	1.24	1.29

Note: The indices for *T. gelada* were calculated using measurements taken from Swindler (1976), those for the Shungura specimens from Eck & Jablonski (1987). 'Koobi Fora' includes specimens from the upper Burgi, KBS and Okote Members; 'Others' includes specimens from upper Bed II, Beds III, IV and Masek at Olduvai and Kapthurin.

mens from the Shungura Formation were recovered, represents about 18,000 generations, assuming a generation time of six years on average, which is rather longer than that of 4.5 years recorded for *T.* *gelada* (Harvey, Martin & Clutton-Brock, 1987). Given appropriate stimulus, significant morphological change could take place in this span of time and would lead to the high variability sampled. Great individual variability, such as the contrasting degree of development of the anterior zygomatic arches of the skulls L345-287 from Unit C-9 and L338y-2257 from Unit E-3, might be expected from a rapidly evolving population but thorough investigation requires additional material and more precise stratigraphic control.

Taxonomy

The specimens from the Lake Turkana Basin represent a single evolving lineage in which the earliest and latest examples are clearly distinct and in which material of intermediate horizons displays variably intermediate morphology. *Theropithecus baringensis* was proposed (as *Papio baringensis* Leakey, 1969) when little was known of the morphology and evolution of *T. brumpti*. Similarly *T. quadratirostris* was considered to differ from *T. baringensis* although its affinities with the latter were recognized (Iwamoto, 1982). Study of numerous well-preserved skulls of *T. brumpti* from the Shungura Formation indicated a lineal relationship between these three species (Eck & Jablonski, 1984), but there is still some doubt as to the generic status of the Usno cranium (see Delson & Dean, chapter 4). The recovery of additional material from Pliocene sites east of Lake Turkana tends to confirm the relationship between *T. brumpti* and *T. baringensis* but it has become increasingly difficult to differentiate stages in this lineage. Specimens from the Tulu Bor Member of the Koobi Fora Formation and the lower and middle Lomekwi Member of the Nachukui Formation show considerable variability, some

resembling *T. brumpti* and others *T. baringensis*. In view of these problems, it is suggested that the distinctive latest and earliest examples be referred to the two subspecies *T. brumpti brumpti* (from the Members C to F of the Shungura Formation and from the upper Lomekwi Member of the Nachukui Formation) and *T. brumpti baringensis* (from the Tulu Bor and Lokochot Members of the Koobi Fora Formation and the lower Lomekwi Member of the Nachukui Formation) respectively, and that specimens of intermediate age that are not clearly attributable to either subspecies be left as *T. brumpti* subspecies indeterminate.

Distribution

Specimens representing the *T. brumpti* phyletic lineage have only been recognized from sites in the adjacent Turkana and Baringo basins dated between 3.5 and 2.0 Ma. This may reflect a restricted distribution for this species or a paucity of sites of this age. With the exception of Hadar in Ethiopia and Makapansgat in South Africa, few other African sites of this time interval are known. *Theropithecus brumpti* has not been recognized at Hadar (G. Eck, pers comm.). The Makapansgat *Theropithecus* has been referred to *T. darti* (Broom & Jensen, 1946) but one mandible, M626 (?M621), is described as differing from the others by its deep fossa (Freedman, 1957; Dechow & Singer, 1984) and the preserved morphology of this specimen is similar to that of *T. b. baringensis*. (Unfortunately the bone is missing at the mental ridges but the remaining portion indicates that they were

well developed.) If Makapansgat is approximately 3 Ma, this specimen correlates
with the Tulu Bor Member of the Koobi
Fora Formation from which *T. b.*
baringensis is known. Makapansgat may
represent another site at which *T. brumpti*
occurred.

Theropithecus oswaldi lineage

In contrast to *T. brumpti*, *T. oswaldi* has
been recovered from Pliocene and Pleistocene deposits in northern, eastern and
southern Africa and is common at sites of
about 2.4 to 0.5 Ma. The many specimens
recovered from the Turkana Basin
(Leakey, 1976; Eck, 1987; Harris *et al.*,
1988) provide a relatively continuous
record of this species from 2.4 to 1.3 Ma,
an interval also represented by sites in
southern Africa (Freedman, 1957, 1976;
Maier, 1970, 1972), in Beds I and II at
Olduvai Gorge in Tanzania (Jolly, 1972;
Leakey & Leakey, 1973) and at Kanjera in
Kenya (Andrews, 1916; Leakey, 1943;
Jolly, 1972; Plummer & Potts, 1989). Later
Pleistocene horizons at Hopefield in South
Africa (Dechow & Singer, 1984), Ternifine
in Algeria (Van Den Brink, 1980; Geraads,
1987), Olduvai Gorge in Tanzania (Hop-
wood, 1934; Leakey & Whitworth, 1958;
Jolly, 1972; Leakey & Leakey, 1973) and
Olorgesailie (Isaac, 1977; Koch, 1986) and
Kapthurin in Kenya, document more pro-
gressive examples of this species.
Theropithecus oswaldi appears to have
become extinct at some time after 700,000
years.

Theropithecus oswaldi is not well
represented in the lower Omo Valley. Only
25 specimens from Units E-3 to G-14 of

the Shungura Formation can presently be
identified as *T. oswaldi*. Fragmentary
Theropithecus specimens recovered from
Members H, J, K and L probably also
belong to *T. oswaldi*. An older fragmentary specimen from Unit C-6 may
represent the earlier *T. darti* (Eck, 1987).
Nine specimens have been recovered from
the Kalochoro, Kaitio, Natoo, and Nariokotome Members of the Nachukui Formation. Numerous specimens have been
recovered from the upper Burgi, KBS, and
Okote Members of the Koobi Fora Formation (2.0–1.3 Ma). Well-preserved crania,
numerous mandibles, and some partial
skeletons document evolutionary traits
previously recognized in specimens from
other sites (Jolly, 1972; Dechow & Singer,
1984). Some of the Turkana specimens are
illustrated in Figs 3.8–3.10.

Stratigraphic correlations within the
Turkana Basin are illustrated in Fig. 3.11:
Units E3 to G14 of the Shungura Formation from which *T. oswaldi* has been
recovered correlate with the Kalochoro
Member of the Nachukui Formation from
which only two specimens of *T. oswaldi*
have been recovered. Apart from upper
Member G, this interval of time is not
documented in the Koobi Fora Formation.

The upper Burgi Member of the Koobi
Fora Formation, which has yielded many
specimens of *T. oswaldi*, correlates with
the upper part of the Kalochoro Member of
the Nachukui Formation and upper Member G of the Shungura Formation.

The KBS and Okote Members of the
Koobi Fora Formation which have yielded
numerous specimens of *T. oswaldi* correlate with the Kaitio and Natoo Members of
the Nachukui Formation and upper Mem-

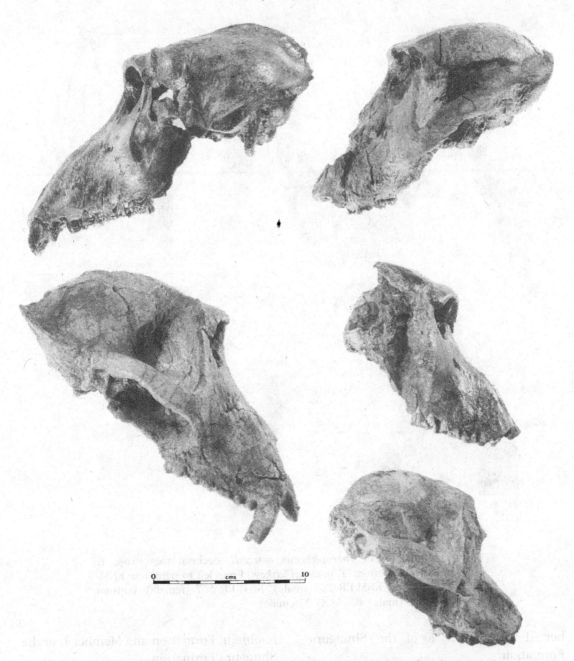

Fig. 3.8. Skulls of *Theropithecus oswaldi*. Left: male; right: female. Anti-clockwise from top left: KNM-ER 1531, KNM-ER 18925, KNM-WT 17435, KNM-ER 180, KNM-ER 971.

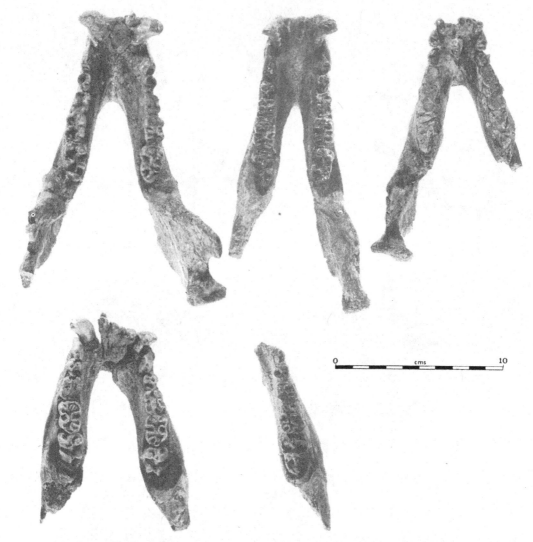

Fig. 3.9. Mandibles of *Theropithecus oswaldi*, occlusal view. Top: *T. oswaldi oswaldi*; bottom: *T. oswaldi leakeyi*. From left to right, top: KNM-ER 18925 (male), KNM-ER 865 (male), KNM-ER 567 (female); bottom: KNM-BK 22638 (male), KNM-OG 2 (female).

ber H to Member K of the Shungura Formation.

The Chari Member of the Koobi Fora Formation which has only yielded one postcranial specimen of *T. oswaldi* correlates with the Nariokotome Member of the

Nachukui Formation and Member L of the Shungura Formation.

Upper Burgi Member *Theropithecus oswaldi* is well represented in the upper Burgi Member of the Koobi Fora Forma-

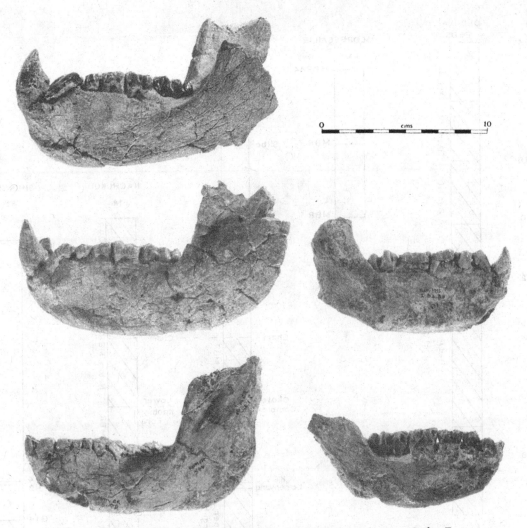

Fig. 3.10. Mandibles of *Theropithecus oswaldi*, lateral view. Left: *T. oswaldi oswaldi*; right: *T. oswaldi leakeyi*. From top left anticlockwise: KNM-ER 865 (male), KNM-ER 18925 (male), KNM-ER 567 (female), KNM-OG 2 (female), KNM-BK 22638 (male).

tion by crania and mandibles which display the characteristic morphological features of this species as described from Kanjera (Jolly, 1972). Among the crania recovered, one, KNM-ER 18925 (Fig. 3.8), was found *in situ* with the mandible articulated. This adult male specimen is remarkably complete. In contrast to *T. brumpti*, *T. oswaldi* has a short muzzle without marked development of the maxillary ridges or fossae, and weakly developed zygomatics which do not protrude anteriorly or superiorly. The supraorbital tori are relatively thin and not bow-shaped, and

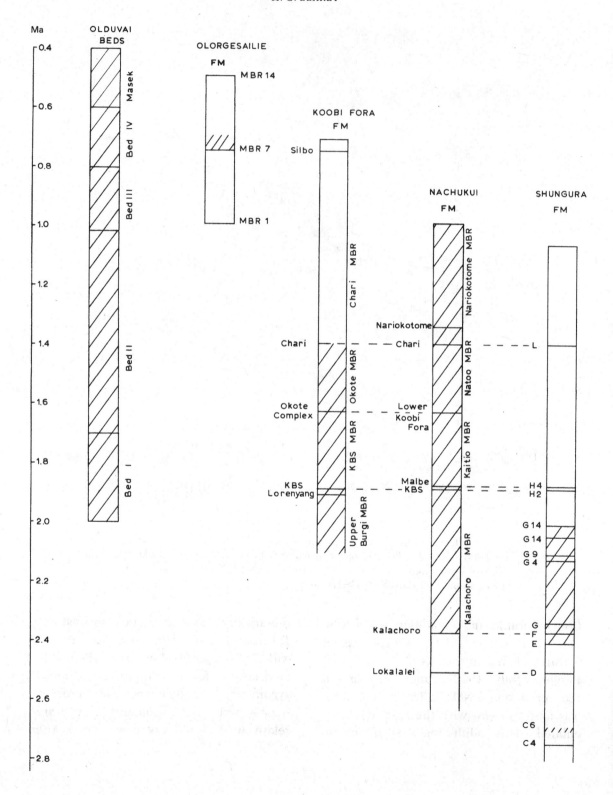

the calvarium is positioned well above the occlusal plane with a high hafting to the face. The outline of the articular eminence differs from that of *T. brumpti* in that it has a clearly defined and relatively long lateral margin and in that the glenoid process is relatively large.

Two fragmentary female specimens have been recovered from the slightly older horizons in the Kalochoro Member, and 25 specimens have been described from the Shungura Formation Units E-3 to G-14 (Eck, 1987).

KBS and Okote Members Numerous specimens from the KBS and Okote Members of the Koobi Fora Formation display more progressive features than those from the underlying upper Burgi Member. These include increasing body size and increasing length of the posterior cheek teeth, particularly M2 and M3 (Tables 3.4 and 3.5, Figs 3.12 to 3.15). The incisors are small and in many of the mandibles were lost during life.

Only six specimens have been recovered from the Natoo and Kaitio Members of the Nachukui Formation. Several fragmentary *Theropithecus* specimens from Members H to K in the Shungura Formation probably represent *T. oswaldi*.

Chari Member No cranial specimens of *T. oswaldi* have been recovered from the Chari Member of the Koobi Fora Formation but two fragmentary specimens were recovered from the Nariokotome Member

of the Nachukui Formation and some fragmentary specimens from Member L of the Shungura Formation most probably represent this species.

Earlier horizons A small mandibular fragment with M3 preserved from Unit C-6 of the Shungura Formation may represent *T. darti* (Eck, 1987).

Evolutionary Trends

Most of the Turkana Basin *T. oswaldi* specimens have been recovered from the time interval represented by the upper Burgi Member through the Okote Member, approximately 2.0–1.4 Ma. Only slight evolutionary progression may be observed in these specimens, but they contrast markedly with later specimens from other sites at which the culmination of these evolutionary trends may be observed. The male type specimen (068/6516) of *T. jonathani* from the Masek Beds at Olduvai Gorge (M.D. Leakey, pers. comm.) has such reduced canines that for some time it was thought to be female. The recent discovery of a well preserved male mandible from Kapthurin, KNM-BK 22638 (Figs 3.9 and 3.10), with teeth of comparable size but less worn than those of Olduvai mandible 068/6516, and comparison with the mandibular (Figs 3.9 and 3.10) and dental specimens from Olorgesailie confirms that the holotype of *T. jonathani* is actually male. These late mandibles are large and have very reduced canines, reduced incisors, and relatively large and com-

Fig. 3.11. Correlation diagram for strata in the Olduvai Beds, and in the Shungura, Koobi Fora, Nachukui and Olorgesailie Formations from which *Theropithecus oswaldi* has been recovered.

Table 3.4. *Measurements of the lower dentition of the total sample of Theropithecus oswaldi Measurements for specimens from the Shungura Formation taken from Eck & Jablonski (1987).*

Site		I₁ L	I₁ B	I₂ L	I₂ B	P₄ L	P₄ B	M₁ L	M₁ AB	M₁ PB	M₂ L	M₂ AB	M₂ PB	M₃ L	M₃ AB	M₃ PB	M₂ LxB	M₃ LxB
Olorgesailie	mean	5.50	6.40	4.60	6.30	11.30	9.30	16.20	11.40	11.40	20.30	14.60	13.30	25.60	16.80		296.38	430.08
	n	26	26	21	21	44	44	36	36	36	30	30	30	21	21		30	21
	max	5.80	7.00	5.00	7.10	13.50	10.20	17.50	13.00	11.80	22.50	17.50	13.80	28.40	19.00		393.75	539.60
	min	5.30	6.10	4.40	5.50	10.20	8.00	14.00	10.00	11.10	18.00	13.00	12.80	23.50	15.40		234.00	361.90
Masek, Kapthurin	mean	4.80	6.20	4.07	5.50	10.48	10.13	14.70	11.90		18.25	14.60	13.30	24.90	16.00	14.00	265	398
	n	1	1	1	1	2	2	1	1		2	2	2	2	2	2	2	2
	max					10.66	10.13				19.00	15.90	13.80	25.00	16.70	14.20	278	414
	min					10.30	9.56				17.50	13.30	12.80	24.80	15.30	13.85	253	382
Olduvai upper Bed II and III, & IV	mean	6.67	6.95	5.53	7.05	11.14	9.5	14.70	11.15	11.40	18.68	13.92	13.44	24.61	15.24	13.84	260	377
	n	4	4	2	2	3	3	3	2	3	6	5	5	7	5	6	5	5
	max	8.07	7.30	5.66	7.20	12.10	10.10	15.20	11.70	11.80	20.00	15.20	15.30	27.00	16.26	14.80	304	439
	min	5.16	6.80	5.40	6.90	10.10	8.76	14.30	10.60	11.10	16.80	13.90	13.28	21.47	14.27	11.76	228	306
Okote	mean	4.40	6.00	4.90	6.40	9.39	8.12	14.30	10.63	10.47	17.46	12.63	12.60	20.88	13.18	12.05	221	277
	n	1	1	2	2	8	6	4	3	3	12	9	9	13	13	13	7	13
	max			4.90	7.00	10.60	8.70	14.40	11.20	10.90	20.40	13.50	14.90	25.00	15.90	14.00	275	398
	min			4.90	5.80	7.80	7.60	14.20	9.90	9.70	15.25	11.50	11.60	17.40	11.50	10.50	172	212
Olduvai Bed I and lower Bed II	mean	5.00	6.10			9.23	7.90	12.30	9.75	9.55	17.00	12.80	12.40	20.90	14.07	12.32	218	294
	n	1	1			3	2	2	2	2	1	1	1	3	3	3	1	3
	max					9.90	8.10	13.20	9.90	9.80				21.20	15.40	13.55		323
	min					8.30	7.70	11.40	9.60	9.30				20.50	13.30	11.15		277
KBS	mean	4.71	5.93	4.87	5.87	9.24	8.41	12.81	10.32	10.75	16.99	13.36	12.71	22.16	13.95	12.65	229	310
	n	2	2	3	3	5	5	5	4	2	10	9	9	13	13	11	9	13
	max	4.83	6.10	5.20	6.00	9.70	8.80	13.40	10.90	11.10	18.30	14.90	13.60	24.90	15.20	13.60	273	371
	min	4.60	5.77	4.40	5.70	8.50	7.60	12.40	9.80	10.40	15.80	12.30	12.10	18.80	12.00	10.50	202	227
Upper Burgi	mean	4.90	6.91	5.20	7.20	9.28	8.10	12.16	9.74	9.65	15.53	12.00	11.57	20.99	13.53	12.02	185	289
	n	3	3	1	1	5	4	7	5	6	19	14	14	12	21	21	14	20
	max	5.70	7.60			10.60	8.60	13.20	10.70	10.50	18.10	13.00	12.70	24.20	15.70	14.10	214	369
	min	3.70	6.00			8.60	7.10	11.40	9.00	8.40	14.20	10.20	10.10	17.20	11.80	10.10	162	203

Table 3.5. Measurements of the lower dentition of male and female T. oswaldi. Sources of measurements as for Table 4. Specimens from upper Bed II, Bed III, Bed IV and Masek at Olduvai as well as those from Kapthurin are treated together as 'Others'. Specimens from the Koobi Fora Formation and Bed II at Olduvai are treated together as 'Koobi Fora and Olduvai Bed I'. Measurements from Olorgesailie include males and females except for the canine and P3.

Site		I_1 L	I_1 B	I_2 L	I_2 B	C L	C B	P_3 L	P_3 B	P_4 L	P_4 B	M_1 L	M_1 AB	M_1 PB	M_2 L	M_2 AB	M_2 PB	M_3 L	M_3 AB	M_3 PB	M_2 L×B	M_3 L×B	C L×B
Site																							
Olorgesailie (males and females except C and P3)	mean	5.50	6.40	4.60	6.30	11.3	15.40	17.70	7.60	11.30	9.30	16.20	11.40		20.30	14.60		25.60	16.80	14.28	296	430	176
	n	26	26	21	21	14	15	16	12	44	44	36	36		30	30		21	21		30	21	14
	max	5.80	7.00	5.00	7.10	13.3	17.00	21.00	8.80	13.50	10.20	17.50	13.00		22.50	17.50		28.40	19.00		394	540	226
	min	5.30	6.10	4.40	5.50	9.9	13.80	15.50	6.50	10.20	8.00	14.00	10.00		18.00	13.00		23.50	15.40		234	362	137
Others	mean	4.98	6.75	4.70	6.35	11.00	15.91	17.73	9.30	11.06	10.04	14.16	11.53	11.40	11.88	14.80	14.34	24.26	16.00	14.28	277	398	180
	n	2	2	2	2	3	3	3	3	4	4	3	3	2	4	3	5	3	2	3			
	max	5.16	7.30	5.40	7.20	13.60	20.00	21.00	9.60	12.10	10.70	14.70	11.90	11.80	20.00	15.90	15.30	25.00	16.70	14.80	304	414	272
	min	4.80	6.20	4.07	5.50	9.50	13.60	15.30	9.00	10.30	9.56	13.50	11.00	11.00	17.50	13.30	12.80	23.00	15.30	13.85	252	382	134
Koobi Fora and Olduvai Bed I	mean	4.15	6.61	4.40	5.70	10.41	16.88	19.46	7.55	9.55	8.37	12.87	10.02	10.08	16.931	12.47	12.32	21.35	13.53	12.11	209	291	180
	n	2	2	2	1	8	8	12	11	8	6	6	5		14	12		11	11		11	9	8
	Olduvai	4.60	7.12	4.40	5.70	12.30	18.26	23.50	8.58	10.60	8.80	14.00	10.90	11.10	20.40	13.50	14.90	24.00	15.70	14.00	275	369	221
	Bed I	3.70	6.10	44.00	5.70	9.20	15.10	16.20	6.60	9.00	8.10	12.00	9.00	9.00	14.50	11.20	10.70	18.50	12.20	10.10	162	226	139
Females																							
Olorgesailie (males and females except C and P3)	mean	5.50	6.40	4.60	6.30	6.70	9.88	12.50	6.60	11.30	9.30	16.20	11.40		20.30	14.60		25.60	16.80		296	430	66
	n	26	26	21	21	28	28	15	11	44	44	36	36		30	30		21	21		30	30	28
	max	5.80	7.00	5.00	7.10	7.50	11.90	13.60	7.10	13.50	10.20	17.50	13.00		22.50	17.50		28.40	19.00		394	540	82
	min	5.30	6.10	4.40	5.50	6.26	9.10	10.70	6.30	10.20	8.00	14.00	10.00		18.00	13.00		23.50	15.40		234	362	59
Others	mean	5.40	8.80	5.66	6.90	6.90	9.80	12.10	8.16	10.10	8.76	14.60			18.55	12.60	13.00				234		68
	n	1	1	1	1	1	1	1	1	1	1	1			1	1	1				1		1
	max														18.55								
	min														17.60								
Koobi Fora and Olduvai Bed I	mean	4.98	6.37	5.10	9.17	5.82	9.17	12.29	6.50	9.18	8.03	12.86	10.18	9.92	15.84	12.03	11.62	20.60	13.25	11.84	190	276	53
	n	4	4	4	10	10	10	7	7	8	6	7	5	5	10	6	7	11	12	11			
	Olduvai	5.70	7.60	5.20	7.20	7.20	10.30	13.40	7.90	10.60	8.70	14.40	10.80	10.80	18.90	13.20	12.50	24.20	15.40	13.55	249	356	61
	Bed I	4.40	5.77	4.90	5.80	5.40	8.06	11.30	5.50	7.80	7.10	11.40	9.50	8.40	14.30	10.20	10.10	17.20	11.50	10.40	146	203	44

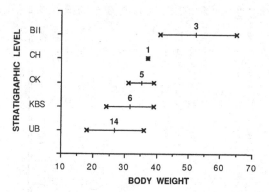

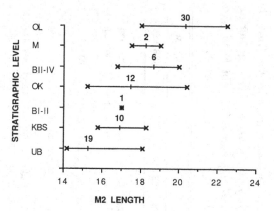

Fig. 3.12. Estimated body weights of *Theropithecus oswaldi* at different intervals in time calculated using dimensions of the femur head as described by Ruff (1988). The mean of the measurements is indicated with the number of individuals in the sample and the length of the bar indicates the range of measurements. UB = upper Burgi Member Koobi Fora Formation; KBS = KBS Member Koobi Fora Formation; OK = Okote Member Koobi Fora Formation; CH = Chari Member Koobi Fora Formation; BII = upper Bed II Olduvai.

Fig. 3.14. Mesiodistal length of the lower M_2 of *Theropithecus oswaldi* at different intervals in time. See Figs. 3.12 and 3.13 for abbreviations.

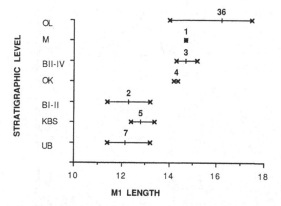

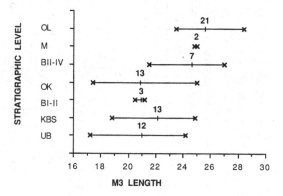

Fig. 3.15. Mesiodistal length of the lower M_3 of *Theropithecus oswaldi* at different intervals in time. See Figs. 3.12 and 3.13 for abbreviations.

Fig. 3.13. Mesiodistal length of the lower M_1 of *Theropithecus oswaldi* at different intervals in time. See Fig. 3.12 for abbreviations and BI-II = Bed I and lower Bed II Olduvai; BII-IV = Upper Bed II, Bed III and Bed IV Olduvai; M = Masek Beds Olduvai and Kapthurin; OL = Olorgesailie.

(Tables 3.4–3.5, Figs 3.13 to 3.15), the relative size of the canines decreases (Table 3.6, Figs 3.16 to 3.18) and the length of the sectorial cusp of the P3 decreases (Fig. 3.19) particularly relative to the increasing size of M_3 (Table 3.6, Fig. 3.20).

plex posterior cheek teeth particularly M3 (Fig. 3.8). With time the complexity and the length of the molar teeth increases

Body weights for *T. oswaldi* were estimated by Jolly (1972) from the robusticity quotient, an expression of the humeral and femoral cross sectional area. Jolly estimated female individuals from

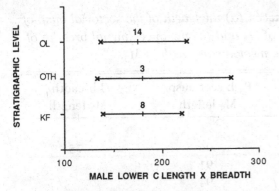

Fig. 3.16. Mesiodistal length multiplied by buccolingual breadth of the male lower canine of *Theropithecus oswaldi* at different intervals in time. KF = Upper Burgi, KBS and Okote Members of the Koobi Fora Formation and Bed I Olduvai; OTH = Kapthurin and Upper Bed II, Bed III, Bed IV and Masek at Olduvai; OL = Olorgesailie.

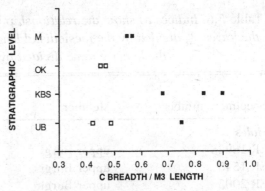

Fig. 3.18. Bucolingual breadth of the canine divided by the mesiodistal length of the M₃ of *Theropithecus oswaldi*. Black squares = males; open squares = females. UB = Upper Burgi Member Koobi Fora Formation; KBS = KBS Member Koobi Fora Formation; OK = Okote Member Koobi Fora Formation; M = Masek Beds Olduvai.

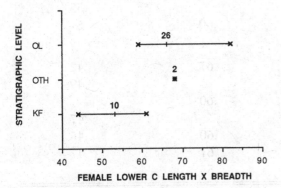

Fig. 3.17. Mesiodistal length multiplied by buccolingual breadth of the female lower canine of *Theropithecus oswaldi* at different intervals in time. See Fig. 3.16 for abbreviations.

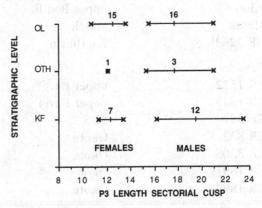

Fig. 3.19. The length of the sectorial cusp of the lower P₃ of *Theropithecus oswaldi* at different intervals of time. Males separated from females. See Fig. 3.18 for abbreviations.

Kanjera to have weighed about 20 kg and the males 35 kg while the Olorgesailie males probably weighed 65 kg, as much as a large female gorilla. Dimensions of the femur head described by Ruff (1988) were used to estimate the body weights of specimens from the Koobi Fora Formation and from Olduvai (Table 3.7, Fig. 3.12). Regressions were drawn for a modern sam-

ple of male *Papio hamadryas cynocephalus* for which individual body weights were known and these regressions used to estimate body weights of the fossil specimens. Using this method the large skeleton from site MCK, upper Bed II, Olduvai Gorge was estimated to be 65 kg. Estimates were not obtained for the Olorgesailie specimens.

Table 3.6. *Indices to show the relationship between (a) the length of the sectorial cusp of the lower P$_3$ divided by the mesiodistal length of M$_3$ and (b) the buccolingual breadth of the lower canine divided by the mesiodistal length of M$_3$.*

Specimen number	Member	P$_3$ L sect cusp/ M$_3$ length	C breadth/ M$_3$ length
Males			
ER 18925	upper Burgi	.96	.75
ER 1545	upper Burgi	.91	
ER 2003	upper Burgi	.94	
ER 577	KBS	.84	.68
ER 865	KBS	.96	.83
ER 985	KBS		.90
ER 1573	KBS		.68
ER 566	Okote	.92	
ER 864	Okote	.90	
3366	upper Bed II	.90	
Masek	Masek		.55
BK 22638	Kapthurin	.60	.57
Females			
ER 1572	upper Burgi	.67	.42
ER 1528	upper Burgi		.49
ER 557	KBS	.60	
ER 833	Okote		.47
ER 5308	Okote	.60	.45
ER 18919	Okote	.67	.47
ER 18924	Okote	.55	

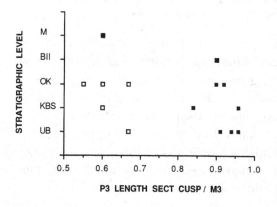

In summary, the evolutionary trends characteristic of *T. oswaldi* include: (1) increase in the complexity of the molars and in their size relative to the anterior dentition, (2) decrease in the size of the canines and the sectorial cusp of the lower

Fig. 3.20. The length of the sectorial cusp of the lower P$_3$ divided by the mesiodistal length of the lower M$_3$ of *Theropithecus oswaldi* at different intervals in time. Males separated from females. See Fig. 3.18 for abbreviations, and Bed II = upper Bed II Olduvai.

Table 3.7. *Estimated body weights of*
T. oswaldi *calculated using dimensions of*
the femur head described by Ruff (1988).

Horizon	n	mean	range
upper Bed II Olduvai	3	52.3	41–65
Chari	1	37	37
Okote	5	35.2	31–39
KBS	6	31.7	24–39
Upper Burgi	14	26.8	18–36

P_3s, (3) development of a reversed curve of
Spee, (4) increase in the size of the glenoid
process, and (5) increase in body size.

Sexual dimorphism is seen both in body
size and in the size of the canines and the
lower P_3 (Table 3.3). Although the canines
and lower P_3 sectorial cusp are reduced
there is still a significant difference in size
between the sexes at equivalent time
horizons. Variability is also high. Crania
from the Koobi Fora Formation show particular variability in the development of
the maxillary ridges and fossae and the
length of the muzzle, the shape of the
anterior zygomatic arch, the thickness
and depth of the zygomatic process at its
midpoint, the thickness and depth of the
supraorbital tori, and the size of the postglenoid process.

Taxonomy

Theropithecus oswaldi was first described
by Andrews (1916) on specimens collected by Dr Felix Oswald at Kanjera. The
larger specimens from upper levels at
Olduvai were described as a separate species, *T. leakeyi* by Hopwood (1934, 1936).
T. oswaldi was divided into several

subspecies by Leakey & Whitworth
(1958). Differentiation was largely geographic. Specimens from Olorgesailie were
assigned to *T. oswaldi mariae*, those from
Olduvai to *T. o. olduvaiensis* and those
from Kanjera to *T. o. oswaldi*. *T.
jonathani* (Leakey & Whitworth, 1958)
was described as a distinct species. Leakey
and Whitworth's scheme was later revised
by Jolly (1972) when increased samples
provided additional information about
evolutionary changes. Additional subspecies were added and earlier ones redefined
or abandoned: *T. o. leakeyi* for specimens
from Bed IV Olduvai and ?Hopefield
(including *T. jonathani* (cf.), *T. o. mariae*
for those from Olorgesailie and ?Ternifine
(cf.), *T. o. darti* for those from
Makapansgat and *T. o. danieli* (Freedman)
1957 for those from Swartkrans, Olduvai I
and/or lower Bed II.

After extensive comparisons of
Theropithecus from Hopefield with those
from other African Pliocene and Pleistocene sites, Dechow & Singer (1984) concluded that only one species, *T. oswaldi*
is represented in all the material. It is
generally agreed that later material,
including progressive specimens referred
to *T. jonathani*, are best attributed to a
single species *T. oswaldi*. Eck & Jablonski
(1987) prefer to retain the species *T. darti*
for the more primitive samples from Hadar
and Makapansgat. They consider that
T. darti differs sufficiently from
T. oswaldi in certain primitive character
states to warrant specific distinction.
These characters have not been described
in detail but the authors mention 'unreduced' incisors, unreduced canines and
P_3s, short postglenoid processes separated

from the external auditory meati by grooves, no triangular depression on the mandibular condyles and small body size as significant (see Eck, chapter 2).

Like *T. brumpti*, *T. oswaldi* represents an evolving lineage in which the early and late stages are clearly definable but intermediate stages grade into each other. A similar solution of subspecific differentiation is appropriate for *T. oswaldi*, the early stages being referred to *T. oswaldi darti* (from Hadar and Makapansgat) the intermediate stages to *T. o. oswaldi* (from Kanjera, the Koobi Fora Formation, Beds I and lower Bed II Olduvai and the Shungura Formation) and the later stages to *T. o. leakeyi* (from Olorgesailie, Kapthurin and upper Bed II, III, IV, and the Masek Beds at Olduvai).

Opposing adaptive strategies of *Theropithecus brumpti* and *Theropithecus oswaldi*

The single remaining extant *Theropithecus* species, *T. gelada*, is unique among the cercopithecoids in its almost exclusively graminivorous diet. Similarities in the dentition between *T. gelada* and its extinct relatives have been interpreted to relate to this unusual dietary adaptation. While *Theropithecus* displays the standard dental formula and characteristic cercopithecid bilophodont dentition, modifications occur in the relatively small size of the incisors and the distinctive morphology and wear patterns of the molars (Eck & Jablonski, 1987). Many of the exposed enamel ridges on the worn molar occlusal surface run longitudinally, indicating that lateromedial excursion of

the mandible was important. Numerous microscopic scratches aligned transversely to the long axis of the jaw may be seen under a light microscope on the worn enamel of the molars of both *T. brumpti* and *T. oswaldi*. The contrasting cranial morphologies of the two lineages *T. brumpti* and *T. oswaldi* appear to relate to contrasting masticatory adaptations both of which emphasize the lateral movement of the jaw. The contrasting adaptations may relate to constraints imposed by a difference in the length of the muzzle which is already apparent in early forms of each lineage before there is any significant development of high crowned complex molars which must have developed in parallel subsequent to the separation of the two lineages.

Alternatively the contrasting masticatory adaptations of the two species could be interpreted to indicate differences in the diet (see Teaford, chapter 12). The tooth morphology of *T. b. brumpti* and *T. o. oswaldi* are closely similar but the more evolved *T. o. leakeyi* molars show more extreme adaptations to graminivory and are clearly distinguishable. As well as being more complex, the occlusal surface of *T. o. leakeyi* molars wears more rapidly to a planar surface in contrast to earlier *Theropithecus* in which the molar cusps retain some shearing function for longer. In addition there appears to be a greater differential in the amount of wear of M1 and M3 in later *T. oswaldi*. As well as feeding on the lusher and less abrasive grasses found in forest glades and river banks, *T. brumpti* may have been at least partially folivorous or may have been supplementing its diet with other herbivorous foods (see Teaford, chapter 12). In contrast the masticatory adaptations of *T. oswaldi* like those of *T. gelada* may relate to a more exclusively graminivorous diet.

Eck & Jablonski (1987) have discussed the masticatory adaptations of *T. brumpti* and compared the relative roles of the muscles in *T. gelada* and *T. brumpti*. The functional significance of the masticatory adaptations of the three species of gelada is further explored by Jablonski in this volume (chapter 11). *Theropithecus gelada* has a large temporalis active in mastication and high occlusal pressures are maintained because the lines of action of the temporalis and masseter are roughly perpendicular. The increased muzzle length of *T. brumpti* alters the lines of action of the temporalis and masseter muscles so that if their origins remained as in *T. gelada* they would not produce the same high occlusal pressures, nor would the temporalis be able to play a significant role in lateral movement. Eck & Jablonski suggest that the flaring anterior zygomatic arches of *T. brumpti* relate to the increased importance of the masseter muscle and provided an arrangement which gave the ability to maintain high occlusal pressures and at the same time to magnify the lateral component of the mandibular movements. The large flaring anterior zygomatic arch moves the origin of the masseter muscle anteriorly and laterally. Details on the inferior surface of the anterior zygomatic arch which define the area of attachment of the superficial masseter muscle indicate that this was probably a large muscle able to produce a large proportion if not the vast majority of

lateral movement of the masticatory cycle and to maintain high occlusal pressures. The pronounced triangular depression on the superolateral surface of the mandibular ramus of *T. brumpti* provides an attachment for fleshy fibres of the deep masseter muscle and tendon. The long symphysis and the marked development of the mental ridges are probably adaptations to counter bending moments on the balancing side of the mandible since biting is always unilateral and with high occlusal pressures these bending moments would be significant.

Theropithecus oswaldi on the other hand has a cranial morphology more similar to that of *T. gelada* and its masticatory musculature was probably also similar. As in *T. gelada* the temporomandibular joint was elevated providing increased mechanical advantage for the masseter muscles and the temporalis muscle can be inferred to have been correspondingly active in mastication. Eck & Jablonski (1987) have described the relative functions of the masticatory muscles of *T. gelada* and shown how the temporalis is not only capable of producing mandibular elevation but lateral mandibular excursion as well. Because the muzzle of *T. oswaldi* was relatively short, like *T. gelada* both the pterygoideus and temporalis muscles were able to produce lateral movement in the jaw, the latter function shared by the pterygoideus. As the molar teeth of *T. oswaldi* became progressively more complex and the M3s enlarged, to maintain efficiency, it would have been necessary to increase the lateral excursion of the mandible. Reduction in the length of the muzzle would have increased the efficiency of the temporalis and pterygoideus muscles but a balance had to be met to maintain a jaw length great enough to accommodate increasingly large molars and short enough to maintain the efficiency of the temporalis muscles. This was in part achieved by reduction in the size of the anterior dentition.

Jolly (1972) noted that *Theropithecus* species have delayed eruption of the molars such that M1 is well worn before M2 fully erupts and M2 worn before M3 erupts, a mechanism which he interpreted as extending the longevity of the teeth. This is most apparent in the later *T. o. leakeyi*. Individuals with very worn teeth such as are seen on the mandible 067/5603 and maxilla S.133 from Olduvai remain with the posterior portion of M3 relatively unworn. Erupting molars of *Theropithecus* tend to be steeply inclined relative to the occlusal plane so that the anterior portion of the tooth comes into occlusion before the posterior portion. The delayed eruption of the molars, the steep angle of the erupting molar, and the reversed curve of Spee which Eck & Jablonski (1987) note acts to concentrate the force produced by the muscles of mastication in a relatively small and mesially situated area of the functionally occluding cheek teeth, together ensure that for most of the animal's life there is some part of an unworn or little worn molar remaining.

To summarize the masticatory adaptations: despite superficial similarity to *Papio* and *Mandrillus* in the long muzzle and long canines, *T. brumpti* developed its own unique masticatory mechanism in which the masseter muscles provide the lateral movement of the mandible. *T. oswaldi* probably processed its food in much the same way as the gelada baboon with the temporalis and pterygoideus producing the lateral excursion of the lower jaw. Because of its larger size or more abrasive food, progressive increase in the size of the posterior cheek teeth led to a reduction in the anterior dentition. The delayed eruption of the molars, the oblique angle of the erupting molars and the reversed curve of Spee collectively extend the longevity of the molars and insure the availability of at least part of a relatively unworn molar until a late stage of wear.

Derivation of the genus *Theropithecus*

Evidence from the Turkana Basin suggests that *Theropithecus* probably derived from the ancestral papionin prior to 3.5 Ma. The earliest known *Theropithecus* are specimens of *T. brumpti* from the Usno Formation and the Lokochot Member of the Koobi Fora Formation. The earliest *T. oswaldi* are from Hadar (see Eck, chapter 2) and from the Tulu Bor Member of the Koobi Fora Formation. An isolated *Theropithecus* tooth has been recovered from Lothagam 3 (Smart, 1976). It has been suggested that Lothagam 3 may be between 4.0 and 4.5 Ma (Hill & Ward, 1988) but if that date is correct it extends the range of this genus by 0.5 to 1.0 Ma. Elsewhere no *Theropithecus* remains have been identified from horizons earlier than 3.5 Ma although small colobines and papionins were relatively common. In the Upper Laetolil Beds at Laetoli in Tanzania

(3.65–4.0 Ma), *Parapapio ado* was the most common cercopithecoid and a larger species (cf. *Papio*) was also present together with a colobine (cf. *Paracolobus* sp.) (Leakey & Delson, 1987). Numerous isolated cercopithecine teeth representing at least three species of papionins and one colobine have recently been recovered from 4 Ma horizons in the Lonyumun Member of the Koobi Fora Formation. The oldest unit at Lothagam, Lothagam 1, which may be as old as 5.0 to 5.5 Ma (Patterson *et al.*, 1970) has yielded a number of small cercopithecoid specimens including both colobines and papionins. Colobines and papionins were thus fairly common at these early Pliocene sites but, with the possible exception of the single isolated lower molar from Lothagam 3, *Theropithecus* was absent.

Cercopithecoid evolution in Africa during the late Miocene and early Pliocene is not well understood. The initial divergence of the Cercopithecinae from the Victoriapithecinae probably took place between 12 and 9 Ma and possibly earlier. The first radiation appears to have been of colobines and this took place mainly in Europe. The first true colobine from Africa was recovered from Ngeringerowa, in Kenya (Benefit & Pickford, 1985), a site dated between 8.5 and 9.3 Ma (A. Deino, pers. comm.). An isolated colobine upper molar was recovered from Nakali, a site of closely similar age (Benefit & Pickford, 1985). In North Africa, nine teeth of ?*Colobus flandrini* from Marceau, Algeria, may be 7 ± 2 Ma (Szalay & Delson, 1979). Colobines must have migrated out of Africa somewhat earlier than this time. The earliest colobine known from Europe was

an upper premolar from Wissberg, Germany, which is probably 11–10 Ma (Szalay & Delson, 1979; Tobien, 1986). The better known *Mesopithecus pentelici* was found in deposits between 9 and 8 Ma (Szalay & Delson, 1979). The later *M. monspessulanus* and *Dolichopithecus ruscinensis* are well represented in the European fossil record (Delson, 1975) and indicate considerably warmer climatic conditions during this time. There is less evidence of early cercopithecines. One tooth from the late Miocene site of Ongoliba in Zaire (Hooijer, 1963, 1970) and another from Ngorora in Kenya (Bishop & Chapman, 1970) may have cercopithecine affinities. The North African cf. *Macaca* (about 9 Ma) and a macaque-like cercopithecoid (about 6 Ma) from Wadi Natrun (Delson, 1975) are the only early cercopithecines known from North Africa.

In East Africa the first evidence of cercopithecoid diversity appeared in the Pliocene at sites such as Lothagam, East Turkana, and Laetoli where papionins were the dominant cercopithecoids. *Theropithecus* appears to originate from this radiation and during the late Pliocene and early Pleistocene, when significant evolutionary change is seen in *Theropithecus*, there was increased cercopithecoid diversity. At least four species of colobines have been recognized in deposits just older than 2 Ma in the Turkana Basin. They occupied niches ranging from forest to open woodland. Some attained larger body size than is found today in the few remaining African colobine species and some displayed terrestrial adaptations (Leakey, 1982; Birchette, 1981, 1982). Cercopithecoid diversity is also seen in the

South African Pliocene. At Makapansgat (about 3 Ma), three species of *Parapapio*, cf. *Cercocebus*, and two colobine species have been recorded (Delson, 1984) as well as *T. o. darti* (and possibly *T. brumpti* as discussed above). At Sterkfontein, a similar diversity is seen with an additional two species of *Papio* but without *Theropithecus* or cf. *Cercocebus* (Delson, 1984). *Theropithecus brumpti* and *T. oswaldi* are part of this Pliocene and Pleistocene cercopithecoid radiation.

The papionins are relatively generalized but with high variability in skull morphology. *Theropithecus* differs from *Parapapio* and *Papio* by its small incisors, its marked postorbital constriction, its long sagittal crest, and by inferiorly divergent lateral margins of the frontal processes of its zygomatic arch. The early stages that have been recognized of the two lineages have relatively unspecialized molar teeth: the molar crowns are only slightly taller than specimens of *Parapapio* and *Papio* although the incisors are already reduced and the differences in muzzle length are already apparent. The ancestral *Theropithecus*, although presently undocumented, was probably a small papionin with a moderately long muzzle, small incisors but relatively unspecialized molar teeth.

Palaeoenvironments, replacement and extinction

The evolution of the two lineages of *Theropithecus* in the Turkana Basin, the success of each and its subsequent demise, is conceivably related to changes in the palaeoenvironment. The two lineages appear to have diverged from a common ancestor prior to 3.5 Ma, and to have evolved in parallel, each developing a dental morphology similar to that of the graminivorous *T. gelada*. In the Turkana Basin the *T. brumpti* lineage is initially common in the older members while there is little contemporaneous evidence of early examples of the *T. oswaldi* lineage. *T. oswaldi* replaced *T. brumpti* at 2 Ma and is thereafter common throughout the younger members of the Koobi Fora Formation. The extinction of *T. oswaldi* probably occurred sometime after 0.7 Ma. The presence of *T. brumpti* in the earlier horizons in the lower Omo Valley and its replacement by *T. oswaldi* in the later horizons has been related to inferred environmental change. It has been suggested that the earlier horizons were sampling faunas from more forested horizons than occurred later (Eck 1987; Eck & Jablonski, 1987).

Palaeogeographic reconstruction of the Turkana region (Brown & Feibel, 1987; Feibel, 1988) indicates that between 4.1 and 2.0 Ma deposition was dominated by sedimentation from a large perennial river. There were two short lacustrine intervals at 3.6 Ma and 3.2 Ma that would have temporarily disrupted the vegetation pattern and from 2.5 to 2.0 Ma there was apparently no deposition in the Koobi Fora region. The ancient Omo River probably supported a riparian vegetation similar to the gallery forest found in the lower Omo Valley today and away from the river more xeric conditions would have prevailed. This fluvial phase of deposition ended at about 2 Ma when the course of the river was blocked tectonically and a large stable lake inundated the basin destroying a 200 km stretch of riverine plant communities. The lake subsequently received sediment deposited by the Omo River flowing in from the north. During the subsequent interval, 1.8 to 1.3 Ma, alternating fluvial and lacustrine regimes prevailed. The lake persisted through the Okote Member in the western part of the basin, periodically expanding to the Koobi Fora region. During the fluvial phases the rivers were smaller than those that existed previously and the lake basin was probably not well drained. Diatoms indicate that the water in the major lake phases was fresh to moderately concentrated (Feibel, 1988).

The cercopithecoids recovered from the Turkana Basin appear to show habitat preferences. *Theropithecus brumpti* seems to have preferred the woodland and gallery forest of the major river system which prevailed prior to 2 Ma. Its disappearance from the fossil record coincides with the change from a fluvial to a lacustrine dominated regime. The tectonic formation of the Member G (upper Burgi Member lake) would have removed the forest habitat from south of the Ethiopian border but *T. brumpti* may have been able to survive higher up the Omo or in the Nile drainage. *Theropithecus oswaldi* became common in the Turkana Basin when lacustrine conditions prevailed indicating that it preferred the more open floodplain and lake margin settings.

The papionins appear to be associated with forest and woodland habitats. In the Lonyumun Member of the Koobi Fora Formation several species are associated with fluvial deposits and probably riverine forest. Those from the upper Burgi Member are found in deltaic deposits and probably inhabited woodland around the Omo Delta. To the west of the lake *Parapapio ado* and *P. whitei* are found in the basin margin deposits suggesting that they inhabited the woodland which probably occurred on the basin margin hills, and *P. whitei* also inhabited the gallery forest of the tributary river at Lomekwi (Harris *et al.*, 1988). These papionins were probably confined to forested and woodland habitats because *T. oswaldi* was dominating the open habitats. Today *Papio* is widespread, inhabiting a variety of environments including forest, woodland, and open grassland. Presumably the demise of *T. oswaldi* left the open habitats available.

The large colobines were associated with both forest and woodland habitats that prevailed prior to 2 Ma. Most of these large colobines became extinct when the upper Burgi Member lake destroyed their habitat. Only the more terrestrial *Cercopithecoides kimeui* is found in number in the overlying KBS Member of the Koobi Fora Formation although two specimens, which are probably *Rhinocolobus*, have been recovered in this member. There is evidence of a large colobine in the overlying Okote Member at Koobi Fora and in Members K and L at the Omo (Leakey, 1982) suggesting that at least one species of large colobine reinvaded the area briefly. The persistence of the large colobines in the Koobi Fora Formation after the disappearance of *T. brumpti* indicates that they were more open country adapted while *T. brumpti* was confined to the gallery forest associated with the major river.

Cerling, Bowman & O'Neill (1988) have recently analysed pedogenic carbonates preserved in the Koobi Fora succession. They interpreted their results to indicate that prior to 1.8 Ma C4 plants constituted 25 to 40 per cent of the flora while after 1.8 Ma C4 plants comprised 60 to 80 per cent of the flora. At corresponding latitudes today C4 floras comprise largely grasses. Trees and shrubs are nearly always C3. Thus the increase in C4 at 1.8 Ma presumably represented an increase in grass cover which would be consistent with the change from a river-dominated regime with extensive gallery forest, to a lake dominated regime with far fewer trees. An analysis of the stable carbon isotopes of tooth fragments of the two *Theropithecus* species, similar to that carried out by Lee-Thorp & van der Merwe (1989) on fragments of tooth enamel of *Papio* and *Theropithecus* from Swartkrans, would be of interest to determine if a dietary difference can be detected. *T. brumpti* may have been feeding largely on C3 vegetation and *T. oswaldi* on C4.

Theropithecus oswaldi is common throughout the later members when a progressive increase in size is displayed together with reduction in canines and lower P_3s and increase in size of the posterior cheek teeth. The increase in body weight may have been in response to a general reduction in tree cover as a deterrent to predators. The increase in molar size may be allometric. A larger body size would have required more efficient processing of larger quantities of vegetation.

Theropithecus oswaldi did not become extinct until after 0.7 Ma. It is last recorded in sites in northern, southern, and eastern Africa that are probably between 0.7 and 0.4 Ma. In northern Africa, *Theropithecus* is last recorded at Ternifine which may be 0.7 Ma and Thomas Quarries which may be 0.4 Ma (Geraads, 1987). In southern Africa, *Theropithecus* from Hopefield may be 0.4 Ma. In eastern Africa specimens of *T. o. leakeyi* have been recovered from Member 7 of the Olorgesailie Formation dated at 0.7 Ma (Potts, 1989; Bye *et al.*, 1987), Kapthurin which is not securely dated and site MRC in the Masek Beds at Olduvai Gorge dated between 0.7 and 0.4 Ma (Hay, 1976). The timing of the extinction of *T. oswaldi* is uncertain because of the incompleteness of the Middle and Late Pleistocene fossil record. *Theropithecus oswaldi* had obviously successfully adapted to a graminivorous diet in open savanna habitats. Climatic oscillations during the terminal phases of the Pleistocene Ice Age may have contributed to the demise of many large savanna herbivores (*Elephas*, *Megalotragus*, *Pelorovis*, *Sivatherium*) and adversely affected others (e.g. *Ceratotherium*) but the timing of this event (or these events) must await further substantiation from the fossil record.

Acknowledgements

I thank the Government of Kenya and the Governors of the National Museums of Kenya. The recovery of specimens from the Koobi Fora and Nachukui Formations was the result of field work coordinated by Richard Leakey and funded by the National Geographic Society, Washington, D.C., the Garland Foundation, the William Donner Foundation, and the National Museums of Kenya. The curatorial staff of the National Museums of Kenya have willingly assisted me in many ways. I am grateful to the European Museums and curators who allowed me access to their specimens in particular the British Museum (NH), London. I particularly thank Kamoya Kimeu and his field crew who were responsible for the discovery of the majority of specimens on which this paper is based, John Harris who provided invaluable assistance and encouragement both in stimulating discussions and in improvements to the manuscript and Peter Kibanga who provided the body weight estimates for *T. Oswaldi*. I am grateful to Eric Delson, Nina Jablonski, and an anonymous reviewer for comments on the manuscript. Lastly, I thank Nina Jablonski and Robert Foley for inviting me to participate in the *Theropithecus* symposium.

References

ANDREWS, C.W. (1916). Note on a new baboon (*Simopithecus oswaldi*, gen et sp. n.) from the (?) Pliocene of British East Africa. *Annals and Magazine of Natural History*, 18, 410–19.

ARAMBOURG, C. (1947). *Mission scientifique de l'Omo 1932–1933, 1(III), Geologie – Anthropologie*. Museum National d'Histoire Naturelle, Paris.

BENEFIT, B. & PICKFORD, M. (1985). Miocene fossil cercopithecoids from Kenya. *American Journal of Physical Anthropology*, 69, 441–64.

BIRCHETTE, M.G. (1981). Postcranial remains of *Cercopithecoides*. *American Journal of Physical Anthropology*, 54, 201.

BIRCHETTE, M.G. (1982). *The Postcranial Skeleton of Paracolobus chemeroni*. PhD. Thesis, Harvard University.

BISHOP, W.W. & CHAPMAN, G.R. (1970). Early Pliocene sediments and fossils from the northern Kenya Rift Valley. *Nature* 226, 914–18.

BROOM, R. & JENSEN, J.S. (1946). A new fossil baboon from the caves at Potgeitersrust. *Annals of the Transvaal Museum*, 20, 337–40.

BROWN, F. & FEIBEL, C. (1987). 'Robust' hominids and Plio-Pleistocene paleogeography of the Turkana Basin, Kenya and Ethiopia. In *Evolutionary history of the 'robust' australopithecines*, ed. F.E. Grine, pp. 325–41. New York: Aldine de Gruyter.

BROWN, F.H., MCDOUGALL, I., DAVIES, I. & MAIER, R. (1985). An intergrated Plio-Pleistocene chronology for the Turkana basin. In *Ancestors: The hard evidence*, ed. E. Delson, pp. 82–90. New York: Alan R. Liss Inc.

BYE, B.A., BROWN, F., CERLING, T. & MCDOUGALL, I. (1987). Increased age estimate for the Lower Paleolithic hominid site at Olorgesailie, Kenya. *Nature*, 329, 237–9.

CERLING, T.E., BOWMAN, J.R. & O'NEIL, J.R. (1988). An isotopic study of a fluvial-lacustrine sequence: the Plio-Pleistocene Koobi Fora sequence, East Africa. *Palaeogeography, Palaeoclimatology, Palaeoecology*, 63, 335–56.

CROOKE, J.H. & ALDRICH-BLAKE, P. (1968). Ecological and behavioural contrasts between sympatric ground dwelling primates in Ethiopia. *Folia Primatologica*, 8, 192–227.

DECHOW, P.C. & SINGER, R. (1984). Additional fossil *Theropithecus* from Hopefield, South Africa: a comparison with other African sites and a reevaluation of its taxonomic status. *American Journal of Physical Anthropology*, 63, 405–35.

DELSON, E. (1975). Evolutionary history of the Cercopithecidae. In Approaches to primate palaeobiology, ed. F.S. Szalay. *Contributions to Primatology*, 5, 167–217.

DELSON, E. (1984). Cercopithecid biochronology of the African Plio-Pleistocene: correlation among eastern and southern hominid-bearing localities. *Courier Forschungs-Institut Senckenberg*, 69, 199–218.

DUNBAR, R.I.M. (1977). Feeding ecology of gelada baboons: a preliminary report. In

Primate ecology: studies of feeding and ranging behaviour in lemurs, monkeys and apes, ed. T.H. Clutton-Brock, pp. 251–73. New York: Academic Press.

DUNBAR, R.I.M. & DUNBAR, E.P. (1974). Ecological relations and niche separation between sympatric terrestrial primates in Ethiopia. *Folia Primatologia*, 21, 36–60.

ECK, G. (1987). *Theropithecus oswaldi* from the Shungura Formation. Lower Omo Basin, southwestern Ethiopia. In *Les Faunes Plio-Pléistocènes de la Basse Vallée de L'Omo (Éthiopie)*. Tome 3, Cercopithecidae de la Formation de Shungura, ed. Y. Coppens, & F.C. Howell, pp. 122–40. Cahiers de Paléontologie. Paris: Editions du Centre National de la Recherche Scientifique.

ECK, G. & JABLONSKI, N. (1984). A reassessment of the taxonomic status and phyletic relationships of *Papio baringensis* and *Papio quadratirostris* (Primates: Cercopithecidae). *American Journal of Physical Anthropology*, 65, 109–34.

ECK, G. & JABLONSKI, N. (1987). The skull of *Theropithecus brumpti* compared with those of other species of the genus *Theropithecus*. In *Les Faunes Plio-Pléistocènes de la Basse Vallée de L'Omo (Éthiopie)*, Tome 3, Cercopithecidae de la Formation de Shungura, ed. Y. Coppens & F.C. Howell, 11–122. Cahiers de Paléontologie. Paris: Editions du Centre National de la Recherche Scientifique.

FEIBEL, C.S. (1988). *Palaeoenvironments of the Koobi Fora Formation, Turkana Basin, northern Kenya*. PhD Thesis, University of Utah.

FEIBEL, C.S., BROWN, F.H. & MCDOUGALL, I. (1989). Stratigraphic context of fossil hominids from the Omo Group deposits: northern Turkana basin, Kenya and Ethiopia. *American Journal Physical Anthropology*, 78, 595–622.

FREEDMAN, L. (1957). The fossil Cercopithecoidea of South Africa. *Annales of the Transvaal Museum*, 23, 121–262.

FREEDMAN, L. (1976). South African fossil Cercopithecoidea: a re-assessment including a description of new material from Makapansgat, Sterkfontein and Taung. *Journal of Human Evolution*, 5, 297–315.

GERAADS, D. (1987). Dating the northern African cercopithecoid fossil record. *Human Evolution*, 2, 19–27.

HARRIS, J.M. (Ed.) (1983). *Koobi Fora Research project*, vol. 2, *The fossil Ungulates: Proboscidea, Perissodactyla and Suidae*. Oxford: Clarendon Press.

HARRIS, J.M., BROWN, F.H. & LEAKEY, M.G. (1988). Stratigraphy and paleontology of Pliocene and Pleistocene localities west of Lake Turkana, Kenya. *Contributions in Science*, 399, 1–128.

HARVEY, P.H., MARTIN, R.D. & CLUTTON-BROCK, T.H. (1987). Life histories in comparative perspective. In *Primate Societies*, ed. B.B. Smuts, R.W. Wrangham, D.L. Cheney, T.T. Struhsaker, & R.M. Seyfarth, pp. 181–96. Chicago: University of Chicago Press.

HAY, R.L. (1976). *Geology of the Olduvai Gorge: a study of sedimentation in a semiarid basin*. Berkeley: University of California Press.

DE HEINZELIN, J. (Ed.) (1983). The Omo Group: archives of the International Omo Research Expedition. *Annales Musée Royal de l'Afrique Centrale, S. 8, Sciences Geologiques*, (85).

HILL A. & WARD, S. (1988). Origin of the Hominidae: The record of African large hominoid evolution between 14 and 4 My. *Yearbook of Physical Anthropology*, 31, 49–83.

HOPWOOD, A.T. (1934). New fossil mammals from Olduvai, Tanganyika Territory. *Annals and Magazine of Natural History*, 10, 546–50.

HOPWOOD, A.T. (1936). New and little known fossil mammals from the Pleistocene of Kenya Colony and Tanganyika Territory. *Annals and Magazine of Natural History*, 10, 636–41.

HOOIJER, D.A. (1963). Miocene mammalia of Congo. *Annales Musée Royal de l'Afrique Centrale, S. 8, Sciences Geologiques*, 46, 1–77.

HOOIJER, D.A. (1970). Miocene mammalia of Congo, a correction. *Annales Musée Royal de l'Afrique Centrale, S. 8, Sciences Geologiques*, 67, 163–7.

ISAAC, G.L. (1977). *Olorgesailie. Archeological*

studies of a Middle Pleistocene lake basin in Kenya. Chicago: University of Chicago Press.

IWAMOTO, T. (1979). Feeding ecology. In *Ecological and socialogical studies in gelada baboons*, ed. M. Kawai, *Contributions to Primatology*, 16, 279–330.

IWAMOTO, M. (1982). A fossil baboon skull from the Lower Omo Basin, southwest Ethiopia. *Primates*, 23, 533–41.

JOLLY, C.J. (1972). The classification and natural history of *Theropithecus* (*Simopithecus*) (Andrews, 1916) baboons of the African Plio-Pleistocene. *Bulletin of the British Museum (Natural History)*, 22, 1–123.

KOCH, C.P. (1986). *The Vertebrate Taphonomy and Paleoecology of the Olorgesailie Formation (Middle Pleistocene, Kenya)*. PhD Thesis, University of Toronto.

LEAKEY, L.S.B. (1943). Notes on *Simopithecus oswaldi* Andrews from the type site. *Journal East Africa Natural History Society*, 17, 39–44.

LEAKEY, L.S.B. & WHITWORTH, T. (1958). Notes on the genus *Simopithecus* with description of a new species from Olduvai. *Occasional Papers Coryndon Memorial Museum*, 6, 1–14.

LEAKEY, M.G. (1976). Cercopithecoidea of the East Rudolf succession. In *Earliest Man and Environments in the Lake Rudolf Basin*, ed. Y. Coppens, F.C. Howell, G. Isaac, & R.E. Leakey, pp. 345–50. Chicago: University of Chicago Press.

LEAKEY, M.G. (1982). Extinct large colobines from the Plio-Pleistocene of Africa. *American Journal of Physical Anthropology*, 58, 153–72.

LEAKEY, M.G. & DELSON, E. (1987). Fossil Cercopithecidae from the Laetolil Beds. In *The Pliocene Site of Laetoli, Northern Tanzania*, ed. M.D. Leakey & J.M. Harris, pp. 91–107. Oxford: Clarendon Press.

LEAKEY, M.G. & LEAKEY R.E. (1973). Further evidence of *Simopithecus* (Mammalia, Primates) from Olduvai and Olorgesailie. *Fossil Vertebrates of Africa*, 3, 101–20.

LEAKEY, R.E. (1969). New Cercopithecoidea from the Chemeron Beds, Lake Baringo, Kenya. *Fossil Vertebrates of Africa*, 1, 53–73.

LEE-THORP, J.A. & VAN DER MERWE, N.J. (1989). Isotopic evidence for dietary differences between two extinct baboon species from Swartkrans. *Journal of Human Evolution*, 18, 183–90.

MAIER, W. (1970). New fossil cercopithecoidea from the lower Pleistocene deposits of the Makapansgat Limeworks, South Africa. *Palaeontologia Africana*, 13, 69–107.

MAIER, W. (1972). The first complete skull of *Simopithecus darti* from Makapansgat, South Africa, and its systematic position. *Journal of Human Evolution*, 1, 395–405.

PATTERSON, B., BEHRENSMEYER, A.K. & SILL, W.D. (1970). Geology and fauna of a new Pliocene locality in northwestern Kenya. *Nature*, 226, 918–21.

PLUMMER, T.W. & POTTS, R. (1989). Excavations and new findings at Kanjera, Kenya. *Journal of Human Evolution*, 18, 269–76.

POTTS, R. (1989). Olorgesailie: new excavations and findings in Early and Middle Pleistocene contexts, southern Kenya rift valley. *Journal of Human Evolution*, 18, 477–84.

RUFF, C. (1988). Hindlimb articular surface allometry in Hominoidea and *Macaca*, with comparisons to diaphyseal scaling. *Journal of Human Evolution*, 17, 687–714.

SHIPMAN, P. & HARRIS, J.M. (1988). Habitat preferences and paleoecology of *Australopithecus boisei* in eastern Africa. In *Evolutionary History of the 'Robust' Australopithecines*, ed. F.E. Grine, pp. 343–81. New York: Aldine de Gruyter.

SMART, C. (1976). The Lothagam 1 fauna: its phylogenetic, ecological, and biogeographic significance. In *Earliest Man and Environments in the Lake Rudolf Basin*, ed. Y. Coppens, F.C. Howell, G. Isaac & R.E. Leakey, pp. 361–9. Chicago: University of Chicago Press.

SWINDLER, D.R. (1976). *Dentition of Living Primates*. London: Academic Press.

SZALAY, F.S. & DELSON, E. (1979). *Evolutionary History of the Primates*. London: Academic Press.

TOBEIN, H. (1986). An early upper Miocene (Vallesian) *Mesopithecus* premolar from Rheinhessen, FRG. *Primate Report*, 14, 49.

4   Are *Papio baringensis* R. Leakey, 1969, and *P. quadratirostris* Iwamoto, 1982, species of *Papio* or *Theropithecus*?

ERIC DELSON AND DAVID DEAN

Summary

1. Eck & Jablonski (1984, 1987) have argued that the holotype crania of *Papio baringensis* (from the Chemeron Formation) and *Papio quadratirostris* (from the Omo Usno Formation) are actually specimens of *Theropithecus* which represent early stages of the *T. brumpti* lineage. These fossils are among the oldest members of the African Papionini which can be allocated to modern genera – the Usno Formation dates to about 3.4 Ma, while the Chemeron Formation is probably slightly younger, although its age is less definitely known.

2. Despite many morphological studies to the contrary, but in line with a variety of molecular analyses, we accept *Papio* (including *Mandrillus* as a subgenus) as the closest known relative of *Theropithecus* among the African Papionini. A mandrill-like cranial morphology (with relative lack of klinorhynchy even with a long face; only moderately deep midface; extended or 'pointed' rather than rounded external occipital protuberance; bulbous, not high-crowned, molars; large front teeth; and ovoid, closely-spaced orbits) is hypothesized to be close to the ancestral morphotype of the baboon–mandrill–gelada clade. Within *Theropithecus*, there are three major lineages: *T. gelada*, *T. oswaldi* (and almost surely *T. darti*) and *T. brumpti*. Based on previous research by others, we take as a working hypothesis the closer phyletic relationship of the first two contrasted with *T. brumpti*.

3. Eck & Jablonski discussed nine characters which were said to differentiate *Theropithecus* from *Papio* and which occurred in the two fossils on which this chapter focuses. We here review these features, showing that several of them in fact are

closely correlated as parts of functional complexes, rather than independent features; most of these, as is well known, relate to the feeding adaptations of *Theropithecus*, but the relatively small brain size of this genus is also important.

4. What Eck & Jablonski did not discuss, however, is the relative airorhynchy of *Theropithecus*, a radical departure from the ancestral moderately klinorhynch condition of African papionins (perhaps comparable to that seen in mandrills today). Although the two fossil skulls have small neurocrania, they are klinorhynch, like *Papio*. The teeth of *Papio quadratirostris* are basically not gelada-like, while those of the Chemeron skull are only possibly so.

5. Other features of *Theropithecus* which differentiate them from *Papio* and the two focal fossils include: the inferior buttressing of the mandible, laterally bowed zygomatics, inferiorly narrowed orbits (often short inferosuperiorly), increased midface height, extension of anterior temporalis origin (anterior temporal line) laterally on the posterior wall of the malar, production of a malar 'visor' and large maxillary tuberosity, a wide, short basiocciput with rounded rather than extended or 'pointed' external occipital protuberance, and a more anterior origin of the root of the zygomatic buttress (anterior to M^3 rather than generally posterior to M^2). Most of these features are directly linked to the great extent of anterior temporalis and masseter in *Theropithecus* species.

6. Although the *baringensis* skull is not convincingly like *Theropithecus*, it may be tentatively referred to the *T. brumpti* lineage on the basis of dental similarities to another Chemeron mandible with less worn teeth (BC 1647) and in turn to definite *Theropithecus* jaws from early horizons in the Turkana Basin (see Leakey, chapter 3).

7. This allocation implies that early members of the *T. brumpti* lineage were conservative in the development of the *Theropithecus* dental pattern and reversed curve of Spee, as well as in retaining a more ancestral (relatively klinorhynch), *Papio*-like skull, compared to contemporaneous members of the *T. oswaldi* lineage. The latter, in turn, shared developing airorhynchy, orthognathy, reversed curve of Spee and greater molar complexity with *T. gelada*, suggesting that together they form the sister clade to the *T. brumpti* lineage. These two clades within *Theropithecus* evinced parallel development of the diagnostic molar pattern and the reversed curve of Spee.

8. Finally, fossils of *Papio* (*Dinopithecus*) from South Africa, Angola, and Ethiopia briefly described here are morphologically identical or closely similar to the *P. quadratirostris* skull and demonstrate that it (and the species) are best referred to that subgenus.

Introduction

In 1969, Richard Leakey reported the recovery of a partial cranium and associated mandible (KNM-BC 2) of a large papionin from the Chemeron Beds of the Baringo Basin, Kenya, which he named *Papio baringensis*. Leakey & Leakey (1976) reported a second specimen, KNM-BC 1647 (originally listed as BC 3, which is in fact the catalogue number of the holotype of *Paracolobus chemeroni* Leakey, 1969), from the same locality. This specimen, which was not illustrated, preserves only a partial lower dentition and some postcranial fragments. In 1982, Mitsuo Iwamoto described a nearly complete cranium without mandible from the Usno Formation of the Omo Group, Ethiopia, which he named *Papio quadratirostris*. This cranium is stored in the NME (see abbreviations at end of 'Introduction') collections but apparently has not yet received a formal catalogue number – it will be termed NME USNO here.

Eck & Jablonski (1984; also Eck, 1983) restudied these fossils in the context of an evaluation of the *Theropithecus* sample from the Omo Group and suggested that both crania were actually *Theropithecus* specimens, probably on the *T. brumpti* (Arambourg, 1947) clade. Delson (1984) rejected this view, combining the Usno skull with other Omo Group fossils previously termed *Papio* (Eck, 1977). Delson (1984) suggested that this species was morphologically closest to one from Swartkrans (South Africa) named *Dinopithecus ingens* Broom, 1937 and that *Dinopithecus* was best ranked as a subgenus of

Papio. Eck & Jablonski (1987) described the Omo *Theropithecus* sample in detail and reiterated their view of the *Theropithecus* status of the USNO and BC 2 crania. These two fossils (and more fragmentary remains allocated to the same species from East and West Turkana) are the earliest members of whichever genus they represent. Moreover, their generic allocation has important implications for the patterns of morphological transformation and phylogenetic geometry for these genera, which together dominate the fossil record of larger cercopithecines in the Pliocene and Pleistocene of eastern Africa. Here we assess their systematic position and their relationships to lineages of *Theropithecus* and *Papio*.

Abbreviations used in this paper for collections include: DGUNL: Departamento de Geologia, Universidade Nova de Lisboa, Portugal. KNM: Department of Palaeontology, National Museums of Kenya, Nairobi, Kenya. NME: National Museums of Ethiopia, Addis Ababa, Ethiopia. TMP: Transvaal Museum, Pretoria, South Africa.

Chronological position of specimens discussed

Although the basis for our study is morphology, the accurate determination of geological age for fossils is important in their overall evaluation. Eck & Jablonski (1984, pp. 127–8) carefully reviewed the available data bearing upon the ages of these specimens, concluding that the Chemeron fossils (from locality JM 90/91) were probably *c.* 4 Ma old, while the Usno skull (from the White Sands locality, probably Unit U8-U9) dated to *c.* 3.3–3.4 Ma.

Recent studies in the Turkana Basin (summarized by Feibel, Brown & McDougall, 1989) confirm the latter estimate, altering the probable range to between 3.35–3.50 Ma (Usno unit U-7 is equivalent to the Lokochot Tuff, with an estimated depositional age of 3.5 ± 0.1 Ma; unit U-10 is equivalent to Tuff B, with an estimated depositional age of 3.36 ± 0.04 Ma). On the other hand, new work in the Baringo Basin employing single-crystal laser fusion argon–argon dating and regional–faunal correlation suggests that locality JM 90/91 is somewhat younger, probably c. 3.2 Ma; although some questions remain about the relationship of this date to the fossils, unless the area of the locality lies within a fault-bounded block (for which there is no evidence), it is unlikely that it is older than 3.2 or younger than 2.5 Ma (A. Hill, pers. comm.). The age range of Theropithecus brumpti, by comparison, is c. 2.95–2.0 Ma (Eck & Jablonski, 1987; same estimates in Feibel et al. 1989).

M. Leakey (see chapter 3) discusses specimens allocated to all three taxa of the putative T. brumpti lineage from the Turkana Basin, including mainly fragmentary materials from as old as the USNO skull. The only described Theropithecus fossils older than the Omo T. brumpti are those of T. darti from the Hadar Formation, Ethiopia (see Eck, chapter 2), perhaps 3.35–2.90 Ma. Unpublished Theropithecus specimens from the Middle Awash Valley, Ethiopia, and a tooth from Lothagam-3 mentioned by Patterson, Behrensmeyer & Sill (1970; KNM-LT 417; see Delson, chapter 5) have an estimated age c. 4 Ma. According to T.D. White (pers. comm.), both of these assemblages could be as young as the oldest Hadar material on faunal grounds but are likely to be somewhat older. No fossils undoubtedly identified as Papio are older than the Usno and Chemeron specimens. In the Omo sequence, material allocated to Papio (Dinopithecus) ranges precisely as does T. brumpti: submembers B11-G13, c. 2.95–2.0 Ma. Specimens from Sterkfontein (South Africa) and Leba (Angola) are probably younger than 2.7 Ma, and other occurrences younger still – e.g. Swartkrans c. 1.9–1.6 Ma (Delson, 1984, 1989).

Lineages of Theropithecus and its relationship to Papio

It is widely accepted that there are three major clades within the Theropithecus group: the living T. gelada, the widespread Plio-Pleistocene T. oswaldi, and the geographically restricted T. brumpti (see Szalay & Delson, 1979; Eck & Jablonski, 1984, 1987; Eck, Leakey, and Jablonski – chapters 2, 3, and 7, respectively). The Pliocene T. darti (see Eck, chapter 2) is less well-known but generally accepted as closest to T. oswaldi. The central question of this paper is the relationship of the two distinctive Pliocene fossils discussed above, but in order to approach that problem, not only the internal phyletic geometry of Theropithecus, but also the place of this genus within the Cercopithecinae must be briefly examined.

Jolly (1967), Maier (1970), and Szalay & Delson (1979), among others, recognized three subtribes within the cercopithecine tribe Papionini: Macacina, Papionina and Theropithecina (sometimes with differing ranks or names). It was generally thought by these morphologists that the numerous distinctive, derived features of *Theropithecus* indicated an ancient origin for this genus. Cronin & Meikle (1979) presented new molecular data indicating a close relationship between *Papio* and *Theropithecus* among the Papionini. They argued that the derived features of *Theropithecus* might well have originated quickly from a *Papio*-like ancestry, perhaps in a punctuated equilibrium mode of evolutionary change. Delson (e.g. 1988; Strasser & Delson, 1987) recognized this argument and included *Theropithecus* within the subtribe Papionina, but continued to illustrate cladograms showing *Theropithecus* as the probable sister-taxon of all other papioninans. Few other workers have addressed this question recently. Fleagle (1988), for example, presented without discussion the cladogram of Strasser & Delson, the phylogram of Cronin & Sarich (1976, which agreed with Cronin & Meikle, 1979) and a third phylogram which placed three lineages of *Theropithecus* nearest to one for *Papio*.

Discussion during the Cambridge conference (summarized in Jablonski, chapter 7) and the results of our analysis here lead us to accept the probable sister-taxa relationship of *Papio* and *Theropithecus*. Disotell, Honeycutt & Ruvolo (1992) came to this conclusion from their new mtDNA studies of papionins as well. This grouping allows a better reconstruction of the ancestral morphotype (list of inferred character states) for *Theropithecus* and thus a better determination of the polarity of character transformation within the genus.

In terms of cranial and dental proportions, facial hafting and mandibular shape, we suggest that the most conservative living form, the one most similar to the last common ancestor of all baboons, may be *Papio* (*Mandrillus*) species (cf. Jolly, 1970). Holding aside the skeletal synapomorphies of this group (especially the swollen muzzle ridges), mandrills and drills present a suite of cranial character states which are similar to those seen in *T. brumpti*, the two focal fossils, and varieties of *P.* (*Papio*). These features, which will be discussed below in detail, include: a relative lack of cranial flexion, even with a long face; only moderately deep midface; extended or 'pointed' rather than rounded external occipital protuberance; bulbous, not high-crowned, molars; large front teeth; and ovoid, closely-spaced orbits. Some of these conditions (such as the low degree of flexion) are unusual for large African papionins but found elsewhere in the tribe.

Eck & Jablonski (1987, Table 42) produced a list of 51 cranial features which varied among the three well-known species of *Theropithecus*, in addition to the seven cranial (and two dental) shared-derived characters which for them diagnosed the genus. Of this set of 51, the three species all differed in 13 characters (their tabulation of Table 43 is incorrect in the 'Skull' or summation column, but accurate elsewhere). In 21 features, *T. oswaldi* is similar to *T. gelada*, in 11 *T. oswaldi* is similar to *T. brumpti* and in only five do *T. gelada* and *T. brumpti* share the same character state. Unfortunately, polarity was not assessed for these 51 characters, so it is not possible to tell whether the shared similarities are derived or plesiomorphic.

As a working hypothesis, to be examined and supported further below, we consider that *T. oswaldi* and *T. gelada* are sister-taxa, the pair being in turn the sister-taxon of *T. brumpti*. Following Eck and Leakey (see chapters 2 and 3, respectively) among other workers, *T. darti* from Makapansgat and Hadar appears to be closest to *T. oswaldi*, although less derived in certain dental and cranial characteristics. From this baseline, we now proceed to an evaluation of the systematic position of the Usno and Chemeron holotype crania. We first consider those points raised by Eck & Jablonski (1984, 1987), then examine other features of these two crania and finally evaluate relevant evidence obtained from other specimens.

Eck and Jablonski's arguments

In their 1984 paper, Eck & Jablonski (p. 111) listed seven features which were said to distinguish *Theropithecus* from *Papio* crania and dentitions: '(1) strong elevation of the temporomandibular joint above the occlusal plane and a tall, upright mandibular ramus; (2) anterior union of the temporal lines and sagittal crest long, if present; (3) great postorbital constriction; (4) inferiorly divergent lateral margins of the frontal processes of the zygomatic bones; (5) a long mandibular symphysis; (6) reduced incisors and a shallow maxillary incisive arc; and (7) molars with high, pinched cusps and well-developed longitudinal crests, M³s with large mesial and distal foveae, and M₃s with con-sistently large hypoconulids.' All of these features were said to occur in BC 2, while all but number 5 were reported for NME USNO. In 1987, the same authors (p. 100) added two other features (here numbered): (8) a reversed curve of Spee (which had been briefly mentioned in 1984); and (9) large temporal fossae. We shall evaluate each of these in turn, with a special focus on the details of the relevant morphology in order to determine whether the suggested similarities are homologous or analogous (=convergent), as well as to ascertain their condition in the two early fossils.

Facial lengthening and klinorhynchy

Character one relates the mandible to the cranium and (as is true for most of these characters) to the masticatory musculature, although there are some questions about the relative value of the different aspects combined here. The tendency for *Theropithecus* mandibles to have a relatively high, long, and vertically oriented ramus was briefly mentioned by Leakey & Whitworth (1958) and discussed at greater length by Jolly (1972). The height of the ramus is directly linked to the elevation of the temporomandibular joint, so that the pattern may be examined on either the cranium or the mandible. Eck & Jablonski (1984, p. 112) plotted three measurements of the cranium on a triangular graph to examine this feature, finding that *Papio* and *Theropithecus* did separate (with a narrow 'empty' zone between them), but that neither fossil specimen fell within the spread of extant (or clearly allocated extinct) species;

instead, both fossils clustered away from all others, apparently due to relatively high values of Eck & Jablonski's variable 1, the distance between the anterior base of the postglenoid process and the posterior root of M^3.

The angulation of ramus to corpus is quite variable in papionins (see Plates 1–2 in Jolly, 1972, noting especially the verticality of the ramus in *Mandrillus*). Therefore, corporamal angulation is less important as a means of distinguishing between *Papio* and *Theropithecus*. This angle, often referred to as the gonial angle, is related to several areas of the masticatory apparatus which are part of somewhat autonomous (developmentally and functionally) regions. The appearance of verticality can be due to a vertical coronoid process, which is related to anterior temporalis development, or to the development of the gonial area, which is related functionally to the masseter-pterygoid 'sling', an area of great expansion in grazing ungulates. The final outcome in papionins is probably mostly the result of the mandible essentially 'tracking' the growing palate ontogenetically (Enlow, 1990, see p. 50) without much canalization of verticality.

Some aspects of the relationship between the corpus and ramus do distinguish between *Papio* and *Theropithecus* on average, but a simplistic analysis contrasting the two by analogy to a carnivore–herbivore distinction (cf. Maynard Smith & Savage, 1959; Eck & Jablonski, 1987; see Jablonski, chapter 11) is not sufficient. Perhaps the most telling problem with the grazing analogy is the concentration on the lateral view of the *Theropithe-*

cus skull. This projection does show a tall, often steeply angled mandible, but we must ask what has brought this about. In ungulates much of the great height of the ramus is contributed by the inferior extension of the gonial region. The expansion of gonion is linked to the anteroposterior elongation of the origin of the masseter onto the maxilla, but in most ungulates the zygoma rarely bows laterally.

The increase in the height of the midface in *Theropithecus* is also not seen in grazing ungulates. Ramal height increase in *Theropithecus* is due to midface height increase, not to elevation of the mandibular condyle as Eck & Jablonski claim. The thicker palate and nasomaxillary complex has probably developed to resist very high occlusal forces. *Theropithecus brumpti*, possibly less derived along these lines, does not show the same level of orthognathy or midface height increase seen in other members of the genus. However, its dentition and the anterior placement of the maxillary root of the zygoma (and thus the masseter) are perhaps a parallel solution to the adaptation presaged by its high-crowned, crenulated teeth.

The dorsal and frontal views of the cranium (Fig. 4.5 and 4.6) may be more telling than the lateral view of the mandible. The anterior approximation of the temporal lines (seen in the dorsal view) is the result of the medial enlargement of anterior temporalis and the relative orthognathy evidenced by the decreased importance of the anterior dentition in the apprehension of food. In order to maintain oblique leverage on the shearing crests of the teeth, the zygomatic arch has bowed out laterally as seen from the frontal view (see Davis, 1964; Turner, 1970; Du Brul, 1974; Ward & Brown, 1986; also D. Dean, 1986, unpublished M.A. thesis, Temple Univ.). This great lateral expansion of the masseter muscle and anterior zygomatic arch is seen in *T. brumpti*. By analogy when the masseter has increased its anterior attachment area in an ungulate or rodent species it has most often taken root on the maxilla rather than producing the visor-like extensions seen in *T. brumpti* or the enlarged masseteric tubercle seen in other species of *Theropithecus* (see below). This 'reinforcement' can best be associated with high occlusal forces. It can reasonably be assumed that the bowing of the zygomatic, resulting in the lateral extension of the masseter's insertion, is an adaptation to putting high occlusal forces on obliquely angled surfaces.

Theropithecus dental morphology also belies the grazing ungulate model. While it is true that *Papio* and *T. brumpti* show elongate muzzles, no specimens of fossil or living *Theropithecus* show the incisor expansion or the canine loss and extensive premolar molarization associated with grazing in ungulates. It is likely that the long muzzle and relative lack of either klinorhynchy or airorhynchy seen in *T. brumpti* is a retention from a mandrill-like condition in these features predicted for the common ancestor of *Theropithecus* and *Papio*. There is no reason to assume that this conservative retention was a newly derived adaptation to large gape and large object feeding (see Jablonski, chapter 11).

Eck & Jablonski (1987; see Jablonski,

chapter 11) concentrated on the lateral cranial view of *Theropithecus* in making their analogy with grazing ungulates. However, this comparison invites a closer look. While most ungulates show superficially similar high-crowned teeth with ribbon-like crenulations and steeply angled power strokes (Hiiemae & Crompton, 1985, p. 278), they do not tend toward airorhynchy or increased midface height.

In *Papio*, as we shall discuss further below, the major distinction from other cercopithecines is increased klinorhynchy (downward bending of the face on the neurocranium and base) in order to maintain the leverage needed by posterior temporalis to act on the anterior dentition. The klinorhynch modifications seen in large macaques or the extinct macacinan *Paradolichopithecus* are different, in that the same result is attained by vertical height increases in the mid-face region.

To document this pattern of facial angulation, Fig. 4.1 presents histograms of the basal angle (between a line connecting basion and sella and the line of the [anterior] hard palate) in a number of extant baboon taxa of both sexes, based on data from Frick (1960). His small samples of drill and chacma baboon are grouped with mandrills and anubis baboons, respectively, and his specimens younger than subadult (without M^3 erupting at all) are not included. It is clear that most *T. gelada* specimens have a very low basal angle, making them nearly airorhynch (with the face upwardly bent). Although a few *Papio* specimens overlap with them, the *Papio* show great variation while *Theropithecus* is restricted to the low end

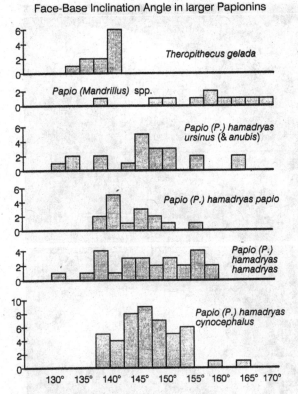

Face-Base Inclination Angle in larger Papionins

Theropithecus gelada

Papio (Mandrillus) spp.

Papio (P.) hamadryas ursinus (& anubis)

Papio (P.) hamadryas papio

Papio (P.) hamadryas hamadryas

Papio (P.) hamadryas cynocephalus

130° 135° 140° 145° 150° 155° 160° 165° 170°

Fig. 4.1. Histograms of basal angle (see text) for six groups of baboons, data from Frick (1960), adults and subadults only, mixed sex, wild-shot and captive animals lumped. Note that *Theropithecus* is restricted to the low end of the range, while various *Papio* species are more variable but mainly in the center and high end.

of the range. It is likely that *Theropithecus* would have an even lower angle were it not for the increased height of the midface in this taxon. This height increase may be in part an adaptation to resist strain from powerful postcanine chewing, but perhaps even more important is the need to produce high occlusal pressure on high-crowned teeth which, even when extremely worn, maintain obliquely oriented shear edges and masticatory muscle lines of action. We suggest that the rela-

tionship of the face to the braincase is less variable in *T. gelada* than in other papioninans due to selection for the application of high occlusal forces to the postcanine teeth at a consistent angle.

P. Dechow (unpublished 1980 Ph.D. dissertation, Univ. Chicago) also discussed the relative airorhynchy of geladas compared to other baboons, noting that it was due to the upturning of the face (especially seen at prosthion) relative to the neurocranium, combined with an extremely short face for its overall size. This places the masticatory muscles closer to the tooth row and allows the development of greater occlusal forces than in other baboons of comparable size (see also Jablonski, chapter 11). Dechow also found that *Papio* (*Mandrillus*) (both mandrills and drills) presented somewhat upturned anterior muzzles, but combined with long faces which are at least moderately klinorhynch.

It is not possible for us to duplicate Frick's measurements (based on X-rays) on the fossil casts available to us, but in order to demonstrate the relative airorhynchy of at least *T. oswaldi* and to place the two questioned fossils in context, we present lateral views of a number of extant and fossil crania (Fig. 4.2 and 4.3). The USNO skull is similar to *P.* (*Mandrillus*) in a number of features, such as the moderate angulation of the face and the apparent elevation of the TMJ. It is also interesting to note that the female *T. darti* from Hadar (see Fig. 4.3, bottom) appears less airorhynch than other specimens (including the Hadar male – see Eck, chapter 2), which may reflect variation in the development of this pattern in the earliest

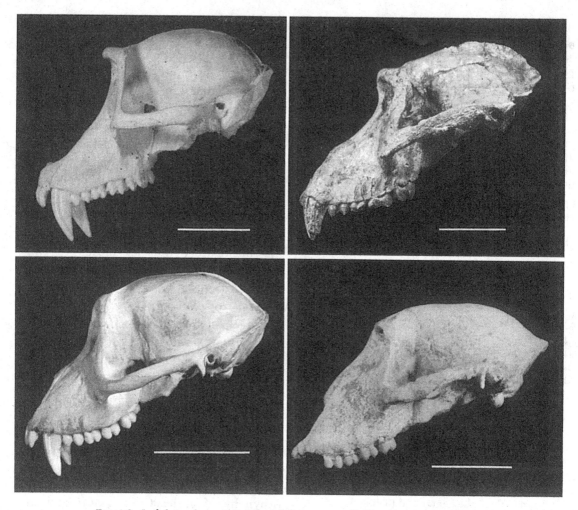

Fig. 4.2. Left lateral views (in Frankfurt horizontal) of male crania of *Papio*, *Theropithecus* and other papionin species. Left page, top row, left to right: *T. gelada*, *T. oswaldi oswaldi* (Kanjera); bottom row: *Macaca thibetana* (one of the largest macaques), *Parapapio ?whitei* (Makapansgat). Right page, top row: *T. brumpti* (NME L. 345–287, Shungura, cast), *P. hamadryas kindae*; middle row: *?T. baringensis* (BC 2, cast), *P. hamadryas ursinus*; lower row: *P. (Dinopithecus) quadratirostris* (USNO), *Papio (Mandrillus) sphinx*. All brought to similar size, scale bars = 5 cm.

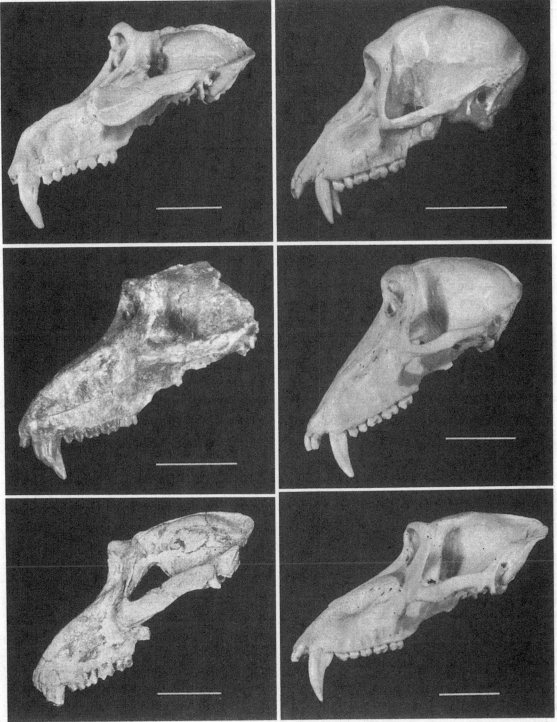

Fig. 4.2 – *contd.*

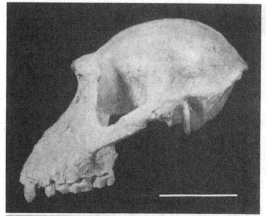

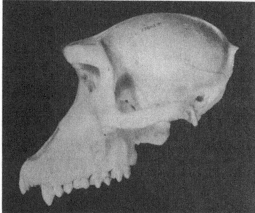

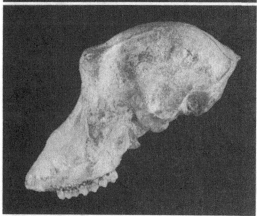

Fig. 4.3. Left lateral views (in Frankfurt horizontal) of female crania of *Theropithecus* species. Top to bottom: *T. oswaldi oswaldi*, Kanjera; *T. gelada*; *T. darti*, Hadar. Scale bar = 5 cm.

members of the genus. *T. brumpti* is also less klinorhynch than other members of the genus, appearing more like (*P.*) *Papio*, as does BC 2.

Relative brain size and chewing muscle orientation

In our opinion, characters 2, 3, and 9 are all part of a second functional complex which may be related to relative brain and masticatory muscle size irrespective of taxonomic differentiation. Given comparable development of the anterior temporalis, in a relatively small-brained individual (or species), this musculature would spread more widely over the vault than in a larger-brained relative; thus, the fusion of the temporal lines would lie anteriorly, the postorbital constriction would be tighter and the temporal fossae larger (the latter two are aspects of the same phenomenon).

It appears from inspection that both BC 2 and USNO have smaller braincases (and presumably brains) than extant baboons of comparable skull size. Moreover, Dechow (unpublished 1980 Ph.D. dissertation, Univ. Chicago) reported that *Theropithecus gelada* has a significantly smaller endocranial volume than does any living baboon, including the rather smaller *P. hamadryas kindae*. Martin (see chapter 10) confirms the relatively small brain size of *Theropithecus gelada* and finds the same result for *T. oswaldi*, evidence that *Theropithecus* is conservative for brain size. We propose that small brain size (and strong postorbital constriction) is symplesiomorphous for *Papio* and *Theropithe-*

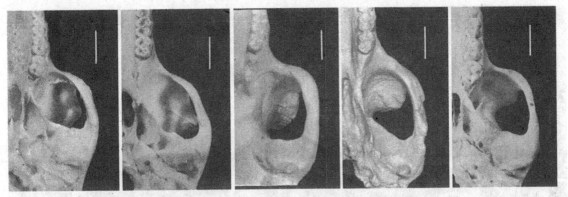

Fig. 4.4. Inferior views, in occlusal plane orientation, of left temporal fossa in selected *Papio* and *Theropithecus* males. Left to right: *P. hamadryas kindae*, *P. h. ursinus*, *P. (D.) quadratirostris* (USNO, cast), *?T. baringensis* (BC 2, cast), *T. gelada*. All brought to similar size, scale bars = 5 cm.

cus. Thus the presence of these two features in BC 2 and USNO is irrelevant in terms of discerning their phylogenetic relationships.

Although it is not one of their nine diagnostic features, Eck & Jablonski (1984, 1987) discussed in some detail the shape of the temporal fossa (especially its inferior outline) in living and extinct baboons. We have considered this question in the light of our analysis of the relationship between temporalis musculature and structures surrounding the temporal fossa. Eck & Jablonski (1984, p. 124) cited a nearly right-angle between the posterior margin of the temporal fossa and the sagittal plane in inferior view in *T. brumpti* and USNO, but not in BC 2. The last two specimens are compared to individuals of modern *Papio* and *Theropithecus* in Fig. 4.4, which reveals that only USNO is even vaguely 'squared-off' (like *T. brumpti*), while BC 2 is most like large, klinorhynch *Papio*. Smaller *Papio* also has a nearly squared posteromedial corner like USNO

(but cf. Fig. 4.7, in which a large *Papio* displays almost the same shape as USNO – the overlapping of the pterygoid plates also interferes with ready observation of this condition in undamaged specimens). The high degree of variation in this feature suggests that the similarity in this simplistic character is not necessarily homologous between USNO and *T. brumpti*.

Concentration on the shape of this posterointernal corner, moreover, misses the key fact of an expanded anterior temporalis above the fossa which encroaches into the posterior surface of the orbital cone and out onto the posterior surface of the malar. It is actually this bowing out of the zygomatic arch that distinguishes *Theropithecus* from *Papio*, not the quadrangular shape. The temporal fossa of *Papio* is like an oval from the inferior view. *Theropithecus* appears less ovoid because of the lateral bowing and anteroposterior shortening of the temporal fossae. Again, this shape is related to an increased development of the anterior

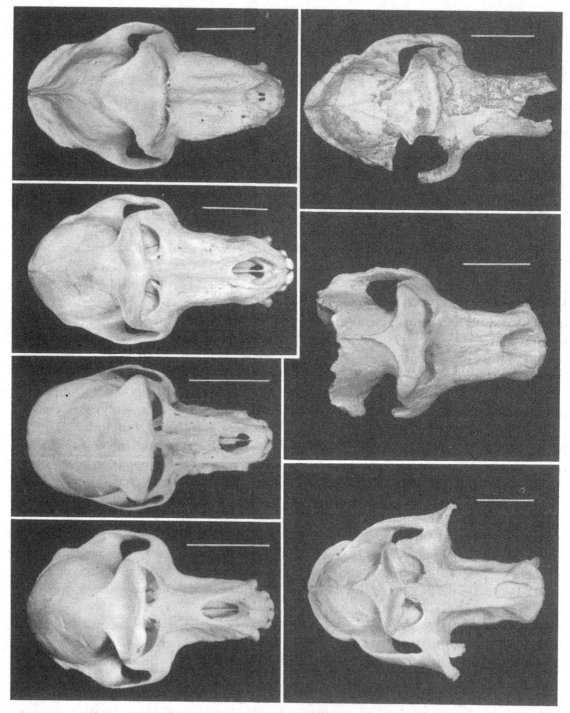

Fig. 4.5. Dorsal views (with occlusal plane horizontal) of male papionin crania. Left column, top to bottom: *Papio (Mandrillus) sphinx, P. (P.)*

temporalis and the masseter-pterygoid sling. In *Papio*, the fossa is extended anteroposteriorly to ensure that posterior temporalis can pull powerfully on the coronoid at an angle such that its line of action is directly in line with the incisive edges of the front teeth. The great anteroposterior length of the opening seen in *Papio* is cut off by the malar visor (see below) in *Theropithecus*. Neither BC 2 nor USNO have zygomatic arches that are bowed or extended out from the skull as far as they are in *Theropithecus*, nor are the arches as robust; by implication, their anterior temporalis (and masseter) muscles are less well developed also. Fig. 4.5 illustrates the two focal fossils and *T. brumpti* as well as a variety of modern baboon crania in dorsal (occlusal plane) view. The relative bowing of the zygomatic arches is obvious, and moreover the superior view of the posterior part of the temporal fossa shows the great similarity of BC 2 to larger *Papio*.

Malar and orbit shape

Character 4 may be more clearly stated as a tendency, in *Theropithecus* but not *Papio*, for the malar to widen inferolaterally. This forms a 'visor' (*sensu* Rak 1983; see pp. 99–102) that serves as a buttress for the laterally bowed zygomatic arch. Eck & Jablonski (1984, 1987) did not pursue their interpretation far enough

to include the three-dimensional angulation of this visor and its buttressing role. As noted above, the orbital cone is narrowed on the posterior surface of the malar because of the expanded anterior temporalis muscle. This bowing of the zygomatic is similar to the pattern seen in *Pongo*, where strong airorhynchy has occurred (Ward & Brown, 1986). In airorhynch taxa (of which *Alouatta* is perhaps the most extreme) the premaxilla bends upward, as it does also in most *Theropithecus gelada* and *T. oswaldi*. Eck & Jablonski (1987, p. 54) described the orbits of *T. brumpti* as ovoid, but as in other *Theropithecus*, they are more accurately said to be elongate and narrower inferiorly than superiorly. By comparison, in *Papio*, the klinorhynchy which is probably an ancestral feature for African papionins is taken to an extreme. The upper face projects forward, but the orbits are of roughly equal width superiorly and inferiorly, the zygomatic arch does not bow laterally and there is no 'visor' (see Fig. 4.6). Eck & Jablonski (1984, p. 124) noted that the zygomatic arch is not flared in either BC 2 or USNO; moreover, there is no 'visor' and the lower part of the orbit is about as wide as the upper.

This malar widening and zygomatic bowing in *Theropithecus* result in the anterior concentration of the origin of the masseter-temporalis muscle complex. Two other aspects of this complex are also

---

Caption to Fig. 4.5 – *contd.*

hamadryas ursinus, P. (P.) hamadryas kindae, T. gelada; right column: *P. (D.) quadratirostris* (USNO), ?*T. baringensis* (BC 2, cast), *T. brumpti* (NME L. 345–287, Shungura, cast). All brought to similar size, scale bars = 5 cm.

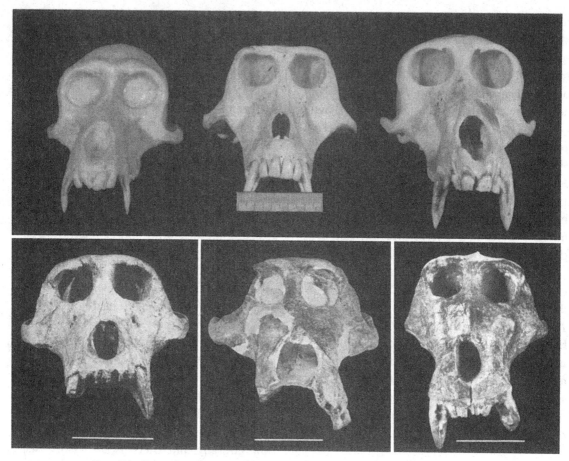

Fig. 4.6. Frontal views, mainly in Frankfurt horizontal, of male crania of *Papio* and *Theropithecus* species. Top row, left to right: *P. hamadryas kindae* (cast), *T. gelada*, *P. h. cynocephalus*; bottom row: *T. oswaldi oswaldi* (Kanjera), *P. (D.) quadratirostris* (USNO), ?*T. baringensis* (BC 2, cast). Scale = 5 cm.

important in distinguishing *Theropithecus* from *Papio*. First, the area of origin of anterior temporalis expands medially and laterally; medially it encroaches upon the posterior surface of the orbital cone, while laterally it extends nearly to the lateral margin of the posterior surface of the malar. In *Papio*, there is far less lateral expansion, and the ridge marking the lateral extent is well medial to the malar edge.

Second, a maxillary tubercle (anterior-most attachment of superficial masseter) can be seen in all *Theropithecus* as a small eminence on the anteroinferior margin of the maxillary process of the malar. This tubercle essentially enlarges to form the extreme 'handle-bar' development seen in *T. brumpti*. The lateral bowing of the malar and zygomatic arch in *Theropithecus* moves the superficial masseter laterally, so that its line of action pulls obliquely rather than vertically. In most other primates, including *Papio*, the zygomatic process of the maxilla and the frontal process of the zygomatic and the supporting mid- and upper-facial structures can resist the pull of the superficial masseter. However, the lateral position of the maxillary tubercle in *Theropithecus* requires the strengthening of the origin, as an 'anchor'. The lateral bowing and increased robusticity of the zygomatic arch produce a structure which can better resist such shearing forces than the relatively straight arch of *Papio*. Moreover, it permits the masseter to supply force at the necessary angle to the obliquely angled surfaces of the high-crowned, diagnostic *Theropithecus* dentition.

Mandibular buttressing

Eck & Jablonski's character 5 was reported as a long mandibular symphysis in *Theropithecus*, comparing the infradentale-menton distance to canine-molar toothrow length. This ratio was generally higher in *Theropithecus* than in *Papio*, but with much overlap; the value for BC 2 fell within the overlap range, at the upper end of *Papio* values (the lack of a mandible with the USNO specimen precludes assessment of this feature). More important, however, is the apparent conflation, once again, of more than one functional feature in this 'character'. Examination of the morphology around the symphysis in most *Theropithecus* mandibles reveals three features: the planum alveolare is concave, leading into a weakly developed superior transverse torus; the inferior torus is more strongly developed, but the greatest buttressing is along the base of the corpus, which is thickened posteriorly well back toward gonion; and anteriorly, the mental ridges continue from this thickened base superiorly toward (but not reaching) infradentale, resulting in a high or tall superior half of the anterior symphysis. By comparison, in *Papio*, the planum is steeper and less hollowed, ending at a more sharply defined superior torus; the inferior torus is weaker, and there is no sign of thickening along the base of the corpus; and the superior part of the anterior symphysis is relatively less tall. The measurement taken by Eck & Jablonski combines the superior and inferior heights, which vary inversely in the two genera, but it does not reflect the toral and basal morphology (just the lengthening) which differ strongly between them. KNM-BC 2 clearly has no basal thickening, but does have a relatively long planum alveolare and stronger superior torus, all as in *Papio*.

Based on the work of Hylander and colleagues, Greaves (1988) and others, it is clear that the shape of the symphysis is controlled by stress-reduction adaptations. Unlike frugivorous cercopithecines and hominoids, the posterior temporalis muscle and the concomitant superior transverse torus are not enlarged in *Theropithecus*. However, it is likely that the posterior extension of the torus (almost comparable to a simian shelf) in *Theropithecus gelada* is most likely to be functionally related to the postcanine dentition. The posterior extension of the symphyseal shelf can resist 'wishboning' (simultaneous lateral pull on both sides) of the mandible due to the lateral placement of the chewing muscles in *Theropithecus gelada* and the concomitant eversion of the lower border of the mandible at the beginning of the power stroke (Hylander, 1984). Hylander, Johnson & Crompton (1987) concluded that the wishboning of the symphysis is maintained or may actually increase after the powerful adductor force has been applied during phase I in anthropoid primates. It is therefore reasonable to speculate that, even though the masticatory muscle geometry in *Theropithecus* differs from that of *Macaca*, *Theropithecus* would have an increased wishboning of the mandible throughout the postcanine power stroke and the resultant posterior extension of the inferior torus (compared to *Papio*) would resist this.

Dental features

In terms of the dentition, characters 6 and 7 of Eck & Jablonski refer to the reduced incisors and high-crowned/high-relief molars of *Theropithecus*, respectively. These features of the teeth, as well as others noted below, are widely accepted as diagnostic of *Theropithecus* and at the origin of its adaptations. Moreover, character 8, the reversal of the curve of Spee, appears to be an adaptation to promote constancy of occlusal pressure along a toothrow in which the first molar is heavily worn while the third molar is still erupting; these two teeth are thus less elevated in the mandible than the intervening M_2, resulting in the reversed curve, which is clearly related to this dental pattern.

However, the cheek teeth of both USNO and BC 2 are worn to the point where differentiation between *Papio* and *Theropithecus* is nearly impossible, because similar molar wear occurs on aged individuals of both genera. On the other hand, the bases of the molar clefts (upper lingual and lower buccal) in *Theropithecus* are flattened, whereas those of *Papio* grade smoothly onto the sidewall of the crown (see illustrations in, for example, Szalay & Delson, 1979, and Delson & Hoffstetter, chapter 6). As far as can be discerned, the latter pattern occurs on USNO, while the situation for BC 2 is less clear.

Eck & Jablonski (1984, pp. 113 and 115) argued instead that the moderately worn third molars of both specimens allow the inference of high crowns and pinched cusps, as in *Theropithecus* but not *Papio*. However, neither the correlation between the observed features (acute angulation between buccal and lingual tooth walls and short but broad worn cusps) and the inferred characters, nor the correlation (or overlap) between those inferred features and generic distinction was demonstrated, rather merely stated.

The small incisors of *Theropithecus* are part of its adaptation for nibbling small hard objects brought to the mouth by hand. KNM-BC 2 clearly has small upper and lower incisors, as reported by all observers from Leakey (1969) onward. However, there are no incisors preserved in NME USNO, and there is some difference of opinion about its original condition. Iwamoto (1982) reported that the preserved alveolus for I^2 is smaller than in modern *Papio* of comparable cranial size, and this appears reasonable. But Iwamoto went on to estimate the width of the incisor arc as 34–39 mm, within the range for the same modern *Papio*. Eck & Jablonski (1984) tentatively accepted this value, which they noted would be very large for *Theropithecus*, but indicated that one adult male *T. darti* from Hadar combines small alveoli and a large arc width due to a midline diastema. This is no more than a possibility for the Usno fossil, so that it was premature for Eck & Jablonski (1984) to assume that small incisors were present in this individual. Extrinsic evidence (see below) tends to contravene this suggestion.

Finally, in terms of their character 8, the curve of Spee shape, Eck & Jablonski (1984, p. 124) accepted a flat curve of Spee for KNM-BC 2 and 1647, but argued that there is a reversed curve typical of *Theropithecus* in the Usno specimen. We do not agree with that determination. This feature does characterize most *Theropithecus* (although not some of the earliest specimens, such as at Hadar), but we do not see it in NME USNO (compare lateral views in Figs 4.2 and 4.3).

Additional characters and their interpretation

As a result of the foregoing analysis, several of the features utilized by Eck & Jablonski (1984, 1987) both to diagnose *Theropithecus* as compared to *Papio* and especially to identify BC 2 and USNO as specimens of *Theropithecus* do not support their arguments. Their characters 1 and 5 combine mandibular conditions which are secondarily related to stress reduction and face size with elevation

above the occlusal plane of the temporomandibular joint and cranial base, of which only the mandibular conditions are truly diagnostic. Characters 2, 3, and 9 may be linked to relative brain as well as masticatory muscle size, especially in the two questionable fossils, and the details of temporal fossa shape have not been carefully analyzed. Feature 4 is poorly phrased, but is related to gelada airorhynchy and the impact of temporalis musculature on the orbit, all of which will be discussed below. Finally, characters 6 and 7 (and 8) relate to the diagnostic dentition of *Theropithecus*, but some of the clearest features (e.g. flattened bases of the clefts and relative incision of the notches) are not mentioned, and we strongly question their application to USNO. KNM-BC 2 does have small incisors and perhaps a hint of the *Theropithecus* molar pattern, although it is so worn that certainty is not possible. In the following section, we consider several additional cranial features we have discerned which may help to distinguish these two genera.

Part of the generalized dolichocephalic profile of papioninans is due to their posterior neurocranium. A posteriorly extended posterior temporalis origin, often associated with a posteriorly pointed compound temporonuchal crest, is a corollary of the front tooth–posterior temporalis complex of all papionins (cf. Fig. 4.2 and 4.5 above). The superoinferior height of the temporalis origin forms an acute angle as it extends posteriorly. Finally, the distance between opisthion and the external occipital protuberance compared to posterior occipital breadth is also great in *Papio*. Figure 4.7 compares

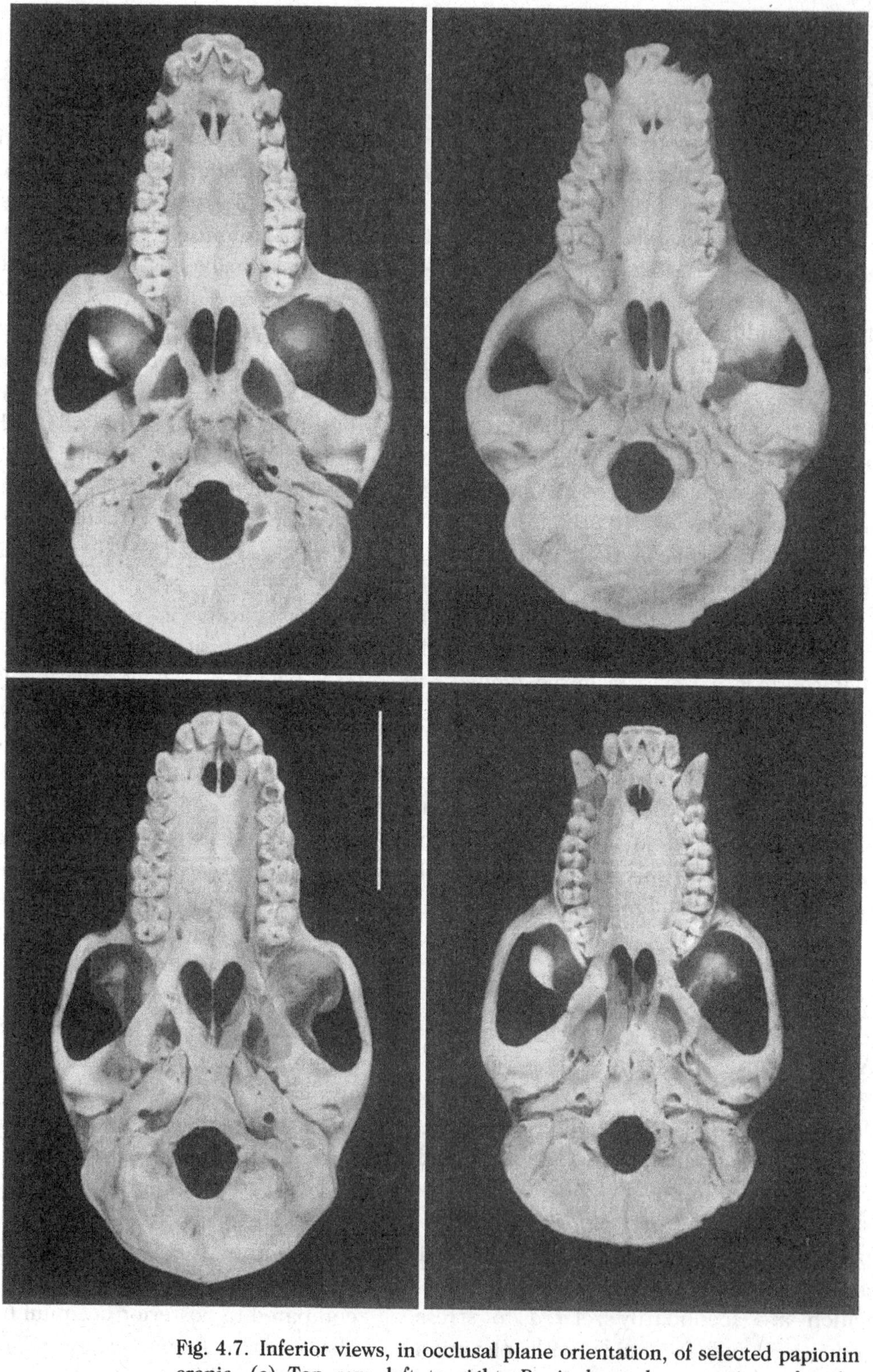

Fig. 4.7. Inferior views, in occlusal plane orientation, of selected papionin crania. (a) Top row, left to right: *Papio hamadryas ursinus*, female; *Theropithecus gelada*, female; bottom row: *Papio hamadryas kindae*,

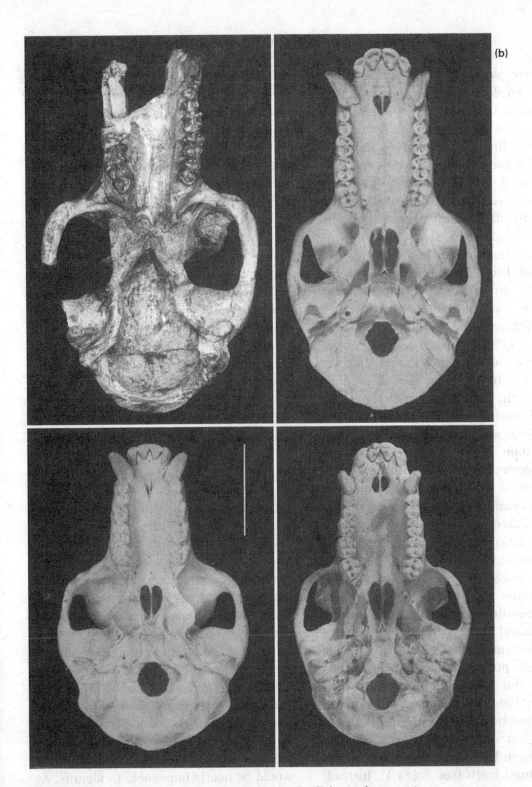

female; *Macaca thibetana*, male. (b) Top row: *P. (D.) quadratirostris* (USNO), male; *P. h. ursinus*, male; bottom row: *T. gelada*, male; *P. h. kindae*, male. Scale bars = 5 cm.

this latter area in living and extinct *Macaca*, *Papio* and *Theropithecus*.

Extant and extinct *Theropithecus* are characterized by a more brachycephalic skull, including the neurocranium, which is more globular and rounded at the posterior end; *T. brumpti* is somewhat intermediate between other *Theropithecus* and *Papio*. In distinction to the typically elongate papioninan skull base, *Theropithecus* shows wide basioccipital and bimastoid breadth. This overall brachycephaly is a reversal which is convergent on *Macaca* and its large extinct relative *Paradolichopithecus*. Swindler, Sirianni & Tarrant (1973) found that *Papio hamadryas cynocephalus* and *Macaca nemestrina* have similar growth profiles, but that the differences between them primarily result from the narrower, more klinorhynch skull of the former species.

All extant and extinct *Theropithecus* show a derived suite of features related to expansion of the anterior temporalis and pterygoid-masseter sling. The expansion of the anterior temporalis muscle has resulted in an expanded and vertical coronoid, stronger post-orbital constriction and anteriorly convergent temporal lines; the anterior temporalis also expanded mediolaterally into the posterior cone of the orbit and far laterally onto the superoposterior surface of the malar. The expanded pterygoid-masseter sling has caused the zygomatic arch to bow laterally. The great bi-euryon breadth and the vertical height of the malar visor combine to cut off the posterior temporalis muscle from having a direct line of action on the front teeth (see Fig. 4.4); instead, posterior temporalis must act on the

incisors indirectly. This is to be expected, as it is well known that *Theropithecus* has smaller incisors and larger, more complex molars than do species of *Papio*.

Another indication of the greater emphasis on postcanine chewing in *Theropithecus* is the position of its maxillary zygomatic process root origin above or immediately posterior to M^2, whereas the root of the malar buttress is usually above or just posterior to M^3 in other papionins. This position of the malar buttress supports the hypothesis that *Theropithecus* is secondarily orthognathous, as well as brachycephalic and airorhynch. Other factors confirming this condition are: the actual facial angle indicating airorhynchy, the raised premaxillary segment of the palate and the bulbous, wide-based neurocranium.

Extrinsic evidence for systematic placement of both fossils

It thus appears that both of the focal fossil specimens are broadly conservative of character states which probably typified the common ancestor of *Papio* and *Theropithecus*. In such cases, even a few (perhaps just one) robust derived conditions may be enough to convincingly allocate a plesiomorphic fossil to one or the other clade. As yet, we do not find any such clear synapomorphy of either BC 2 or USNO with *Theropithecus*; the problem is that *Papio* is generally plesiomorphic with respect to its common ancestor with *Theropithecus*, so that potential derived features linking one of the fossils to *Papio* would be nearly impossible to identify. As argued elsewhere (Strasser & Delson,

1987, p. 93), the lack of derived characters by comparison to a sister taxon does not invalidate or make less distinct any diagnosable taxon, but it does render more difficult the identification of the earliest members of that clade as opposed to a common ancestor or 'stem' taxon. Additional material from other sites may help to better interpret both holotypes and their respective species.

?*Theropithecus baringensis*

Eck & Jablonski (1984, p. 113) and M.G. Leakey (pers. comm. to them) agreed that the teeth of a second Chemeron mandible fragment (KNM-BC 1647) are *Theropithecus*-like, although they are also moderately worn; Delson's notes indicate this as well. This resemblance is greatly strengthened by the presence at a comparable date of definite *Theropithecus*, albeit with relatively low-crowned and less complex teeth, in the Lokochot horizons of the Turkana Basin (see Leakey, chapter 3). If we accept the Eck–Jablonski–Leakey argument that it is unlikely that two papionin taxa of the same size would be represented in the few known specimens from locality JM 90/91, then in light of the new Turkana specimens, we are forced to accept hesitantly that BC 2 indeed represents an early member of the *Theropithecus brumpti* lineage. The implications of this view for *Theropithecus* phylogeny, if indeed correct, are far greater than indicated by the cited authors.

As noted above, Eck & Jablonski (1984, p. 124) accepted a flat curve of Spee for KNM-BC 2 and 1647, and because some

Hadar *Theropithecus* have a relatively flat curve, which is generally accepted as the ancestral condition, they were forced to infer that a reversed curve evolved independently three times in the three *Theropithecus* lineages. Moreover, even accepting that the dentition of ?*T. baringensis* is incipiently gelada-like, the contemporaneous forms from Hadar present fully developed *Theropithecus* molars (as do also the probably significantly older teeth from Lothagam [Delson, chapter 5, this volume] and perhaps the Awash). This would require the parallel development of the full molar pattern at least twice, depending upon the relationships of *T. gelada*. Finally, the same is true for the partial airorhynchy common at least to the *T. oswaldi* lineage and to *T. gelada*, but of uncertain development in the relatively long-faced *T. brumpti* group.

At the outset, we suggested that *T. brumpti* might be the sister taxon of the combined *T. gelada* and *T. oswaldi–T. darti* clades. The several derived characters seen in the first species (above and in Eck & Jablonski, 1987) are combined with a number of conservative features, including facial angulation (degree of klinorhynchy) and the height of the midface perhaps comparable to that expected for the common ancestor of *Papio* and *Theropithecus* and moderate-size incisors. On the other hand, *T. gelada* and *T. oswaldi* share the development of airorhynchy, compared to the ancestral African papionin condition, as well as a tendency toward orthognathy, a deeper midface and a shortened, rounded neurocranium. Although *T. oswaldi* itself shows progressive reduction in incisor size and an increase in molar size and complexity through time, the early members of that sublineage (*T. darti* from Hadar and Makapansgat) share with *T. gelada* a moderate molar complexity and with *T. gelada* and *T. brumpti* moderate-size incisors (Eck & Jablonski, 1987; see Eck, chapter 2). The reversed curve of Spee and moderate molar complexity shared by *T. brumpti* and other species appear to have developed in parallel, as neither feature is present in early members of the *T. brumpti* lineage, such as BC 2 and BC 1647 from Chemeron or the various fragmentary remains from the Koobi Fora Formation (see Leakey, chapter 3). Thus, if KNM-BC 2 and 1647 are both early members of the *T. brumpti* lineage, the major division within *Theropithecus* must be between that clade and the remaining taxa. This view was tentatively supported on general character similarity by Eck & Jablonski (1984 and especially 1987), but Jablonski (see chapter 7) illustrates a split between the *T. brumpti–T. oswaldi* clades and the living *T. gelada*, which she considers the most conservative species of the genus. Our findings of shared airorhynchy in *T. gelada* and the *T. oswaldi* lineage rejects that interpretation. A further taxonomic evaluation of this finding is presented by Delson (see chapter 5).

Papio quadratirostris

It is less possible to be certain about the affinities of the USNO fossil. Leakey (see chapter 3) allocates several fragmentary specimens to the same taxon, but she then synonymizes it with parts of the *T. brumpti* lineage, and identification of

those fossils with Iwamoto's cranium is not clear. Delson (1984) referred *P. quadratirostris* to *Dinopithecus*, which was in turn argued to be a subgenus of *Papio*. Material allocated to *P. (Dinopithecus)* is known only from four other site groups: Schurweburg and Swartkrans (both South Africa), Leba (Angola) and the lower Shungura sequence (Omo Basin, Ethiopia). The fossils from the latter two samples are indistinguishable from each other, but significantly smaller than the Schurweburg and Swartkrans sample of *P. (D.) ingens* (see Figs 4.8–4.11).

The Shungura collection includes a male palate (NME L. 185-6), parts of three male mandibles, most of a female mandible (Omo 47-1970-2008) and parts of two fragmented female skulls (of which NME Omo 42–1972–1 is more complete) which have been tentatively reconstructed, as well as numerous smaller fragments and isolated teeth (E. Delson, unpub. data). The Leba sample (all numbered DGUNL LEBA02-19) includes a partial male maxilla (06), a partial frontal bone (probably male, 05), a female mandible (03) and a well-preserved female skull (lacking the superior region, 02) which are all nearly identical to their Shungura counterparts, along with other fragments (M.T. Antunes & E. Delson, unpub. data); other than a single mandible of *Cercopithecoides williamsi*, no other cercopithecid species is represented among the 20 known specimens. The Swartkrans specimens (Freedman, 1957; Delson, unpub. data) include a partial female skull (TMP SK 553), a male neurocranium without face (SK 599, recently reconstructed by Dr. Ron Clarke), several

Table 4.1. *Measurements (in mm) of male upper teeth of* Papio (Dinopithecus) *specimens.*

Species: Site: Specimen Number	P. (D.) quadratirostris			P. (D.) ingens			
	Usno NME USNO	Omo E NME L. 185–6	Leba DGUNL LEBA06	Schurw. TMP SB 7	SK 1 TMP SK 546	SK 1 TMP SK 578	SK 1 TMP SK 577
C^1W	12.0	<14	17.4		13.5	15.8	
C^1L	13.0	<17	17.8		21.0	20.1	
P^3W	8.0	10.0	11.4		10.3	11.3	9.8
P^3L	6.3	8.9			8.3	9.0	8.7
P^4W	9.5	12.3	11.5		10.9	12.9	11.0
P^4L	7.3	9.2	8.4		8.0	9.5	9.7
M^1AW	11.1				11.8	13.3	
M^1PW	11.5				11.8	12.5	
M^1L	11.7	13.0	11.9		13.5	14.1	
M^2AW	13.4	14.3	15.2	17.1			
M^2PW	13.1	13.2	14.2	16.4			
M^2L	14.3	15.1	15.8	17.9			
M^3AW	13.8	15.0	16.0	16.6			
M^3PW	12.1	12.5	12.7	13.7			
M^3L	15.5	15.9	15.2	19.7			

Notes: NME, National Museum of Ethiopia, Addis Ababa; DGUNL, Departamento de Geologia, Universidade Nova de Lisboa, Portugal; TMP, Transvaal Museum, Pretoria, South Africa. Schurw., Schurweburg; SK 1, Swartkrans, 'Hanging Remnant' of member 1.

For canines and premolars, L(ength) is always maximum mesiodistal, W(idth) is maximum buccolingual, taken perpendicular to length. For molariform teeth, AW and PW, respectively, are taken across the mesial and distal loph(id)s usually at the cervix; L, however, is taken at interdental contact points, often estimated due to wear, and decreases significantly in worn teeth.

partial male maxillae and other fragments, while the Schurweburg holotype mandible (TMP SB 7) may be associated with a large damaged cranium bearing M^{2-3} (SB 3).

Table 4.1 presents measurements of $C^1–M^3$ for the holotype of *P. (D.) quadratirostris* along with those for male specimens from the four mentioned sites. Dental metrics are quite similar for the two Omo specimens and the Leba maxilla

(and see Fig. 4.8), while the South African molars, at least, are rather larger. M.G. Leakey (pers. comm.) suggests that the premolars of NME USNO are relatively smaller compared to its molars than those of the other specimens, being similar to those of *T. brumpti* which are usually shorter than half the length of M^2. The situation observed for USNO may reflect its advanced state of wear, but without a

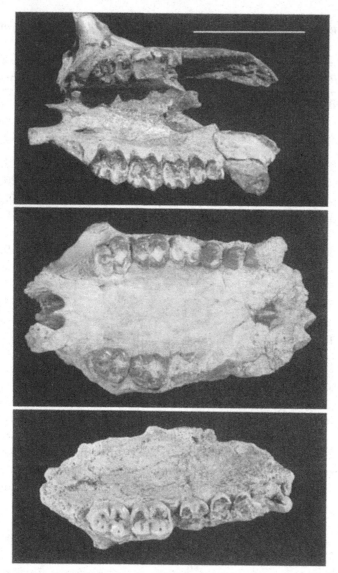

Fig. 4.8. Occlusal views of male maxillae referred to *Papio (Dinopithecus) quadratirostris*. Top to bottom: holotype (USNO), Shungura palate NME L. 185–6, Leba half palate DGUNL LEBA06. Scale bar = 5 cm.

sample of more individuals, such comparisons are difficult to evaluate; the purpose of these data is only to demonstrate overall similarity of size.

Table 4.2 lists several craniofacial measurements for the same material. The Leba frontal (Fig. 4.9) is slightly smaller than the USNO cranium in nasion-bregma distance and biorbital width, but its temporal lines are fused farther anteriorly. The apparently adult Swartkrans skull (SK 599), which is unlikely to be *T. oswaldi*

Table 4.2. *Measurements (in mm) of selected cranial dimensions of male* Papio (Dinopithecus) specimens.

Species: Site unit: Specimen Number	P. (D.) quadratirostris			P. (D.) ingens
	Usno NME USNO	Omo E NME L. 185–6	Leba DGUNL LEBA05/6*	SK 1 TMP SK 599
Nasion-Bregma	68.5		66.0	79.0
Nasion-Min. Temporal width pt.	85.0		50.0	72.0
Temporal Line width at level of postorbital constriction	32.5		13.0 (est.)	33.0 (est.)
Biorbital width	106.5		90.0	115.0+
Orbit width	29.0		28.0	30.0
Length M^3–C^1	72.5	82.0	77.0	
Length M^3–P^3	53.0	63.0		
Length M^3–M^1	41.5	45.0	44.0	
Length P^4–C^1	32.0	36.5	34.0	
Length P^4–P^3	13.0	18.0		
Alveolar process width at M^2	17.5	19.0	19.0	
Palate depth at M^2	10.0	12.0	11.0	

* Measurements on the frontal taken on LEBA05, dental values from LEBA06.

rather than *P.* (*Dinopithecus*) because of its short postglenoid processes (Fig. 4.10 and see Delson, chapter 5), is larger than either of the other two, and its relative point of temporal line fusion would appear to be intermediate. Moreover, the posteromedial corner of its temporal fossa is rather squared in inferior view. As it is most unlikely that this specimen belongs to the *T. brumpti* lineage, the short postglenoid and right-angled temporal fossa corner of that species are probably non-homologous with those described here.

The posterior neurocranium of each of the crania referred to *P.* (*Dinopithecus*), both male and female (Fig. 4.10), has the narrowed and extended external occipital protuberance typical of *Papio* (including USNO), as discussed above. In addition, the female mandibles from Leba and Shungura (Fig. 4.11) preserve large incisors, while all the mandibles have a typically *Papio* symphyseal morphology, as also discussed above. The Usno cranium occludes as well or better with the most complete male mandible from Shungura (unfortunately lacking incisors) than does the Shungura male palate. It appears to us highly unlikely that the holotype of *P. quadratirostris* had small incisors.

Taken together, the Leba and Shungura specimens document the presence in the 3–2 Ma range of a species of *P.* (*D.*) smaller

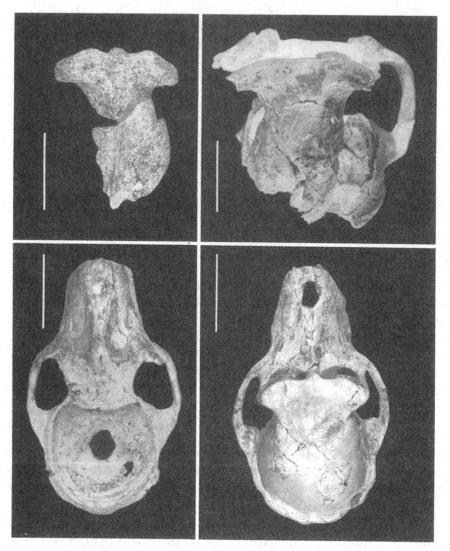

Fig. 4.9. Dorsal views (with occlusal plane horizontal) of crania referred to
Papio (Dinopithecus) quadratirostris. Top row, left to right: Leba male?
frontal DGUNL LEBA05, Swartkrans male cranium TMP SK 599 [*P. (D.)
ingens*]; bottom row: Leba female cranium DGUNL LEBA02, Shungura
female cranium NME Omo 42–1972–1. Cf. Fig. 4.5. Scale bar = 5 cm.

than the type species from Schurweburg
and Swartkrans. The lack of any teeth with
Theropithecus morphology at Leba suggests that the frontal does belong to the
same species as the other cercopithecines,
using the same parsimony argument as
was employed above for BC 2 and BC
1647. Since the frontal (LEBA 05) is too
large for the female skull from Leba (02), it
is probably male. Dentally, the species

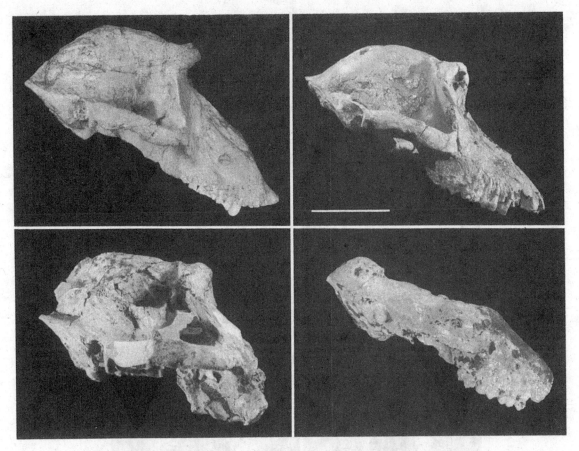

Fig. 4.10. Right lateral views, in Frankfurt horizontal, of crania referred to *Papio (Dinopithecus)* species. Left column, *P. (D.) ingens* from Swart-krans: female (TMP SK 553) above male (TMP SK 599). Right column, *P. (D.) quadratirostris*, Shungura female (NME Omo 42–1972–1) above Leba female (DGUNL LEBA02). Cf. Fig. 4.2. Scale bar = 5 cm.

known at Leba is comparable in size to NME USNO (and the Shungura fossils), and based on the Leba frontal, the temporal lines in males joined to form a sagittal crest quite far anteriorly, as was also probably the case for the Swartkrans population. In sum, the most likely con-clusion is that the Usno fossil represents the same species as the specimens from the Shungura Formation and from Leba, a species which is best termed *P. (D.)* *quadratirostris* Iwamoto, 1982. A rela-tively small brain size and strong postorbi-tal constriction, as well as a flat muzzle dorsum and especially the lack of maxil-lary or mandibular corpus fossae, appear to be among the few diagnostic features of this basically conservative subgenus. The possibility that *P. (Dinopithecus)* is in some sense ancestral to *P. (Papio)* or to other extant baboons is intriguing but beyond the scope of this paper. Modern

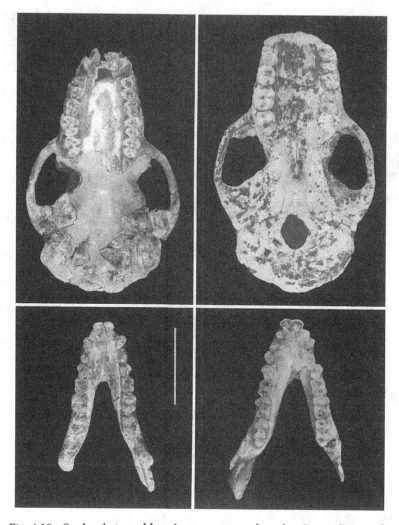

Fig. 4.11. Occlusal view of female specimens referred to *Papio (Dinopithecus) quadratirostris*. Left column Shungura cranium NME Omo 42–1972–1, mandible Omo 47–1970–2008; right column Leba cranium DGUNL LEBA02, mandible DGUNL LEBA03. Scale bar = 5 cm.

baboons (including mandrills), as well as all other Plio-Pleistocene species of *Papio*, have both larger brains and larger incisors than do species of *P. (Dinopithecus)* or *Theropithecus*, which is perhaps related to the exploitation of high-quality energy sources requiring preparation by the front teeth but needing less grinding. It is possible that this distinction from *Theropithecus* lies at the root of the eventual replacement of that genus throughout its previous pan-African range by *Papio* in the later Pleistocene.

Acknowledgements

We thank Nina Jablonski for the invitation to participate in the Cambridge meeting and for her patience during the preparation of this manuscript; Lorraine Meeker for printing and arranging the photographic figures, which were taken by her and the authors; Gerald Eck, Nina Jablonski, Cliff Jolly, Meave Leakey and other symposium participants for fruitful discussions of *Theropithecus* evolution; and Drs. Eck, Jablonski, Leakey, Peter Andrews, and William Kimbel for reviews and comments on the manuscript. Study of modern and fossil specimens was greatly facilitated by the curators of the collections of paleontology and/or mammalogy at numerous museums, especially the American Museum of Natural History, New York; and National Museums of Kenya, Nairobi. Research for this study was financially supported, in part, by grants to ED from the US National Science Foundation (most recently BNS 84-19939) and the PSC-CUNY Faculty Research Award Program (13453, 667370 and 669381).

References

ARAMBOURG, C. (1947). *Mission scientifique de l'Omo 1932–1933, 1 (111). Geologie – Anthropologie*. Paris: Museum National d'Histoire Naturelle.

BROOM, R. (1937) On some new Pleistocene mammals from limestone caves of the Transvaal. *South African Journal of Science*, 33, 750–68.

CRONIN, J.E. & MEIKLE, W.E. (1979). The phyletic position of *Theropithecus*: congruence among molecular, morphological and paleontological evidence. *Systematic Zoology*, 28, 259–69.

CRONIN, J.E. & SARICH, V.M. (1976). Molecular evidence for dual origin of mangabeys among Old World Monkeys. *Nature*, 260, 700–2.

DAVIS, D.D. (1964). The giant panda. A morphological study of evolutionary mechanisms. *Fieldiana: Zoology* Memoir 3.

DELSON, E. (1984). Cercopithecid biochronology of the African Plio-Pleistocene: correlation among eastern and southern hominid-bearing localities. *Courier Forschungs-Institut Senckenberg*, 69, 199–218.

DELSON, E. (1988). Studies on the northern Pleistocene. *Journal of Human Evolution*, 17, 643–5.

DELSON, E. (1989). Chronology of South African australopith site units. In *Evolutionary History of the 'Robust' Australopithecines*, ed. F.E. Grine, pp. 317–24. New York: Aldine de Gruyter.

DISOTELL, T.R., HONEYCUTT, R.L. & RUVOLO, M. (1992). Mitochondrial DNA phylogeny of the Old World monkey tribe Papionini. *Molecular Biology and Evolution*, 9, 1–13.

DU BRUL, E.L. (1974). Origin and evolution of the oral apparatus. *Frontiers of Oral Physiology*, 1, 1–30.

ECK, G.G. (1977). Diversity and frequency distribution of Omo Group Cercopithecoidea. *Journal of Human Evolution*, 6, 55–63.

ECK, G.G. (1983). The six species of *Theropithecus*. *American Journal of Physical Anthropology*, 60, 190–1.

ECK, G.G. & JABLONSKI, N.G. (1984). A reassessment of the taxonomic status and phyletic relationships of *Papio baringensis* and *Papio quadratirostris* (Primates: Cercopithecidae). *American Journal of Physical Anthropology*, 65, 109–34.

ECK, G.G. & JABLONSKI, N.G. (1987). The skull of *Theropithecus brumpti* compared with those of other species of the genus *Theropithecus*. In *Les faunes Plio-Pléistocènes de la Basse Vallée de l'Omo (Éthiopie)*. Tome 3, Cercopithecidae de la Formation de Shungura, pp. 11–122. Cahiers de Paléontologie, Travaux de Paléontologie

Est-Africaine. Paris: Editions du Centre National de la Recherche Scientifique.

ENLOW, D. (1990). *Facial Growth*, 3rd ed. Philadelphia: Saunders.

FEIBEL, C.S., BROWN, F.H. & McDOUGALL, I. (1989). Stratigraphic context of fossil hominids from the Omo Group deposits: northern Turkana Basin, Kenya and Ethiopia. *American Journal of Physical Anthropology*, 78, 595–622.

FLEAGLE, J.G. (1988). *Primate Adaptation and Evolution*. San Diego: Academic Press.

FREEDMAN, L. (1957). The fossil Cercopithecoidea of South Africa. *Annals of the Transvaal Museum*, 23, 121–257.

FRICK, H. (1960). Uber die Variabilität der präbasalen Kyphose bei Pavianschädeln. *Zeitschrift für Anatomie und Entwicklungsgeschichte*, 121, 446–54.

GREAVES, W.S. (1988). A functional consequence of an ossified mandibular symphysis. *American Journal of Physical Anthropology*, 77, 53–6.

HIIEMAE, K.M. & CROMPTON, A.W. (1985). Mastication, food transport, and swallowing. In *Functional Vertebrate Morphology*, ed. M. Hildebrand, D.M. Bramble, K.F. Liem & D.B. Wake, pp. 262–90. Cambridge: Harvard University Press.

HYLANDER, W.L. (1984). Stress and strain in the mandibular symphysis of primates: a test of competing hypotheses. *American Journal of Physical Anthropology*, 64, 1–46.

HYLANDER, W.L., JOHNSON, K.R. & CROMPTON, A.W. (1987). Loading patterns and jaw movements during mastication in *Macaca fascicularis*: a bone-strain, electromyographic and cineradiographic analysis. *American Journal of Physical Anthropology*, 72, 287–314.

IWAMOTO, M. (1982). A fossil baboon skull from the lower Omo basin, southwestern Ethiopia. *Primates*, 23, 533–41.

JOLLY, C.J. (1967). The evolution of the baboons. In *The Baboon in Medical Research*, vol. 2, ed. H. Vagtborg, pp. 427–57. Austin: University of Texas Press.

JOLLY, C.J. (1970). The large African monkeys as an adaptive array. In *Old World Monkeys*, ed. by J.R. Napier & P.H. Napier, pp. 141–74. New York: Academic Press.

JOLLY, C.J. (1972). The classification and natural history of *Theropithecus* (*Simopithecus*) (Andrews, 1916), baboons of the African Plio-Pleistocene. *Bulletin of the British Museum (Natural History), Geology*, 22, 1–123.

LEAKEY, L.S.B. & WHITWORTH, T. (1958). Notes on the genus *Simopithecus* with a description of a new species from Olduvai. *Coryndon Memorial Museum Occasional Papers*, 6, 3–14.

LEAKEY, M.G. & LEAKEY, R.E.F. (1976). Further Cercopithecinae (Mammalia, Primates) from the Plio-Pleistocene of East Africa. *Fossil Vertebrates of Africa*, 4, 121–46.

LEAKEY, R.E.F. (1969). New Cercopithecidae from the Chemeron Beds of Lake Baringo, Kenya. *Fossil Vertebrates of Africa*, 1, 53–69.

MAIER, W. (1970). Neue Ergebnisse der Systematik und der Stammesgeschichte der Cercopithecoidea. *Zeitschrift für Säugetierkunde*, 35, 193–214.

MAYNARD SMITH, J. & SAVAGE, R.J.G. (1959). The mechanics of mammalian jaws. *School Science Review*, 40, 289–301.

PATTERSON, B., BEHRENSMEYER, A.K. & SILL, W.D. (1970). Geology and fauna of a new Pliocene locality in northwestern Kenya. *Nature*, 226, 918–21.

RAK, Y. (1983). *The Australopithecine Face*. New York: Academic Press.

STRASSER, E. & DELSON, E. (1987). Cladistic analysis of cercopithecid relationships. *Journal of Human Evolution*, 16, 81–99.

SWINDLER, D., SIRIANNI, J. & TARRANT, L. (1973). The topography of the premaxillary-frontal region in non-human primates. *Folia primatologica*, 19, 18–23.

SZALAY, F.S. & DELSON, E. (1979). *Evolutionary History of the Primates*. New York: Academic Press.

TURNER, W.D. (1970). Mammalian Masticatory Apparatus. *Fieldiana: Geology*, 18, 147–355.

WARD, S.C. & BROWN, B. (1986). The facial skeleton of *Sivapithecus indicus*. *Comparative Primate Biology*, vol. 1, *Systematics, Evolution and Anatomy*, ed. D.R. Swindler & J. Erwin, pp. 413–52. New York: Alan R. Liss.

5 *Theropithecus* fossils from Africa and India and the taxonomy of the genus

ERIC DELSON

Summary

1. The female *Theropithecus* skull from Swartkrans, South Africa, SK 561, is described for the first time after reconstruction by R. Clarke. It is closely similar to female crania from Kanjera, but the anterior dentition is better preserved. The probable male cranium SK 599 lacks a face or teeth, but is most likely referable to *Papio (Dinopithecus) ingens* on the basis of its short postglenoid processes. Several isolated teeth from Brain's recent excavations at Swartkrans demonstrate no significant size increase from members one to three; this suggests only a short span of time (perhaps less than 0.25 Ma) separated these horizons, rather than the 1+ Ma originally suggested by Brain.

2. The first fossil *Theropithecus* specimen ever published, a single lower molar from Ain Jourdel, Algeria, is described. It is a typical M_1 of a small species whose only distinction is the acute angle formed at the base of its median lingual notch. The distal humerus from Garaet Ichkeul, Tunisia, is probably from a macaque (rather than a *Theropithecus*, as suggested by Geraads).

3. The isolated *Theropithecus* molar from Lothagam-3, Kenya, is described. It is a slightly damaged and moderately worn tooth, probably M_2, of a size comparable to several Plio-Pleistocene species of the genus. Its crown complexity and relief is high, and as it probably dates to between 4.0–3.3 Ma, it is unlikely to be a member of the *T. brumpti* lineage, whose early members have weakly complex molar crowns.

4. Two *Theropithecus* molars are described from Kanam East, Kenya. Although heavily worn, they compare most favorably with the Hadar sample of *T. darti*, which corresponds to a suggested earlier Pliocene age for the deposits.

5. Two isolated *Theropithecus* lower molars (probably M_1 and M_2) have

been recovered from the Senga 5A archeological site in the Semliki region of eastern Zaire. The larger, unworn tooth is comparable in size to those from a wide range of other sites, but it is most similar to populations from the Turkana Basin dating between 2.4–2.0 Ma, as suggested for other elements of the fauna. It is not possible to distinguish between isolated teeth of *T. oswaldi* and *T. brumpti* in this time range.

6. The only specimen of a *Theropithecus* ever described from outside Africa was recovered by E. Khan, not V.J. Gupta (the provenance of some of whose specimens has proven spurious). This maxilla with M^{2-3} is also typical of the genus, but of quite large size; it probably dates to between 1.0 and 0.1 Ma.

7. It is generally accepted that there are three main lineages within *Theropithecus*. Of these, the *T. brumpti* lineage is argued to be the sister to the *T. gelada* + *T. oswaldi* sublineages, based especially on Delson & Dean (see chapter 4). To recognize this phyletic pattern, the new subgenus *T. (Omopithecus)* is described for the *T. brumpti* lineage. Formal synonymies and diagnoses are given for the genus and the two accepted subgenera, but the subgenus *Simopithecus* is no longer recognized.

8. Within the *T. oswaldi* lineage, the earliest site samples (from Hadar and Makapansgat) are tentatively separated as the species *T. darti*. The Ain Jourdel molar cannot be readily allocated to a known species, and the ICZN will be petitioned to suppress the nomen previously applied to it, 'Cynocephalus' (= *Theropithecus*) *atlanticus*, which might otherwise be a senior synonym.

9. Within *T. oswaldi*, I follow the suggestion of M. Leakey (see chapter 3) that two African subspecies be recognized: *T. o. oswaldi* and the younger, larger *T. o. leakeyi* which presents greater reduction of the incisors, canines and P_3. The subspecies time boundary is unclear but can broadly be drawn between most Turkana Basin specimens and those from Kanjera, Swartkrans, Peninj, and Olduvai Beds I–lower II on the one hand and those from Olduvai upper Bed II and above, Olorgesailie, Kapthurin, Hopefield, Ternifine, and Thomas Quarries on the other. Outside Africa, the Mirzapur specimen (from India) is reduced in rank to a subspecies, *T. o. delsoni*. The questions of subspecies in modern savannah baboons and the use of this category in the fossil record are briefly discussed.

Introduction

Since the first description of extinct *Theropithecus* from Kanjera (Andrews, 1916), numerous fossils have been reported from a great number of site units. It is useful to discuss here a variety of such specimens which have never been described sufficiently to permit colleagues to interpret or refer to them. The collection from Swartkrans, South Africa, is large and well known, but additional specimens have become available in recent years. Isolated teeth or fragments from Ain Jourdel (Algeria), Mirzapur (India) and other sites are also in need of comparative analysis. This conference volume provides an excellent opportunity to present both the new and older material. The important sample from Ternifine, Algeria, which has never been described since its recovery in the 1950s, is discussed separately by Delson & Hoffstetter (see chapter 6).

Specimens discussed here are housed in the collections of numerous institutions, whose standard acronyms are as follows:

BM(NH) British Museum (Natural History), London (Department of Palaeontology)

IMNZ Institut des Musées nationales de Zaire, Kinshasa

KNM National Museums of Kenya, Nairobi

MNHN-P Muséum national d'Histoire Naturelle, Paris (Institut de Paléontologie)

NMT National Museum of Tanzania, Dar-Es-Salaam

PUC-GM Panjab University, Chandigarh, Geology Museum

TMP Transvaal Museum, Pretoria, South Africa (Department of Palaeontology)

UWMA University of the Witwatersrand Medical School (Department of Anatomy)

These acronyms will be employed the first time specimens are mentioned but not thereafter unless there is the possibility of confusion. Data presented in other papers in this volume will not be repeated here.

Swartkrans, South Africa

The sample of *Theropithecus* from Swartkrans was described in some detail by Freedman (1957), with additional notes by Freedman & Brain (1977). In 1957, Freedman noted that the best preserved cranium was that of a female, SK 561, described as 'almost complete but very badly crushed' (p. 208; see Fig. 5.1). In 1980, with financial support from the Wenner-Gren Foundation, I arranged for Dr Ron Clarke to prepare a number of South African cercopithecid fossils in the Transvaal Museum, including SK 561. Through the 1970s and 1980s, Dr C.K. Brain of that museum continued detailed excavation at Swartkrans, reinterpreting the local stratigraphy and collecting a large number of new cercopithecid fossils (Brain *et al.*, 1988). Brain asked me to study these new specimens as part of my general work on southern African cercopithecid palaeontology.

By way of brief review, Brain *et al.* (1988; also Brain, 1989) reported a local stratigraphy in which Member 1 at Swartkrans was heavily eroded after deposition in the chamber. The main fossiliferous patch excavated by Broom and other early workers was termed the 'Hanging Remnant', as it was emplaced high on the north wall. A second patch of Member 1 remained on the floor, separated from the former by a large gap; it was called the 'Lower Bank' or 'Orange Breccia'. All horizons above this were originally

Fig. 5.1. Crushed female *Theropithecus oswaldi* cranium SK 561, before reconstruction. Scale = 5 cm.

referred to as Member 2 or b, but they have now been subdivided into four distinct Members. Member 2, termed the 'Orange' breccia, and Member 3 (the 'Black' breccia) have yielded hominid fossils and cercopithecids discussed here.

As reconstructed by Clarke, the maxilla of TMP SK 561 (from the Member 1 'Hanging Remnant') is nearly complete, lacking only a small area on the right above P^4–M^1, the base of the left orbit and the midline above the nasal aperture (Fig. 5.2). The upper dentition is complete and moderately worn, with the incisors slightly crushed together and the buccal surface of the left M^3 cracked away. The middle of the palate is lacking, as is most of the sphenoid complex, although the pterygoid

plates are preserved in the basisphenoid region. The frontal bone is nearly complete and unwarped, but the internal portions of the orbits are damaged. Part of the right malar is lacking in a strip from the superior contact with the frontal down to the base of the zygomatic arch, as is the middle third of the right arch itself. The parietals were badly crushed and warped and are still missing large areas of bone, but the shape of the vault has been restored as fully as possible. The occipital and temporals are mainly complete,

Fig. 5.2. Standard views of reconstructed female *Theropithecus oswaldi* cranium SK 561, from the Swartkrans 'Hanging Remnant' of Member 1; scale bar = 5 cm.

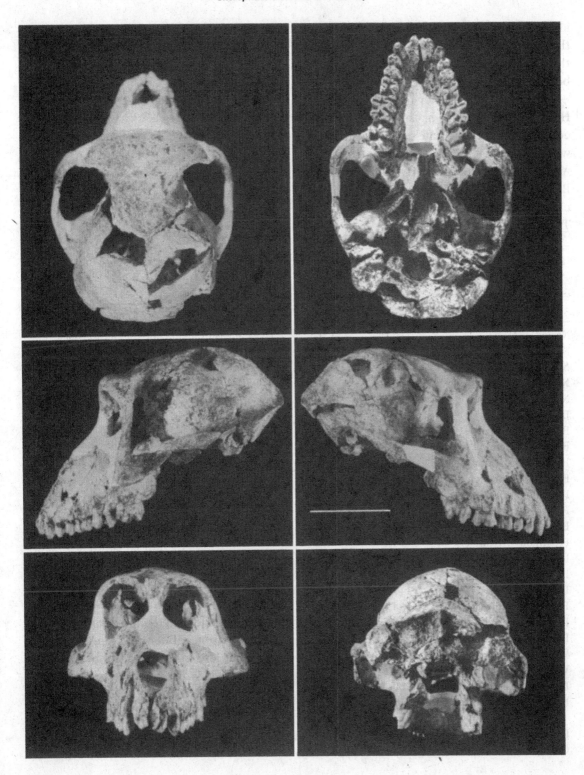

although lacking small areas of bone on the basal surfaces, especially the middle third of the right nuchal crest and most of both postglenoid processes.

I shall use Eck's paper on *T. darti* from Hadar (see chapter 2) as the model for my comparison of the female cranium SK 561, which is in fact little different from its well-preserved counterpart from Kanjera, BM(NH) M14936. Measurements which can be taken accurately are presented in Table 5.1.

Table 5.1. *Measurements (in mm) of female* Theropithecus *cranium SK 561.*

Variable		Eck*
Nasion – Inion	116.0	[12]
Nasion – Basion	83.5	
Nasion – Bregma	71.0	
Nasion – Prosthion	85.5	
Nasion – Staphylion	52.6	
Basion – Inion	51.0	
Basion – Bregma	66.0	[15]
Basion – Vertex	67.5	[15]
Basion – Prosthion	126.5	
Basion – Staphylion	60.0	
Prosthion – Inion	175.0	3
Prosthion – Bregma	148.0	
Prosthion – Staphylion	65.0	[9]
Prosthion – Orbitale	74.0	
Prosthion – Infraorb. for.	61.0	1
Inion – Bregma	61.0	
Inion – Orbitale	117.0	4
Bi-postglenoid width	84.0	
Bizygomatic width	119.0	11
Postorbital constriction	45.0	14
Width temporal lines (min.)	2.5	
Width " at postorb. const.	49.0	
Biorbital width	83.0	
Facial height (top supra- orbital torus to alv. plane)		
Interorbital width	10.8	
Orbit height	25.0	

Table 5.1. *Contd.*

Variable		Eck*
Orbit width	28.5	
Nasal aperture max. width	18.5	
External palatal width: M^2	56.0	5
External palatal width: P^4	42.0	
External palatal width: C^1	34.0	7
Alveolar process width: M^2	15.5	
Palatal depth at M^2	9.3	
Toothrow (partial) lengths:		
$M^3 – I^1$	72.0	
$M^3 – C^1$	67.5	
$M^3 – P^3$	60.0	21
$M^3 – M^1$	45.0	6
$P^4 – C^1$	23.0	
$P^4 – P^3$	15.0	
Alveolar width of incisors	19.0 (est.)	8

* These measurements correspond to those described by Eck (chapter 2, this volume), under the indicated number; a number in brackets, e.g. [12], is not identical to that of Eck.

The Swartkrans skull lacks both maxillary ridges and maxillary fossae. The nasal aperture is well defined by the premaxillary wings, although the midline is damaged both superiorly and inferiorly. The aperture is rather wider than in the Kanjera specimen, but of similar height. Perhaps due to this greater width, the length of the muzzle dorsum as measured by Eck is shorter than in M14936, but the overall muzzle length and cranial length are comparable. Midcranial length is again somewhat less in the Swartkrans fossil. As Eck notes, it is not possible to determine on SK 561 whether the nasal bones were raised slightly in the midline. The curve of Spee is rather flat, as in M14936, although the premaxilla appears to rise somewhat, perhaps resulting in the crowding of the upper incisors.

The root of the zygomatic buttress arises just distal to the mesial loph of M^3, while in M14936 it is placed just mesial to that loph. Although both specimens are damaged in the suborbital region, they appear to have been quite comparable in shape, in terms of the slope of the maxilla onto the zygoma. In both crania, the zygomatic arches are placed well out from the vault, although not strongly bowed beyond their anterolateral extent; the temporal fossae are capacious. The Kanjera specimen differs in having a much stronger inferior expansion at the anteroinferior corner of the facial surface of the zygomatic arch, while SK 561 is smoother in that area (although broken just medial to it on both sides). The posterior surfaces of the postorbital plates are damaged, so that it is difficult to discern clearly the extent of the origin of anterior temporalis, but it seems to have been similar to the condition in M14936.

Although much of the orbital margin is damaged in SK 561, the orbits appear to be quite tall, especially compared to M14936. As reconstructed, they are much larger, but there is no clear bony evidence to support this condition. The supraorbital tori are lightly constructed, as in the Kanjera and perhaps Hadar females, but somewhat more robust than in UWMA MP 222 from Makapan. In superior view, they curve convexly anteriorly and present a shallow ophryonic groove, as in M14936. The temporal lines curve sharply up and back from the posterior margin of the zygomatic processes of the frontal. They are at first a bit more obtuse than in M14936, and project laterally over the postorbital constriction (rather than almost vertically), but as they converge toward bregma, they appear more elevated and sharper than in the Kanjera fossil. The region at bregma is damaged in SK 561, but the temporal lines appear to have met near or just posterior to that landmark, continuing as a low ridge which meets the nuchal crest at inion. The nuchal plate is strongly muscle-marked, and the crest may extend up to 7 mm beyond the vault superiorly. Basally, SK 561 is badly damaged, but the mastoid and temporomandibular joint regions appear quite similar to those of M14936, although the postglenoid processes are not as well developed; despite some damage, they appear to have been relatively small. It is worth mentioning here that the partial adult male (?) cranium TMP SK 599 was also reconstructed by Clarke and is illustrated in Delson & Dean (this volume, chapter 4, Figs. 4.9 and 4.10). That specimen appears to have been correctly identified by Freedman (1957) as a 'Dinopithecus' rather than a *Theropithecus*, in part based upon the rather weak postglenoid processes, apparently less developed in this large male than in SK 561.

From Brain's more recent collections, seven isolated teeth and a partial mandibular corpus have been identified as *Theropithecus* (most are illustrated in Fig. 5.3). TMP Skx 9579 from the Orange Breccia of Member 1 is a partial right corpus with worn and damaged P_4–M_3. The distal half of the last molar appears to identify the specimen as a *Theropithecus*. The body is reasonably well preserved and the buccinator channel small. Skx 38,376 is an

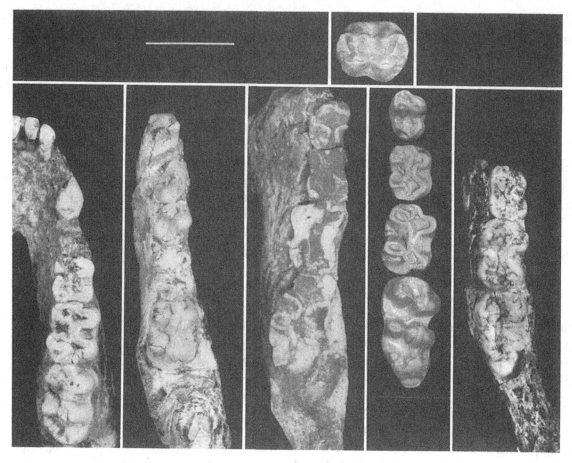

Fig. 5.3. Swartkrans dentitions in occlusal view (of *Theropithecus oswaldi* unless otherwise noted). Left to right: from 'Hanging Remnant', SK 411 (female–right I_{1-2}, P_3, M_{1-3}), SK 426 (subadult male–right P_3-M_2); from Member 1, Skx 9579 (sex indeterminate–heavily worn right P_4-M_3); (top) Skx 32148 (right $M^{2?}$, cast); casts of isolated unsexed teeth, arranged as a toothrow: Skx 38376 (right P_4, Member 1), Skx 2996 (right $M_{1?}$, Member 2), Skx 28812 (right $M_{2?}$, Member 3), Skx 27586/27587 (left M_3, photographically reversed, Member 3); *Papio (Dinopithecus) ingens* (for comparison), from 'Hanging Remnant', SK 404 (right M_{1-3}). Scale bar = 2 cm.

isolated, nearly unworn right P_4 from the Orange Breccia which is tentatively identified as *Theropithecus*. Skx 2996, from the Brown Breccia of Member 2, is an M_1 (or possibly M_2) presenting typical *Theropithecus* trefoil wear, in a corpus fragment. From the Black Breccia of Member

3, there are a number of less worn teeth: Skx 27586/27587 is a nearly unworn, possibly unerupted rootless crown of a left M_3; Skx 28490 is a slightly worn left M_3 crown with some damage to the mesial end (so that length must be estimated); Skx 37323 is a heavily worn right M_3; Skx

28812 is an isolated right M_2 (or M_1); and Skx 32148 is an unworn upper molar crown, probably of a right M^2, which appears referable to *Theropithecus*.

Table 5.2 presents dental data on the new Swartkrans teeth, as well as on the original 'Hanging Remnant' sample of Member 1. The new specimens from Member 3 appear indistinguishable from the earlier sample, which suggests that there is

Table 5.2. *Measurements (in mm) of* Theropithecus *teeth from Swartkrans*

(a) *Swartkrans 'Hanging Remnant' – older Member 1 finds.*

Variable	N	Minimum	Maximum	Mean	SE	CV
I_1W	1			5.7		
I_1LA	1			2.3		
I_1L	1			2.8		
I_2W	1			5.6		
I_2LA	1			2.2		
I_2L	1			3.8		
C_1L FEMALE	1			5.2		
C_1H FEMALE	1			9.9		
C_1W MALE	1			8.5		
C_1H MALE	1			31.5		
P_3W FEMALE	1			5.9		
P_3L FEMALE	1			10.2		
P_3FH FEMALE	1			10.6		
P_3W MALE	2	6.20	7.40	6.80	0.60	12.48
P_3L MALE	2	15.10	15.40	15.25	0.15	1.39
P_3FH MALE	2	21.50	24.50	23.00	1.50	9.22
P_4W	4	8.00	8.80	8.38	0.19	4.61
P_4L	4	9.00	11.00	10.13	0.43	8.43
M_1AW	3	9.40	10.00	9.70	0.17	3.09
M_1PW	3	9.60	10.50	9.97	0.27	4.74
M_1L	3	11.40	12.90	11.93	0.48	7.03
M_2AW	4	12.00	12.80	12.35	0.17	2.69
M_2PW	4	11.50	12.40	11.90	0.20	3.29
M_2L	4	14.60	17.00	15.78	0.58	7.33
M_3AW	5	12.90	16.30	14.04	0.63	10.02
M_3PW	4	11.80	14.80	13.08	0.63	9.62
M_3L	6	18.00	26.60	21.85	1.39	15.60
dP_4AW	1			7.6		
dP_4PW	1			7.8		
dP_4L	1			12.5		
I^1W	2	6.10	6.20	6.15	0.05	1.15
I^1LA	2	4.70	4.80	4.75	0.05	1.49

Table 5.2. *Contd.*

Variable	N	Minimum	Maximum	Mean	SE	CV
I^1L	2	5.60	7.00	6.30	0.70	15.71
I^2W	2	5.70	5.80	5.75	0.05	1.23
I^2LA	2	3.40	3.90	3.65	0.25	9.69
I^2L	2	5.50	5.60	5.55	0.05	1.27
C^1W FEMALE	2	8.30	8.40	8.35	0.05	0.85
C^1L FEMALE	2	8.40	9.00	8.70	0.30	4.88
C^1H FEMALE	1			9.7		
C^1W MALE	3	12.20	12.80	12.50	0.17	2.40
C^1W MALE	3	16.60	17.70	17.17	0.32	3.21
P^3W	5	8.50	10.30	9.34	0.32	7.59
P^3L	6	7.70	9.90	8.38	0.32	9.22
P^3H	6	7.10	9.00	8.23	0.34	10.15
P^4W	5	9.60	10.70	10.26	0.20	4.39
P^4L	7	7.90	10.40	9.21	0.36	10.42
P^4H	5	7.70	11.60	9.94	0.64	14.31
M^1AW	6	10.70	11.70	11.30	0.18	3.84
M^1PW	6	10.00	12.00	10.93	0.28	6.33
M^1L	8	11.90	15.20	13.63	0.45	9.39
M^2AW	10	11.90	16.00	14.12	0.37	8.25
M^2PW	7	11.10	13.70	12.69	0.34	7.17
M^2L	10	14.20	20.00	17.38	0.59	10.65
M^3AW	1			13.9		
M^3PW	1			11.8		
M^3L	1			17.2		
dP^4AW	1			9.0		
dP^4PW	1			8.8		
dP^4L	1			9.8		

(b) Swartkrans 'Orange' – newer Member 1 finds

P_4W	1			8.9		
P_4L	2	11.50	11.70	11.60	0.11	1.34
M_3PW	1			13.1		
M_3L	1			23.3		

(c) Swartkrans 'Brown' – Member 2

M_1AW	1			9.9		
M_1PW	1			9.4		
M_1L	1			12.7		

Table 5.2. *Contd.*

Variable	N	Minimum	Maximum	Mean	SE	CV
M²AW	1			14.3		
M²PW	1			13.2		
M²L	1			17.2		

(d) Swartkrans 'Black' – Member 3

M₂AW	1			12.3		
M₂PW	1			12.3		
M₂L	1			16.2		
M₃AW	3	13.11	14.57	14.01	0.45	5.60
M₃PW	3	11.53	13.76	12.62	0.64	8.85
M₃L	3	22.00	26.38	24.47	1.30	9.17
M²AW	1			14.3		
M²PW	1			13.2		
M²L	1			17.2		

Notes: N: number of measurable specimens; SE: standard error of the mean; CV: coefficient of variation. Values are given separately by sex for canines and P_3s.

For incisors, canines and premolars, L(ength) is always maximum mesiodistal, W(idth) is maximum buccolingual, taken perpendicular to length; therefore, for lower canines, because of their turned placement in the mandible, W is greater than L. Incisor LA is taken at the alveolar plane (=cervical level), as maximal length is only available on unworn teeth whose incisal edges have not been reduced due to attrition. P_3 FH (flange height) is the distance from the cusp apex to the most mesial extent of the enamel along the mesial flange; it is equivalent to L(h) of Freedman (1957) or L of Singer (1962); L here is the maximum mesiodistal length of the tooth, as for Freedman (1957). For molariform teeth, AW and PW, respectively, are taken across the mesial and distal loph(id)s usually at or just apical to the cervix; L, however, is taken at interdental contact points, often estimated due to wear, and decreases significantly in worn teeth.

little time difference between the lower three members (Delson, 1989). Brain (1989) originally suggested that as much as 500 Ka might have separated each of these members, based in part upon preliminary thermoluminescence dating; more recently, however (pers. comm.), he indicated that he and other faunal analysts now agreed with the shorter time frame, implying less than 250 Ka for the total span of Member 1–3 time. The additional cercopithecids from the new Swartkrans sample, mainly *Papio* but also including a large colobine, will be described elsewhere.

Ain Jourdel, Algeria

History of study

The first fossil specimen of a *Theropithecus* ever reported was described by Phillippe Thomas (1884) in a monograph on the freshwater Neogene of Algeria. From the locality of Ain Jourdel, in the Constantine region, Thomas reported three lithologic units: a clayey silt, then a sandy conglomerate, and finally a molasse. In the middle conglomerate, he recovered a single tooth of a cercopithecid, which he briefly described as follows (my translation from his French, p. 14): '. . . a single lower posterior molar . . . This tooth, following M. Gaudry who was kind enough to examine it, indicates a monkey much larger than the modern Barbary Ape of Algeria; moreover, it presents characters which precisely recall those of *Cynocephalus porcarius*, Desm[arest], today restricted to southern Africa. We designate this monkey, provisionally at least, under the name of *Cynocephalus atlanticus*'. Plate IV included a drawing (Fig. 4) of the lateral view of this specimen at natural size; the occlusal view stated to be present also is not seen.

The horizon yielding this specimen was said to be of Pliocene age by Thomas. Arambourg (1969, 1979) discussed the site briefly (without mention of the primate tooth), indicating that it could not be relocated. Geraads (1987) has re-examined the fragmentary materials and suggested that the fauna of the conglomerate level is of later Pliocene age, perhaps about 2.5 Ma. The presence of equids and especially several bovids suggests an open-country environment.

Few researchers before 1974 commented upon the cercopithecid specimen. Hill (1970) discussed the tooth as a lower third molar and reproduced the illustration alongside those of M_3s of a baboon and a mandrill. Although he did not mention the clear lack of a hypoconulid (thus precluding its identification as an M_3), Hill (p. 358) did note the relatively great height of the crown and 'deep vertical grooves between mesial and distal pairs of cusps. This suggests an allocation to the Theropithecini would be more appropriate'. Nonetheless, he listed the taxon as *Papio atlanticus*, as had Romer (1928) in a faunal list. Delson (1974) listed the material as *Theropithecus atlanticus*, but did not provide further details there or in later papers. Geraads (1980), in a report on the fauna (including *Theropithecus*) from the Thomas quarries of Morocco, mentioned that as *T. atlanticus* was the first named species of the genus, it should be considered the senior synonym of *T. oswaldi*. Later, Geraads (1987) discussed the Ain Jourdel specimen as *T. atlanticus*, while identifying the Thomas quarries finds as *T. oswaldi*, without further taxonomic evaluation.

Geraads also questioned the allocation of a partial humerus from the slightly older Tunisian locality of Garaet (or Lac) Ichkeul. Delson (1974) had mentioned the presence of *Macaca* sp. at that site, based upon the humeral fragment in the MNHN-P collections, identified as a canid by Arambourg but never published. Geraads (1987, p. 22) stated that he was 'unable to find any significant difference from *Theropithecus gelada*', but the development of the medial epicondyle closely resemble that of macaques, being far less retroflected than in any humerus of *Theropithecus*. Pickford (see chapter 8) makes the important observation that although both *Macaca* and *Theropithecus* occur in the Pliocene and Pleistocene of the Maghreb, they do not appear to overlap at any site yet known, which probably is due to their differing environmental requirements. This observation was not made explicitly by Geraads (1987), but it is clear from his concluding table. More recently, however, Raynal *et al.* (1990) reported on a faunal assemblage from Ahl-Al Oughlam, near Casablanca (Morocco). This assemblage, estimated to date between 2.5–2.0 Ma, is said to include *Macaca* and a *Theropithecus* whose teeth are smaller and lower-crowned than those of Ternifine.

Description

The Ain Jourdel specimen (see Figs 5.4 and 5.5) is a nearly unworn isolated lower molar, lacking roots and perhaps not fully erupted (but apparently completely developed). It is typical of modern and fossil *Theropithecus* teeth in almost all features,

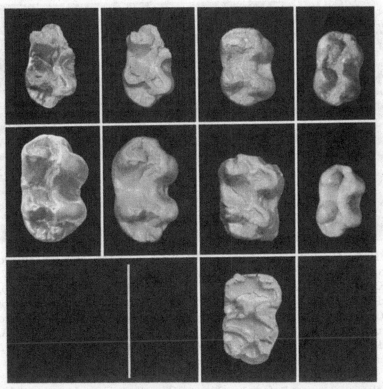

Fig. 5.4. Occlusal views of various *Theropithecus* lower molars. Top row, left to right: *T.* cf. *oswaldi* IMNZ Sn 5A-520, Senga 5A, right $M_{1?}$, original and cast; MNHN-P [AJO 001], *T.* sp. indet. from Ain Jourdel (holotype of '*Cynocephalus atlanticus*') right $M_{1?}$, cast and original. Middle row, left to right: *T.* sp. indet IMNZ Sn 5A-405, Senga 5A, right $M_{2?}$, original and cast; *T.* sp. indet., KNM-LT 417, Lothagam-3, right $M_{2?}$, cast; *T. gelada*, right M_2. Bottom row: *T.* cf. *darti*, KNM-KE 237, Kanam East (West), R M_3, cast. Scale bar = 2 cm.

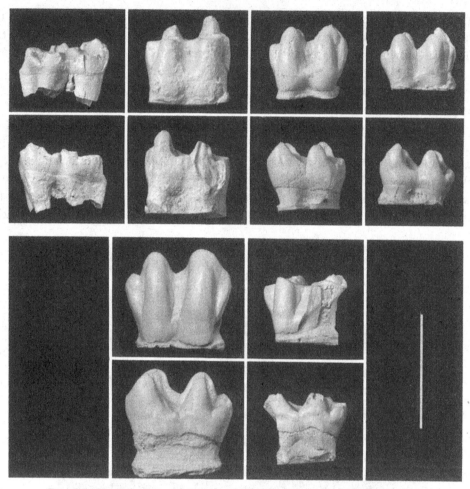

Fig. 5.5. Lateral views of various *Theropithecus* lower molars. Top two rows left to right (buccal above, lingual below): *T.* cf. *darti*, KNM-KE 237, Kanam East (West), R M_3, cast; *T.* sp. indet., KNM-LT 417, Lothagam-3, right $M_{2?}$, cast; MNHN-P [AJO 001], *T.* sp. indet., Ain Jourdel (holotype of '*Cynocephalus atlanticus*') right $M_{1?}$, cast; *T. gelada*, right M_2. Bottom two rows, left to right (buccal above, lingual below): *T.* sp. indet., Senga 5A, IMNZ Sn 5A-405, right $M_{2?}$, cast; Sn 5A-520, right $M_{1?}$, cast. Scale bar = 2 cm.

such as high crown relief; deeply excavated foveas, lingual notches and buccal clefts; trigonid basin (mesial fovea) somewhat short mesiodistally; and median buccal cleft with a flattened base, rather than flowing smoothly onto the buccal surface. In addition, the lophids are angled slightly oblique to the tooth's long axis.

One distinctive feature is that the base of the median lingual notch (trigonid notch) forms an acute angle, rather than having a nearly flat base as in all other *Theropithecus*.

Measurements of this tooth are provided in Table 5.3, along with some comparative data on living and fossil material. The most

Table 5.3. *Measurements (in mm) of Ain Jourdel, Senga, Lothagam and Kanam lower molars and comparative* Theropithecus *lower teeth.*

Species/site	Source	Tooth	Note	AW	PW	Len	AW/L	Wear
Ain Jourdel	Delson	$M_{1?}$		9.4	9.3	14.1	0.67	None
Senga 5A	Delson	$M_{1?}$	broken	mes	10.0	15+		High
Senga 5A	Delson	$M_{2?}$		12.6	12.2	19.3	0.65	None
Lothagam-3	Delson	$M_{2?}$		11.3	10.6	15.1	0.75	Med
Kanam East	Delson	M_3	est.	10.8	9.8	16.0		High
T. gelada	Delson	M_2		8.6	8.6	13.3	0.65	None
	Delson	M_1		6.8	7.2	10.8	0.63	Low
	Delson	M_1		7.0	7.7	9.9	0.71	Med.
	Delson	M_1		7.4	7.5	9.8	0.76	High
	Delson	dP_4		5.3	6.0	8.2	0.65	Med.
	G. Eck,	M_1	Mean	7.4	7.6	10.6		
	1980		Max.	8.0	8.6			
	unpubl.		Min.	6.8	7.1			
	Ph.D.	M_2	Mean	9.2	8.9	13.4		
	thesis		Max.	10.6	9.7	14.4		
			Min.	8.3	7.9	11.9		
T. darti	Delson	M_1		8.7	8.5	12.8	0.68	Low
Hadar	Delson	M_1	same	8.3	8.1	11.2	0.74	Med.
	Delson	M_2	jaw	11.2	9.7	14.3	0.78	Low
	Delson	M_3		9.6	8.4	15.2		Med.
	Delson	M_3		10.8	10.1	15.8		High
	Delson	M_3		10.8	9.1	16.6		Med.
Makapan Grey	Delson	M_3	Min.	12.1	10.8	18.2		High
Kanjera	Delson	M_3		11.7	10.8	19.7		High
T. oswaldi	Leakey,	M_1	Min.	9.9	9.7	14.2	0.70	??
Okote horiz.	chapt. 3	M_1	Max.	10.9	11.2	14.2	0.76	??
T. brumpti	Eck &	M_1		9.4	9.6	12.9	0.73	Med-
Shung. C	Jablonski, 1987							High
Shung. G		M_1						High
" Omo 75-C40	Delson	M_2	cast	13.0	12.3	18.6	0.70	Low
KF 102	Delson	M_1		10.3	10.2	14.4	0.72	N0 ?
(horizon?)								

Notes: Delson observations are on original specimens only, unless otherwise indicated. See Table 5.2 for measurement definitions. The first four specimens are fossils whose allocation to species is still uncertain.

striking feature of the tooth metrically is its relative narrowness, even when compared to specimens which are nearly as little worn (and thus not shortened by interproximal wear). The proportions of the specimen are most comparable to an unworn M_2 and little-worn M_1 of *T. gelada*, but the size is greater; dP_4s are especially narrow across the protolophid, while M_2s of the other fossils are relatively wider. Overall size and proportions are comparable to those of selected M_1s of later Pliocene (and early Pleistocene?) *Theropithecus* species, but the Ain Jourdel tooth is distinguished, as noted, by the shape of its median lingual notch and to some degree its narrowness. It is compared in Fig. 5.4 with the unworn M_2 of *T. gelada* and a variety of other lower molars of the fossils discussed below. It is not readily possible to allocate this single tooth to any of the known species, although it appears most similar to early *T. darti*. Perhaps when the Ahl-Al Oughlam specimens are described, they will include material closely comparable to the Ain Jourdel tooth, but until then the latter must be considered indeterminate as to species.

Lothagam Hill, Kenya

Patterson, Behrensmeyer & Sill (1970) first reported a faunal assemblage from Lothagam Hill; *Simopithecus* sp. was included in the faunal list. Smart (1976) discussed the fauna of the lower horizon, Lothagam-1, in some detail, and Behrensmeyer (1976) provided an overview of the whole sequence, in regional context. The upper fossiliferous horizon,

Lothagam-3, which included the '*Simopithecus*', was estimated on faunal grounds to date around 4 Ma, perhaps slightly older than Ekora and equivalent to Kanapoi. A volcanic sill intruded underneath the Lothagam-3 fauna was dated to about 3.73 Ma, while the basalt apparently lying stratigraphically between Kanapoi and Ekora was dated between 2.5–4.0 Ma; the latter date (although incompletely published) was generally accepted. Hill & Ward (1988) recalculated the sill date to 3.8 ± 0.2 Ma, but otherwise followed the earlier interpretation. T.D. White (pers. comm.) indicates that the fauna of Lothagam-3 appears older than that of Hadar (i.e. >3.35 Ma), but is unlikely to be older than about 4 Ma.

A single isolated lower molar, catalogued as KNM-LT 417, represents *Theropithecus* from this locality. This tooth (see Figs 5.4 and 5.5) is moderately worn, the roots are damaged, and there is a large flake of enamel missing from the entoconid and a sliver of enamel from the mesial face of the metaconid. A large mesial contact facet compresses the trigonid basin, but the distal contact facet is faint if present at all. The wear pattern, deep and flattened lingual notch, and somewhat flattened base of the buccal cleft all combine to identify the tooth as belonging to a member of the genus *Theropithecus*. Its proportions (see Table 5.3) suggest that it is probably a M_2, although it is possible that it might be a M_1; it is too broad for a dP_4. The size is roughly comparable to those of M_2 from Hadar allocated to early *T. darti* (see also Eck, chapter 2), and its crown pattern is at least as complex as those. This is interesting, in light of the low degree of crown complexity seen in the earliest members of the *T. brumpti* lineage farther north in the Turkana Basin (see Leakey, chapter 3, and discussion in Delson & Dean, chapter 4). Eck has suggested that molar size reflects age in *Theropithecus* samples, but the LT 417 basal length (Eck's variable 68) is about 15 mm, a value equalled by M_2s from Pliocene through early Middle Pleistocene sites; clearly the technique is designed to work only with a sample of several teeth. Until more than a single tooth is known (and the age and even toothrow position clarified), it is not feasible to assign this specimen to a named species. Its crown complexity suggests that it does not belong to an early member of the *T. brumpti* lineage, however.

Kanam East, Kenya

The Kanam localities in western Kenya were first discovered by Louis Leakey in the early 1930s, and their geology, faunas, and age have been the subject of much discussion since (see review in Pickford, 1987). Several cercopithecid teeth and jaw fragments collected by Leakey were deposited in the BM(NH); they were mentioned by Szalay & Delson (1979) and Delson (1984) and are being described by Harris & Harrison (1991). In the KNM collections are several additional specimens, including two molars identified as *Theropithecus*, which M.G. Leakey has kindly permitted me to describe here. Both are catalogued as KNM KE 237, indicated as from Kanam East (West). The exact age of these deposits is still uncertain, although at Kanam West

(whose relationship to the Kanam East subsites is uncertain) Pickford (1987) indicated the presence of a late Late Pleistocene horizon (the Apoko Beds) and an earlier Pliocene one (the Kanam Beds). More recently, Plummer & Potts (1989, who also provided a good location map of the region) reported that at nearby Kanjera, the Apoko Beds may range back as far as latest Early Pleistocene, while the overlying Black Cotton Soil horizon is late Late Pleistocene. It seems most likely that the cercopithecids from Kanam East are of earlier Pliocene age, and perhaps the *Theropithecus* teeth can throw light on that question.

The two molars are heavily worn and might possibly derive from a single individual. One tooth is clearly a right M_3, whose cusps have been worn down so that only large transverse 'lakes' of dentin are visible (see Fig. 5.4). The deep bases of the two lingual notches are still unworn, as is the flattened base of the median buccal cleft (Fig. 5.5). This wear has left the lingual face of the tooth much higher above the cervix than the buccal side, on which enamel has been worn away around the base of the protoconid and hypoconid, as well as from a large area at the distal base of the hypoconulid. The outline of the tooth as preserved is rather 'blocky': nearly quadrangular, rather than elongate, with no obvious narrowing of the hypoconulid heel. Some Hadar and Makapan teeth are similar in shape. Measurements are provided in Table 5.3.

The second tooth is even more worn, but I identify it as an upper ?first molar, probably of the left side (Fig. 5.6). Enamel is flaked off around most of the crown

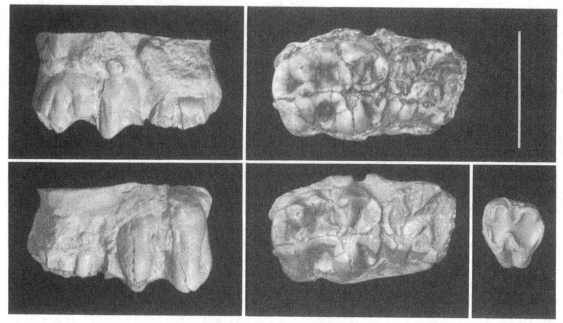

Fig. 5.6. Occlusal and lateral views of various *Theropithecus* upper molars. *T. oswaldi delsoni* from Mirzapur, India (PUC-GM A/643, holotype of '*T. delsoni*'), right M$^{2-3}$, buccal above lingual views of cast, occlusal view of original above cast. Lower right, *T.* cf. *darti* KNM-KE 237, Kanam East (West), R M^1, cast, occlusal.

base, except at the median buccal and lingual areas and buccally at the ?metaconid. This renders measurement most difficult, and only rounded estimates of width and length are provided in Table 5.4.

The overall size of the teeth is small. The M$_3$ falls well within the range for Hadar specimens as measured by me or by Eck (see chapter 2). Makapan teeth are larger, even at a similar wear stage, as are those from Kanjera, Swartkrans, and later populations. The basal length of the tooth is about 14.7 mm, a value matched by Eck only in Hadar Sidi Hakoma (older) specimens, and still well below the mean. The upper molar is less readily placed. As seen from Table 5.4, several modern *T. gelada*

are only slightly smaller, while selected Hadar and Makapan specimens are slightly larger. It is unlikely that the specimens represent *T. gelada*, given the blocky shape of M$_3$, and thus an identification as *T.* cf. *darti* appears most reasonable. In turn, this tends to confirm the earlier Pliocene age.

Senga 5, Semliki, Zaire

Harris *et al*. (1987) described the geology, paleontology, and archeology of the Senga 5A site, in the Lusso Beds of the Upper Semliki Group, along the Semliki River in western Zaire. Various lines of faunal evidence · suggested an age between 2.35–2.0 Ma. A more detailed analysis is

Table 5.4. *Measurements (in mm) of* Theropithecus *upper molars from Kanam, Mirzapur and Comparisons.*

Locality	Source	M^1	AW	PW	L	
Modern	Delson		9.3	8.3	11.3	
Modern	Delson		9.6	8.3	11.3	
Kanam East	Delson	(?)	10.	10.	11.	(all estimated)
Hadar	Delson		10.0	9.9	11.7	
Mak* Mb 4	Delson		9.8	9.5	11.9	

Locality	Source	M^2	AW	PW	L	M^3	AW	PW	L
Mirzapur	Delson		14.0+	13.3	14.6+		17.3	16.3	20.8
Olorge-	Jolly,	Min	15.0	13.5	18.7	Min	15.7	13.8	18.6
sailie	1972	Max	19.0	17.4	22.3	Max	18.0	15.9	22.5

* Makapansgat Member 4 (grey breccia)

being prepared for publication by Boaz *et al*.

The original faunal list for Senga 5A included *Theropithecus* sp., and Dr Noel Boaz has recently requested that I describe the two teeth identified as that taxon. Although they are currently identified only by field numbers, they will eventually be catalogued in the collections of a museum under the authority of the IMNZ. Specimen Sn 5A 405 is a nearly complete, unworn crown of an M_2 (or possibly M_1), lacking the roots and some of the enamel above the cervix except on the buccal side. It is of typical *Theropithecus* morphology, with high occlusal relief (tall cusps and deep basins and lingual notches), projecting mesial and distal shelves and a small cuspule tending to flatten the base of the median buccal cleft (see Figs 5.4 and 5.5). Sn 5A 520 is smaller, more worn, and more heavily damaged: the mesial quarter is broken away, and enamel is missing

buccally almost to the median cleft. Wear has exposed subcircular dentin pits on all four major cusps, but this confirms the identification of the tooth as *Theropithecus*. Neither tooth appears to be elongate enough to be identified as a dP_4, and as they are from the same general horizon, it is likely that the larger specimen is an M_2, while the smaller is an M_1. It is conceivable that they are from the same individual, but this is not significant here. Measurements of the teeth are provided in Table 5.3.

Comparison with the data for Swartkrans teeth in Table 5.2 reveals an overall similarity of size, especially in width; the unworn Senga tooth is longer than its generally more worn Swartkrans counterparts, and both are larger overall than the Kanjera specimens summarized by Jolly (1972). Leakey (see chapter 3) reported metric data for a variety of *T. oswaldi* specimens, Delson & Hoffstetter (see chapter 6) give measurements for the

Ternifine sample and Eck (1987) provided them for the few Shungura specimens. Clearly the Middle Pleistocene samples (Olorgesailie and Ternifine, especially) average much larger than the Senga teeth, although there is some overlap with the smallest individuals. Teeth of *T. oswaldi* from the Koobi Fora Okote horizon are perhaps closest overall, but those from the KBS and Upper Burgi levels (and the correlative Omo E–G) are also similar, if not as long for their breadth. Some of this discrepancy may be due to the unworn condition of the larger Senga tooth, but it may also be slightly more elongate than those from other sites.

Eck & Jablonski (1987) and Leakey (see chapter 3) also provide data for Turkana Basin *T. brumpti*, some of which show close similarity in size to the Senga teeth. I have measured a cast of one Shungura juvenile female, tentatively identified as *T. brumpti*, which is also morphologically similar (Omo 75-C40, Table 5.3). Given the difficulties in distinguishing isolated teeth of the two species (Eck & Jablonski, 1987), no further attempt is made here. Eck (see chapter 2; also 1987) does suggest that the basal (cervical) length of lower molars can be used as a guide for comparing samples, but the value obtained for Sn 5A 405 (16.1 mm) falls into the overlap zone for all populations between Makapansgat and Olorgesailie. Overall, it is likely that a time range equivalent to the middle Shungura Formation, perhaps 2.4–2.0 Ma, would be reasonable for the Senga 5A teeth; they may be relatively elongate compared to their contemporaries.

Mirzapur, India

History of Study

V.J. Gupta (1977) first mentioned the possibility of *Theropithecus* from India in his description of a fragment of right maxilla with two teeth. The specimen, PUC-GM A/643, was reported to have come from 'the Lower Boulder Conglomerate Formation exposed 1.4 km WWS [sic] of Mirzapur Rest House (76° 43′ 44″ : 30° 54′ 10″), Kharar Tehsil, Dist. Ropar, Punjab' (p. 450). Although papers by E. Khan discussing the geology of this area were cited, there was no indication as to who had actually found the fossil. Gupta identified the teeth in the maxilla as questionably M^{1-2} and indicated that they closely resembled Kenyan *Theropithecus* as described by Jolly (1972).

In 1981, Gupta & Sahni reported that the maxilla was undoubtedly of a large *Theropithecus*, which they named *T. delsoni*. The teeth were reidentified as M^{2-3}, based upon 'further examination of the maxillary fragment' (p. 70), without specifying details. The new species was diagnosed mainly as a large-sized cercopithecine, with M^3 robust and slightly larger than M^2, overall size significantly larger than *T. darti* and *T. oswaldi*, approaching the dimensions of *T. brumpti*; and accessory cuspules better developed than in the African species.

Pickford (see chapter 8) has added to this history by reporting that it was he who first located the specimen in the artiodactyl collections in Chandigarh and brought it to Gupta's attention. I can further add that in 1980 Sahni visited me in New York to discuss the fossil but was unable to show it to me because it had been packed preparatory to his return to India. On examining a number of casts of cercopithecid dentitions, he indicated that the Indian fossil (of which I was unaware at the time) was most similar to *T. oswaldi* specimens. I was astounded by this idea but convinced by the cast he sent me on his return and much honoured by his nomenclature.

As discussed by Pickford (see chapter 8), doubt has recently been raised about the correct provenance of numerous fossils described by V.J. Gupta and a variety of international colleagues (see Talent, 1989; Krishtalka, 1989, who focused on this specimen). In the light of this problem, I contacted Prof Ashok Sahni who most graciously provided both a copy of the original catalogue entry for this specimen and the specimen itself for further study. The fossil is one of a group collected in May, 1959, by Dr Ehsanullah Khan, a respected scientist whose collections have never been questioned; the date was some six years prior to Gupta's earliest publications. The catalogue entry indicates that the maxilla (identified as a suid molar) was recovered seven-eighths of a mile west-west-south (sic) of Mirzapur Resthouse. On the same page are listed a suid ramus from half mile south and two *Equus* skulls from one mile south-east of the Resthouse. Thus, the provenance of this fossil does not appear in further doubt.

The age of the specimen however, is less clear. It is indicated as Middle Pleistocene by Gupta & Sahni, but it is not even certain what definition they are using. Azzaroli & Napoleone (1979) discussed

paleomagnetic chronology of the Chandigarh region, reporting the possible presence of the Jaramillo subchron near the Pinjor/Boulder Conglomerate 'boundary', which is poorly defined in any case. Unfortunately, the Mirzapur area is just off the map (to the west) given by Azzaroli & Napoleone and even farther away from the more eastern region mapped by Tandon *et al.* (1984), in their paleomagnetic study. Combining the magnetic and stratigraphic data suggests that the Mirzapur area might be younger than 1 Ma. If the Middle Pleistocene is accepted to begin roughly at 0.75 Ma (see Delson, 1988), this specimen is most likely of later Early Pleistocene or Middle Pleistocene age. Azzaroli (1985) discussed faunal interchange in the Siwalik sequence and noted that bovids, suids, and other taxa of African affinity appeared in India mainly after about 2.5 Ma, with movement probable in both directions. He negated later Pleistocene connections, but this specimen may serve to suggest that those were indeed possible.

Description

Figure 5.6 illustrates the original Mirzapur maxilla and a cast (to avoid the loss of detail due to discoloration). Measurements of the Mirzapur teeth are provided in Table 5.4, with comparative data on Olorgesailie upper teeth taken from Jolly (1972; more extensive samples from Olorgesailie could not be included because Leakey & Leakey [1973] only gave proportional width values, not raw measurements, while Leakey [this volume, chapter 3] only provided measurements of lower dentitions). There is no sign of a distal facet on the larger tooth, which may have been the reason for Gupta & Sahni (1981) to suggest that the Mirzapur specimen likely contains M^{2-3}. The anterior tooth is heavily worn, and as they noted, the mesial margin is completely worn away, so that only a minimum estimate of both length and breadth is possible. I therefore agree with their identification.

The description of the specimen given by Gupta & Sahni is quite accurate and detailed and need not be repeated. Their suggestion that it presents more accessory cuspule development than in African *Theropithecus* does not seem however, to be valid. There are several small cracks in the enamel of both teeth which have led to some expansion in dimensions. Nonetheless, in terms of size, the Mirzapur teeth are clearly among the largest upper molars known for *Theropithecus*, within the range for Olorgesailie but larger than the few Ternifine teeth (Delson & Hoffstetter, see chapter 6) or other samples reported by Jolly (1972). They are also far larger than the teeth of *T. brumpti* reported by Eck & Jablonski (1987). Arambourg's (1947) original description of the 'type series' of *T. brumpti* from Omo gave 20 mm as the length for the incompletely erupted M^3 of that Eck & Howell (1982) chose as the species lectotype, and it may be this high value to which Gupta & Sahni (1981) referred in their diagnosis (see above). In fact, however, my estimate of M^3 maximum length is only 18.8 mm for this specimen, although my other measurements for it accord exactly with those given by Arambourg. The systematic position of the Mirzapur specimen will be considered further in the final section of this paper.

Taxonomy of *Theropithecus*

Genus-group systematics

The problems of specific and generic taxonomy for *T. gelada* and its extinct relatives have burdened researchers for over 40 years. Jolly (1972) provided a major advance by demonstrating the generic identity of *Theropithecus* and *Simopithecus*, the name which had previously been applied to extinct species only and which Jolly relegated to subgeneric status. Szalay & Delson (1979) accepted this system, recommending that a new subgenus might prove necessary to receive the highly distinctive *T. brumpti*. Dechow & Singer (1984) also formally accepted two subgenera. On the other hand, Maier (1972; commenting upon preliminary work by Jolly) and Freedman (1976) retained *Simopithecus* and *Theropithecus* as distinct genera, citing mainly potential problems with phylogeny, rather than evaluating the set of clearly derived features shared by the two. More recently, Eck & Jablonski (1987, p. 100) argued that all of the relevant species should be included in *Theropithecus* without further genus-group subdivision, this being 'a small and tightly-knit group, ... already overburdened with taxonomic terms'. They further elaborated upon their previous arguments (Eck & Jablonski, 1984) that three phyletic lineages can be discerned within *Theropithecus*: *T. gelada*, *T. oswaldi* (with *T. darti*) and

T. brumpti (with *T. baringensis* and
T. quadratirostris). The generic unity of
Theropithecus appears to have been
accepted by Leakey (see chapter 3) as
well, as no mention of the term *Simopithecus* appears in her paper.

Delson & Dean (see chapter 4)
recognized the same three lineages within
the genus. However, they argued that the
holotype crania of *Papio baringensis* R.
Leakey, 1969 and *Papio quadratirostris*
·Iwamoto, 1982 were essentially of conservative, *Papio*-like morphology. After
considering extrinsic evidence as well,
they retained Iwamoto's species in *Papio*,
including it in the subgenus *P. (Dinopithecus)*. However, on the basis of dental
evidence, including similarity of both the
holotype and another Chemeron toothrow
to newly described jaws of early *Theropithecus* from the Turkana Basin (Leakey,
see chapter 3), they accepted the possible
placement of ?*T. baringensis* as an early
member of the *T. brumpti* lineage. This
species would still have retained a conservative *Papio*-like cranium and molars
of relatively low complexity; only later
T. brumpti developed complex, high-
crowned molars, but never a relatively airorhynch facial skeleton such as is seen in
T. oswaldi and *T. gelada*. Such a relationship, combined with the more complete
development of *Theropithecus*-like dental
and cranial features in Hadar *T. darti*,
implied to them that the main division
within *Theropithecus* was between the
T. brumpti lineage and all other species,
including the modern form.

Delson & Dean had suggested this pattern as a working hypothesis from the outset, because it had been recognized earlier
by Eck & Jablonski (1984, 1987). Eck (this
volume, chapter 2) differentiates *T. darti*
from *T. oswaldi* and that lineage from
other species (see below), but he does not
comment on inter-lineage relationships,
nor does Leakey (see chapter 3). It is
important to realize that retention of a
suite of conservative features (as in the
cranium of *T. gelada* or the dental
similarities between *T. gelada* and
T. brumpti) or a set of autapomorphies (as
is true for each of the three lineages and
their terminal species at least) does not aid
in the recognition of synapomorphies
between clades which can be used to
determine phylogeny in a three-taxon
problem (Hennig, 1966; Eldredge & Cracraft, 1980).

Delson & Dean further suggested that
within Papionini, the closest relative of
Theropithecus is *Papio*, including three
lineages within that genus: savannah
baboons (subgenus *Papio*), mandrills and
drills (subgenus *Mandrillus*) and an
extinct group of small-brained but
relatively mandrill-like species (subgenus
Dinopithecus). Many of the features seen
in mandrills are shared also by *Theropithecus* species, and it was suggested that
the former are cranially the most conservative living 'baboons', setting aside
their obvious autapomorphies.

If the relationships as determined by
Delson & Dean (and by Eck & Jablonski,
1987) are accepted, there are three main
branching nodes within the *Theropithecus*
clade: first, between a mandrill-like common ancestor and *Theropithecus*; second,
between the *T. brumpti* lineage and the

two others; and third, between the *T. gelada* and *T. oswaldi* lineages. The first node would be characterized by the development of relative molar complexity and incisor reduction. The second node would document the beginning of air-orhynchy and orthognathy (increasing midface height, upturned premaxilla, rounded neurocranium) in the *T. gelada/oswaldi* clade. The third node would be characterized by a 'canalization' of further anterior tooth reduction and size increase in the *T. oswaldi* lineage. The development of a reversed curve of Spee and increased molar complexity in *T. brumpti* and *T. oswaldi* (but not the earliest members of the former lineage especially) would presumably have been convergent. This hypothesis of relationships is depicted in Fig. 5.7. It can be tested in more detail through analysis of these characters and others discussed in Delson & Dean (see chapter 4), Jablonski (see chapter 7) and Eck & Jablonski (1987), but many of these characters have not been subjected to polarity analysis, and the necessary work is beyond the scope of this chapter. Other traits characteristic of each of the three main lineages are summarized by Jablonski (see chapter 7).

Either or both of the intra-*Theropithecus* nodes could be named at supra-generic level. If both are recognized, it would be best to place the *brumpti* lineage in a new genus, while recognizing *T. (Simopithecus)* for the *oswaldi* lineage. In view of most previous arguments, this would seem to be 'taxonomic overkill'. However, in order to emphasize the closer relationship between the *T. gelada* and

T. oswaldi lineages argued here and by Delson & Dean, I propose that a new subgenus be recognized formally for the *T. brumpti* lineage, as originally suggested but not named by Szalay & Delson (1979). The genus-group systematics of *Theropithecus* would thus be as follows:

Theropithecus I. Geoffroy, 1843

(= or including *Macacus* auctorum: Rüppell, 1835, in part. *Gelada* Gray, 1843. *Simopithecus* Andrews, 1916. *Theropythecus* Vram, 1922, lapsus?. *Papio* Erxleben, 1777: Broom & Jensen, 1946; Buettner-Janusch, 1966; in part. *Dinopithecus* Broom, 1937: Arambourg, 1947; Broom & Hughes, 1949; in part. *Brachygnathopithecus* Kitching, 1952, in part. *Gorgopithecus* Broom & Robinson, 1949: Kitching, 1953, in part.)

Included subgenera: *T. (Theropithecus)* and *T. (Omopithecus)*, new.

Type species: *T. gelada* (Rüppell, 1835).

Distribution: Pliocene to later Middle Pleistocene; South Africa, East African Rift Valleys, Maghreb. ?Later Early Pleistocene or Middle Pleistocene; northern India. Modern; central Ethiopian plateau.

Diagnosis: A medium-sized to large papionin cercopithecine generally characterized by such features as: anterior union of the temporal lines and sagittal crest long, if present; marked postorbital constriction and relatively small brain size; at least some development of airorhynchy* relative to African papionin common ancestor; inferiorly divergent lateral margins of the frontal processes of the zygomatic bones and some lateral encroachment of the malar into the orbital cone posteriorly; large temporal fossae and

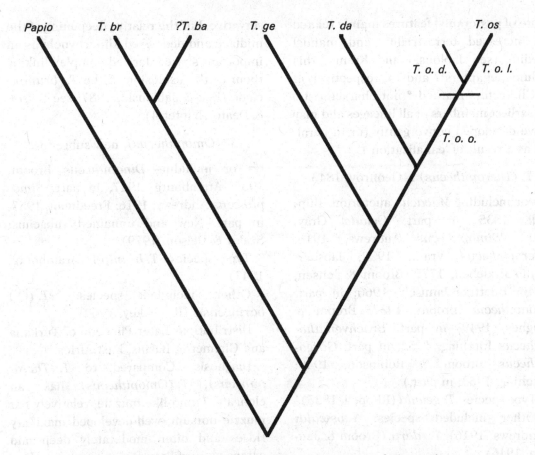

Fig. 5.7. Cladogram of relationships among the species of *Theropithecus*. *T. br* = *Theropithecus brumpti*; *?T. ba* = *?T. baringensis*; *T. ge* = *T. gelada*; *T. da* = *T. darti*; *T. os* = *T. oswaldi*. At upper right, thin horizontal line represents 'transition' between *T. oswaldi oswaldi* (*T. o. o.*) and *T. oswaldi leakeyi* (*T. o. l.*), while thin oblique line represents Asian *T. oswaldi delsoni* (*T. o. d.*).

bowed zygomatic arch, with the root of its maxillary buttress usually posterior to M^2; rounded neurocranium; strong elevation of the temporomandibular joint above the occlusal plane, enlarged postglenoid process and a tall, upright mandibular ramus; a long mandibular symphysis, with buttressed anterior inferior margins of the ramus; relatively reduced incisors (and a shallow maxillary incisive arc); molari-

form teeth with relatively high relief* (deep notches and clefts, tall, pinched cusps and well-developed mesial and distal foveae), trigonid basin (mesial fovea) somewhat short mesiodistally, median buccal cleft with a flattened base, rather than flowing smoothly onto the buccal surface, and M_3s with consistently large hypoconulids; a reversed curve of Spee*; delayed eruption of distal molars; and a suite of postcranial features mainly related to increased terrestriality and manual feeding (see Jablonski and Krentz, this volume, chapters 7 and 14, respectively).

Characters marked * not characteristic of earliest members of all lineages and may have developed convergently (or in parallel as a result of 'canalization').

T. (Theropithecus) I. Geoffroy, 1843

(= or including *Macacus* auctorum: Rüppell, 1835, in part. *Gelada* Gray, 1843. *Simopithecus* Andrews, 1916. *Theropythecus* Vram, 1922, lapsus?. *Papio* Erxleben, 1777: Broom & Jensen, 1946; Buettner-Janusch, 1966; in part. *Dinopithecus* Broom, 1937: Broom & Hughes, 1949; in part. *Brachygnathopithecus* Kitching, 1952, in part. *Gorgopithecus* Broom & Robinson, 1949: Kitching, 1953, in part.)

Type species: *T. gelada* (Rüppell, 1835).

Other included species: *T. oswaldi* (Andrews, 1916); *T. darti* (Broom & Jensen, 1946).

Distribution: Pliocene to later Middle Pleistocene; South Africa, East African Rift Valleys, Maghreb. ?Later Early Pleistocene or Middle Pleistocene; northern India. Modern; central Ethiopian plateau.

Diagnosis: Compared to *T. (Omopithecus)*, *T. (Theropithecus)* has: a relatively airorhynch skull; less elongate muzzle, but deep midface; weak maxillary ridge development and rounded muzzle dorsum; superior margin of supraorbital torus with shallow depression; zygomatic arch only moderately robust; neurocranium rounded, especially in posterior region.

Many of the shared features of the *T. gelada* and *T. oswaldi* lineages are conservative, but the relative deepening of the midface and decreased klinorhynchy is an important shared derived complex linking them to the exclusion of the *T. brumpti* clade (Eck & Jablonski, 1987; see Delson & Dean, chapter 4).

T. (Omopithecus), new subgenus

(= or including *Dinopithecus* Broom, 1937: Arambourg, 1947, in part. *Simopithecus* Andrews, 1916: Freedman, 1957, in part. New and unnamed subgenus: Szalay & Delson, 1979).

Type species: *T. brumpti* (Arambourg, 1947).

Other included species: ?*T. (O.) baringensis* (R. Leakey, 1969).

Distribution: Later Pliocene of Turkana and Chemeron Basins, East Africa.

Diagnosis: Compared to *T. (Theropithecus)*, *T. (Omopithecus)* has: an elongate, *Papio*-like muzzle, relatively flat muzzle dorsum, well-developed maxillary ridges and often moderately deep and extensive maxillary fossa; convex surface of the nasal aperture; generally large and broad malar with anteroinferior prolongation; robust zygomatic arch, triangular in cross-section; generally deep mandibular corpus fossa; well-developed, sinusoidal mental ridge and strongly rugose mental protuberance; and postcranial features reflecting a trend (in *T. brumpti*) to increased elbow stability but shoulder flexibility (detailed by Jablonski and Krentz, chapters 7 and 14, respectively).

The development of facial fossae is similar to that seen in *T. gelada*, while the mandibular mental protuberance is similar to that of *T. oswaldi*; other similarities and differences are discussed by Eck & Jablonski (1984, pp. 126–7; 1987, pp. 102–4). The reversed curve of Spee and complex molars seen in later forms are not present in early members of this lineage.

Species and subspecies in the *T. oswaldi* clade

Eck (see chapter 2) describes the sample of *Theropithecus* from Hadar for the first time and formally includes it in *T. darti*, a species which he distinguishes from an unsubdivided *T. oswaldi*. Leakey (see chapter 3) takes a more conservative view, recognizing only one species on this clade, *T. oswaldi*, with three temporal subspecies. Earlier workers (e.g. Leakey & Whitworth, 1958; Jolly, 1972; Freedman, 1976; Szalay & Delson, 1979) grouped the species variably, often recognizing subspecies on the basis of differing combinations of morphology (especially size), time, and geography.

As a result of my own study of the Hadar (and Makapansgat) *Theropithecus*, I tend to agree with Eck that these two samples are broadly similar (and more conservative than *T. oswaldi*) in relative incisor and canine reduction, enlargement of the postglenoid process, and a number of other, less marked features. Many of these, if not all, are probably closely related to the body size increase which characterizes the history of the *T. oswaldi* lineage. From a theoretical or methodological standpoint, Leakey's argument is more valid, to unite the members of a lineage unless there is evidence of a cladogenetic speciation event, but as yet there is no overlap between the latest *T. darti* and the oldest *T. oswaldi*. None of the alternative taxonomic possibilities – such as recognizing each sample or group of contemporaneous populations as a distinct species; separating only Hadar sample, while including the Makapansgat one in *T. oswaldi* as it already presents more derived conditions than the Hadar monkeys; or including all the fossils in a monolithic *T. oswaldi* – appear as reasonable in this case as the proposals of Eck or Leakey. For the moment, I accept the specific distinction of *T. darti*, but I realize this may be arbitrary and look forward to data on *Theropithecus oswaldi* in the 3.0–2.5 Ma time range in order to test the proposition further.

The Ain Jourdel tooth might fall near the end of this range, and the difficulties found above in allocating the specimen to one of the known taxa may suggest that some speciation occurred in the latest Pliocene. Details on the Ahl-Al Oughlam sample of apparently similar age might clarify this situation. If the Ain Jourdel specimen is referred to any of the known, named species, then Geraads' (1980) comment about the name *Theropithecus atlanticus* having priority must be evaluated. The name was originally proposed provisionally (see above), but that only makes a name unavailable if proposed after 1930, according to the International Code of Zoological Nomenclature. It seems unlikely (although conceivable) that the minor distinction of the shape of the base of the median lingual notch would be reflective of species distinction for the Ain Jourdel tooth and the population from which it was drawn. Rather than upsetting broadly accepted nomenclature, I will request that the International Commission on Zoological Nomenclature suppress the nomen *Papio atlanticus* Thomas, 1884, for the purposes of the Law of Priority. Pursuant to the Code, current terminology may be continued pending a ruling.

In the taxonomy accepted here, *T. oswaldi* is distinguished from *T. darti* on the basis of larger size, taller post-glenoid processes, and (greater) relative reduction of incisors and canines. Within *T. oswaldi* as defined in this way, Leakey (see chapter 3) has recognized (but not diagnosed) two time-successive subspecies: *T. o. oswaldi* and the later, larger *T. o. leakeyi*. Although previous workers (see above) have recognized many more subspecies for the same group of populations, it now seems more reasonable to accept a reduction to two in Africa. Present evidence reveals that the East African representatives of *T. oswaldi* were widespread at any given time, and due to the strong similarity in morphology and size of contemporaneous populations (see also Delson & Hoffstetter, chapter 6), it is likely that the South African and North African populations were in contact with the better-known eastern forms. The distinctions (in dental size and proportions, for example) seen among such penecontemporaneous populations as those from Swartkrans and Kanjera, or those from Hopefield, Olorgesailie, and Ternifine, appear to be less than those discernable today among the varieties of savannah baboons which I recognize as subspecies. Thus, the extinct gelada subspecies would seem to have had larger ranges than modern savannah baboon subspecies.

The recognition of temporal subspecies among fossil samples always raises special problems (cf. Tattersall, 1986 and Delson, 1990, for example). I argue that such a step should proceed from a modern analogy involving closely related forms, in this case the savannah baboons. Jolly & Brett (1973) have recommended that all living baboons be classified as subspecies of the single species *Papio hamadryas*, and Szalay & Delson (1979) have gone further to recognize the small 'Rhodesian' population usually considered as a variety of the yellow baboon as the distinct *P. hamadryas kindae*. The six modern subspecies thus recognized vary in their degree of distinction from one another, but at most of the population contact areas, there is a hybrid zone. Szalay & Delson also recognized at least one extinct subspecies, the South African Plio-Pleistocene *P. hamadryas robinsoni*, reducing it from its original species rank.

More recently, Hayes, Freedman & Oxnard (1990) have reviewed this problem (although they ignored Szalay & Delson's alternative), utilizing multivariate analysis of dental measurements to seek a solution. As they argued for full species separation of at least the three studied populations (*Papio* [*h.*] *ursinus* from southern Africa, *P.* [*h.*] *cynocephalus* from Malawi and Zambia, and *P.* [*h.*] *anubis* from Kenya), a brief survey of their results is in order. In a principal components analysis, it was recognized that the strong component 1 (69 per cent of variation in males, 74.6 per cent in females) was size-dependent, with the large *anubis* and *ursinus* samples separating from the smaller *cynocephalus*, and loadings showed that other components depended heavily on incisor and canine dimensions. A canonical variates analysis yielded less readily interpretable differences, but a dendrogram based on the intergroup distance coefficients again clustered the *ursinus* and *anubis* populations to the exclusion of *cynocephalus*. This suggests that the results were based almost entirely on size discrimination, not on tooth shape, as desired. There is no question that size may be different among the (sub)species of this group, but at what rank are such differences to be recognized. The small number of hybrids used in this study makes that portion of it less important, and their discussion of hybrid zones and interbreeding is not convincing. It is possible, and in fact likely, that at least some of these baboon populations are undergoing evolutionary change now, but the overall pattern seems to be one of partial interbreeding at the boundaries between neighbouring subspecies. Subspecies are expected to be temporary phenomena, compared to the geographic and chronological range of a species, and the modern baboon populations appear to fit that model well, as do the extinct samples of *Theropithecus*.

Leakey (see chapter 3) does not precisely diagnose the subspecies that she recognizes, but her data suggest that a division is possible roughly between (1) most Turkana Basin specimens and those from Kanjera, Swartkrans, Peninj, and Olduvai Beds I-lower II on the one hand and (2) those from Olduvai upper Bed II and above, Olorgesailie, Kapthurin, Hopefield, Ternifine and Thomas Quarries on the other. The youngest reasonably extensive sample from the Turkana Basin

sequence, that from the Okote member and its correlatives, is harder to allocate to subspecies on the basis of the data presented by Leakey (see chapter 3), although it is probably referable to the earlier, smaller form. This uncertainty is especially obvious because Leakey graphed only lower molar lengths, whose variation is greatly increased by the effects of wear. Plotting molar width might not yield a clearer picture, but at least the data would all be comparable. Nonetheless, such features as overall tooth size, the ratio of canine breadth (or P_3 flange length) to molar length (or breadth), and estimated body mass (from long bone diameters) all suggest a similar division between the two groups of site samples listed.

The single specimen from Mirzapur discussed above appears to fit readily into the second of these groups, *T. o. leakeyi*. If the specimen had been recovered from an African locality, I would have no difficulty referring it to that subspecies. However, given its location some 5000 km from the nearest locality yielding *T. oswaldi* (more nearly 8000 km along a reasonable dispersal route), it is perhaps best not to include it in either of the named African subspecies. Instead, its specific nomen could be used at the subspecies level: *Theropithecus oswaldi delsoni*. It is of course conceivable that the population involved was a late member of the *T. brumpti* lineage (no known member of which presents such large and complex molars) or belonged to as yet unknown species or even lineage, but these alternatives are less likely and not otherwise supported at present. This leaves the *T. oswaldi* lineage with two species, of which the younger has three temporogeographic subspecies.

Acknowledgements

Once again, I thank Nina Jablonski for the invitation to participate in the Cambridge meeting and for her patience during the preparation of this manuscript; Lorraine Meeker for printing and arranging the photographic figures, which were taken by her, ED and David Dean; and Drs Gerald Eck, Nina Jablonski, Bill Kimbel and Meave Leakey for fruitful discussions of *Theropithecus* evolution and for reviews and comments on the manuscript, although they each (not always for the same reasons) disagree with my proposed taxonomy. Study of modern and fossil specimens was greatly facilitated by the curators of the collections of paleontology and/or mammalogy at numerous museums, especially BM (NH); KNM; MNHN-P; TMP; UWMA; American Museum of Natural History, New York; and South African Museum, Cape Town. Special thanks are due Professor Ashok Sahni of Panjab University Department of Geology, Chandigarh, for loan of the holotype of *T. o. delsoni*, and Drs Bob Brain of TMP and Noel Boaz for inviting me to study the new material from excavations at Swartkrans and Senga 5A, respectively. Research for this study was financially supported, in part, by grants from the US National Science Foundation (most recently BNS 84-19939) and the PSC-CUNY Faculty Research Award Program (13453, 667370, and 669381). Preparation of the Swartkrans fossils and development of the cataloguing system used in preparing the specimen lists (in Appendix I) were supported by the Wenner-Gren Foundation for Anthropological Research.

References

ANDREWS, C.W. (1916). Note on a new baboon (*Simopithecus oswaldi*, gen. et sp. n.) from the (?) Pliocene of British East Africa. *Annals and Magazine of Natural History*, 18, 410–19.

ARAMBOURG, C. (1947). *Mission Scientifique à l'Omo, vol. I, Géologie, Anthropologie* [Fasc. 3]. Paris: Muséum national d'Histoire Naturelle.

ARAMBOURG, C. (1969). Les vertébrés du Pléistocène de l'Afrique du Nord. *Archives du Muséum national d'Histoire Naturelle* (Paris) Ser. 7, 10, 1–126.

ARAMBOURG, C. (1979). *Vertébrés Villafranchiens d'Afrique du Nord (Artiodactyles, Carnivores, Primates, Reptiles, Oiseaux)*. Paris: Fondation Singer-Polignac.

AZZAROLI, A. (1985). Provinciality and turnover events in late Neogene and early Quaternary vertebrate faunas of the Indian Subcontinent. *Contributions to Himalayan Geology*, 3, 27–37.

AZZAROLI, A. & NAPOLEONE, G. (1979). Magnetostratigraphic investigation of the Upper Sivaliks near Pinjor, India. *Rivista Italiana di Paleontologia*, 87, 739–62.

BEHRENSMEYER, A.K. (1976). Lothagam Hill, Kanapoi and Ekora: a general summary of stratigraphy and faunas. In *Earliest Man and Environments in the Lake Rudolf Basin*, ed. Y. Coppens, G.L. Isaac, F.C. Howell & R.E.F. Leakey, pp. 163–70. Chicago: University of Chicago Press.

BRAIN, C.K. (1989). New information from the Swartkrans cave of relevance to 'robust' australopithecines. In *Evolutionary History of the 'Robust' Australopithecines*, ed. F.E. Grine, pp. 311–16. New York: Aldine De Gruyter.

BRAIN, C.K., CHURCHER, C.S., CLARK, J.D.,

GRINE, F.E., SHIPMAN, P. SUSMAN, R.L. TURNER, A. & WATSON, W. (1988). New evidence of early hominids, their culture and environment from the Swartkrans cave, South Africa. *South African Journal of Science*, 84, 828–35.

BROOM, R. (1937). On some new Pleistocene mammals from limestone caves of the Transvaal. *South African Journal of Science*, 33, 750–68.

BROOM, R. & HUGHES, A.R. (1949). Notes on the fossil baboons of the Makapan caves. *South African Science*, 2, 194–6.

BROOM, R. & JENSEN, J.S. (1946). A new fossil baboon from the caves at Potgietersrust. *Annals of the Transvaal Museum*, 20, 337–40.

BROOM, R. & ROBINSON, J.T. (1949). A new type of fossil baboon, *Gorgopithecus major*. *Proceedings of the Zoological Society, London*, 119, 374–83.

BUETTNER-JANUSCH, J. (1966). A problem in evolutionary systematics: nomenclature and classification of baboons, genus *Papio*. *Folia Primatologica*, 4, 288–308.

DECHOW, P.C. & SINGER, R. (1984). Additional fossil *Theropithecus* from Hopefield, South Africa: a comparison with other African sites and a reevaluation of its taxonomic status. *American Journal of Physical Anthropology*, 63, 405–35.

DELSON, E. (1974). Preliminary review of cercopithecid distribution in the circum-Mediterranean region. *Mémoires de Bureau de Recherches Géologiques et Minières (France)*, 78, 131–5.

DELSON, E. (1984). Cercopithecid biochronology of the African Plio-Pleistocene: correlation among eastern and southern hominid-bearing localities. *Courier Forschungs-Institut Senckenberg*, 69, 199–218.

DELSON, E. (1988). Studies on the Northern Pleistocene. *Journal of Human Evolution*, 17, 643–5.

DELSON, E. (1989). Chronology of South African australopith site units. In *Evolutionary History of the 'Robust' Australopithecines*, ed. F.E. Grine, pp. 317–24. New York: Aldine De Gruyter.

DELSON, E. (1990). [Species and species concepts in paleoanthropology.] Commentary [on paper by D. Pilbeam]. *Evolutionary Biology at the Crossroads*, ed. M.K. Hecht, pp. 141–5. New York: Queens College Press.

ECK, G.G. (1987). *Theropithecus oswaldi* from the Shungura Formation, Lower Omo Basin, southwestern Ethiopia. In *Les Faunes Plio-Pléistocènes de la Basse Vallée de l'Omo (Éthiopia)*, Tome 3, Cercopithecidae de la Formation de Shungura, ed. Y. Coppens & F.C. Howell, pp. 123–39. Cahiers de Paléontologie, Travaux de Paléontologie Est-Africaine, Editions du Centre National de la Recherche Scientifique, Paris.

ECK, G.G. & HOWELL, F.C. (1982). Un primate des formations Plio-Pléistocènes d'Afrique orientale: *Theropithecus brumpti* (Arambourg), (Primates, Cercopithecidae). *Comptes Rendus hebdomadaires de l'Academie des Sciences (Paris)*, Sér. II, 295, 397–400.

ECK, G.G. & JABLONSKI, N.G. (1984). A reassessment of the taxonomic status and phyletic relationships of *Papio baringensis* and *Papio quadratirostris* (Primates: Cercopithecidae). *American Journal of Physical Anthropology*, 65, 109–34.

ECK, G.G & JABLONSKI, N.G. (1987). The skull of *Theropithecus brumpti* compared with those of other species of the genus *Theropithecus*. In *Les faunes Plio-Pléistocènes de la Basse Vallée de l'Omo (Éthiopie)*. Tome 3, Cercopithecidae de la Formation de Shungura, ed. Y. Coppens & F.C. Howell, pp. 11–122. Cahiers de Paléontologie, Travaux de Paléontologie Est-Africaine. Paris: Editions du Centre National de la Recherche Scientifique.

ELDREDGE, N. & CRACRAFT, J. (1980) *Phylogenetic Patterns and the Evolutionary Process*. New York: Columbia University Press.

ERXLEBEN, J.C.P. (1777). *Systema regni animalis ... Classis I, Mammalia*. Lipsiae: Weygand.

FREEDMAN, L. (1957). The fossil Cercopithecoidea of South Africa. *Annals of the Transvaal Museum*, 23, 121–257.

FREEDMAN, L. (1976). South African fossil Cercopithecoidea: A re-assessment including a description of new material from Makapansgat, Sterkfontein and Taung.

Journal of Human Evolution, **5**, 297–315.

FREEDMAN, L. & BRAIN, C.K. (1977). A re-examination of the cercopithecoid fossils from Swartkrans (Mammalia: Cercopithecidae). *Annals of the Transvaal Museum*, **30**, 211–18.

GEOFFROY SAINT-HILAIRE I. (1843). Description des mammifères nouveaux ou imparfaitement connus . . . Famille des Singes. *Archives du Muséum d'Histoire Naturelle, Paris*, **2**, 486–592.

GERAADS, D. (1980). La faune des sites à *Homo erectus* des carrières Thomas (Casablance, Maroc). *Quaternaria*, **22**, 65–94.

GERAADS, D. (1987). Dating the northern African cercopithecid fossil record. *Human Evolution*, **2**, 19–27.

GRAY, J.E. (1843). *List of the specimens of Mammalia in the collection of the British Museum*. London: British Museum.

GUPTA, V.J. (1977). Fossil cercopithecoid from the Lower Boulder Conglomerate Formation (Middle Pleistocene) of Mirzapur, Kharar Tehsil, District Ropar, Punjab. *Recent Researches in Geology* (Chandigarh), **3**, 450–2.

GUPTA, V.J. & SAHNI, A. (1981). *Theropithecus delsoni*, a new cercopithecine species from the Upper Siwaliks of India. *Bulletin of the Indian Geological Association*, **14**, 69–71.

HARRIS, E.E. & HARRISON, T. (1991). Undescribed fossil cercopithecids from the Plio-Pleistocene of Kanam East in Western Kenya (abstract). *American Journal of Physical Anthropology*, Supplement 12, 88.

HARRIS, J.W.K., WILLIAMSON, P.G., VERNIERS, J., TAPPEN, M.J., STEWART, K., HELGREN, D., DE HEINZELIN, J., BOAZ N.T. & BELLOMO, R.V. (1987). Late Pliocene hominid occupation in Central Africa: the setting, context and character of the Senga 5A site, Zaire. *Journal of Human Evolution*, **16**, 701–28.

HAYES, V.J., FREEDMAN, L. & OXNARD, C.E. (1990). The taxonomy of savannah baboons: an odontomorphometric analysis. *American Journal of Primatology*, **22**, 171–90.

HENNIG, W. (1966). *Phylogenetic Systematics*. Chicago: University of Illinois Press.

HILL, A. & WARD, S.C. (1988). Origin of the Hominidae: The record of African large hominoid evolution between 14 My and 4 My. *Yearbook of Physical Anthropology*, **31**, 49–83.

HILL, W.C.O. (1970). *Primates: Comparative Anatomy and Taxonomy (Volume VIII: Cynopithecinae: Papio, Mandrillus, Theropithecus)*. Edinburgh: The University Press.

JOLLY, C.J. (1972). The classification and natural history of *Theropithecus (Simopithecus)* (Andrews, 1916), baboons of the African Plio-Pleistocene. *Bulletin of the British Museum (Natural History), Geology*, **22**, 1–123.

JOLLY, C.J. & BRETT, F.L. (1973). Genetic markers and baboon biology. *Journal of Medical Primatology*, **2**, 85–99.

KITCHING, J.W. (1952). A new type of fossil baboon: *Brachygnathopithecus peppercorni*, gen. et sp. now. *South African Journal of Science*, **49**, 15–17.

KITCHING, J.W. (1953). A new species of fossil baboon from Potgietersrust. *South African Journal of Science*, **50**, 66–9.

KRISHTALKA, L. (1989). Missing links: tales of a tub. *Carnegie Magazine*, **59**, (4), 11, 42–43.

LEAKEY, L.S.B. & WHITWORTH, T. (1958). Notes on the genus *Simopithecus* with a description of a new species from Olduvai. *Coryndon Memorial Museum Occasional Papers*, **6**, 3–14.

LEAKEY, M.G. & LEAKEY, R.E.F. (1973). Further evidence of *Simopithecus* (Primates, Mammalia) from Olduvai and Olorgesailie. *Fossil Vertebrates of Africa*, **3**, 101–20.

LEAKEY, R.E.F. (1969). New Cercopithecidae from the Chemeron Beds of Lake Baringo, Kenya. *Fossil Vertebrates of Africa*, **1**, 53–69.

MAIER, W. (1972). The first complete skull of *Simopithecus darti* from Makapansgat, South Africa, and its systematic position. *Journal of Human Evolution*, **1**, 395–405.

PATTERSON, B., BEHRENSMEYER, A.K. & SILL, W.D. (1970). Geology and fauna of a new Pliocene locality in northwestern Kenya. *Nature*, **226**, 918–21.

PICKFORD, M. (1987). The geology and palaeontology of the Kanam erosion gullies (Kenya). *Mainzer geowissenschaftlicher Mitteilungen*, **16**, 209–26.

PLUMMER, T.W. & POTTS, R. (1989). Excavations and new findings at Kanjera, Kenya. *Journal of Human Evolution*, 18, 269–76.

RAYNAL, J.-P., TEXIER, J.-P., GERAADS, D. & SBIHI-ALAOUI, F.-Z. (1990). Un nouveau gisement paléontologique Plio-Pléistocène en Afrique du Nord: Ahl-Al Oughlam (ancienne carrière Deprez) à Casablanca (Maroc). *Comptes Rendus hebdomadaires de l'Academie des Sciences (Paris)*, Sér. II, 310, 315–20.

ROMER, A.S. (1928). Pleistocene mammals of Algeria. *Bulletin of the Logan Museum, Beloit*, 1, 80–163.

RÜPPELL, E. (1835). *Neue Wirbeltiere zu der fauna von Abyssinien gehörig, vol. 1 Säugetiere* Schmerber: Frankfurt.

SINGER, R. (1962). *Simopithecus* from Hopefield, South Africa. *Bibliotheca Primatologica*, 1, 43–70.

SMART, C. (1976). The Lothagam-1 fauna: its phylogenetic, ecological and biogeographic significance. *Earliest Man and Environments in the Lake Rudolf Basin*, ed. Y. Coppens, G.L. Isaac, F.C. Howell & R.E.F. Leakey, pp. 361–9. Chicago: University Chicago Press.

SZALAY, F.S. & DELSON, E. (1979). *Evolutionary History of the Primates*. New York: Academic Press.

TALENT, J. (1989). The case of the peripatetic fossils. *Nature*, 338, 613–15.

TANDON, S.K., ROHTASH KUMAR, KOYAMA, M. & NIITSUMA, N. (1984). Magnetic polarity stratigraphy of the Upper Siwalik Subgroup, east of Chandigarh, Punjab Sub-Himalaya, India. *Journal of the Geological Society of India*, 25, 45–55.

TATTERSALL, I. (1986). Species recognition in human paleontology. *Journal of Human Evolution*, 15, 165–75.

THOMAS, P. (1884). Quelques formations d'eau douce de l'Algérie. *Mémoires de la Societé Géologique de France*, 3(2), 1–53.

VRAM, U.G. (1922). Sul genere *Theropythecus*. *Archivos di Zoologia Italiana*, 10, 169–214.

6 *Theropithecus* from Ternifine, Algeria

ERIC DELSON AND ROBERT HOFFSTETTER

Summary

1. The Ternifine *Theropithecus*, referred to *T. oswaldi leakeyi*, is here described for the first time.
2. Published analyses of the site and its fauna suggest that it may date to about 700 Ka, that the paleoenvironment was quite open and arid around the central spring-fed lake, and that the bone accumulation was partly of human cultural origin.
3. Three well-preserved specimens of the male mandible document all of the morphology of this element, revealing minor differences from described Olduvai jaws.
4. The dental sample includes 68 teeth (not counting definite antimeres), of which five are deciduous.
5. The degree of sexual dimorphism in lower canine length and width is strong, as known for other populations of this subspecies; this is one additional line of evidence supporting the interpretation of male gender for the holotype of '*Simopithecus jonathani*' from the Olduvai Masek

Beds – its P_3 mesial flange was small because it received a small C^1.
6. In both cheek tooth width (length is a poor comparator because it decreases greatly with advanced wear) and canine length and width, the Olorgesailie sample has the highest mean among all *T. o. leakeyi* populations, and its variation range encompasses all of them.
7. Molar size does not increase monotonically with time, but varies somewhat, with a possible decrease in the youngest fossils (e.g. Masek and Thomas Quarries).

Introduction

The largest sample of *Theropithecus* from North Africa is that collected from Ternifine (once Palikao, now Tighenif), Algeria, by Camille Arambourg and Hoffstetter in 1955–6. This sample has never been described nor illustrated in any detail (although it was analyzed by S. van den Brink in an unpublished 'mémoire de D.E.A.', 1980, Univ. Paris VI), and we thus take this opportunity to document it here, as well as to make some comparisons with other late *Theropithecus* populations. Following M.G. Leakey and Delson (chapters 3 and 5, respectively, this volume), we include the Ternifine sample in *T. oswaldi leakeyi* Hopwood, 1934 (a mainly Middle Pleistocene subspecies typified by increased molar size and canine/premolar reduction).

Historical review of studies of Ternifine and its primates

Fossil bones were discovered outside the village of Palikao by local quarrymen as early as 1872 (and recognized as such by Tommasini). The locality was first reported in 1879 by Pomel, who described a new species of proboscidean. Tommasini (1883) was the first to report Paleolithic artifacts. In 1885, the Association française pour l'Avancement des Sciences held a meeting in Algeria, and in three abstracts Pomel (1886a, b, c) discussed the site's mammalian fauna and archeology. These were followed by a longer report of a field trip (1888a) and then a review of the geology and fossil mammals (1888b). Camille Arambourg visited the region briefly in 1931 and recovered additional fossils.

In 1954, 1955, and 1956, Arambourg and Hoffstetter undertook extensive new excavations at Ternifine. Arambourg & Hoffstetter (1954) reported the first results of this work, and later Arambourg (1954a) described two hominid mandibles as *Atlanthropus mauritanicus* (now *Homo erectus*). Specimens of *Theropithecus* were first mentioned by Arambourg (1954b) as 'a large baboon' (un grand Cynocéphale), then as 'a large baboon similar to the giant fossil forms of eastern and southern Africa' (1955). Following the second field season, Arambourg & Hoffstetter (1955) reported the discovery of a third hominid mandible and a parietal, and included among the fauna 'a large baboon recalling certain giant forms from South Africa, as well as a cercopithecine close to the macaques', but this putative second form was not mentioned again. A long monograph on the fossil hominid remains (Arambourg, 1963) was introduced by a brief review of the history of study and geology of Ternifine (Arambourg & Hoffstetter, 1963). They described the excavation techniques and stratigraphy, but did not even provide a complete faunal list, merely indicating once again the presence of a large baboon. Cooke (1963, p. 70) included '*Simopithecus* cf. *major*' in a tabulation of North African fossil mammals, but this taxon is a *lapsus*, perhaps based on a personal communication from Arambourg or a confusion with *Gorgopithecus major*, a South African large papionin previously thought related to *Theropithecus* (cf. Freedman, 1957).

In 1966, Arambourg briefly showed the Ternifine hominids and cercopithecids to Delson in Paris, and they discussed these specimens again in 1969, just weeks before Arambourg's death. In 1970, Hoffstetter permitted Delson to study the *Theropithecus* collection from Ternifine but indicated that he planned to give the fossils to a student to describe in due course. Jolly (1972) did not discuss the material in any detail in his revision of the fossil *Theropithecus*, but he tentatively included the Ternifine population in the new Olorgesailie subspecies, *T. (S).* *oswaldi mariae*. Delson (1975, Fig. 19) published photographs of two mandibular fragments and nine isolated teeth, citing the taxon as *Theropithecus (Simopithecus)* aff. *oswaldi*. Szalay & Delson (1979) recognized numerous subspecies within *T. (S). oswaldi*, placing the Ternifine sample as 'subspecies indet. A' without diagnosis. S. van den Brink analyzed the Ternifine *Theropithecus* in her unpublished 1980 'mémoire de D.E.A.'. More recent studies of the locality and its fauna (see below) have discussed the cercopithecids little if at all. Delson (see chapter 5) has formally rejected the use of *Simopithecus* even as a subgenus, following the suggestions of several recent authors.

Age of the locality

The most recent overall analysis of the stratigraphy and geochronology of Ternifine is that by Geraads *et al.* (1986). They described the first results of a renewed collection effort in 1982–3, on the 110th anniversary of the finding of the site. Faunal studies summarized there, and those by Geraads (1987) and Tong (1986, 1989) concluded that Ternifine is intermediate in age between Sidi Abdallah (Morocco) or Aïn Hanech (Algeria) and the Thomas quarries (Morocco), probably closer to the latter. The former are probably Early Pleistocene and the latter later Middle Pleistocene in age, which leaves Ternifine close to the Early-Middle Pleistocene boundary, as originally suggested by Arambourg & Hoffstetter (1963) and by Jaeger (1975). Preliminary paleomagnetic

studies indicated normal polarity at the base of the section, thus either early Brunhes chron (*c.* 750–600 Ka) or Jaramillo subchron (*c.* 950 Ka). Geraads (1987) accepted the younger of these dates as most likely.

Paleoenvironment and taphonomy of Ternifine

The fauna from Ternifine includes a number of taxa which provide an indication of the paleoenvironment. The fossils themselves were recovered from a lacustrine deposit fed by artesian springs. Geraads (1981) described 11 species of ruminant artiodactyls, of which the most common were gazelles and alcelaphines, indicating a generally open and dry habitat. The small mammals (Jaeger, 1975; Tong, 1986, 1989) appear to confirm the essentially sub-desertic habitat, with gerbillids dominating the assemblage. On the other hand, the large number of hippopotamid remains and the presence of the anuran *Discoglossus* indicate that the most proximal animal community might be hydrophilous. Geraads noted that *Ursus* and *Theropithecus* also might indicate a different environment, which is supported by the common occurrence of *T. oswaldi* in localities tied to a well-watered landscape (Jolly, 1972).

In terms of the taphonomy of the Ternifine accumulation, Denys, Patou & Djemmali (1984) have reported cutmarks, mainly on hippopotamid bone recovered during the 1982–3 campaign. These indications of marrow extraction and flensing, as well as a possible bone tool, led them to suggest that the assemblage was at least partly of human origin, rather than solely a natural or carnivore accumulation. Denys *et al.* (1987) extended this work, suggesting the likelihood of a mammalian carnivore or avian raptor having been responsible for at least part of the accumulation of small mammals.

Description and comparisons

Mandible

The Ternifine fossils are housed in the Institut de Paléontologie, Muséum national d'Histoire Naturelle, Paris (MNHN-P), catalogued under the prefix TER; the museum acronym will not be repeated in the descriptions. There are three specimens of the male mandible in the Ternifine sample, which among them preserve most of the mandibular morphology. Measurements on these three specimens are given in Table 6.1.

A male symphysis preserving all the teeth from I_1–P_4 on both sides (TER 1702, Fig. 6.1) reveals more symphyseal morphology than any other male of the species yet described. There is little indication of any mandibular corpus fossa, not even as much as in Old 067/5603 (from Bed II), which is about the same size overall. Eck (see chapter 2, following Eck & Jablonski, 1987) has discussed fossa development in terms of the ratio of fossa depth (indentation) to corpus thickness. In TER 1702, this ratio is about 0.05 (difficult to determine because the indentation is so slight), which is appreciably less than in any other *T. oswaldi* and comparable only to the least excavated *T. darti* (see Eck, chapter 2, Table 2.12).

The mental ridges begin at the break below P_4, converging around a broad flattened region which extends up to the middle of the anterior symphyseal surface. About 20 mm below the anterior alveolar margin (infradentale), the ridges end in a shallow depression. The surface bulges slightly up to infradentale. The area between the ridges is not rugose or raised, as in the two large Olduvai mandibles. The planum alveolare is only slightly concave, flatter than in 067/5603. The tooth rows diverge posteriorly, with an angle of 60° from infradentale to the centers of each P_4. Presumably, the cheek tooth rows became subparallel after that point.

A left corpus (TER 1703, Figs 6.2 and 6.4), probably of an adult male, lacks the superior half of the ramus but extends forward to P_4 (below M_1 basally). This individual was slightly smaller and probably older than the preceding, as the P_4 is worn more than those in the symphysis 1702. Again, there is little indication of a mandibular corpus fossa, although the surface bone below P_4 is broken away, exposing matrix and spongy bone within. Gonion is smoothly rounded below, curving up to the point of breakage about the level of the alveolar plane. The corpus shallows slightly from M_1 to M_3, with a ratio of these two depths of 1.21, comparable to values reported by Eck (see chapter 2) for Olduvai mandibles.

A third specimen (TER 1815, Fig. 6.2) preserves most of a right ramus, whose large size suggests it may also be male. Its maximum mesiodistal length is about 58 mm, while height from the base of the notch between coronoid and condyle to

Table 6.1. *Measurements (in mm) of Ternifine mandibles.*

Variable (Eck number*)	TER 1702	TER 1703
Symphyseal length (infradentale–gnathion)	66.0 (est.)	
Symphyseal height (perpendicular to occlusal plane)	36.0 (est.)	
Length of planum alveolare (infradentale to inferodistal most point on superior transverse torus, 'supertorion')	39.0	
Height of planum alveolare (gnathion–'supertorion')	33.0 (est.)	
Depth of the incisive plane (26)	38.2 (est.)	
Width of the incisive plane (27)	33.9	
Alveolar width across I_2s (28)	16.8	
Maximum width across I_2s	17.7	
Alveolar width across I_1s	7.8	
Maximum width across I_1s	9.8	
External width across C_1s (29)	35.0	
Internal width between C_1s	12.5	
Internal width between P_3s	29.5	
Toothrow (partial) lengths:		
$M_3 – I_1$ (reconstruction, est.)	100	
$M_3 – C_1$ (reconstruction, est.)	97	
$M_3 – P_4$		64.5
$M_3 – M_1$ (30)		54.6
$P_4 – C_1$	42.2	
$P_4 – I_1$	45.3	
Alveolar thickness at M_2		
Corpus depth, below middle of		
P_4 (buccal side)	43.6	
M_1 (buccal side) (24)		45.2
M_2 (buccal side)		42.3
M_3 (buccal side) (25)		37.4
M_1 (lingual side)		46.9
M_3 (lingual side)		42.3
Depth mandibular corpus fossa (22)	<1.0	
Thickness of corpus at fossa (23)	17.8	

* Numbers in parentheses correspond to measurements described by Eck (chapter 2, this volume).

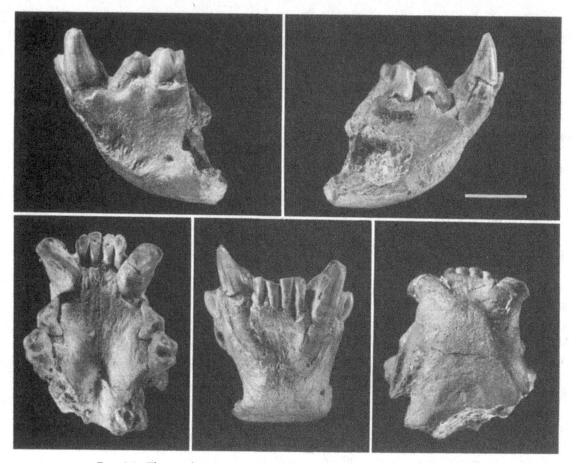

Fig. 6.1. *Theropithecus oswaldi leakeyi* from Ternifine, Algeria. Young adult male symphysis, MNHN-P TER 1702, scale bar = 2 cm. Top row, left and right lateral views; bottom row, occlusal, anterior and basal views; all in occlusal plane orientation.

the base of the ramus (anterior to gonion) is about 96 mm.

Figure 6.3 is a photographic reconstruction of the mandible based on the preceding three specimens. The symphysis is 5–10 per cent larger than the corpus fragment and was enlarged by 7 per cent to better connect with it.

Dentition

In addition to the three large fragments, 53 isolated permanent teeth were identified as *Theropithecus*. There is also a small fragment of juvenile corpus with right dP_4-M_1 (TER 1704), as well as (one each) isolated right and left dP_4, right dP^4 and partial left dP^3. The teeth are in general typical of large members of the genus; measurements are presented in Table 6.2, and selected examples are illustrated in Figs 6.4–6.6. A full listing of Ternifine specimens is included in Appendix I to this volume.

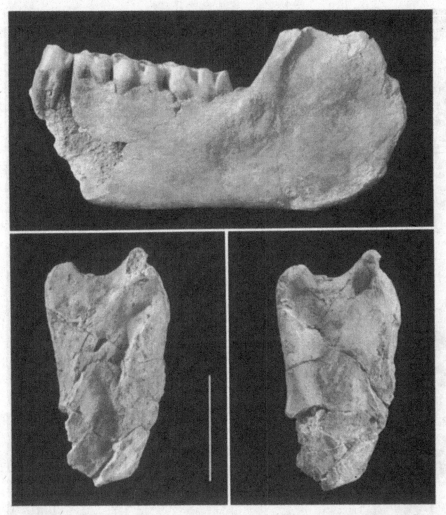

Fig. 6.2. *Theropithecus oswaldi leakeyi* from Ternifine, Algeria. Above, older adult male left mandibular corpus, MNHN-P TER 1703, lateral view; below, right partial ?male mandibular ramus, MNHN-P TER 1815, internal and (photographically reversed) external views; scale bar = 5 cm.

Among the most interesting of these teeth are the lower canines, which document strong sexual dimorphism in the Ternifine population. Two male lower canines (one in the symphysis, another isolated [TER 1725]) are robust in length and width, but not very tall above the cervix. A single much smaller specimen

(TER 1726) is identified as a female. It is relatively less thick mesiodistally than the male, with a lightly built root (Fig. 6.6). This degree of dimorphism is comparable to that seen in the more extensive sample of slightly larger *Theropithecus* from Olorgesailie (Jolly, 1972; Leakey & Leakey, 1973; and especially, Leakey,

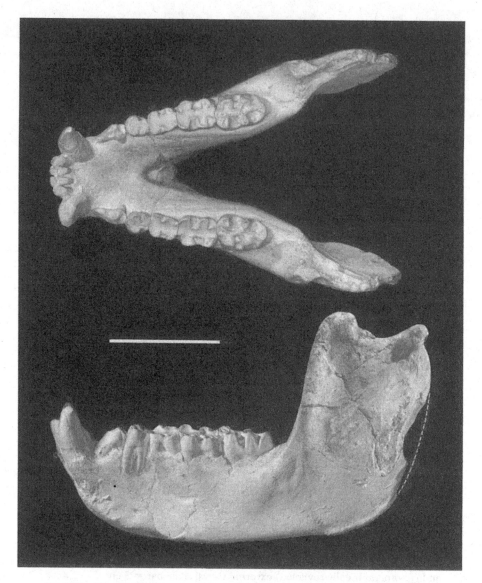

Fig. 6.3. *Theropithecus oswaldi leakeyi* from Ternifine, Algeria. Photo-graphic reconstruction of male mandible, based on MNHN-P TER 1702, 1703 and 1815, scale bar = 5 cm.

chapter 3). Along with the clear sex dif-ference in root length and width, this level of crown dimorphism further serves to confirm the male gender of the Olduvai Masek Beds mandible 068/6516 (previously Old. 1472, 57), as argued by several authors. That specimen's canine roots are almost as strong as in the two Ternifine male canines, but the crown is quite low, which in turn explains the short mesiobuccal flange of the P_3: it must have received a very small upper canine crown.

Table 6.2. *Measurements (in mm) of Ternifine Theropithecus teeth.*

Variable	N	Minimum	Maximum	Mean	SE	CV
I_1W	3	6.2	8.7	7.17	0.78	
I_1LA	3	3.8	5.5	4.93	0.57	
I_1L	3	4.6	7.4	5.63	0.89	
I_2W	3	5.4	6.8	6.23	0.43	
I_2LA	3	3.5	5.2	4.10	0.55	
I_2L	3	4.2	6.3	5.13	0.62	
C_1W FEMALE	1			10.8		
C_1L FEMALE	1			6.0		
C_1H FEMALE	1			14.4		
C_1W MALE	2	14.3	14.7	14.50	0.20	
C_1L MALE	2	10.0	11.0	10.50	0.50	
C_1H MALE	2	20.3	23.4	21.85	1.55	
P_3W MALE	4	7.4	9.2	8.18	0.38	
P_3L MALE	3	15.3	20.4	17.87	1.47	
P_3FH MALE	3	15.9	20.5	17.97	1.35	
P_4W	7	8.9	9.8	9.53	0.13	3.62
P_4L	8	9.7	11.8	11.09	0.28	7.04
M_1AW	6	9.5	11.3	10.57	0.27	6.21
M_1PW	6	10.2	11.0	10.58	0.14	3.35
M_1L	4	12.7	16.5	15.08	0.85	
M_2AW	6	13.0	15.0	13.45	0.31	5.68
M_2PW	6	12.5	14.6	12.97	0.34	6.35
M_2L	6	17.0	19.8	18.10	0.39	5.28
M_3AW	8	14.3	16.2	15.11	0.26	4.80
M_3PW	8	12.7	14.5	13.33	0.22	4.74
M_3L	8	21.5	27.7	24.06	0.73	8.59
dP_4AW	3	7.5	8.3	7.97	0.24	
dP_4PW	3	8.2	9.0	8.57	0.23	
dP_4L	3	11.4	12.1	11.70	0.21	
I^2W	1			9.0		
I^2LA	1			5.4		
I^2L	1			7.5		
C^1W MALE	2	11.3	14.4	12.85	1.55	
C^1L MALE	3	15.2	17.1	15.83	0.63	
C^1H MALE	2	33.0	36.5	34.75	1.75	
$P^{?3}W$	1			9.1		
$P^{?3}L$	1			8.2		
$P^{?3}H$	1			10.0		
$P^{?4}W$	3	10.4	11.7	10.97		

Table 6.2. *Contd.*

Variable	N	Minimum	Maximum	Mean	SE	CV
$P^{?4}L$	3	9.1	10.3	9.70	0.35	
$P^{?4}H$	3	9.4	11.2	10.57	0.58	
M^1AW	2	12.5	13.0	12.75	0.25	
M^1PW	2	11.8	13.3	12.55	0.75	
M^1L	2	13.2	15.9	14.55	1.35	
M^2AW	2	15.5	17.6	16.55	1.05	
M^2PW	2	14.4	16.5	15.45	1.05	
M^2L	2	19.4	19.7	19.55	0.15	
M^3AW	4	15.5	17.6	16.50	0.55	
M^3PW	4	13.6	16.5	14.58	0.65	
M^3L	4	19.7	21.1	20.45	0.29	
dP^3AW	1			7.7		
dP^4AW	1			10.0		
dP^4PW	1			10.4		
dP^4L	1			10.9		

N: number of measureable specimens; SE: standard error of the mean; CV: coefficient of variation. Values are given separately by sex for canines and P_3s. For incisors, canines and premolars, L(ength) is always maximum mesiodistal, W(idth) is maximum buccolingual, taken perpendicular to length; therefore, for lower canines, because of their turned placement in the mandible, W is greater than L. Incisor LA is taken at the alveolar plane (=cervical level), as maximal length is only available on unworn teeth whose incisal edges have not been reduced due to attrition. P_3FH (flange height) is the distance from the cusp apex to the most mesial extent of the enamel along the mesial flange: it is equivalent to L(h) of Freedman (1957) or L of Singer (1962); L here is the maximum mesiodistal length of the tooth, as for Freedman (1957). For molariform teeth, AW and PW, respectively, are taken across the mesial and distal loph(id)s usually at the cervix; L, however, is taken at interdental contact points, often estimated due to wear, and decreases significantly in worn teeth.

Fig. 6.4. Occlusal views of dentition of *Theropithecus oswaldi leakeyi* from Ternifine, Algeria, scale bar = 2 cm. (a) Upper teeth, left to right (and top to bottom within arranged 'toothrows'), all MNHN-P TER: left column, right dP^4 (1733), right $M^{2?}$ (1762), right $M^{3?}$ (1758); center column, right P^3 (1775), right P^4 (1774), right M^1 (1764), right $M^{2?}$ (1761), right $M^{3?}$ (1759); right column (photographically reversed), left P^4 (1765), left $M^{2?}$ (1763), left M^3 (1760). (b) Lower teeth, left to right (and top to bottom within arranged 'toothrows'), all MNHN-P TER: 1703, left male? P_4-M_3 from corpus; left $M_{1?}$ (1720), left $M_{2?}$ (1757), left M_3 (1708); right $M_{1?}$ (1715), right $M_{2?}$ (1711), right M_3 (1707); right dP_4-M_1 (1704).

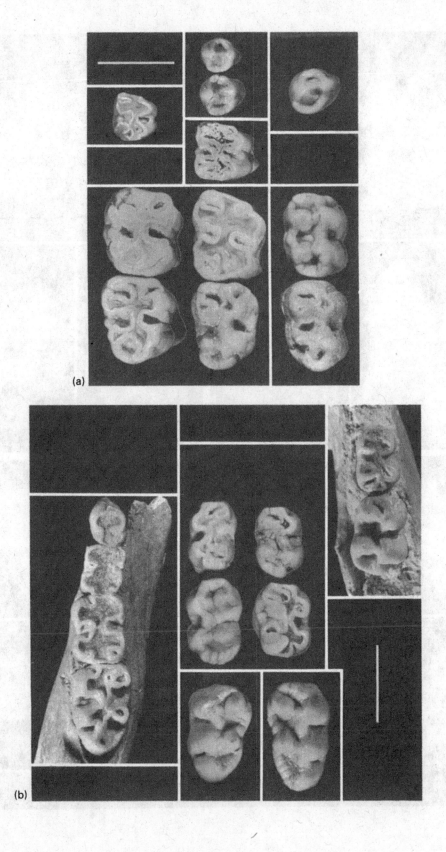

(a)

(b)

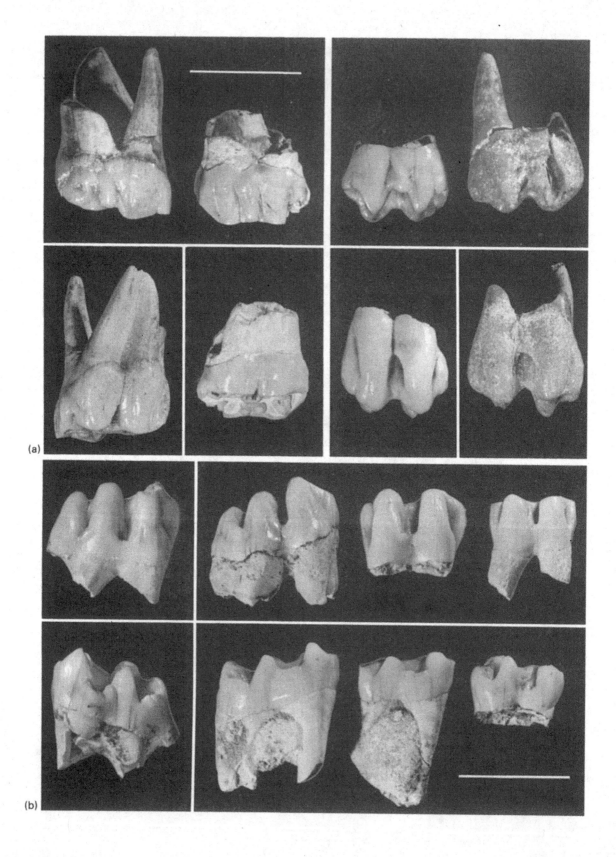

(a)

(b)

Fig. 6.6 Canines of *Theropithecus oswaldi leakeyi* from Ternifine, Algeria, in buccal view, scale bar = 2 cm. Left to right: MNHN-P TER 1725, male left C_1; 1726, female left C_1; 1780, male right C^1.

Although Eck (see chapter 2) continues to interpret this specimen as female (following Eck & Jablonski, 1984 and Leakey & Whitworth, 1958), Leakey (see chapter 3)

Fig. 6.5. Lateral views of dentitions of *Theropithecus oswaldi leakeyi* from Ternifine, Algeria, scale bar = 2 cm, all MNHN-P TER. (a) Upper teeth, top row buccal views, second row lingual views, left to right: right $M^{3?}$ (1758), right $M^{2?}$ (1762), left $M_{2?}$ (1763), left $M^{3?}$ (1760). (b) Lower teeth, top row buccal views: right M_3 (1708), right M_3 (1713), right $M^{2?}$ (1757), right $M_{1?}$ (1720); second row lingual views: right M_3 (1708), left M_3 (1706), left $M_{2?}$ (1710), left $M_{1?}$ (1715) (last three photographically reversed).

has provided additional evidence of its being male (see also Jolly, 1972; Leakey & Leakey, 1973; Szalay & Delson, 1979; Dechow & Singer, 1984), based on a new specimen from Kapthurin.

It is interesting to compare the overall dental measurements of the samples of *Theropithecus* from Ternifine, Olorgesailie, and Hopefield, as these localities are of comparable age and span the whole of the African continent (see Potts, 1989; Gibbons, 1990. Klein & Cruz-Uribe, 1991, suggested an age for the main Hopefield fauna between 0.7–0.4 Ma). Measure-

Table 6.3. *Measurements (in mm) of Hopefield* Theropithecus *teeth*.

Variable	N	Minimum	Maximum	Mean	SE	CV
P_3W FEMALE	3*	6.8	8.0	7.43	0.35	
P_3L FEMALE	1	12.1	12.1	12.10		
P_3FH FEMALE	3*	11.4	12.9	12.27	0.45	
C_1W MALE	2*	18.1	19.7	18.90	0.79	
C_1L MALE	2*	11.0	13.7	12.35	1.32	
C_1H MALE	2*	23.9	29.0	26.45	2.55	
P_3L MALE	1			17.8		
P_3FH MALE	1			25.4		
P_4W	4*	8.8	9.9	9.47	0.24	
P_4L	4*	10.1	11.1	10.67	0.23	
M_1AW	2*	9.7	11.8	10.75	1.05	
M_1PW	2*	9.9	11.6	10.75	0.83	
M_1L	3*	13.3	13.8	13.55	0.15	
M_2AW	2	14.0	14.9	14.45	0.49	
M_2PW	2	12.7	14.6	13.61	0.95	
M_2L	3	17.6	19.9	18.75	0.67	
M_3AW	4	15.0	18.2	16.69	0.66	
M_3PW	3	14.7	15.6	15.7	0.27	
M_3L	5	25.0	27.1	26.03	0.36	3.10
C^1W FEMALE	1*			10.2		
C^1L FEMALE	1*			9.3		
C^1H FEMALE	1*			18.0		

See Table 6.2 for measurement definitions.

* Indicates at least one specimen's measurements taken from Singer (1962) [SAM 6174, 6882 and 13905] or Dechow & Singer (1984) [16650 and 16680 P_3s].

ments of the Olorgesailie sample are provided by Jolly (1972), Leakey & Leakey (1973), and especially Leakey (see chapter 3), although no single published set is complete. Singer (1962) and Dechow & Singer (1984) published measurements on the Hopefield specimens, but these differ by up to 10 per cent from measures taken by Delson[1]. The latter are reported in Table 6.3 when available, along with

[1] Although Dechow & Singer (1984, p. 406) imply that their measurement techniques follow Dechow (unpublished Ph.D. Thesis, 1980), the latter did not give definitions for his dental metrics; instead, they follow Singer (1962), who in turn used the definitions of Freedman (1957), which are comparable to Delson's. It is thus unclear why there may be significant differences in values reported by these authors. Given the extensive wear on most Hopefield teeth and the uncertain condition of the Olorgesailie specimens, and in light of the great reduction in length of papionin cheek teeth with high wear, comparisons of length are less useful than those of width (although most workers emphasize the former).

published values for specimens unavailable to Delson.

In general, the Olorgesailie range of variation encompasses both Hopefield and Ternifine, with the means decreasing in that order. Various authors (e.g. Delson, 1983; Eck & Jablonski, 1987; see Eck and Leakey, chapters 2 and 3, respectively) have shown that the size of *Theropithecus* molar teeth increases through time. But attempts at more detailed correlation of size and age are not feasible – populations of comparable age vary too much both within and between samples for such precise comparisons to be meaningful. In fact, it appears that molar size may have decreased in the apparently youngest samples – Olduvai Masek Beds, Kapthurin (see Leakey, chapter 3) and Thomas Quarries (Geraads, Beriro & Roche, 1980) – but that may be due to small sample sizes.

Male canine length and width also show variation among the Middle Pleistocene samples without close correlation with age. The Hopefield specimen is extremely robust, while the Ternifine and Thomas Quarry canines are more gracile, and the Masek and Kapthurin fossils are smaller still. Once again, variation within the Olorgesailie sample appears to encompass all of the preceding individuals.

Leakey (see chapter 3, Table 3.6) also examines the relative canine-complex reduction in *T. oswaldi* by means of ratios comparing P_3 flange height and C_1 width to M_3 length in numerous specimens. Anterior teeth and M_3 do not co-occur in any Ternifine jaw, but if the two mandibles are combined, ratios of 0.65 and 0.63 are obtained. The P_3/M_3 ratio is very low, exceeded (or approached) only by the new Kapthurin mandible. The canine ratio places Ternifine between the Kapthurin and Masek jaws and two of four from the KBS member; two other KBS specimens have higher ratios (larger canines) than a single upper Burgi jaw, however, indicating less consistency of interpretation. Based on Delson's measurements, one Hopefield specimen (South African Museum, Cape Town, No. 16680) has a premolar index of 0.74. Another Hopefield jaw (16648) preserves all three teeth in damaged condition; Delson's measurements yield premolar and canine indices of 1.02 and 0.72, respectively. Not only are the two premolar values far apart, the second is higher than for any specimen measured by Leakey. Despite potential inconsistencies, this indicates again that the Ternifine canine/P_3 complex is reduced by comparison to the penecontemporaneous Hopefield sample.

In conclusion, the Ternifine sample of *Theropithecus oswaldi leakeyi* is reasonably typical of those populations placed in this taxon. It shows high sexual dimorphism, a reduced canine complex and a strongly buttressed mandibular corpus. Further comparison with additional specimens from younger North African localities, if they become available, might throw light on the apparent decoupling of size changes with time in different regions of Africa late in the history of this species.

Acknowledgements

E.D. thanks Nina Jablonski for the invitation to participate in the Cambridge meeting, for her patience during the preparation of this manuscript and for comments on early drafts; Chester Tarka for preparing Fig. 6.4; Lorraine Meeker for printing and/or compiling the figures; Drs Gerald Eck and Meave Leakey for fruitful discussions of *Theropithecus* evolution; and Drs Eck, Leakey, Jablonski, and Bill Kimbel for reviews and comments on the manuscript. Study of modern and fossil specimens was greatly facilitated by the curators of the collections of paleontology and/or mammalogy at numerous museums, especially the Muséum national d'Histoire Naturelle, Institut de Paléontologie, Paris; American Museum of Natural History, New York; South African Museum, Cape Town; and National Museums of Kenya, Nairobi. The collection of fossils at Ternifine was supported by the Algerian government and the French Centre national pour la Recherche Scientifique (see Arambourg & Hoffstetter, 1963). Research for this study was financially supported, in part, by grants to E.D. from Columbia University, the US National Science Foundation (most recently BNS 84-19939) and the PSC-CUNY Faculty Research Award Program (13453, 667370 and 669381).

References

ARAMBOURG, C. (1954a). L'hominien fossile de Ternifine (Algérie). *Comptes Rendus hebdomadaires de l'Académie des Sciences (Paris)*, 239, 893–5.

ARAMBOURG, C. (1954b). L'Atlanthrope de Ternifine, un chaînon complémentaire de l'ascendance humaine, fabriquait des bifaces chelléens. *La Nature*, Paris, No. 3235, 401–4.

ARAMBOURG, C. (1955). Le gisement de Ternifine et l'*Atlanthropus*. *Bulletin de la Société Préhistorique de France*, 52, 94–5.

ARAMBOURG, C. (1963). Le gisement de Ternifine. Deuxième partie, L'*Atlanthropus mauritanicus*. *Archives de l'Institut de Paléontologie Humaine*, Mém. 32, 37–190.

ARAMBOURG, C. & HOFFSTETTER, R. (1954). Découverte, en Afrique du nord, de restes humains du Paléolithique inférieur. *Comptes Rendus hebdomadaires de l'Académie des Sciences (Paris)*, 239, 73–4.

ARAMBOURG, C. & HOFFSTETTER, R. (1955). Le gisement de Ternifine. Résultat des fouilles de 1955 et découverte de nouveaux restes d'*Atlanthropus*. *Comptes Rendus hebdomadaires de l'Académie des Sciences (Paris)*, 241, 431–3.

ARAMBOURG, C. & HOFFSTETTER, R. (1963). Le gisement de Ternifine. Première partie, Historique et Géologie. *Archives de l'Institut de Paléontologie Humaine*, Mém. 32, 1–36.

COOKE, H.B.S. (1963). Pleistocene mammal faunas of Africa, with particular reference to southern Africa. In *African Ecology and Human Evolution*, ed. F.C. Howell & F. Bourlière, pp. 65–116. Chicago: Aldine.

DECHOW, P.C. & SINGER, R. (1984). Additional fossil *Theropithecus* from Hopefield, South Africa: a comparison with other African sites and a reevaluation of its taxonomic status. *American Journal of Physical Anthropology*, 63, 405–35.

DELSON, E. (1975). Evolutionary history of the Cercopithecidae. *Contributions to Primatology*, 5, 167–217.

DELSON, E. (1983). Evolutionary tempos in catarrhine primates. In *Modalités, Rythmes et Mécanismes de l'Evolution Biologique: Gradualisme Phyletique ou Equilibres Ponctués?* ed. J. Chaline, pp. 101–6. *Colloques Internationaux du CNRS No. 330.* Paris: CNRS.

DENYS, C., GERAADS, D., HUBLIN, J-J. & TONG, H. (1987). Méthode d'étude taphonomique des microvertébrés. Application au site pléistocène de Tighenif (Ternifine, Algérie). *Archaeozoologia*, 1, 53–82.

DENYS, C., PATOU, M. & DJEMMALI, N. (1984). Tighennif (Ternifine, Algérie). Premiers résultats concernant l'origine de l'accumulation du material osseux de ce gisement pléistocène. *Comptes Rendus hebdomadaires de l'Académie des Sciences (Paris)*, Sér. II, 299, 481–6.

ECK, G.G. & JABLONSKI, N.G. (1984). A reassessment of the taxonomic status and phyletic relationships of *Papio baringensis* and *Papio quadratirostris* (Primates: Cercopithecidae). *American Journal of Physical Anthropology*, 65, 109–34.

ECK, G.G. & JABLONSKI, N.G. (1987). The skull of *Theropithecus brumpti* compared with those of other species of the genus *Theropithecus*. In *Les faunes Plio-Pléistocènes de la Basse Vallée de l'Omo (Éthiopie)*. Tome 3, Cercopithecidae de la Formation de Shungura, ed. Y. Coppens & F.C. Howell, pp. 11–122. Cahiers de Paléontologie, Travaux de Paléontologie Est-Africaine. Paris: Editions du Centre National de la Recherche Scientifique.

FREEDMAN, L. (1957). The fossil Cercopithecoidea of South Africa. *Annals of the Transvaal Museum*, 23, 121–257.

GERAADS, D. (1981) Bovidae et Giraffidae (Artiodactyla, Mammalia) du pléistocène de Ternifine (Algérie). *Bulletin du Muséum national d'Histoire Naturelle, Paris*, Sér. 4, 3, Sec. C, 47–86.

GERAADS, D. (1987). Dating the northern African cercopithecid fossil record. *Human Evolution*, 2, 19–27.

GERAADS, D., BERIRO, P. & ROCHE, H. (1980). La faune et l'industrie des sites à *Homo erectus* des carrières Thomas (Maroc). Précisions sur l'âge de ces Hominidés. *Comptes Rendus hebdomadaires de l'Académie des Sciences (Paris)*, Sér. D, 291, 195–8.

GERAADS, D., HUBLIN, J.-J., JAEGER, J.-J., TONG, H., SEN, S. & TOUBEAU, P. (1986). The Pleistocene hominid site of Ternifine, Algeria: new results on the environment, age and human industries. *Quaternary Research*, 25, 380–6.

GIBBONS, A. (1990). Paleontology by bulldozer. *Science*, 247, 1407–9.

HOPWOOD, A.T. (1934). New fossil mammals from Olduvai, Tanganyika territory. *Annals and Magazine of Natural History*, Ser. 10, 14, 546–7.

JAEGER, J.-J. (1975). The mammalian fauna and hominid fossils of the Middle Pleistocene of

the Maghreb. In *After the Australopithecines*, ed. by K.W. Butzer & G.L. Isaac, pp. 399–418. The Hague: Mouton.

JOLLY, C.J. (1972). The classification and natural history of *Theropithecus* (*Simopithecus*) (Andrews, 1916), baboons of the African Plio-Pleistocene. *Bulletin of the British Museum (Natural History), Geology*, 22, 1–123.

KLEIN, R. & CRUZ-URIBE, K. (1991). The bovids from Elandsfontein, South Africa, and their implications for the age, palaeoenvironment, and origins of the site. *African Archaeological Review*, 9, 21–79.

LEAKEY, L.S.B. & WHITWORTH, T. (1958). Notes on the genus *Simopithecus* with a description of a new species from Olduvai. *Coryndon Memorial Museum Occasional Papers*, 6, 3–14.

LEAKEY, M.G. & LEAKEY, R.E.F. (1973). Further evidence of *Simopithecus* (Primates, Mammalia) from Olduvai and Olorgesailie. *Fossil Vertebrates of Africa*, 3, 101–20.

POMEL, A. (1879). Ossements d'éléphants et d'hippopotames découverts dans une station préhistorique de la plaine d'Eghis (province d'Oran). *Bulletin de la Société géologique de France*, 3 (7), 44–51.

POMEL, A. (1886a). Sur la station préhistorique de Ternifine, près Mascara. *Comptes rendus de l'Association française pour l'avancement de Science*, Paris, 14 (1), 128.

POMEL, A. (1886b). Station préhistorique de Ternifine, près de Mascara (Algérie). *Comptes rendus de l'Association française pour l'avancement des Sciences*, Paris, 14 (1), 164.

POMEL, A. (1886c). Station préhistorique de Ternifine (Mascara). *Comptes rendus de l'Association française pour l'avancement des Sciences*, Paris, 14 (2), 504.

POMEL, A. (1888a). Visite faite à la station préhistorique de Ternifine (Palikao) par le groupe excursionniste D. *Comptes rendus de l'Association française pour l'avancement des Sciences*, Paris, 17 (1), 208–12.

POMEL, A. (1888b). La station quaternaire de Palikao (Alger). II. Note géologique et paléontologique. *Matériaux pour l'histoire primitive et naturelle de l'Homme*, Paris, 22, 224–32.

POTTS, R. (1989). Olorgesailie, new excavations and findings in Early and Middle Pleistocene contexts, southern Kenya rift valley. *Journal of Human Evolution*, 18, 477–84.

SINGER, R. (1962). *Simopithecus* from Hopefield, South Africa. *Bibliotheca Primatologica*, 1, 43–70.

SZALAY, F.S. & DELSON, E. (1979). *Evolutionary History of the Primates*. New York: Academic Press.

TOMMASINI, P. (1883). Gisement chelléen de Ternifine, en Algérie. *Bulletin de la Société d'Anthropologie, Paris*, Sér. 3, 6, 426–9.

TONG, H. (1986). The Gerbillinae (Rodentia) from Tighennif (Pleistocene of Algeria) and their significance. *Modern Geology*, 10, 197–214.

TONG, H. (1989). Origine et évolution des Gerbillidae (Mammalia, Rodentia) en Afrique du nord. *Mémoires de la Société Géologique de France*, Nouv. Sér., 155, 1–120.

7 The Phylogeny of *Theropithecus*

NINA G. JABLONSKI

Summary

1. In this paper, the phylogenetic relationships of *Theropithecus* to other cercopithecid genera are considered and the phyletic relationships within the genus are examined.
2. *Theropithecus* is distinguished from other Papionini by a large number of derived gross anatomical features associated mostly with its adaptations to a highly terrestrial habitus and 'manual grazing'.
3. Available neontological and palaeontological evidence suggests that, among the Papionini, *Theropithecus* shared a common ancestor most recently with *Papio*, between 7 and 3.5 Ma ago.
4. *Theropithecus* comprises three distinct species groups: one including the modern gelada, one connecting *T. darti* and *T. oswaldi*, and one containing *T. brumpti* and its predecessors. It is thought that these lineages arose within a relatively short space of time, between 4.0 and 3.5 Ma. Delson (chapter 5) has suggested that the anatomical distinctions of *T. brumpti* and its putative forebears are sufficient to warrant their placement in the subgenus *Omopithecus*, leaving the subgenus *Theropithecus* to accommodate *T. gelada*, *T. darti*, and *T. oswaldi*. This scheme is not universally accepted, however, and two alternative views are presented.
5. *Theropithecus gelada* exhibits a mixture of plesiomorphic and autapomorphic features, some of the latter possibly related to its adaptation to the cold, dry environment of the Ethiopian highlands. This combination of features is not particularly useful in determining the origin of its lineage relative to others.
6. *Theropithecus darti* and *T. oswaldi* comprise a single continuous lineage of animals adapted to a highly terrestrial, grass-eating existence in the lowland grasslands of Africa. The lineage is characterized by several trends, including marked increase in body size and marked diminution of incisor and canine size. The absence of morphological discontinuity within the lineage suggests that the differences between *T. darti* and *T. oswaldi* may best be expressed as subspecific designations within *T. oswaldi*.

7. *Theropithecus brumpti* was the only species of *Theropithecus* that appears to have been a forest dweller. The species was relatively short-lived, restricted in its geographical distribution, and characterized by a highly specialized masticatory apparatus. The origin of its lineage is not completely clear, but the form originally referred to as *Papio baringensis* may have been its predecessor.

Introduction

At the Cambridge conference, the phylogeny of *Theropithecus* was discussed at some length in a workshop specially devoted to the topic. The discussions were lively and productive, having been started off by the principal discussants – Eric Delson, Gerald Eck, Clifford Jolly, and Meave Leakey – each placing their preferred cladograms or phylogenetic trees of the genus on the front board. Thanks to the large collection of fossil specimens and casts available to discussants, morphological details relevant to the arguments at hand could be verified or questioned on the spot, with the result that much idle and contentious speculation was averted. While the workshop failed to produce complete agreement among discussants concerning the precise course of evolution of *Theropithecus*, a concurrence of opinion on several critical points was reached. As will become clear in the course of this chapter, the points that remained unresolved in the discussions were those that harkened back to some of the eternal problems of palaeobiology,

such as the chronospecies concept, on which workers tended to agree to disagree. This chapter represents a distillation of the discussions at that workshop to which has been added some information relating to the subject concerning chromosomal and molecular evolution in the Papionini that was not mentioned at the conference.

The study of the phylogeny of *Theropithecus* is perhaps most useful to evolutionary biology because of the problems it poses. These problems, so irksome to the would-be taxonomist, present us face to face with multifarious workings of the evolutionary process and the farcical nature of strict rules in biology.

The position of *Theropithecus* within the Cercopithecidae

At present it is generally agreed that *Theropithecus* belongs to the tribe Papionini of the family Cercopithecidae. In addition to *Theropithecus* the Papionini comprise *Papio*, *Mandrillus*, *Cercocebus*, *Lophocebus* (if considered generically distinct from *Cercocebus*), *Parapapio*, *Dinopithecus*, *Gorgopithecus*, *Macaca*, *Procynocephalus*, and *Paradolichopithecus*. These genera are united within the Papionini by a series of craniodental and pedal characteristics that have been discussed at length elsewhere (Szalay & Delson, 1979; Strasser & Delson, 1987; Strasser, 1988). The living Papionini also share the possession of 42 chromosomes. *Theropithecus* is widely recognized as the most highly derived of the papionins (Szalay & Delson, 1979; Strasser & Delson, 1987), and in some taxonomic schemes the distinctions of the genus have been recognized through erection of the monotypic subtribe Theropithecina (Szalay & Delson, 1979), or the tribe Theropithecini (Jolly, 1966; Maier, 1972b) that also included *Dinopithecus* and *Gorgopithecus*.

The position of *Theropithecus* within the Cercopithecidae has been most extensively examined from the point of view of comparative gross anatomy; information bearing on the problem derived from various biochemical and karyological approaches has generally been less plentiful and helpful.

Gross anatomical, karyological, and molecular distinctions of the genus *Theropithecus*

Theropithecus is distinguished at the genus level by a large number of dental and skeletal characteristics, and the genus is considered highly autapomorphic in features of gross anatomy. Of these, the features of the cranium and dentition have been most thoroughly documented, largely because these elements constitute the major fraction of the fossil record of the genus. With few exceptions, these derived features are characteristics of the masticatory apparatus or are features associated indirectly with the demands or stresses of mastication. The distinctive postcranial characteristics of the genus have been less exhaustively examined than those of the skull in part because of the paucity of postcranial skeletons and postcranial fossil remains in clear association with *Theropithecus* crania or dentitions. It is clear, however, that the species of *Theropithecus* share several features of the postcranial skeleton related to a terrestrial habitus and a distinctive mode of feeding requiring increased manual dexterity for the harvesting of grass parts. The dental and skeletal specializations of *Theropithecus* are listed in Table 7.1.

Studies of the chromosomes of the Papionini have generally not yielded results useful to phylogenetic studies because of the profound karyological homogeneity of the group as revealed by classical methods of staining (Stanyon *et al.*, 1988). The introduction of chromosomal banding techniques has revealed some evidence of inter- and intrageneric variation among the Papionini, but the phylogenetic significance of this variation has not always been clear. In their study of the banded karyotypes of extant papionins, Stanyon *et al.* (1988) showed the presence in *Papio* and *Cercocebus* of a large clear staining band on chromosome 2 by G-banding techniques. This band is absent in macaques and is variably present in *T. gelada*, with some individuals being heterozygous for the presence or absence of the band. This was interpreted by Stanyon *et al.* (1988) as weak evidence of *Theropithecus* being more closely linked to *Macaca*, by what may be a shared primitive characteristic.

Molecular criteria, on the other hand, have demonstrated a particular closeness between *Theropithecus* and *Papio*. Cronin, Cann & Sarich (1980) indicated that the molecular picture of the Papionini is one of a single primary adaptive radiation with only *Papio* and *Theropithecus* sharing a substantial period of common ancestry. On the basis of results of albumin plus transferrin immunological

Table 7.1. *Features of the dentition and skeleton distinguishing* Theropithecus *from other genera of Cercopithecidae*

Feature	Reference
Dentition:	
Incisors small relative to cheek teeth	Jolly, 1972; Hylander, 1975
Cheek teeth high crowned with steep sides and increased relief: foveae highly excavated and notches deeply incised; columnar cusps; large distal accessory cuspules on M_2 and M_1; large M_3 hypoconulid	Jolly, 1972; Delson, 1975; Szalay & Delson, 1979
Cheek teeth show a distinctive pattern of wear with the exposure of complexly curved infoldings of enamel and dentinal 'lakes' as cusps become worn	Jolly, 1972; Meikle, 1977
Cheek teeth exhibit extreme interproximal wear, especially anteriorly	Jolly, 1972; Meikle, 1977; Jablonski, 1981
Mandibular cheek teeth oriented in a reversed curve of Spee	Jablonski, 1981; Eck & Jablonski, 1984, 1987
Skull:	
Large infratemporal fossa; anteriorly set temporalis musculature; marked postorbital constriction	Jolly, 1972; Szalay & Delson, 1979; Jablonski, 1981; Eck & Jablonski, 1987
Deep posterior maxilla; short and convex premaxilla; relatively steep anteorbital drop	Szalay & Delson, 1979
Temporomandibular joint set high above occlusal plane; upright mandibular ramus	Jolly, 1972; Szalay & Delson, 1979; Jablonski, 1981; Eck & Jablonski, 1984, 1987
Deep mandibular body and symphysis	Hylander, 1979; Eck & Jablonski, 1984, 1987
Postcranial skeleton:	
Narrow bicipital groove; concave medial border of deltopectoral crest	Krentz, chapter 14
Distinct groove on distal end of deltoid tuberosity	Krentz, chapter 14
Superior border of medial epicondyle horizontal	Krentz, chapter 14
Medial surface of olecranon highly concave	Krentz, chapter 14
Relative elongation of forearm and greater flexibility of elbow and wrist joints	Krentz, 1992
Relative elongation of thumb (especially pollical metacarpal)	Napier & Napier, 1967;

Table 7.1. *Contd.*

Feature	Reference
and relative shortening of index finger (especially proximal and middle phalanges); high opposability index	Maier, 1972a; Etter, 1973; Jablonski, 1986
Robust proximal phalanges	Jolly, 1967
Greater trochanters reduced in length, but with laterally extended articular surfaces	Krentz, 1988, 1992
Femoral shaft angled superiolaterally	Krentz, 1988, 1992

distance, plasma protein electrophoresis, and two types of DNA hybridization, Cronin & Meikle (1979, 1982) concluded that *Theropithecus* differed from *Papio* by little more than did species (*sensu* Cronin & Meikle) within *Papio*. The length of the period of common ancestry of *Papio* and *Theropithecus* is, however, a matter of some debate among specialists in molecular evolution. Cronin & Meikle (1982) estimated that, according to the molecular clock hypothesis, the time of divergence of the two genera was 3.2 Ma, with a standard deviation of 1.3 Ma and a standard error of 0.46 Ma. Working from the results of electrophoretic examination of red blood cell enzymes, however, Lucotte (1982) placed the divergence of *Theropithecus* and *Papio* at 7 Ma.

Although the results of the karyological and molecular studies appear at first glance to be at variance, it is more likely that both sets of results indicate a high level of karyological and molecular conservatism in the Papionini. The karyological evidence may be interpreted to indicate that the ancestral papionin was polymorphic for certain chromosomal variants, including the presence or absence of the large clear-staining band on chromosome 2, and that one form or another became fixed in various evolutionary lines (Stanyon *et al.*, 1988). Karyological conservatism, it is argued, has been promoted in the Papionini because of a demographic structure characterized by extensive gene flow due to male transfer between large, multi-male groups (Stanyon *et al.*, 1988).

Phylogenetic hypotheses for the Papionini

The weight of neontological evidence indicates that, within the Papionini, *Theropithecus* exhibits many gross anatomical distinctions, but far fewer chromosomal or molecular ones. Cronin & Meikle (1982) argued that the 'traditional conclusion' that *Theropithecus* diverged early in papionin history before the *Papio*, *Mandrillus*, and *Cercocebus* lineages separated from one another is based on the 'dubious assumption' that the gross morphological distinctiveness of the genus is indicative of its early branching (p. 473). They contended that the derived gross anatomical features of *Theropithecus* could have arisen at any time and do not necessarily imply early separation.

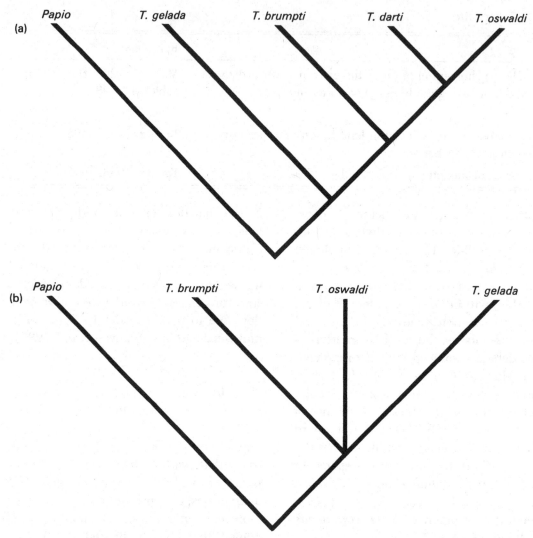

Fig. 7.1. Phylogenetic hypotheses for *Theropithecus* discussed in the text.
(a) According to this view, *T. gelada* would be viewed as the most primitive
species of the genus. A version of this hypothesis presented at the conference by Jolly also showed *T. jonathani* as the most derived species (and
a sister species to *T. oswaldi*). (b) This hypothesis would depict the origin
of the three major lineages of *Theropithecus* as an unresolved trichotomy
based on the absence of compelling morphological information concerning
the affinity of *T. gelada* to either of the other two species groups. According
to Leakey (chapter 3) this scheme could also recognize *T. baringensis* as a
subspecies of *T. brumpti* and *T. darti* as a subspecies of *T. oswaldi*. (c)
According to this scheme, the primary split in the genus is seen as that
between *T. (Omopithecus)* lineage and the *T. (Theropithecus)* lineage. For
details consult the text and Delson (chapter 5). Species and subspecies
abbreviations as in Fig. 5.7, chapter 5.

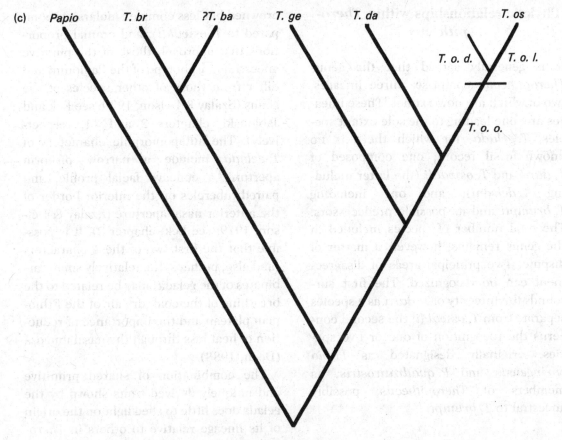

Fig. 7.1 – *cont.*

The 'traditional conclusion' to which they referred is actually an umbrella term for several phylogenetic schemes that placed the divergence of *Theropithecus* just before (Jolly, 1967) or just after (Simons, 1970; Jolly, 1972; Delson, 1975) the separation of macaques from the rest of the papionins, but in all cases before the divergence of the putative common ancestor of *Papio*, *Mandrillus*, and *Cercocebus*.

The phylogenetic hypotheses presented at the Cambridge conference by the proponents of some of these early schemes suggest that the molecular evidence has influenced their phylogenetic hypotheses. Their evolutionary schemes reflected the consensus that, among living papionins, *Theropithecus* is most closely related to *Papio* (Fig. 7.1). The time of divergence of *Theropithecus* and *Papio* is more problematical, but must predate the earliest known fossils of the *T. brumpti* lineage dated at about 3.5–4.0 Ma (Eck & Jablonski, 1987; see Leakey, chapter 3) and the earliest fossils of *T. darti*, now known to be at least 3.3 Ma old (see Eck, chapter 2).

Phyletic relationships within *Theropithecus*

It is generally agreed that the genus *Theropithecus* comprises three lineages, two of which are now extinct. These lineages are: one leading to the sole extant species, *T. gelada*, for which there is no known fossil record; one composed of *T. darti* and *T. oswaldi* (the latter including *T. delsoni*); and one including *T. brumpti* and its possible predecessors. The total number of species included in the genus remains, however, a matter of dispute. Two principal areas of disagreement can be recognized. The first surrounds the integrity of *T. darti* as a species separate from *T. oswaldi*; the second concerns the recognition of one or two species, originally designated as *Papio baringensis* and *P. quadratirostris*, as members of *Theropithecus*, possibly ancestral to *T. brumpti*.

The position of *Theropithecus gelada*

The extant gelada of the Ethiopian highlands has been widely recognized as 'morphologically conservative' (Szalay & Delson, 1979, p. 379) and there is general agreement that, in Jolly's words (1972, p. 115), 'A stock of primitive *Theropithecus* probably became isolated in the Ethiopian highlands, and gave rise to the modern gelada, while the other forms continued to specialize in the lowlands.'

The gelada exhibits several gross anatomical characters that appear to be primitive for the genus, including small body size, slightly reduced incisors, high-

crowned but less complex molars (as compared to *T. oswaldi*), and cranial proportions, that approach those of the putative ancestral morphotype of the Papionini and differ from those of other species of the genus (Szalay & Delson, 1979; see Eck and Jablonski, chapters 2 and 11, respectively). The autapomorphic characters of *T. gelada* include a narrow piriform aperture, a concave facial profile, and paired tubercles on the inferior border of the anterior nasal aperture (Szalay & Delson, 1979; see Eck, chapter 2). It is possible that the first two of these characters (and also, perhaps, the relatively small turbinates of the gelada) may be related to the breathing of the cold, dry air of the Ethiopian plateau and the importance of reduction of heat loss through the nasal mucosa (Dean, 1988).

The combination of shared primitive and uniquely derived traits shown by the gelada does little to shed light on the origin of its lineage relative to others in *Theropithecus*. Furthermore, in the absence of any fossil evidence relating to the origin of the gelada, the time of origin of its lineage can only be inferred by process of elimination from the fossil records of the other lineages. The weight of current evidence suggests the gelada lineage retains a large number of characteristics considered primitive for *Theropithecus*. Its origin must, therefore, have occurred very close to the origins of the *T. darti*–*T. oswaldi* and *T. brumpti* lineages. Some workers (e.g. Jolly) have maintained that the lineage leading to the extant gelada is the most primitive of the genus, implying that gelada lineage split off prior to the division of the remainder of the genus into the

T. darti–T. oswaldi and *T. brumpti* lines. Others (e.g. Leakey) have indicated that there is insufficient evidence to resolve the question of the origin of *T. gelada* lineage relative to the others within the genus, and that it is premature to alter the taxonomy to express phyletic relationships that are not strongly supported by morphological evidence. Eck & Jablonski (1987) suggested that the gelada originated after the divergence of the ancestor of the *T. brumpti* lineage, a hypothesis that would make *T. gelada* the sister taxon of *T. darti*. In this volume, Delson takes this argument further by proposing a formal unification of the *T. gelada* and *T. darti–T. oswaldi* lineages within the subgenus *Theropithecus*. (For further discussion see next section.) These alternative views are reflected in the three phylogenetic hypotheses of the genus presented in Fig. 7.1. Jolly (1972) offered a date of 3.5 Ma for the origin of the gelada lineage, and this would appear to be an acceptable estimate considering that the earliest remains assignable to the *T. brumpti* lineage are 3.5–4.0 Ma in age (Eck & Jablonski, 1987; see Leakey, chapter 3).

The *Theropithecus darti–Theropithecus oswaldi* lineage

The lineage comprising *Theropithecus darti* and *T. oswaldi*[1] has frequently been referred to as the most successful *Theropithecus* lineage because it was extremely long-lived and the most geographically widespread. *Theropithecus darti* has been recognized from two widely separated sites, Hadar in Ethiopia and Makapansgat in South Africa, of mid- to late Pliocene age (see Eck, chapter 2), and possibly also from the Omo (Eck, 1987) in Ethiopia. *Theropithecus oswaldi* is known from a multitude of Plio-Pleistocene sites in northern, eastern, and southern Africa. The North African sites that have yielded *T. oswaldi* are Ain Jourdel, Thomas Quarry, and Ternifine (see Delson, and Delson and Hoffstetter, chapters 5 and 6, respectively). The East African sites that have produced *T. oswaldi* include Olduvai (all beds), East Turkana, West Turkana, Lothagam, Kaiso, Omo, Kanjera, Kapthurin, and Olorgesailie (Jolly, 1972; Eck, 1987; see Leakey, chapter 3). It is interesting that only two of the South African sites, Swartkrans and Hopefield, have yielded remains of the species. Perhaps most remarkable is the fact that *T. oswaldi* may have radiated into Eurasia in Middle Pleistocene times, as demonstrated by the single, well-documented specimen originally referred to as *T. delsoni* from Mirzapur, India, that Delson (chapter 6) has assigned to *T. oswaldi*. A discussion of the distribution of *T. oswaldi* in relation to climatic and geographic factors during the African Plio-Pleistocene is provided by Pickford (chapter 8).

The time range of the *T. darti–T. oswaldi* lineage spans the interval from approximately 3.3 Ma (its estimated first

---

1 Jolly (1972) suggested that the genus *Simopithecus*, which included the species *S. darti* and *S. oswaldi*, be downgraded to the level of subgenus within *Theropithecus*. This proposition has been accepted by most students, but, as a result, the subgeneric name *Simopithecus* is seldom used.

appearance at Hadar) to probably less than 0.4 Ma (its youngest occurrence, as recorded at Thomas Quarry). This is based on the combined time ranges for *T. darti* (approximately 3.3–2.4 Ma, see Eck, chapter 2) and *T. oswaldi* (approximately 2.4–0.4 Ma, see Delson and Leakey, chapters 5 and 3, respectively).

The *T. darti–T. oswaldi* lineage is characterized by several morphological characteristics and trends – associated with a highly terrestrial habitus and exclusive graminivory in lowland grassland habitats – that clearly distinguish it from the other lineages of *Theropithecus*. These attributes have been described at length elsewhere (Jolly, 1972; Eck & Jablonski, 1987) and so need only be listed here: (1) short, weakly developed (or absent) maxillary ridges and shallow maxillary fossae; (2) a short muzzle dorsum and a relatively short muzzle; (3) an oval piriform aperture with margins defined by a ridge; (4) short and very shallow (or absent) mandibular fossae; (5) weakly developed and slightly curved mental ridges; (6) incisors reduced in absolute and relative size; (7) upper and lower canine teeth low-crowned and thick, and sectorial P_3 shortened; and (8) increased size and complexity of the crowns of the cheek teeth.

Theropithecus darti generally manifests the least developed expression of these trends. In addition, the species – especially as it is known from the earliest specimens from Hadar – is distinguished from *T. oswaldi* by several other characteristics, some of which might be considered primitive for the genus. These characteristics, discussed fully by Eck in this

volume, include: (1) a muzzle dorsum with a concave-convex-concave curvature (as opposed to a simple convex curvature in *T. oswaldi*); (2) a narrow oval piriform aperture (instead of the broad oval of *T. oswaldi*); (3) very broadly bowed zygomatic arches, at least in males; (4) a postglenoid process that is short and is separated from the tympanic portion of the temporal bone by a distinct fissure or groove; and (5) small body size (comparable to the gelada). Eck (chapter 2) submits that in several characters (relative sizes of the incisors and the canine mechanism, size, and orientation of the postglenoid process), *T. darti* is more similar to *T. gelada* and *T. brumpti* than it is to *T. oswaldi*. These similarities, in addition to evidence for small body size in the species, suggest the primitive status of *T. darti* and uphold the concept of *T. darti* as a small, primitive species that was directly ancestral to *T. oswaldi* (Jolly, 1972; see Eck, chapter 2).

The fossil record of *Theropithecus oswaldi* reveals a picture of a species that flourished in Africa for nearly two million years. All evidence points to the fact that *T. oswaldi* evolved from *T. darti* approximately 2.4 Ma, and that it firmly established a niche as the primate grazer of the lowland grasslands of the African Plio-Pleistocene (Jablonski, 1981; Eck, 1987). Over time, this adaptation had profound effects on most aspects of the anatomy of the species as well as on its body size. Among the most remarkable of these specializations were those of the masticatory apparatus for the processing of increasingly large volumes of grass. Impressive also was the tremendous increase in body size over time, from an estimated 35 kg for Early Pleistocene males from Kanjera to 65 kg for late Middle Pleistocene males from Olorgesailie and Olduvai (Jolly, 1972; see Leakey, chapter 3). It is interesting to note that specimens from late populations of *T. oswaldi* demonstrated modifications of the temporomandibular joint that suggest that these animals were unable to perform gape displays to expose the canine teeth, and thus possibly did not rely upon dental displays as a means of social control (see Jablonski, chapter 11). For further details of the anatomical trends shown by *T. oswaldi*, the reader is referred to the contributions in this volume by Leakey, Jablonski, Martin, Krentz, and Lee and Foley (chapters 3, 11, 10, 14, and 19, respectively).

Several students have come forward with the suggestion that *T. darti* and *T. oswaldi* be considered as the single species *T. oswaldi* because there are not sufficient differences between the two forms to warrant their separation at the species level (Singer, 1962; Szalay & Delson, 1979; Dechow & Singer, 1984; Leakey, chapter 3). This view has been supported by two interrelated arguments. The first concerns the sample of *T. darti* from Makapansgat, which in certain individuals and in some characteristics is intermediate between the sample of *T. darti* from Hadar and acknowledged specimens of *T. oswaldi* (see Eck, chapter 2). The second argument pertains to the difficulty of dividing a continuous morphocline or chronospecies into separate species. Because *T. darti* and *T. oswaldi* are clearly part of one continuously evolving

lineage, can one draw a meaningful taxonomic line between them? The answer is, to an extent, a matter of personal taste or prejudice. Eck (Eck & Jablonski, 1987; see also chapter 2) recognizes sufficient unique characters in *T. darti* to warrant its retention as a separate species. Leakey (chapter 3) suggests that the differences between continuous segments of the *T. darti–T. oswaldi* lineage are best expressed as the subspecific distinctions *T. oswaldi darti*, *T. oswaldi oswaldi*, and *T. oswaldi leakeyi*.

Modern populations of *Papio* baboons are now considered by many to belong to various subspecies of one species, *P. hamadryas*, because they are potentially or actually interbreeding and because gene flow between populations (at hybrid zones) is significant. (See, e.g. Hayes, Freedman & Oxnard, 1990, however, for a different interpretation.) If our understanding of the patterns of geographical variation among populations of modern *Papio* baboons can be applied to the pattern of variation in the *T. darti–T. oswaldi* lineage through time, recognition of subspecific rather than specific distinctions between the members of the *T. darti–T. oswaldi* lineage may be a more accurate reflection of biological reality. This is likely to remain a controversial and contentious point for some time.

The *Theropithecus brumpti* lineage

Theropithecus brumpti is perhaps the most highly autapomorphic species of *Theropithecus* and one of the most unusual species of fossil primate known.

Its distribution in time and space was narrower than that of the *T. darti–T. oswaldi* lineage, being limited to the period from approximately 2.7–2.0 Ma (Eck & Jablonski, 1987) and apparently restricted to the Turkana Basin as judged by its presence only at the sites of the Omo, East Turkana, and West Turkana. It should be emphasized that this interval saw riverine forests as the dominant habitat in the Turkana Basin and that the replacement of *T. brumpti* by *T. oswaldi* in the Turkana Basin coincides with the invasion of the lake that destroyed the forest (M. Leakey, pers. comm.). (As discussed below and by Leakey, chapter 3, and Eck & Jablonski, 1987, the time range of the *T. brumpti* lineage may extend back farther in time, perhaps to 3.4 Ma or older, in which case the *T. brumpti* lineage would be the most ancient of the genus.)

Theropithecus brumpti is distinguished by a large suite of cranial and dental features related in part to its unique masticatory apparatus, and also by features of its postcranial skeleton that suggest a partly arboreal habitus. These features have been described at length elsewhere (Eck & Jablonski, 1984, 1987; Jablonski, 1986; see Krentz, chapter 14), and will only be listed here. The cranial and dental features unique to *T. brumpti* among species of *Theropithecus* are: (1) long, strongly developed maxillary ridges and deeply excavated, rectangular-shaped maxillary fossae; (2) long and flat muzzle dorsum and relatively long muzzle; (3) piriform aperture with poorly defined margins and convex profile; (4) broad, flaring (and in some specimens, upward curling) anterior zygomatic arch with a greatly enlarged surface area for the attachment of the superficial masseter muscle; (5) robustly built zygomatic arch that is triangular in cross-section at the zygomaticotemporal suture; (6) long and deeply excavated mandibular fossae; and (7) strongly developed and sinusoidally curved mental ridges. Jablonski (chapter 11) suggests that some of these features reflect a feeding adaptation for the species that was quite different from that of other theropiths, possibly involving the eating of large shoots, fruits, or tubers.

Theropithecus brumpti is also distinguished by a unique suite of postcranial characters related to increased flexibility of the shoulder joint and increased stability of the elbow joint (Krentz, chapter 14). These characters, which are fully described and discussed by Krentz (1991; see also chapter 14), include: (1) well-marked areas for the attachment of the rotator cuff muscles of the shoulder on the proximal humerus; (2) a long olecranon process, a relatively low coronoid process, and a relatively shallow trochlear notch, characters all associated with habitual elbow flexion; and (3) large and flattened ventromedial surface of the shaft of the radius indicative of an extensive area of origin of the flexor pollicis longus muscle. These characters suggest that *T. brumpti* was more arboreal than other species of *Theropithecus*, a suggestion that is supported by palaeoenvironmental evidence that the species inhabited riverine forests (Eck & Jablonski, 1987).

While the uniqueness of *T. brumpti* has never been disputed, the origin and relations of the species have been discussed at length without consensus. This has been due, in part, to disagreements over the presence or absence in the fossil record of other species assignable to the *T. brumpti* lineage. Eck & Jablonski (1984) argued that the species originally referred to as *Papio baringensis* and *P. quadratirostris* were more correctly placed in *Theropithecus* and, further, that these species shared several derived characteristics with *T. brumpti* that indicated a close or perhaps ancestral relationship to it. This suggestion has met with a mixture of approval and disapproval. Delson & Dean originally argued that *P. baringensis* and *P. quadratirostris* should remain within *Papio*, based on the facts that: (1) some of the theropith dental features of *P. baringensis* and *P. quadratirostris* cited by Eck & Jablonski (1984) are not clear on the type specimens of the two species; and (2) that other similarities these species share with *T. brumpti* are homoplasies not indicative of common ancestry. Leakey (chapter 3), on the other hand, has supported the placement of *P. baringensis* and *P. quadratirostris* within the *T. brumpti* lineage on the basis of recently discovered fossils from West Turkana that suggest the *in situ* transition from a '*baringensis*' to a '*brumpti*' morphology with a possible '*quadratirostris*' intermediate.

Discussions at the Cambridge conference did not entirely resolve this disagreement, but they led to a clarification of a few important issues and pointed the way toward an eventual consensus. Inspection of relevant fossils (and pictures of fossils) from West Turkana convinced most discussants (with Delson wavering) that at least '*P.*' *baringensis* was represen-

ted at the site and that it was almost certainly a predecessor of *T. brumpti*. Leakey (chapter 3) suggested that the relationship between '*baringensis*' and '*brumpti*' was best expressed by subspecific designations within *T. brumpti*; hence, *T. brumpti baringensis* for the Chemeron and early Turkana forms and *T. brumpti brumpti* for the later Turkana and Shungura (Omo) specimens. *Papio quadratirostris*, it was conceded, was not yet sufficiently well represented in the fossil record to definitely assess its taxonomic affinity. Delson contended that the type specimen of *P. quadratirostris* probably represented a male *Papio* (*Dinopithecus*), heretofore unknown in the fossil record, while Eck and Leakey appeared willing to accept the temporary verdict of Papionini *incertae sedis* for the species. If the earliest appearance of *T. baringensis* marks the divergence of the *T. brumpti* lineage, then the lineage may have originated about 4.0–3.5 Ma, depending on the age determined for the Chemeron fossils. The dating of the 'Basal Grits' at Chemeron from which the *T. baringensis* fossils were retrieved has proven problematical. All that can be said conclusively about the age of the specimens is that they are younger than the underlying Kaparaina Basalt that has an age of approximately 5.5 Ma, (Al Deino, pers. comm. via G. Eck).

Conclusions

The large amount of well-dated fossil evidence available suggests that *Theropithecus* was clearly established before 3.5 Ma. Less persuasive neontological evidence suggests that, among the Papionini, *Theropithecus* is most closely related to *Papio*, and that its time of divergence from *Papio* may be anywhere during the interval from approximately 7.0 to about 4.0 Ma. It seems highly likely that *Theropithecus* and *Papio* share a common ancestor in *Parapapio*.

The evidence currently available suggests that the principal lineages of *Theropithecus* appear to have diverged from one another by approximately 3.5 Ma. Depending on how the phylogenetic relationships within the genus are viewed, any one of three evolutionary scenarios for the genus can be generated. Implicit in all of these scenarios is the recognition that the three main species groups diverged from one another within a short space of time. Scenario 1 would see the origin of the *T. gelada* lineage as the first lineage-splitting event within the genus (Fig. 7.1a). Scenario 2 would depict the origin of the three lineages as an unresolved trichotomy (Fig. 7.1b) based on the absence of compelling evidence concerning the affinity of the *T. gelada* lineage to either of the others. Finally, Scenario 3 would depict a primary split between the *T. brumpti* species group and the others. This could be expressed taxonomically as the distinction between *T. (Omopithecus) brumpti* and its predecessors on the one hand and the *T. gelada* and *T. darti–T. oswaldi* groups united within the *T. (Theropithecus)* subgenus on the other (see Delson, chapter 5).

Of great interest is the nature of the adaptation of the common ancestor of all *Theropithecus*. Evidence from the postcranium of living and fossil *Theropithecus*

species suggests an essentially terrestrial adaptation for the ancestral morphotype of the genus, but it is uncertain whether an exclusively terrestrial habitus, as displayed by the modern gelada, would be indicated for the common ancestral form (see Krentz, chapter 14). In light of the striking similarities in hand morphology between *T. brumpti* and *T. gelada*, the ancestral theropith morphotype appears to have possessed a hand specialized for 'manual grazing' (Jablonski, 1986). Was this, from the beginning, a specialization for plucking grass parts or was it a modification that facilitated the harvesting of a variety of plant parts of small size? There is no conclusive evidence that permits either of these hypotheses to be eliminated and, clearly, one's view of the nature of the original adaptation of the genus depends on the phylogenetic hypothesis one subscribes to. The least well known lineage, leading to the extant gelada, became isolated in the Ethiopian highlands and gradually evolved the characteristics of a primate grazer adapted to montane conditions. The *T. darti–T. oswaldi* lineage, in contrast, spread widely into lowland grassland habitats in Africa (and probably Eurasia), its gradual but marked increase in body size through time made possible by the rich environments which it favoured and its relatively efficient mode of processing large amounts of vegetation (see Dunbar, and Lee and Foley, chapters 18 and 19, respectively). The most highly apomorphic species of the genus, *Theropithecus brumpti*, exhibited an interesting mixture of characters that reflect an adaptation to forest life and eclectic herbivory that is in stark contrast to the life-styles of the highly terrestrial graminivores that make up the rest of the genus. *Theropithecus brumpti* was, no doubt, one of the most unusual looking primates of any age.

References

BOUVIER, M. (1986). A biomechanical analysis of mandibular scaling in Old World monkeys. *American Journal of Physical Anthropology*, 69, 473–82.

CRONIN, J.E. & MEIKLE, W.E. (1979). The phyletic position of *Theropithecus*: congruence among molecular, morphological, and paleontological evidence. *Systematic Zoology*, 28, 259–69.

CRONIN, J.E. & MEIKLE, W.E. (1982). Hominid and gelada baboon evolution: agreement between molecular and fossil time scales. *International Journal of Primatology*, 3, 469–82.

CRONIN, J.E., CANN, R. & SARICH, V.M. (1980). Molecular evolution and systematics of the genus *Macaca*. In *The Macaques: Studies in Ecology, Behavior and Evolution*, ed. D.G. Lindburg, pp. 31–51. New York: Van Nostrand Reinhold Company.

DEAN, M.C. (1988). Another look at the nose and the functional significance of the face and nasal mucous membrane for cooling the brain in fossil hominids. *Journal of Human Evolution*, 17, 715–18.

DECHOW, P.C. & SINGER, R. (1984). Additional fossil *Theropithecus* from Hopefield, South Africa: a comparison with other African sites and a reevaluation of its taxonomic status. *American Journal of Physical Anthropology*, 63, 405–35.

DELSON, E. (1975). Evolutionary history of the Cercopithecidae. In *Approaches to Primate Paleobiology*, ed. F.S. Szalay, pp. 167–217. Basel: Karger.

ECK, G.G. (1987). *Theropithecus oswaldi* from the Shungura Formation, Lower Omo Basin, southwestern Ethiopia. In *Les Faunes Plio-Pléistocènes de la Basse Vallée de l'Omo (Éthiopie)*, Tome 3. Cercopithecidae de la Formation de Shungura, ed. Y. Coppens &

F.C. Howell, pp. 124–39. Cahiers de Paléontologie, Travaux de Paléontologie Est-Africaine. Paris: Editions du Centre National de la Recherche Scientifique.

ECK, G.G. & JABLONSKI, N.G. (1984). A reassessment of the taxonomic status and phyletic relationships of *Papio baringensis* and *Papio quadratirostris* (Primates: Cercopithecidae). *American Journal of Physical Anthropology*, 65, 109–34.

ECK, G.G. & JABLONSKI, N.G. (1987). The skull of *Theropithecus brumpti* compared with those of other species of the genus *Thero-pithecus*. In *Les Faunes Plio-Pléistocènes de la Basse Vallée de l'Omo (Éthiopie)*. Tome 3. Cercopithecidae de la Formation de Shungura, ed. Y. Coppens & F.C. Howell, pp. 11–122. Cahiers de Paléontologie, Travaux de Paléontologie Est-Africaine. Paris: Editions du Centre National de la Recherche Scientifique.

ETTER, H.F. (1973). Terrestrial adaptations in the hands of Cercopithecinae. *Folia Primatologica*, 20, 331–50.

HAYES, V.J., FREEDMAN, L. & OXNARD, C.E. (1990). The taxonomy of savannah baboons: an odontomorphometric analysis. *American Journal of Primatology*, 22, 171–90.

HYLANDER, W.L. (1975). Incisor size and diet in anthropoids with special reference to Cercopithecidae. *Science*, 189, 1095–8.

HYLANDER, W.L. (1979). The functional significance of primate mandibular form. *Journal of Morphology*, 160, 223–40.

JABLONSKI, N.G. (1981). *Functional analysis of the masticatory apparatus of* Theropithecus gelada *(Primates: Cercopithecidae)*. Ph.D. Dissertation, University of Washington, Seattle.

JABLONSKI, N.G. (1986). The hand of *Thero-pithecus brumpti*. In *Primate Evolution – Selected Proceedings of the Tenth Congress of the International Primatological Society*, vol. 1, ed. J.B. Else & P.C. Lee, pp. 173–82. Cambridge: Cambridge University Press.

JOLLY, C.J. (1966). Introduction to the Cercopithecoidea, with notes on their use as laboratory animals. *Symposium of the Zoological Society, London*, 17, 427–57.

JOLLY, C.J. (1967). The evolution of the baboons. In *The Baboon in Medical Research*, ed. H. Vagtborg, pp. 23–50. Austin: University of Texas Press.

JOLLY, C.J. (1972). The classification and natural history of Theropithecus (Simopithecus) (Andrews, 1916), baboons of the African Plio-Pleistocene. Bulletin of the British Museum (Natural History), Geology. 22, 1–123.

KRENTZ, H. (1988). The femur of Theropithecus: evidence for the appearance of shuffling behavior. American Journal of Physical Anthropology, 75, 234.

KRENTZ, H. (1992). Functional analysis of the postcranial skeleton of the species of Theropithecus. Ph.D. Dissertation, University of Washington, Seattle.

LUCOTTE, G. (1982). Distances electrophorétiques entre les différentes espèces de babouins du genre Papio et la gelada Theropithecus. Biochemical Systematics and Ecology, 10, 99–101.

MAIER, W. (1972a). Anpassungstyp und systematische Stellung von Theropithecus gelada Rüppell, 1835. Zeitschrift für Morphologie und Anthropologie, 63, 370–84.

MAIER, W. (1972b). The first complete skull of Simopithecus darti from Makapansgat, South Africa, and its systematic position. Journal of Human Evolution, 1, 395–405.

MEIKLE, W.E. (1977). Molar wear stages of Theropithecus gelada. Kroeber Anthropological Society Papers, Number 50, pp. 21–5.

NAPIER, J.R. & NAPIER, P.H. (1967). A Handbook of Living Primates. New York: Academic Press.

SIMONS, E.L. (1970). The deployment and history of Old World monkeys (Cercopithecidae, Primates). In Old World Monkeys, ed. J.R. Napier & P.H. Napier, pp. 97–138. New York: Academic Press.

SINGER, R. (1962). Simopithecus from Hopefield, South Africa. Bibliotheca Primatologica, 1, 43–70.

STANYON, R., FANTINI, C., CAMPERIO-CIANI, A., CHIARELLI, B. & ARDITO, G. (1988). Banded karyotypes of 20 Papionini species reveal no necessary correlation with speciation. American Journal of Primatology, 16, 3–17.

STRASSER, E. (1988). Pedal evidence for the origin and diversification of cercopithecid clades. Journal of Human Evolution, 17, 225–45.

STRASSER, E. & DELSON, E. (1987). Cladistic analysis of cercopithecid relationships. Journal of Human Evolution, 16, 81–99.

SZALAY, F.S. & DELSON, E. (1979). Evolutionary History of the Primates. New York: Academic Press.

PART II

Biogeography and evolutionary biology

8   Climatic change, biogeography, and *Theropithecus*

MARTIN PICKFORD

Summary

1. Examination of the detailed chronological and geographical distribution of *Theropithecus* sites throughout Africa has indicated that the diversity and distribution of the genus were greatly influenced by global and regional climatic changes, as well as by interactions with predators and herbivores.

2. The small species range of the extant species *Theropithecus gelada* is not unusual compared with those of other anthropoid primates in Africa. What is unusual and in need of explanation is the very large range occupied by *T. oswaldi* in Africa during different parts of the Pleistocene.

3. If it is accepted that the known chronological ranges of fossil *Theropithecus* are representative of the actual distributions, then it can be determined that *T. oswaldi* occupied huge ranges (two or three biogeographic provinces) for only 32 per cent of the duration of the lineage. *Theropithecus oswaldi* lived for about 2.5 Ma, but only occurred in southern Africa and the Maghreb for about 0.8 Ma of this period. For the remainder of the time it was confined to tropical Africa.

4. The distribution patterns of extinct species of *Theropithecus* appear to be correlated with changes in climate in eastern Africa brought about by the rise of the 'Roof of Africa' during the Plio-Pleistocene. Other changes in species ranges seem to be related to global-scale climatic changes which resulted in latitudinal shifts in the boundary between the Ethiopian and the Palaearctic biogeographic realms, while still others may have been related to the radiations of *Papio* and *Homo* during the Pleistocene. Less likely causes of changes in the distribution of *Theropithecus* are represented by other herbivores, in particular the bovids and suids, although among the latter group the genus *Phacochoerus* may have contributed toward the shrinkage of *Theropithecus* territory.

Introduction

Theropithecus species have been recorded in various parts of Africa from strata ranging in age from 4 Ma to the present day. The genus has also been reported to occur in Indian Pleistocene deposits. Despite the fact that extant *T. gelada* has a very restricted species range (Figs 8.1–8.3), fossil species of the genus have been found at the two latitudinal extremities of Africa, well beyond the limits of all extant African primate species other than humans. Indeed, the only other cercopithecid genus which can compete with fossil *Theropithecus* regarding the large size of its species range is *Macaca*, which occurs widely in the Palaeartic and Oriental regions.

The aim of this chapter is to examine the distribution of *Theropithecus* species in time and space, and to search for the causes which led to fluctuations in its species range.

Theropithecus fossil record

Theropithecus fossils have been reported from numerous sites in northern, eastern, and southern Africa, ranging in age from about 4 Ma to the present day (Eck & Jablonski, 1987). Geraads (1987) provides a detailed history of North African cercopithecids based on mammalian biostratigraphy, and lists *Theropithecus* at Ain Jourdel (2.5 Ma), Ternifine (0.7 Ma) and Thomas Quarries (0.4 Ma) (see also Delson, chapter 5.) The oldest of these occurrences is *T. atlanticus* (Thomas, 1884) while the other two records are probably referable to the species *T. oswaldi* (Andrews, 1916).

The most recent reviews of East African *Theropithecus* species are those of Eck & Jablonski (1984, 1987). The earliest known species is *T. baringensis*, although there has been some confusion about the age of the strata from which the specimen came. This debate is summarized by Eck & Jablonski (1984). In short, the specimen came from the Chemeron Formation, locality JM 90/91, thought to be about 4 Ma old on stratigraphic grounds, but said to be much younger (about 2 Ma) on the basis of an *Elephas recki* molar found nearby. For the purposes of this discussion, I concur with Eck & Jablonski (1984) that the fossil is of lower Pliocene age.

From about 3.4 Ma until about 0.7 Ma, the East African record of *Theropithecus* species is more or less continuous, with fossils from numerous localities arranged along the length of the Gregory Rift Valley from as far north as Djibouti to as far south as northern Tanzania. A few specimens have come from localities outside this rift valley, including a molar from the Moyo Valley, Nyeri (Pickford, 1987) in the Kenya highlands, a molar from Marsabit Road dated about 2.3 Ma and many specimens from the type locality of *T. oswaldi* (Kanjera, Kenya), originally thought to be associated with a Middle Pleistocene

Fig. 8.1. Geographic and geochronologic distribution of *Theropithecus* species. Taxa followed by question marks are of uncertain systematic status, but have been attributed to *Theropithecus* at one time or another. Glacial periods marked as dotted rectangles on left margin.

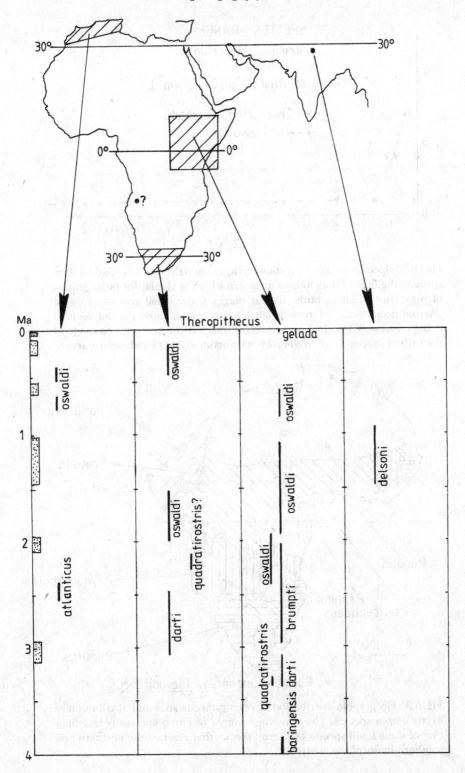

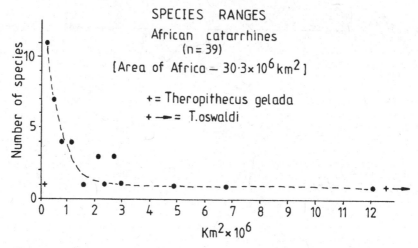

Fig. 8.2. Species ranges of extant African catarrhines. The dashed line through the field of dots follows a pattern which is similar for other groups of organisms, such as birds. In this diagram, the small species range of *Theropithecus gelada* (cross in left hand corner of curve) is not seen as being unusual. What does seem extraordinary is the vast species range of the extinct species *T. oswaldi* (cross with arrow at right hand end of curve).

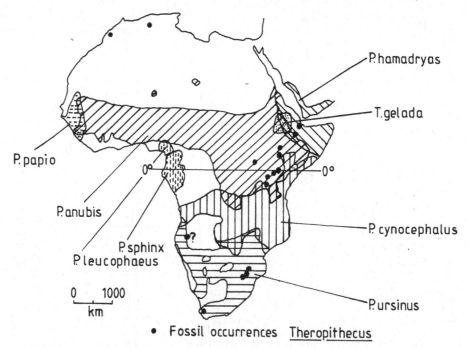

Fig. 8.3. Geographic distribution of *Theropithecus* sites and distribution of extant *Papio* species. The combined ranges of *Papio* species is less than that of some fossil species of *Theropithecus* that reached the northern and southern limits of the continent.

fauna, but now known to be much older (about 2 Ma) (Pickford, 1986). A maxilla fragment from Kaiso Village (Uganda) is probably about 2.3 Ma old, as is a specimen from Senga Zaire (Boaz, 1990).

In contrast, the fossil record of *Theropithecus* in southern Africa is much less continuous than that in East Africa. According to Delson (1984) the genus was present in South Africa from about 3.0 to 2.5 Ma ago (*T. darti*), and then from 2.0 to 1.5 Ma ago (*T. oswaldi*) with a final record about 0.4 Ma old (*T. oswaldi* at Hopefield). Delson (1984) reports that the species *Papio (Dinopithecus) quadratirostris* occurs in Leba, Angola, in association with *Cercopithecoides williamsi*, but until the fossils have been described and illustrated, it will not be possible to determine their relationship to *T. quadratirostris* from Ethiopia.

Judging from the chronological and geographic distribution of fossil theropiths, equatorial Africa seems to have been the source area for all species, and that, from time to time, some species were able to spread towards higher latitudes, on occasion reaching the northern and southern extremities of the continent (Fig. 8.1).

The first of the widespread species was *T. darti*, which appears to have originated in tropical Africa, and then to have spread to southern Africa. Its chronological range in East Africa is older than its South African records, if the correlations are reliable (Delson, 1984).

Theropithecus darti is apparently the species most likely to have given rise to the long-lived species *T. oswaldi*, which occurs throughout the Pleistocene period in East Africa, but only for short periods in northern and southern Africa. It seems unlikely that South African *T. darti* independently gave rise to *T. oswaldi*. It would appear more likely that the transition from one species to the other took place in tropical Africa, and that from an initially restricted species range in the tropics *T. oswaldi* expanded its species range until it reached southern Africa.

It appears that *T. oswaldi* then became extinct in South Africa about 1.5 Ma ago, and that it only re-colonized the southern end of the continent towards the end of the Pleistocene (Hopefield dated about 0.4 Ma ago). An alternative scenario that *T. oswaldi* was present in South Africa throughout the Pleistocene period, but has not been found in sites aged between 1.5 and 0.5 Ma, seems less likely. This is because there was a significant size increase in East African *T. oswaldi* during the Pleistocene period, and the Hopefield specimens resemble the younger of the East African members of the species, suggesting a close genetic relationship between the two groups. Otherwise, contemporaneous parallel evolution or continuous gene flow between eastern and southern Africa must be invoked.

In summary, therefore, it would appear that *Theropithecus* species colonized southern Africa on at least three separate occasions (Late Pliocene, Early Pleistocene and Late Pleistocene).

The North African records of *Theropithecus*, all in the Maghreb, north of 30° north latitude, are much less complete than those from East and South Africa. Nevertheless, it appears that *Theropithecus* species colonized the northern tip of Africa at least twice during the

Pleistocene, along with other characteristically tropical herbivores such as bovids, giraffids, and rhinocerotids, once at its inception (*T. atlanticus*) and once in the Middle Pleistocene (*T. oswaldi*). The remarkable point about the spread of *Theropithecus* to the latitudinal limits of the continent, is that successive colonizations of northern and southern Africa appear to have been contemporaneous within the limits of accuracy of age determinations. Given the uncertainties of making correlations between northern, eastern, and southern Africa, it is surely no coincidence that both time periods during which *Theropithecus* occurs in North Africa should correlate reasonably closely with periods when the genus was present in South Africa. If the spread of *Theropithecus* to the opposite ends of the continent from an equatorial base, did indeed occur simultaneously two or three times, as the admittedly limited data (Fig. 8.1) seems to indicate, then we should look for causes at least at the scale of the continent, and preferably at the scale of the globe. Localized environmental changes, for example, would be unlikely to produce the simultaneous spread of the genus both northwards and southwards by huge distances.

Theropithecus delsoni was reported from a locality 1.4 km west-south-west of Mirzapur Rest House, north of Chandigarh, India. It was collected from the Tavi Formation of Middle Pleistocene age (Gupta, 1977; Gupta & Sahni, 1981).

The holotype, and thus far the only known specimen of *T. delsoni*, was first recognized as belonging to a cercopithecid primate by myself during a visit to Chandigarh in 1977. The specimen was on display in the Museum of the Geology Department of the University of Chandigarh, labelled as belonging to a suid, together with details of its discovery locus. I pointed out the misidentification of the specimen to V.J. Gupta, and discussed the importance of the fossil (only 17 fossil cercopithecids have ever been recorded from the Siwalik succession of the subcontinent). During the conversation which ensued, I mentioned that the specimen should be compared to *Papio subhimalayanus*, but that functional comparison between it and *Theropithecus* would be interesting to do, because the molar cusps were high and infolded, indicating the possibility of similar diets in the Siwalik species and *Theropithecus*. Plans were made for publishing a jointly authored paper after the necessary studies had been done, but these were dropped when Gupta (1977) published a single authored paper within a couple of months of our conversation.

It is of great interest that the genus *Theropithecus* reached the Indian Subcontinent during the Middle Pleistocene (Barry, 1987). (See also Delson, chapter 5.)

Ranges of *Theropithecus* species

It has often been remarked that the present day species range of *T. gelada* is minute when compared to the known fossil distribution of the genus *Theropithecus*. Simons & Delson (1978) commented that the 'extinction of such a successful and widespread taxon is always a problem of some interest' and they list a number of ideas that have been put forward to explain the 'extinction' of the genus over much of its former range. They list human hunting, changing climatic conditions, competition for resources by *Papio* species, and behavioural subordination to *Papio*, as possible factors related to the fact that *Theropithecus* is now only a 'relict and probably ancient population' surviving in upland Ethiopia. Interactions with *Papio* are 'infrequent', presumably because geladas live between 2000 and 5000 m altitude (Haltenorth & Diller, 1977), and *Papio* only ranges up to 2300 m altitude (Wolfheim, 1983) in Ethiopia.

While the present range of *Theropithecus* is strikingly smaller than the ranges of some (but not all) extinct species of the genus, it should be pointed out that its small species range is not unusual when compared with species ranges of other cercopithecoids (Fig. 8.2). Indeed, species ranges as extensive as that of *T. oswaldi* rather than the diminutive range of the gelada represent the unusual situation which requires special explanation. Out of 39 species of catarrhines living in Africa, 22 have species ranges less than 1 million square kilometres in extent, and only three species have ranges greater than 4 million square kilometres. Of these the most widespread species, other than humans, is the vervet monkey with a species range of about 12 million square kilometres. The latter range is about the same size as the combined ranges of the 'savanna' baboons (*P. ursinus*, *P. cynocephalus*, *P. hamadryas*, *P. anubis*, and *P. papio*) (Wolfheim, 1983). The range of *T. oswaldi* was potentially much greater than that of the vervet monkey or of 'savanna' baboons (cross with arrow at the right-hand end of Fig. 8.2) on the basis of fossil discoveries in the Maghreb, East and South Africa (Fig. 8.3). It is this fact which is interesting and unusual, not the small species range of *T. gelada*.

Figure 8.4 summarizes the species range data for *Theropithecus* during the Plio-Pleistocene. Out of seven currently recognized species, only *T. oswaldi* occurred simultaneously in more than one biogeographic province, and only for a very short time did it expand into three biogeographic provinces (northern, eastern, and southern Africa). If *T. atlanticus* is a synonym of *T. brumpti* or *T. darti*, then this species would represent a second lineage with a large species range. *T. darti* has been recorded from two provinces (eastern and southern Africa) but if the biostratigraphic correlations are reliable (Delson, 1984) it appears not to have occupied both at the same time. The presence of the species *T. quadratirostris* at Leba, Angola (Delson, 1984), needs to be verified before it can be considered a widespread species occurring in two provinces (eastern and south-western Africa). (For related discussion on the affinity of this species, see Delson and Dean, chapter 4.)

Lumping of species-range data, as has been done in the left-hand column of Fig. 8.4, yields a very different picture from that in which the species are treated individually. It is preferable to examine the past distribution of organisms at the species level, if one is to understand the factors which led to variation in the size of areas occupied by those organisms.

The right-hand column of Fig. 8.4 reveals that the only convincing evidence

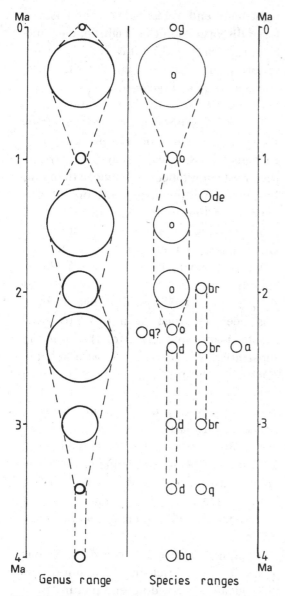

Fig. 8.4. Reconstructed species and genus ranges of *Theropithecus* through time. Small circles: species recorded from one biogeographic province. Medium sized circle: species recorded from two biogeographic provinces. Large circles: species recorded from three biogeographic provinces. Abbreviations: a = *atlanticus**; ba = *baringensis**; br = *brumpti*; d = *darti*; de = *T. delsoni**; o = *oswaldi*; q = *quadratirostris**. * = Species known from single specimens.

for the existence of a widespread species of *Theropithecus* concerns the species *T. oswaldi*. From a centre in tropical Africa it spread to the southern end of the continent by about 2 Ma, where it survived for about 0.5 Ma, before its range shrunk back to the tropics alone. About 0.7 Ma ago, the same species, albeit somewhat larger in body size than were previous populations, experienced a second expansion of species range, this time to include both southern and northern Africa. In so doing, *T. oswaldi* broke all records for size of species range among African catarrhine primates other than that of *Homo*. By about 0.4 Ma the range of *T. oswaldi* shrunk again, until it became extinct.

Interpretation of *T. atlanticus* is difficult in the present context, because it is so poorly known, and because its relationships to other theropiths remain unclear. It appears to be a contemporary of the species *T. darti*, *T. brumpti*, and possibly, early populations of *T. oswaldi*, of which it may be a synonym. Until its systematic position has been determined, it is not possible to be precise, but it is possible to say that about 2.5 Ma ago the genus range of *Theropithecus* included northern, eastern, and southern parts of Africa.

Historical zoogeography of *Theropithecus*

An interesting point about the North African cercopithecid record, is that there appears to have been mutual exclusion between the genera *Theropithecus* and *Macaca* (Geraads, 1987). Macaques are known from strata aged about 3 Ma old

(Ichkeul, Ain Brimba) in the Maghreb, after which they are unknown in the region until about 0.2 Ma ago, probably re-entering Africa from Europe where populations lived throughout the Pleistocene (Ardito & Mottura, 1987). During the period of probable absence of *Macaca* from North Africa, the genus *Theropithecus* occurs there sporadically, presumably spreading northwards from the tropics.

At the present time, the Maghreb comprises part of the Palaearctic biogeographic realm, to which *Macaca sylvanus* belongs. The presence of *Theropithecus* and numerous other tropical mammalian lineages in Pleistocene strata of the Maghreb (Geraads, 1987) suggests that at times the zoogeographic affinities of the region lay with the Ethiopian biogeographic realm rather than with the Palaearctic. These primates, as well as other faunal elements including suids, giraffids, and bovids tell much the same story regarding the biogeographic affinities of the Maghreb. The region seems to have alternated between having Palaearctic and Ethiopian affinities, in phase with global climatic fluctuations.

During glacial periods, the boundaries of the various latitudinally arranged climato-ecological belts (Arctic, Taiga, Tundra, Boreal, Temperate, etc.) shifted equator-wards, while during interglacial periods they moved back towards higher latitudes. The Maghreb region, between 30° and 35°, is located at a latitude which is close to the present-day boundary between the temperate and tropical zones, and during large-scale global climatic changes, such as occurred during the Pleistocene period, its biogeographic affinities changed according to which side of the Palaearctic–Ethiopian boundary zone it lay. It remains to be determined whether the fossil record accords with this scenario.

The three levels at which *Theropithecus* occurs in the Maghreb (Geraads, 1987) all correlate with interglacial periods of Europe, whereas the records of *Macaca* in the same region coincide with periods of glaciation in Europe. Even though the quantity of data is not very great, the little that we do have accords well with the scenario of latitudinally fluctuating biogeographic boundaries. Further studies along these lines, using other mammalian species and plants would be well worth doing. As it is, the past distribution of *Theropithecus* and *Macaca* and other Palaearctic and Ethiopian faunal elements seems to be pointing towards an interesting line of research.

Global scale climatic changes of the sort that gave rise to the series of glacial and interglacial periods in the northern hemisphere, are considered to have had comparable effects in the southern hemisphere. The southern tip of Africa, being at a similar latitude south of the equator as the Maghreb is to the north of it, provides additional evidence related to the scenario. Even though the record is spotty, records of *Theropithecus* in southern Africa are located in parts of the column correlated with interglacial periods, while at least parts of the periods during which this genus is apparently lacking in southern Africa correlate with glacial periods. The evidence, as it is, does not run contrary to the notion that latitudinal fluctuations in the range of *Theropithecus* were to some extent controlled by global scale climatic fluctuations of the Plio-Pleistocene.

Eastern Africa, being located along the equator was, as it were, at the centre of activity, and had a continuous occupation by the genus. Not to say that species ranges stayed constant in the tropics, they did not (Eck & Jablonski, 1987), as shown by variations in the distribution patterns of *T. brumpti* and *T. oswaldi* during the Early Pleistocene (Cooke & Coryndon, 1970; Boaz, 1990).

Regional scale, as opposed to global scale changes

Global scale changes during the Pleistocene produced such impressive results in the high latitudes that there has been a tendency to 'impose' global scale factors on the tropics while interpreting the palaeoclimatology of such regions. In East Africa, three decades of preoccupation with 'pluvial-interpluvial' stratigraphy as a tropical equivalent of the glacial–interglacial stratigraphy of Europe, led to much confused thought and circular reasoning. The climatostratigraphic time scale was abandoned as a tool in East Africa after the definitive proof (Cooke, 1958) that its evidential foundation was better interpreted in terms of local geotectonic and volcanic history, rather than global climatic change.

Following initial suggestions by Soloman (1939), that all the evidence said to support the sequence of pluvial-interpluvial periods, was better interpreted in terms of local geomorphologic history (tectonics related to rifting, volcanicity, erosion, transportation of sediments, and deposi-

tion), the Uganda Palaeontology Expedition has been examining the region of the Western Rift valley with this advice in mind. What has emerged is evidence for geomorphological changes of sufficient magnitude to account, at least in part, for many of the changes reported to have occurred in East Africa's fauna and flora during the Plio-Pleistocene (Pickford, 1990). This evidence does not deny the existence of global-scale climate changes that had effects on East African faunas and floras. What it does is modify these effects, sometimes to such an extent that the 'globally produced' signal is swamped by that generated locally (Pickford, 1991).

Figure 8.5 illustrates tropical Africa up to 10° latitude either side of the equator. There is a strong 'latitudinal' signal (due to global-scale effects) represented by the distribution of tropical rainforest. But this signal is abruptly interrupted in Eastern Zaire/Western Uganda by the 'rise to the rift' culminating in the 'Roof of Africa'. The latter is a chain of mountains (mostly of tectonic origin, but several of volcanic affinities) over 700 km long oriented approximately north–south. The region to the east of this mountain chain (much of Uganda, Kenya, Ethiopia, Somalia, and northern Tanzania) lies in a rain shadow for those climatic systems derived from the west, dependent ultimately on moisture obtained from the Gulf of Guinea (Coppens, 1986). Without this north–south mountain range, which culminates in altitudes of 5000 m (Ruwenzori) covered in permanent glaciers, East Africa's present-day climate would be very different from what it is. The belt of humidity would extend considerably

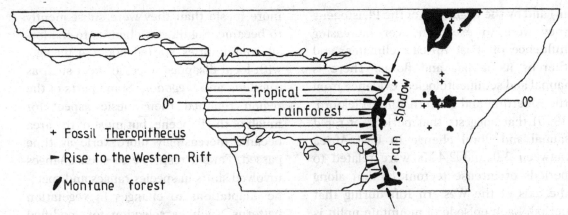

Fig. 8.5. Central Africa: effect of the 'Roof of Africa' on the distribution of rainforest and rainshadow in East Africa.

further to the east, perhaps as far as the Indian Ocean.

The most interesting fact to emerge recently is that the uplift of this mountain chain occurred relatively recently. The earliest evidence for rifting in the Albertine Rift, to which mountain uplift is genetically related, took place during the upper Miocene, perhaps about 8–7 Ma ago (Fig. 8.6). Faulting was episodic,

interspersed with long periods of relative tectonic calm, and it was incremental. Thus during the latest Miocene and the Plio-Pleistocene, the Roof of Africa experienced several episodes of rapid uplift, followed by periods of relative stability. Initially quite low (altitude about 800–1200 m above sea-level) the mountains grew to be climatically important during the Pliocene (altitudes up to 2000

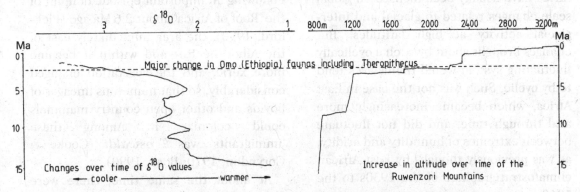

Fig. 8.6. Comparison of global (left frame) and regional (right frame) events upon climatic change. Global climatic changes have been postulated to have caused faunal and floral changes in East Africa (at the Omo, for example: dotted line at 2.4 Ma), but the uplift of the 'Roof of Africa' (jagged curve on the right) is likely to have played an important role, if not an overriding one, in influencing the xerification of East Africa during the Plio-Pleistocene.

m) and by the beginning of the Pleistocene they were to exert an ever increasing influence on East Africa's climates, and thereby its faunas and floras. There is faunal and sedimentological evidence from the Albertine and Eastern Rifts (Pickford, 1990) that suggests that periods of rapid faunal and floral change in East Africa between 3.0 and 2.4 Ma were related to periods of intense tectonic action along the axis of the Western Rift during that period. Each episode of mountain uplift is considered to have led to increased severity of the rain shadow effect in East Africa. The punctuated nature of mountain uplift is recorded as punctuated faunal and floral changes in the area of rain shadow, each episode of uplift leading to increased aridity in East Africa. Many of the faunal and floral changes reported from East African Plio-Pleistocene sites accord with this reconstruction of events (Pickford, 1990).

Previous studies of faunal change in East Africa have usually been focused on global scale changes related to glacial and interglacial activity at high latitudes. But changes brought about by such a cyclically fluctuating system would themselves tend to by cyclic. Such was not the case in East Africa, which became increasingly more arid through time, and did not fluctuate between extremes of humidity and aridity, as was previously thought by East African climatostratigraphers of the 1930s to the 1960s.

Up to 4 Ma ago, much of East Africa was more humid than it is now. Most of the sites of lower Pliocene age, such as Chemeron in Kenya, contain faunas which indicate that the vicinity of the sites was more mesic than they were subsequently to become. Significant changes to the fossil communities started about 4 Ma ago, with local disappearances of taxa such as *Anancus* and *Stegodon*. Some parts of the region retained their mesic aspect for much of the Pliocene, but most of the area became increasingly more xeric as time passed. Faunal response to such changes involved shifts in species ranges and specific adaptations to changes in vegetation patterns, such as selection for modified dento-gnathic or locomotor systems. By about 2.6 Ma ago, there was very little, if any, thick forest left in East Africa, except as small patches on mountain tops and as galleries along river courses.

To some extent the distribution of *Theropithecus* was influenced by these changes. For example, *T. oswaldi* would have been an unlikely inhabitant of the Albertine Basin before 2.6 Ma ago, because the region was most likely surrounded by tropical forest and forest-savanna mosaic. Following an important episode of uplift of the 'Roof of Africa' about 2.6 Ma ago (Pickford, 1990), the area immediately east of the Albertine Rift and within it became more xeric, and the vegetation changed considerably, so that numerous lineages of bovids and other open country mammals could colonize it. Among these 'immigrants' was *T. oswaldi* (Cooke & Coryndon, 1970; Boaz, 1990).

At about the same time, there were changes in the Omo region of Ethiopia, with much shifting of species ranges and increased evolutionary activity. Nearly all groups of herbivores were affected during this period, including *Theropithecus*. Response among theropiths was of two kinds: shift of species ranges and autochthonous evolution. Eck & Jablonski (1987) and Eck (1976) discuss the changes in representation of *T. brumpti* and *T. oswaldi* in the Omo sequence, and postulate their ecological separation (probably incomplete). *Theropithecus brumpti* was present before the change at 2.6 Ma but afterwards survived in reduced frequency, whereas *T. oswaldi* either migrated into the region from pre-existing populations elsewhere, or it evolved locally due to selective forces brought about by the changes. Either way, it would appear that uplift of the 'Roof of Africa' was the 'trigger' for these changes, rather than global scale or other effects that we are unaware of.

Possible competitors: predators

Another primate lineage which responded to these Early Pleistocene changes is represented by the Hominidae, a group that was to become more and more important to *Theropithecus*, especially if the hunting of giant geladas at Olorgesailie actually occurred, but this was to be deep in the future (Olorgesailie is about 0.7 Ma old).

The fossil record reveals that *Theropithecus* and *Homo* survived together in Africa throughout the Pleistocene. *Theropithecus oswaldi* experienced its greatest expansion of species range during the Acheulian period, despite postulated hominid hunting pressures. For most of its history, therefore, the distribution of *Theropithecus* was presumably not greatly affected by hominid activity. It would appear that climatic changes, due to both

global and regional factors were ultimately more influential in this respect. The great latitudinal fluctuations in species ranges were probably directly related to global climatic changes of the glacial–interglacial type. The less spectacular species range changes, and the more important speciation events may well have been responses on the part of theropiths to regional climatic changes due to the episodic uplift of the 'Roof of Africa'.

However, humans may have had an increasing effect on the distribution of *Theropithecus* during the recent past, say from about 0.3 Ma until the present. During this period, technological advances and increasing human populations have led to ever increasing habitat alteration, which have affected virtually all organisms in the continent in one way or another. Improved hunting techniques were invented at about this time, with the development of hafted tools and long distance projectiles, and it is possible, though undemonstrated, that such improvements might have led to the reduction of the geographic range of theropiths or that it led to their extinction in all of Africa except for the Ethiopian highlands.

Possible competitors: herbivores

Competition from other herbivores may have impinged more seriously upon the distribution patterns of theropiths. An obvious candidate in this respect is the genus *Papio* which has one of the largest distribution ranges among African primates (Fig. 8.3). The fossil record of *Papio* is much less well known than that of *Theropithecus*. Many former records of early *Papio* have been based on specimens of *Theropithecus*. It is actually rather difficult to demonstrate the existence of this genus in the Pleistocene of East Africa, several of the supposed East African *Papio* listed by Delson (1984) having subsequently been identified as *Theropithecus*. In southern Africa, the presence of *Papio* in the early Pleistocene is more securely demonstrated (Delson, 1984). It is feasible that the spread of *Papio* throughout much of sub-Saharan Africa took place late in the Pleistocene, and that the spread of this genus led to competition for resources (not necessarily solely dietary ones but possibly sleeping places) with *Theropithecus*, which has only managed to survive in high-altitude grassland which *Papio* species are apparently unable to occupy successfully.

Among non-primate herbivores, there are few direct competitors of *Theropithecus* that would be likely to have affected its distribution throughout its former vast range. There are no bovids, for example, with a distribution pattern that even remotely approaches that of *Theropithecus* at its maximum extension. The only African herbivore that comes anywhere near being such a candidate for resources over a significant proportion of the former range of *Theropithecus*, is the warthog, *Phacochoerus*. Although *Phacochoerus* has been reported from many sites in Africa, from the Mediterranean seaboard to the southern cape of the continent, and from the Atlantic to the Indian Ocean coasts, nearly all the records are thought to be from horizons post-dating the early Pleistocene (Harris & White, 1979). These authors report that most specimens from

Early Pleistocene sites are either misidentified specimens of *Metridiochoerus compactus*, or are likely to be from younger levels overlying, or formerly overlying the Pleistocene exposures where they were collected. Be that as it may, the genus *Phacochoerus* occurs in southern and eastern Africa during the Middle Pleistocene, and throughout the continent in Late Pleistocene deposits. Its rise is to some extent correlated to the decline in diversity and range of *Theropithecus*.

At present, *Phacochoerus* ranges upwards to about 2500 m altitude in East Africa, so there is minor overlap between the upper distribution limits of the warthog, and the lower distributional limits of the gelada (about 2000 m). Over most of their ranges they do not compete, and it could be argued that *Phacochoerus* has displaced *Theropithecus* by dietary competition from all the lower altitude habitats that it occupied during the Middle and Late Pleistocene.

In this respect it is difficult to choose between *Papio* and *Phacochoerus*, because the altitudinal range of both of these herbivores is similar, as are their huge distribution ranges in Africa. It is possible that combined competition from *Papio* and *Phacochoerus* lies behind the shrinkage of *Theropithecus* ranges during the latest Pleistocene to Holocene, but we should not ignore the other possibilities described above, including global and regional climatic changes. (For discussion of other factors that may have led to the shrinkage of *Theropithecus* ranges, see Dunbar, and Lee and Foley, chapters 18 and 19 respectively.)

Summary and Conclusions

Theropithecus species have been collected from many Plio-Pleistocene sites in eastern, southern, and northern Africa. Examination of the detailed chronological and geographic distribution patterns indicates that the diversity and distribution of the genus may have been greatly influenced by global and regional climatic changes, as well as by interactions with predators and by competition from other herbivores.

It is stressed that the present minuscule species range of *T. gelada* appears unusual only if one compares it with the vast range occupied by *T. oswaldi* at different times during the Pleistocene. If, however, it is compared with the ranges of all other anthropoids known from Africa, its small range appears to be quite normal. Viewed from this perspective, it is the huge expansion of the range of *T. oswaldi* that requires explanation, rather than the so-called 'relict' range of the extant gelada.

If the known chronological ranges of fossil *Theropithecus* are approximately representative of their actual distribution, then we can calculate that the species *T. oswaldi* occupied huge ranges, comprising two or three biogeographic provinces, for only 32 per cent of the lifetime of the lineage. The species lived for about 2.5 Ma but it only occurred in southern Africa and the Maghreb for about 0.8 Ma of this period. The rest of the time it appears to have been confined to tropical Africa.

These relatively short periods of time during which *Theropithecus* species occur at latitudes of 30° N and S coincide with interglacial periods of Europe, whereas their apparent absence from these areas tend to coincide with glacial periods. It seems probable that *Theropithecus* distribution patterns were to some extent influenced by global climatic events, one effect of which was to cause latitudinal shifts in global climato-ecological belts. Areas such as the Maghreb are situated at latitudes close to major biogeographic realm boundaries, and during the Pleistocene, their biogeographic affinities fluctuated between tropical and temperate types. During periods when the boundary zone between the Palaearctic and Ethiopian biogeographic realms lay to the north of the Maghreb (during interglacial periods) it experienced an influx of tropical faunal and floral elements, of which *Theropithecus* was twice implicated. Conversely, during glacial periods in Europe, the same boundary zone was displaced southwards far enough for the Maghreb to lie within the boundaries of the Palaearctic, as it does today. During two of these periods, the only cercopithecids in the Maghreb were macaques of Palaearctic affinities.

In equatorial Africa, *Theropithecus* was a constant presence, even though species ranges and diversity patterns varied through time. Some of these changes in distribution patterns were very probably related to regional scale climatic changes brought about by successive episodes of uplift along the Roof of Africa, which parallels the Western Rift Valley, and is genetically related to it. These mountains became increasingly important climate modifiers in eastern tropical Africa during the Pliocene and Pleistocene and are largely responsible for the xerification of East Africa during the Quaternary period. Their effect on floras and faunas of the region throughout the Plio-Pleistocene was incrementally punctuated and widespread. Few mammalian lineages, for example, were unaffected by the changes brought about by the uplift of these mountains. Among these mammals was *Theropithecus*, the distribution patterns and evolution of which were, with little doubt, influenced by these changes.

Other possible factors affecting distribution patterns and diversity of *Theropithecus* include interactions with predators, including humans, and competition for dietary and non-dietary (e.g. sleeping places) resources with other herbivores such as *Papio* and *Phacochoerus*.

Acknowledgements

I particularly wish to thank Nina Jablonski for extending an invitation to participate in the Cambridge Theropithecus Conference. Unfortunately I had made prior arrangements covering the period of the conference, but despite this setback Nina encouraged me to participate *in absentia* and bullied me into producing this paper. I am also anxious to thank Professor Yves Coppens (Collège de France) and Professor Phillipe Taquet (Institut de Paléontologie) for their support. Finally, thanks go to Dr Brigitte Senut (Institut de Paléontologie) for discussions and encouragement, and for permission to include the latest interpretations about the Roof of Africa, part of the results of the Uganda Palaeontology Expedition which she co-directs.

References

ANDREWS, C.W. (1916). Note on a new baboon (*Simopithecus oswaldi*, gen. et sp. n.) from the (?)Pliocene of British East Africa. *The Annals and Magazine of Natural History*, 18, 410–19.

ARDITO, G. & MOTTURA, A. (1987). An overview of the geographic and chronologic distribution of west European cercopithecoids. *Human Evolution*, 2, 29–45.

BARRY, J.C. (1987). The history and chronology of Siwalik cercopithecids. *Human Evolution*, 2, 47–58.

BOAZ, N. (1990). The Semliki Research Expedition: history of investigation, results, and background to interpretation. *Virginia Museum of Natural History Memoir*, 1, 3–14.

COOKE, H.B.S. (1958). Observations relating to Quaternary environments in east and southern Africa. *Geological Society of South Africa Annex*, 60, 1–73.

COOKE, H.B.S. & CORYNDON, S. (1970). Pleistocene mammals from the Kaiso Formation and other related deposits in Uganda. *Fossil Vertebrates of Africa*, 2, 107–224.

COPPENS, Y. (1986). Evolution de l'homme. *La Vie des Sciences Comptes rendus Séries générales*, 3 (3), 227–43.

DELSON, E. (1984). Cercopithecid biochronology of the African Plio-Pleistocene: correlation among eastern and southern hominid-bearing localities. *Courier Forschungsinstitut Senckenberg*, 69, 199–218.

ECK, G. (1976). Diversity and frequency distribution of Omo Group Cercopithecoidea. *Journal of Human Evolution*, 6, 55–63.

ECK, G.G. & JABLONSKI, N.G. (1984). A reassessment of the taxonomic status and phyletic relationships of *Papio baringensis* and *Papio quadratirostris*. *American Journal of Physical Anthropology*, 65, 109–34.

ECK, G.G. & JABLONSKI, N.G. (1987). The skull of *Theropithecus brumpti* compared with those of other species of the genus *Theropithecus*. In *Les Faunes Plio-Pléistocènes de la Basse Vallée de l'Omo (Éthiopie)*. Tome 3. Cercopithecidae de la Formation de Shungura, ed. Y. Coppens & F.C. Howell, pp. 11–122. Cahiers de Paléontologie, Travaux de Paléontologie Est-Africaine. Paris: Editions du Centre National de la Recherche Scientifique.

GERAADS, D. (1987). Dating the northern African cercopithecid fossil record. *Human Evolution*, 2, 19–27.

GUPTA, V.J. (1977). Fossil cercopithecoid from the lower Boulder Conglomerate Formation (Middle Pleistocene) of Mirzapur, Kharar Tehsil, District Ropar, Punjab. *Recent Researches in Geology*, 3, 450–2.

GUPTA, V.J. & SAHNI, A. (1981). *Theropithecus delsoni*, a new cercopithecine species from the upper Siwaliks of India. *Bulletin of the Indian Geological Association*, 14, 69–71.

HALTENORTH, T. & DILLER, H. (1977). *A Field Guide to the Mammals of Africa including Madagascar*. London: Collins.

HARRIS, J.M. & WHITE, T.D. (1979). Evolution of the Plio-Pleistocene african Suidae. *Transactions of the American Philosophical Society*, ns 69, 1–128.

PICKFORD, M. (1986). Caenozoic palaeontological sites of Western Kenya. *Münchener Geowissenschaftliche Abhandlungen*, A8, 1–151.

PICKFORD, M. (1987). The chronology of the Cercopithecoidea of East Africa. *Human Evolution*, 2, 1–17.

PICKFORD, M. (1990). Uplift of the Roof of Africa and its bearing on the evolution of mankind. *Human Evolution*, 5, 1–20.

PICKFORD, M. (1991). Growth of the Ruwenzoris and their impact on palaeoanthropology. In *Primatology Today*, ed. A. Ehara, T. Kimura, O. Takenaka & M. Iwamoto, pp. 513–16. Amsterdam: Elsevier Science Publishers.

SIMONS, E.L. & DELSON, E. (1978). Cercopithecidae and Parapithecidae. In *Evolution of African mammals*, ed. V.J. Maglio & H.B.S. Cooke, pp. 100–19. Cambridge, Mass: Harvard University Press.

SOLOMAN, J.D. (1939). The Pleistocene succession in Uganda. In *The Prehistory of Uganda Protectorate*, ed. T.P. O'Brien, p. 15. Cambridge: Cambridge University Press.

THOMAS, P. (1884). Recherches stratigraphiques et paléontologiques sur quelques formations d'eau douce de l'Algérie. *Memoires de la Société géologique de France*, 3ème Sér., 3 (2), 1–51.

WOLFHEIM, J.H. (1983). *Primates of the World: Distribution, Abundance and Conservation*. Seattle: University of Washington Press.

9   African terrestrial primates: the comparative evolutionary biology of *Theropithecus* and the Hominidae

R. A. FOLEY

Summary

1. The Hominidae and the genus *Theropithecus* overlap chronologically, geographically, and ecologically, and therefore provide an appropriate case study in comparative primate evolutionary biology.
2. By comparing the fossil histories of the various taxa of baboons and hominids – the African terrestrial primates (ATP) – it is possible to draw inferences about evolutionary process and pattern.
3. For baboons and hominids there is a strong relationship between generic diversity and date of origin; there is no relationship between generic duration and diversity.
4. Both baboons and hominids show a successional pattern of evolution with genera replacing each other during the Pleistocene.
5. Overall species longevity of the ATP is very similar to that of other mammalian genera. The earlier taxa of *Theropithecus* – are more long-lasting.
6. While there is no clear trend of complete replacement of *Theropithecus* by *Papio*, the latter do become numerically dominant in the Pleistocene.
7. From a biogeographical perspective East Africa seems to be the focus for most evolutionary novelty and diversity, while other areas of Africa display some characteristics of refugia.
8. There is a close relationship between the frequency of climatic oscillations during the last five million years and species diversity, especially for *Theropithecus*. Other climatic variables do not show such a strong relationship.

Introduction

It has long been recognized that the evolutionary history of *Theropithecus* overlaps with that of the hominids[1]. Jolly (1970) was the first to explore in detail the similarities in functional biology of the two groups, and to use *Theropithecus* as an analogue model for early hominid adaptation, and the underlying functional principles as a basis for interpreting hominid features. Dunbar (1984) looked at the adaptive strategies of *Theropithecus*, in the context of the drier environments of Africa, and argued that contrasting adaptive solutions to similar ecological problems could be identified in the two groups. Foley (1984, 1987) compared the evolutionary histories of the Hominidae and the Papionini, and drew the conclusion that both groups were responding to a general pattern of community change and evolution. Shipman, Bosler & Davis (1981), among others (e.g. Isaac, 1977), analysed the bone accumulations of *Theropithecus* at Olorgesailie to investigate possible predatory interactions between the two groups.

The purpose of this chapter is to extend this work, but from a somewhat different perspective. I intend to examine the two lineages in terms of some general evolutionary issues to see the extent to which such a comparison can throw light on the patterns and processes of primate evolution in general, and in particular upon events in Africa during the later part of the Caenozoic. In other words, hominids, *Theropithecus*, and the baboons in general

– the African terrestrial primates (ATP)[2] – as an example of comparative evolutionary biology.

The particular topics to be considered here include questions relating to speciation and diversity, taxonomic longevity, competition and coevolution, the effect of climate, as well as direct interactions between the two groups.

In the first instance, though, it will be necessary to consider the empirical and methodological basis for adopting this comparative approach.

Similarities, differences, and interactions

Similarities

Theropithecus and hominids are clearly linked by general features, such as the fact that they are both primates and both have a relatively rich fossil record. From an evolutionary point of view, though, they also share some more specific characteristics.

Date of origin The first hominids are known in the fossil record at about 5.0 Ma (Hill & Ward, 1988), although molecular and genetic data would place their origins at about 7.5 Ma ± 1.2 Ma. The first *Theropithecus* occur in the Chemeron Formation, dated (probably) to about 4.0 Ma (Eck & Jablonski, 1987). They may also

1 Taxonomic terminology is inevitably a problem in this chapter, and it is unlikely that the terms used here will please everyone! However, for the sake of linguistic clarity rather than taxonomic accuracy all bipedal apes are referred to the Hominidae rather than the Homininae.
2 For the sake of simplicity the hominids and Papionini discussed here are referred to collectively as the African terrestrial primates, abbreviated in the text to ATP.

occur at Lothagam, for which a date of approximately 5.0 Ma has been suggested but not entirely accepted. Their sister groups, *Papio* and *Parapapio*, also occur in the Late Pliocene (*Parapapio ado* at Sterkfontein (Delson, 1975, 1984, 1988) and *P. jonesi* at Lothagam (Patterson, Behrensmeyer & Sill, 1970)) indicating a somewhat earlier date of origin. The Papionini have yet to be studied as intensively as the Hominoidea from the perspective of molecular evolution, but the data that exist have indicated dates of 3.0 Ma (based on fibropeptides A & B (Nakamura, Takenaka & Takahashi, 1983) or between 2.5 and 3.0 Ma based on albumin and transferrin (Cronin & Sarich, 1976). In a general paper Cronin & Meikle (1979) suggest a slightly older date of 3.2 ± 1.3 Ma.

Geographical distribution At present the distributions of *Theropithecus* and hominids could not be more distinct (an Ethiopian isolate and global, respectively!), but in palaeontological terms the similarities are more pressing. Both groups appear to have African origins, and for the bulk of their evolutionary history were confined to the African continent. More specifically, both groups appear to belong to the 'savanna' faunas of the Late Caenozoic of sub-Saharan Africa. Evidence for hominids outside Africa prior to 1.0 Ma is inconclusive, based largely on undiagnostic artefacts, and evidence for any *Theropithecus* beyond Africa consists of a single specimen of *T. delsoni* from the Siwaliks, in India (see Delson chapter 5). This specimen suggests that *Theropithe-*

cus, like the hominids, expanded out of sub-Saharan Africa during the Pleistocene, but the extent and nature of this are largely unknown.

Evolutionary diversity Both groups are currently represented by one extant species. This contrasts markedly with their earlier evolutionary history. In the fossil record there have been claims for as many as 16 species of hominid (Groves, 1989; Foley, 1991), although a figure closer to half this would be more acceptable to most palaeoanthropologists. The number of *Theropithecus* species is also still under debate, but a figure of six (Eck & Jablonski, 1984; Jablonski, 1986; and see Jablonski, chapter 7) would seem to be appropriate. The main point to be noted here is that both groups underwent an adaptive radiation during the Plio-Pleistocene, and subsequently suffered a major decrease in diversity.

These evolutionary characteristics represent the principal similarities of the two groups. Others are more difficult to define, but are clearly important. For example, it could be argued that both groups display the most extreme locomotor modifications of the cercopithecoids and hominoids respectively; both are the most specialized terrestrialists of their superfamilies, and both display dental specializations. In other words, it may be argued that hominids and *Theropithecus* are among the most specialized catarrhines. The implications of these observations will be discussed below.

Differences

Despite these interesting similarities it is clear that the two groups possess characteristics unique to each one. These should be borne in mind during any comparison.

Phylogeny *Theropithecus* and hominids each belong to one of the two major divisions of the catarrhines – monkey and ape. It is clear from any comparison of these two groups that there are distinct phylogenetic characteristics and evolutionary trajectories displayed.

Body and brain size While there is overlap in body size between the two groups, it is probable that the hominids are on the whole relatively larger than *Theropithecus*. Furthermore, brain size increase is dramatic during the course of hominid evolution, whereas this characteristic has remained relatively stable among *Theropithecus* (see Martin, chapter 10).

Generic diversity Most authorities recognize two (*Homo* and *Australopithecus*) or even three (*Paranthropus*) genera among the Hominidae, whereas (by definition) there is only one genus *Theropithecus*. Whether this is a real distinction or a function of the way taxonomy has been approached in the two groups is unclear (see below, section 'taxonomy').

Evolutionary success Above all, perhaps, the ultimate evolutionary success of the two groups offers the greatest contrast. It may be argued that from a palaeontological perspective the sole survivor of *Theropithecus*, *T. gelada*, is a mere isolate teetering on the edge of extinction, shortly to join its congenerics. While some may argue that the same is true for *Homo sapiens*, it at least currently enjoys a numerical and ecological dominance unsurpassed by any other mammalian species.

Interactions

Apart from these points of comparison in overall evolutionary terms, it is also the case that hominids and *Theropithecus* directly interacted during the course of their evolution, and therefore may have played a part in each other's evolution.

While unambiguous evidence for sympatry is always difficult to document in the fossil record, it does seem, from localities such as Olduvai, Olorgesailie and East Turkana, Kenya, that *Theropithecus* and both *Homo* and *Australopithecus* did occupy the same habitats. Several levels of interaction may therefore have occurred, with selective consequences for each lineage.

Competitive interactions Hominids and *Theropithecus* may have competed for the same resources; in particular, plant resources may have been the focus for direct and indirect competition. Space – such as sleeping and shelter sites – may also have been competed for (Isaac, 1977; Shipman *et al.*, 1981).

Predatory interaction While it is unlikely that *Theropithecus* would have preyed on hominids (or anything else for that matter), it is certainly the case that the hominids may have hunted or scavenged

Theropithecus. Evidence from Olorgesailie certainly points in this direction.

Either of these ecological interactions
would indicate that hominids and *Theropithecus* were involved in a coevolution-
ary process – that is, where adaptations in
one group impinge on the activities, and
hence reproductive success of the other,
in such a way as to impose reciprocal
selective pressures.

It is the purpose of this chapter to
examine the underlying basis for these
similarities, differences, and interactions
in the context of the overall principles of
evolutionary ecology and the specifics of
the African Pliocene and Pleistocene.

Methodological framework

The basic methods employed here are
straightforward – a comparison of the fossil records of the hominids and *Theropithecus*. This will be carried out in terms
of particular evolutionary issues. At the
outset, however, it is necessary to consider briefly two methodological issues.

Taxonomy

As mentioned above, the taxonomic treatment of the two groups has not been consistent. Not surprisingly (with the
exception of this volume) the hominids
have been the focus of far more attention
than is the case with *Theropithecus*.
Historically there has been a tendency for
'splitting' in the former group. A case can
certainly be made that sufficient variation
exists to justify a number of different species, but there may be less of a basis for

splitting the hominids – all linked by the
basic adaptation of bipedalism – into three
genera. If two or three genera are
appropriate then it may be argued that it is
unacceptable to compare these groups
with the single genus *Theropithecus*. A
more appropriate comparison would be
with *Paranthropus* or *Homo*. Conversely,
it might be more appropriate to compare
the hominids as a whole with the Papionini as a whole. While the focus of this
chapter remains the hominids and the
Theropithecus, analytically the Hominidae and Papionini are best treated
together as the ATP.

The actual number of taxa involved in
these analyses, therefore, may be open to
dispute. Focus here will be on those that
are represented in the fossil record, and
assumptions are not made about those
which were probably in existence but are
unknown   palaeontologically   (e.g.
T. gelada or several *Papio* species). The
genera   *Macaca*,   *Cercocebus*,   and
Mandrillus are also excluded. The justification for these decisions lies largely in
the practical need to maintain consistency, which is at least more obtainable
than completeness.

Estimates of the number of both
hominid and *Theropithecus* fossil species
varies according to different authors.
Number of taxa identified in the fossil
record are likely to be underestimated due
to two factors: (a) the incompleteness of
the fossil record; and (b) the insensitivity
of osteological features alone to speciation
processes (Tattersall, 1986). This is contrary to the views of Turner & Chamberlain (1989) who argue that species
recognition features should be employed.

However, for early hominids at least these are unknown. Neither does it take into account the probability that some hybridization between groups may well have occurred. *Species here are taken to mean groups that appear on morphological grounds to have some unique pattern of morphological, geographical, and chronological distribution, such as to imply relatively independent evolutionary trajectories* (see Jablonski's discussion of Jolly, in chapter 1, for a discussion of the validity of the species concept among the Papionini).

For hominids the number of species may be as large as 16 (Groves, 1989), although most authorities would prefer a smaller number. Twelve species appear to be relatively well documented, although many of these may be problematic. *Australopithecus afarensis* may be two species (Olson, 1981; Senut & Tardieu, 1985); there may be a second species at Sterkfontein (Clarke, 1985); the African and Asian forms assigned to *Homo erectus* may be separate taxa (Andrews, 1984), which would in turn raise problems of the assignment of later Pleistocene material throughout the Old World. Some authorities view *H. sapiens* as a polytypic species incorporating all late Middle and Late Pleistocene specimens, others as anatomically modern humans specifically, with other archaic specimens assigned to a number of other species (Tattersall, 1986). 'Minimalists' would in contrast prefer six species (*A. afarensis*, *A. africanus*, *Paranthropus robustus*, *H. habilis*, *H. erectus*, *H. sapiens*).

The Papionini also suffer from systematic difficulties. A relatively 'splitting'

approach (to maintain consistency with the hominids) would yield some 23 species, seven of which are extant. While there is no existing consensus on this matter, for the purposes of this chapter it has been necessary to make firm decisions about taxonomic units that other authorities would dispute (see Delson and Dean, Eck, and Jablonski, chapters 4, 2, and 7, respectively). Such decisions should be treated as experimental assumptions rather than formal taxonomic assignments.

The taxonomic units used in these analyses are shown in Table 9.1.

Table 9.1 *Principal taxonomic units of hominids and baboons used in this paper. See text for discussion of limitations and problems.*

Hominidae
Australopithecus
  A.afarensis (AUA)
  A.africanus (AUF)
Paranthropus
  P.aethiopicus (PRA)
  P.robustus (PRR)
  P.crassidens (PRC)
  P.boisei (PRB)
Homo
  H.habilis (HOH)
  H.ergaster (HOT)
  H.erectus (HOE)
  H.heidlebergensis (HOA)
  H.neanderthalensis (HON)
  H.sapiens (HOS)

Papionini
Parapapio
  P.ado (PAA)
  P.jonesi (PAJ)

Table 9.1 – *cont.*

   P.broomi (PAB)
   P.antiquus (PAN)
   P.whitei (PAW)
Dinopithecus
   D.ingens (DNI)
Gorgopithecus
   G.major (GGM)
Papio
   P.robinsoni (PPR)
   P.hamadryas (PPH)
   P.cynocephalus (PPC)
   P.anubis (PPA)
   P.ursinus (PPU)
   P.papio (PPP)
   P.angusticeps (PPN)

Theropithecus
   T.gelada (THG)
   T.darti (THD)
   T.quadratirostris (THQ)
   T.baringensis (THR)
   T.brumpti (THB)
   T.oswaldi (THO)
   T.atlanticus (THA)

Note: Abbreviations show key used in tables and figures.

Fossil distribution and taphonomy

It may be argued that any comparison of two lineages must assume that their fossil records are equivalent, and not subject to any unique taphonomical bias. If, for example, one record is more complete than the other, then any comparison is likely to be misleading.

However, it is extremely unlikely that the fossil record of *any* two lineages will be exactly the same, for by definition separate species will occupy separate ecological niches and therefore the probability of fos-

silization is liable to vary. All that can be said in justification is that a characteristic shared by both hominids and *Theropithecus* is the relative richness of their fossil history in comparison with any other group of higher primate. While this does not preclude the possibility of distortion through differential preservation, it does perhaps justify this preliminary exercise in comparative evolutionary biology. Furthermore, there may be taphonomical factors operating simultaneously on both groups. This may be particularly significant in the case of geographical and chronological patchiness. It must be remembered, for example, that the sub-Saharan record for both groups comes essentially from only two areas of Africa, and that certain periods, such as that between 1.0 and 0.5 Ma, are relatively poorly represented in the fossil record.

Materials and methods

Within these limitations, the material incorporated here consists of the range of ATP material recovered from the Pliocene and Pleistocene in Africa. These data take the form of temporal and spatial distributions of taxa as recorded in site reports and faunal lists. Assumptions about dating have followed the current estimates, but some may be open to alternative interpretations. For the sake of analytical simplicity these locality data are categorized by broader geographical units that may represent reasonably homogenous biogeographical units, and by half million year units. The analyses carried out in this chapter are based on the distributional data shown in Table 9.2. From these and

Table 9.2 *Geographical and chronological distribution of the African terrestrial primates under consideration here, based on the fossil record.*

Time	Afar/Awash Hominid	Papionine	Turkana Hominid	Papionine	Central/West Kenya Hominid	Papionine	S.Kenya/N.Tanzania Hominid	Papionine	Southern Africa Hominid	Papionine	North Africa Hominid	Papionine
Present	HOS	PPC PPH THG	HOS	PPH PPA	HOS	PPC PPA	HOS	PPC PPA	HOS	PPU	HOS	
0–0.5	HOS HOA	THO PPH THG	HOS HOA	THO	HOE	THO?	HOA	THO	HOA	THO	HOE HON HOS	THO
0.5–1.0				THO			HOE	THO	HOE		HOE	
1.0–1.5			HOE PRB	THO		TH?	HOE PRB	THO	PRC? PRR?	PPR PPN		
1.5–2.0	PRB		HOE HOH HOT PRB	THB THO DNI	PRB	THO THB?	HOH PRB	THO	PRR PRC	THO PPR GGH DNI		THA

(Ma)							
2.0–2.5	HOH	THB THO DNI				AUF	PAJ
2.5–3.0	AUA?	PRA	DNI	PAA THB THO		AUF	PAA PAW THD PAB PAJ
3.0–3.5	AUA	THD		PAA	AUA	AUF	THD
3.5–4.0	THO	AUA?	THB THQ	PAA	AUA		PAJ
4.0–4.5	AUA						
4.5–5.0	AUA	THO	TH? PAJ	AUA			

Sources used: Afar: Johanson, Taieb & Coppens, 1982; Hall *et al.* 1984; Aronson & Taieb 1981. Awash: Kalb *et al.* 1982a, b. Turkana: Walker *et al.* 1986; Patterson *et al.* 1970; Leakey 1976; Harris *et al.* 1988; Eck & Jablonski 1987; Eck 1976; Central Kenya: Pickford 1986. West Kenya: Pickford 1986. Southern Kenya: Potts 1989; Isaac 1977. Northern Tanzania: Leakey & Delson 1987. South Africa: Freedman 1976; Szalay & Delson 1979; Delson 1975, 1984, 1988. North Africa: Delson 1975, 1984; Szalay & Delson 1979.

Note: See Table 9.1 for abbreviations of the taxa.

Table 9.3 *Number of species, date of origin and duration of African terrestrial primate genera. Mean number of species are based on a conservative estimate of diversity; maximum number uses the full range of species that have been proposed.*

	Australopithecus	Paranthropus	Homo	Theropithecus	Parapapio	Papio
Number of genera (mean)	2	3	3	5	5	7
Number of species (max)	4	4	6	7	6	9
Date of origin (Ma)	5	2.7	2.3	4.0	2.0	5.0
Duration (Ma)	2.2	1.7	2.3	4.0	3.0	2.0

the source data it has been possible to estimate dates of origin and extinction, patterns of endemism and distribution, taxonomic longevity, and patterns of covariation between the different groups and with climatic parameters. While no doubt subject to error and the usual inconsistencies of the fossil record they provide at least a preliminary base on which to attempt an analytical approach to primate evolutionary biology.

Results

Diversity of Genera

While both groups are represented by only one extant species, the situation in the Pleistocene was very different. One simple question that arises is whether the number of species per higher taxonomic level – in this case genus – are the same, and if not what factors affect the pattern of diversification. A simple expectation is that generic diversity (number of species per genus) should be a function of time.

Data on generic diversity for hominids and papionines are shown in Table 9.3, along with information about estimated (fossil) dates of origin and temporal longevity. In general there is a relationship between the date of origin and the number of species per genus for all ATP, but no relationship with duration. More specifically, a strong negative correlation exists between minimum hominid generic diversity and time of origin ($r = -0.991$, $p = 0.088$) although this is not significant; and between maximum papionine generic diversity and date of origin ($r = -1.00$, $p = 0.000$). Figure 9.1 shows the same trends for all generic diversity variables, but the regression slopes are steeper and intercepts higher for baboons compared with hominids (see caption to Fig. 9.1 for regression equations).

The conclusions to be drawn from these data are that both baboons and hominids have continued to diverge throughout their evolution, but that the rate for baboons is generally higher and that rates of divergence are more rapid for the

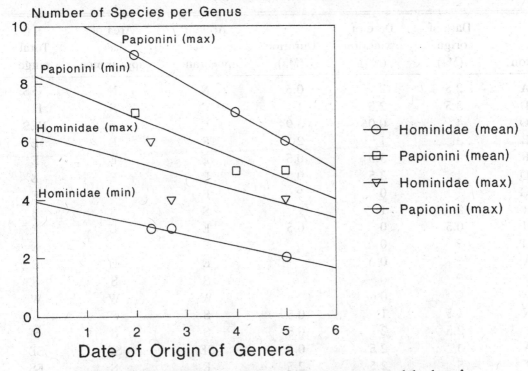

Fig. 9.1. Relationship between number of species per genus and the date of origin of the genera for hominids and Papionini. For both groups the maximum estimated number of species and minimum or more widely accepted number of species are plotted. Linear regression lines are shown for each variable. The most highly correlated relationship is for minimum hominid generic diversity and maximum papionine generic diversity, for which the regression equations are $y = 3.9749–0.3925x$ and $y = 11.000–1.000x$ respectively. *Dinopithecus* and *Gorgopithecus* are included in *Papio*. When plotted separately the papionines show a much more disrupted pattern of generic diversity, with a stronger relationship between number of species and duration of the genus emerging.

Pleistocene than for the Pliocene. In particular the relatively high rate of diversity for *Homo* and *Papio* indicates continued selective or stochastic events into the late Middle and Late Pleistocene.

Temporal pattern of species richness

This may be pursued in more detail by looking at the actual pattern of species richness through time, rather than just the overall situation for genera. An estimate of the date of origin, extinction and geographical distribution of the component species of ATP are shown in Table 9.4. These estimates are drawn from the data used to construct Table 9.2. The tabulated data are shown in bar graph form in Fig. 9.2. The vertical axis represents time, and the horizontal one

Table 9.4 *Dates and location of origin, extinction and length of duration of the African terrestrial primates.*

Taxon	Date of origin (Ma)	Date of extinction (Ma)	Duration (Ma)	Area of first appearance	Area of last appearance	Total range
THA	2.5	2	0.5	N	N	NES
THD	3.5	2.5	1	E	S	ES
THO	4	0.06	3.94	E	S	NES
THB	3	1	2	E	E	E
THR	4	3.5	0.5	E	E	E
THQ	4	3.5	0.5	E	E	E
THG	?	0	?	E	E	E
PPR	2	1	1	S	S	S
PPH	0.5	0	0.5	E	E	E
PPC	?	0		E	E	E
PPA	?	0		E	E	E
PPU	?	0		S	S	S
PPP	?	0		W	W	W
PPN	1.5	1	0.5	S	S	S
PPI	2.5	2	0.5	S	S	S
PAA	3	2.5	0.5	SE	SE	SE
PAJ	5	2.5	2.5	E	S	ES
PAB	3	2.5	0.5	S	S	S
PAN	?	?		S	S	S
PAW	?	?		S	S	S
DNI	2	1.5	0.5	SE	SE	SE
GGM	2	1.5	0.5	S	S	S
AUA	5	2.9	2.1	E	E	E
AUF	3.5	2	1.5	S	S	S
PRA	2.7	2.3	0.4	E	E	E
PRR	2	1	1	S	S	S
PRC	2	1	1	S	S	S
PRB	2	1	1	E	E	E
HOH	2.3	1.7	0.6	E	E	E
HOT	2.1	1.7	0.4	E	E	E
HOE	1.6	0.4	1.2	E	E?	NES
HOA	0.5	0.1	0.4	?	?	?
HON	0.15	0.03	0.12	EU	EU	Eu/ME
HOS	0.14	0	0.14	S	?	Global

See Table 9.2 for sources used. N=northern Africa, S=Southern Africa, E=Eastern Africa, EU=Europe, ME=Middle East.
See Table 9.1 for abbreviations of the taxa.

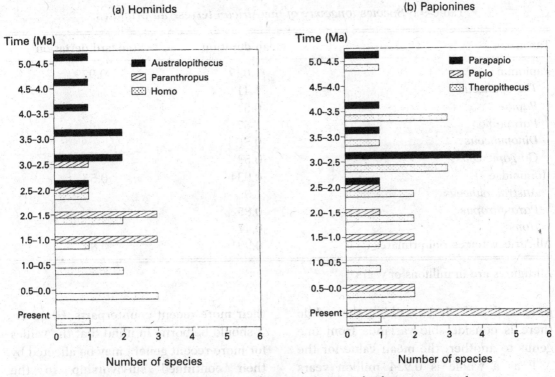

Fig. 9.2. Number of species within the genera of African terrestrial primates for each half million year time unit. Histograms are based on date drawn from Table 9.4. (a) Hominids; (b) Papionines.

the number of species for hominids and baboons. Hominids display an overlapping successional character, with an early diversification and subsequent extinction of *Australopithecus*, followed by the overlapping radiations of *Homo* and *Paranthropus*, of which the former is more prolonged. Papionines, on the other hand, show a successional relationship between *Parapapio* and *Papio* but *Theropithecus* has a complex pattern of constant presence throughout the Pliocene and Pleistocene, punctuated by peaks of diversity. To some extent it can be argued that *Papio* has replaced *Theropithecus* as the principal baboon species.

Species longevity

A further comparison that may be made is in terms of species longevity – that is the length of time from first to last appearance in the fossil record. This parameter has been extensively studied in palaeobiology (Gingerich, 1977; Stanley, 1978, 1979). Stanley in particular has argued that mammals in general show a typical longevity of approximately one million years. Table 9.4 shows the duration in millions of years, as observed in the fossil record, of the various ATP species under consideration here. Table 9.5 shows the mean species longevity of each of the principal

258   R. A. FOLEY

Table 9.5 *Species longevity of the African terrestrial primates.*

Taxon	Mean duration	Standard deviation
Papionini	1.029	0.979
Theropithecus	1.41	
Papio	0.5	
Parapapio	1.67	
Dinopithecus	0.5(?)	
Gorgopithecus	0.5?	
Hominidae	0.934	0.55
Australopithecus	1.8	
Paranthropus	0.85	
Homo	0.47	
All African terrestrial primates	0.991	0.83

All figures are in millions of years.

genera of ATP. It is striking that while there is considerable variance from one genus to another, the mean value for the ATP as a whole is 0.934 million years (SD = 0.83), very close to the mean for mammals (Stanley, 1978, 1979). There is no significant difference between baboons and hominids, but three groups stand out as having particularly high mean longevity times – *Theropithecus*, *Parapapio*, and *Australopithecus*. These are all the more ancient genera. Various potential explanations arise for this observation. It is possible that the earlier taxa lived during a period of relatively stable climate, and were therefore less subject to the risk of extinction (see below). Alternatively, there may be community structure components involved, whereby interspecific interactions were of a different nature during the Late Tertiary compared to the Quaternary (Foley, 1984, 1987). It may be this factor that differentiates the more ancient lineages referred to above from

their more recent counterparts. However, it should be borne in mind that the values for more recent genera may be affected by their continued survivorship to the present day.

Relative frequencies of ATP species

As well as the absolute frequencies of species of ATP, the changing proportions in the fossil record are of some interest. It has long been observed, both in locality occurrences and absolute numbers, that *Theropithecus* outnumber *Papio* throughout the Pleistocene (Jolly, 1972), and indeed it has been argued that *Papio* is completely absent from East Africa during much of the Pleistocene (Eck & Jablonski, 1987). Using the data given in Table 9.6 the relative proportions of three groups – hominids, *Theropithecus*, and *Papio/Parapapio* – are graphically represented in Fig. 9.3.

Surprisingly, perhaps, there is no clear

Table 9.6 *Absolute and relative frequencies of African terrestrial primates (ATP) during the last five million years.*

Time (Ma)	Number of species					Percentage of species		
	Hominids	Papionini	Theropithecus	Papio	All ATP	Hominids	Papio/ Parapapio	Theropithecus
Present	1	6	1	5	7	14.3	71.4	14.3
0–0.5	3	2	1	1	5	60	20	20
0.5–1	1	1	1	0	2	50	0	50
1–1.5	4	4	1	3	8	50	37.5	12.5
1.5–2	6	6	2	4	12	50	33.3	16.7
2–2.5	2	4	2	2	6	33.3	33.3	33.3
2.5–3	3	8	3	5	11	27.3	45.5	27.3
3–3.5	2	2	1	1	4	50	25	25
3.5–4	1	4	2	2	5	20	40	40
4–4.5	1	0	0	0	1	100	0	0
4.5–5	1	2	1	1	3	3.33	33.3	33.3

trend of *Theropithecus* replacing the *Papio/Parapapio* clade, and perhaps it is the case that the numerical (abundance) dominance of *Theropithecus* is not reflected in the diversity data. Clearly ecological dominance is not the same as evolutionary activity, and indeed it may well be the case that they are inversely related – small, low density, isolated populations may be more evolutionarily active than larger, widely distributed ones. This would certainly be in accord with some models of the population genetics of speciation; in other words, the very ecological success of *Theropithecus* provided them with the basis for the relative stability of evolutionary events that we have seen from the data discussed above. The periods when either *Theropithecus* or the hominids are entirely dominant are ones where the fossil data are extremely poor (4.5–4.0 Ma and 1.0–0.5 Ma). Broadly speaking it

seems to be the case that *Theropithecus* and hominids were the dominant ATP during the Pliocene and Pleistocene, with *Parapapio* significant only during the Pliocene and *Papio* only becoming significant during the last part of the Pleistocene. To some extent, though, this chronological pattern hides a more complex geographical situation.

Geographical patterns

Table 9.4 also shows where, as best as the fossil evidence can indicate, the geographical origins of a taxa lie, and where they were last recorded (East, South, and North Africa).

It is clear that most ATP species known in the fossil record have their origins in eastern Africa. This may confirm the notion that equatorial latitudes are the focus of more evolutionary novelty than

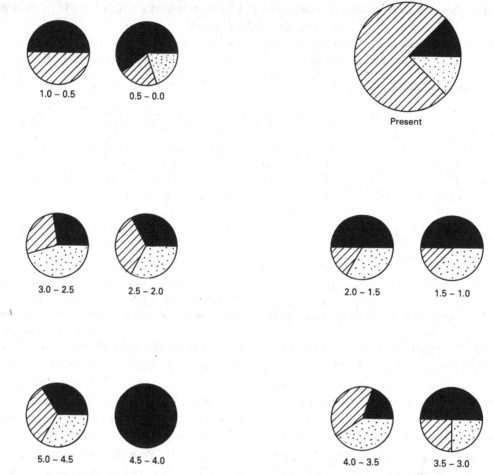

Fig. 9.3. Relative proportions of (a) Hominidae (black); (b) *Papio/Parapapio* (diagonal hatching), and (c) *Theropithecus* (dotted hatching) during the last five million years.

are higher latitudes, with species expanding out from the equator to the north and south. The reasons for this may be the greater productivity of low latitude environments, and hence the greater potential for niche separation, or it may be more closely related to the nature of the environmental gradients through time and space (Foley, 1989). For *Theropithecus*, 85.7 per cent of known taxa have their origins in East Africa; for *Paranthropus*

and *Australopithecus* it is 50 per cent, and for *Homo* it is 75 per cent.

In contrast, while most taxa appear to last longer in the low latitude, equatorial regions, the bias is not so significant; in particular 43 per cent of *Theropithecus* species make their last appearance outside East Africa. This may relate to the existence of refugia.

There is, furthermore, a variable pattern of geographical distribution. Seventy-six per cent of all ATP species – living and fossil – have geographically limited distributions (i.e. are confined to one part of Africa). For *Theropithecus* this figure is 57 per cent, and for hominids 75 per cent. It may be argued that some genera speciate more frequently in response to geographical expansion than others, as for example *Paranthropus*. This is likely to be a function of habitat or dietary tolerance or variability. The precise relationship between spatial distribution is a topic that requires further investigation, but is to some extent hampered by lack of consensus about taxonomic assignment.

Turning to more precise geographical patterns, Table 9.7 shows the distribution of Papionini and hominid taxa through time by biogeographical–palaeontological unit. The units employed follow the major biogeographical units employed in Table 9.2, and should reflect decreasing probability of biological continuity. This Table may well simply reflect taphonomical factors, but it is noticeable that the Turkana Basin shows consistently high levels of diversity for both Papionini and hominids, as to a lesser extent does South Africa.

African terrestrial primates and climatic change

It would not be unexpected that there should be some relationship between the patterns discussed above and climatic change. The onset of cooler conditions have been linked by Vrba (1985) to the speciation and extinction patterns of hominids and bovids. However, climatic change is an extremely complex process and cannot be characterized by a single variable such as change in temperature. To develop more precise analyses of the relationships between evolutionary events and climate it is necessary to quantify climatic change and to characterize its various evolutionarily significant components.

In terms of its effect on evolutionary events four possible climatic parameters can be measured: (1) the amount of climatic change from one period to another, (2) the magnitude of climatic variation within a period, (3) the number of climatic fluctuations or cycles, and (4) the actual climate. Using the composite oxygen isotope curve given by Prentice & Denton (1988), values for these four parameters were calculated for the last five million years in half million year units (Fig. 9.4). Successive climatic change was measured by the difference in modal Delta O^{18} values from one half million year period to another. The magnitude of climatic variation within a half million year unit was measured by the maximum and minimum Delta O^{18} values for each period. The actual climate is indicated by the modal Delta O^{18} value, and the frequency of change (or climatic stability) is measured by the number of major oscillations per half million year period.

Figure 9.4 shows the pattern of change in these three parameters over the last five million years. As can be seen, the trajectories of each is different, underscoring the point that climatic change will not necessarily be linked in a simple way to evolutionary events. For example, the period around three million years shows the greatest amount of climatic instability, but the period around 2.5 Ma ago shows

Table 9.7 Number of species of hominid and Papionini in the principal fossil localities of Africa.

Time	Afar/Awash		Turkana Basin		Central/W. Kenya		S.Kenya/N.Tanz.		South Africa		North Africa	
	Ho	Pap	Ho	Pap	Ho	Pap	Ho	Pap	Ho	Pap	Ho	Pap
Now	1	3	1	2	1	1	1	2	1	1	1	0
0.0–0.5	2	3	2	1	1	1	0	1	1	1	3	1
0.5–1.0	0	0	0	1	0	0	1	1	1	0	1	0
1.0–1.5	0	0	2	1	0	1	2	1	2	2	0	0
1.5–2.0	0	0	4	3	1	2	2	1	2	4	0	1
2.0–2.5	0	0	1	3	0	0	0	0	1	1	0	0
2.5–3.0	1	1	1	3	0	0	0	0	1	5	0	0
3.0–3.5	1	1	0	0	0	0	1	1	1	1	0	0
3.5–4.0	0	1	1	1	0	2	1	1	0	0	0	0
4.0–4.5	1	0	0	0	0	0	0	0	0	0	0	0
4.5–5.0	0	0	1	2	1	0	0	0	0	0	0	0
Mean	0.5	0.6	1.3	1.5	0.3	0.7	0.8	0.6	0.9	1.4	0.4	0.2

Ho = hominid species, Pap = baboon species.

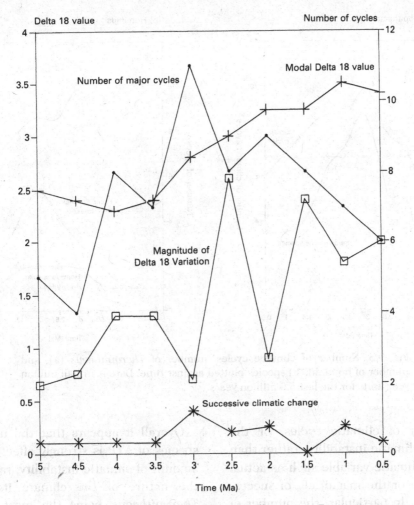

Fig. 9.4. Climatic change during the last 5.0 million years. Four parameters are shown: number of climatic cycles; modal Delta O^{18} value; amount of change from one successive half million year period to another; and magnitude of climatic variation within each half million year period. See text for discussion.

the greatest variation in climate. Furthermore, while the terminal Pliocene is a period of great climatic instability, it is during the Late Pleistocene that the climate deteriorates most rapidly. Given these differences in pattern according to the variable used it is important to distinguish types of climatic change, rather than simply looking at the relationship between a single variable, such as temperature, and evolutionary events.

Correlation coefficients were calculated to examine the relationship between the different patterns of climatic change and the diversity of ATP species. Species diversity was found to relate most strongly to

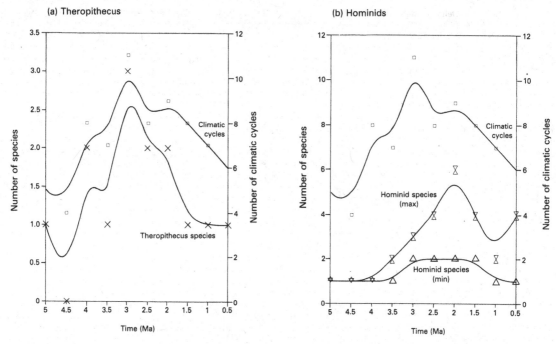

Fig. 9.5. Number of climatic cycles, number of *Theropithecus* (a), and number of hominid (b) species plotted against time. Date is in half million year units for the last 5.0 million years.

the number of climatic cycles, or the amount of climatic instability, rather than any other climatic variable such as actual temperature or the magnitude of successive change. In particular, the number of *Theropithecus* species showed a very strong positive relationship with the number of climatic cycles ($r = 0.9079$, $p < 0.001$). Other baboon species show a less marked relationship ($r = 0.755$, $p < 0.01$), and hominids, using both maximum or minimum number of species, a still weaker relationship ($r = 0.5324$, $p > 0.01$) (Fig. 9.5). The other climatic variables, including those transformed to allow for a time lag in the relationships, do not show any statistically significant relationships with any species diversity data.

Overall it appears that the number of species of ATP is strongly affected by the amount of climatic instability, rather than the nature of the climate itself, with *Theropithecus* being the most sensitive and hominids the least. Furthermore, when looked at in detail (Fig. 9.5), the earlier part of hominid evolution is more closely related to climatic stability than the later parts. This last observation may relate either to the greater geographical range of the hominids by the later parts of the Pleistocene period, or to the extent to which hominids were buffered from the influence of climatic change.

Two initial, albeit tentative inferences may be drawn from this. The first is that if the appearance of new taxa is associated with climatic instability, and in particular the frequency with which the climate changes (whatever the direction of that change), then there may well be a stochastic mechanism involved. Climatic change is likely to lead to both new selective pressures and to population isolation. This may occur during any period of change. However, whether the new 'morphs' that arise during these periods survive and become established species may be a matter of probability. The greater the number of events, the greater the probability. This may be the mechanism by which Vrba's 'effect hypothesis' can occur, with daughter species budding off but only under certain circumstances surviving – those circumstances being dependent upon selective and competitive conditions. The second inference is that particular taxonomic groups of ATP may have been more subject to this process than others. *Theropithecus* may have tracked environmental events so closely because of a unique combination of factors. It possesses, for a primate, an extreme form of dietary adaptation to dependence upon grass, and while this resource is abundant and widely distributed, its local distribution, availability, and abundance would have fluctuated markedly during the last five million years.

Discussion and conclusions

The data and analyses presented in this chapter may be used to draw a series of preliminary conclusions about: (a) the evolution of the ATP; and (b) the nature of evolution among the primates. The first of these is a specific historical event, the second a question of general evolutionary biological principles. It is worth bearing these two different aspects in mind, for ultimately what makes evolutionary biology so challenging is that it combines the explanation of unique events with the development of general theoretical frameworks.

Evolutionary history of the African terrestrial primates

In terms of the specific historical questions, several major points can be made. First, that neither hominid nor *Theropithecus* evolution occurs in isolation. Both groups can be seen to be responding to more general conditions, and, as Dunbar (1984) has pointed out, they show contrasting solutions to similar problems. Both groups appear to be strongly influenced by climatic events, but these are geographically specific and chronologically variable. Both groups are undergoing adaptive radiations. Both groups are extremely successful and appear to have similar species longevity. The differences lie in the extent to which different taxa are geographically variable at different periods; the extent to which they are sensitive to particular climatic (and hence environmental) parameters; and the extent to which they are evolutionarily active at different periods.

In conclusion, hominids appear to be less speciose than the Papionini. This may be a product of a number of ecological and life history variables, such as environmental tolerance, dietary specializations, home range size, body size, and social behaviour. Their evolutionary history shows a more successional character, such that there are really three markedly different parts to the evolution of the family — the establishment and radiation of the early bipeds (the australopithecines), the radiation of low quality plant food specialists (the paranthropines) and the radiation of highly encephalized hominines.

Papionines, in contrast, are probably less demarcated in terms of adaptive radiations. In particular *Theropithecus* appears to have been consistently successful throughout the Pleistocene, becoming reduced only towards the very end of the period. They were especially sensitive to climatic change. In terms of the relative patterns of the different groups of the Papionini, certainly abundance indicates that for most of the Pliocene and Pleistocene *Theropithecus* was the most successful taxon, with *Papio* radiating relatively late.

ATP and evolutionary biology

Certain general points may also be made about the light that the history of these two groups throw on our understanding of evolutionary ecological mechanisms.

Evolutionary tempo When considered on a broad scale, evolution among these higher primates appears, in terms of speciation rates, species longevity, patterns of diversity, etc. to be similar to those found in other mammals (Stanley, 1978, 1979). This, in particular, should indicate the importance of not treating human evolution as a unique process as well as a unique event (Foley, 1987).

Climate Climatic factors are clearly a major driving force in evolution. The analyses presented here illustrate the importance of dissecting climate into parameters that are of evolutionary significance. Furthermore, it is important to stress that climate cannot direct evolutionary change, but merely set the conditions under which competitive interactions are played out (see Lee and Foley, chapter 19). The principal implication to be drawn is that evolutionary patterns occurring in response to climatic changes are likely to be variable in rate and in terms of the responses of different species.

Community evolution The generalities that emerge in ATP evolution indicate that these primates are not evolving in isolation but as part of a larger community. It may well be that the pattern of hominid and Papionini evolution conforms to a model of overall African savanna community evolution (Foley, 1984, 1987). In particular the parallel between frequencies of *Paranthropus/Theropithecus* (the specialists) and *Homo/Papio* is suggestive (Foley, 1987).

Coevolution The data presented here are insufficient to test the extent to which direct coevolution occurred between hominids and papionines. A circumstantial case can be made on the basis of the data presented here, but it would be inappropriate to stress such a case without paying equal attention to the evolutionary patterns of other species of sympatric mammals. Overall it would be expected that interspecific competition will vary in its intensity through time, with such com-

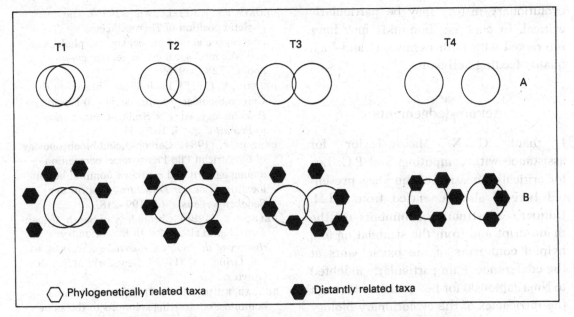

Fig. 9.6. Model of interspecific competition at different evolutionary stages. (A) Degree of niche overlap between two taxa that are diverging from T1 to T4. Character displacement through time will reduce niche overlap and competition. (B) The same pattern, but with more distantly related species shown. As evolutionary divergence increases species will come to compete increasingly with taxa that are more distantly related to them. These taxa will themselves be going through the same process.

petition being between closely related species during periods of speciation and evolutionary novelty, and between more phylogenetically distant groups at other times (Fig. 9.6).

Limitations

It is also worth bearing in mind the (major) limitations on the analyses presented here, and which should be subject to further work. In particular, the basic geographical and chronological distributions that lie at the heart of the discussion here may require further refinement and modification. Two factors not considered here are the actual rates of

morphological change within lineages (i.e. microevolution) as opposed to the macro-level changes associated with the appearance and disappearance of taxa; and the question of abundance, which may be an alternative and complementary measure of evolutionary success. Furthermore, in the same way that hominids and *Theropithecus* are part of the general pattern of ATP evolution and adaptation, so too the ATP are part of a larger mammalian community, and the effects of suids, bovids, and carnivores should certainly be explored. As discussed elsewhere (Lee and Foley, chapter 19) *Theropithecus* may have competed closely with the other graminivores, and their

evolutionary history may be particularly
critical. In contrast, hominids may have
interacted with both carnivores and high
quality feeding herbivores.

Acknowledgements

I thank C.G.N. Mascie-Taylor for
assistance with computing, and P.C. Lee
for critical discussion of the ideas presented here. I also benefitted from R.I.M.
Dunbar's constructive comments on the
manuscript and from the stimulating and
helpful comments of the participants at
the conference. I am particularly indebted
to Nina Jablonski for her guidance through
the intricacies of the evolutionary biology
of *Theropithecus*, and for her encouragement to pursue the issues discussed here.

References

ANDREWS, P.J. (1984). An alternative
interpretation of the characters used to define
Homo erectus. *Courier Forschungsinstitut
Senckenberg*, 69, 167–75.

ARONSON, J.L. & TAIEB, M. (1981). Geology and
paleogeography of the Hadar hominid site,
Ethiopia. In *Hominid Sites: Their Geologic
Settings. AAAS Selected Symposium*, 63,
165–95. Washington: American Association
for the Advancement of Science.

BROWN, F.H., McDOUGALL, I., DAVIES, I. &
MAIER, R. (1985). An integrated Plio-
Pleistocene chronology for the Turkana Basin.
In *Ancestors: the Hard Evidence*, ed. E.
Delson, pp. 82–90. New York: Alan R. Liss.

CLARKE, R.J. (1985). *Australopithecus* and early
Homo in Southern Africa. In *Ancestors: the
Hard Evidence*, ed. E. Delson, pp. 171–7.
New York: Alan R. Liss.

CRONIN, J.E. & SARICH, V.M. (1976). Molecular
evidence for dual origin of mangabeys among
Old World monkeys. *Nature*, 260, 700–2.

CRONIN, J.E. & MEIKLE, W.E. (1979). The
phyletic position of *Theropithecus*:
congruence among molecular, morphological
and paleontological evidence. *Systematic
Zoology*, 28, 259–69.

DELSON, E. (1975). Evolutionary history of the
Cercopithecidae. In *Approaches to Primate
Paleobiology*, ed. F.S. Szalay. *Contributions
to Primatology*, 5, 167–217.

DELSON, E. (1984). Cercopithecid biochronology
of the African Plio-Pleistocene: correlation
among eastern and southern hominid-bearing
localities. *Courier Forschung Institut der
Senckenbergensis*, 69, 199–218.

DELSON, E. (1988). Chronology of South African
Australopith site units. In *Evolutionary
History of the 'Robust' Australopithecines*, ed.
F.E. Grine, pp. 317–24. New York: Aldine de
Gruyter.

DUNBAR, R.I.M. (1984). Theropithecines and
hominids: contrasting solutions to the same
ecological problems. *Journal of Human
Evolution*, 12, 647–58.

ECK, G. (1976). Diversity and frequency
distribution of Omo Group Cercopithecoidea.
Journal of Human Evolution, 6, 55–63.

ECK, G.G. & JABLONSKI, N.G. (1984). A
reassessment of the taxonomic status and
phyletic relationships of *Papio baringensis*
and *Papio quadratirostris* (Primates:
Cercopithecidae). *American Journal of
Physical Anthropology*, 65, 105–34.

ECK, G.G. & JABLONSKI, N.G. (1987). The skull
of *Theropithecus brumpti* compared with
those of other species of the genus *Theropithecus*. In *Les Faunes Plio-Pléistocènes de
la Basse Vallée de l'Omo (Éthiopie)*. Tome 3.
Cercopithecidae de la Formation de
Shungura, ed. Y. Coppens & F.C. Howell, pp.
11–122. Cahiers de Paléontologie, Travaux de
Paléontologie Est-Africaine. Paris: Editions du
Centre National de la Recherche Scientifique.

FOLEY, R.A. (1984). Early man and the Red
Queen: tropical community ecology and
hominid adaptation. In *Hominid Evolution
and Community Ecology*, ed. R.A. Foley, pp.
85–118. London: Academic Press.

FOLEY, R.A. (1987). *Another Unique Species:
Patterns in Human Evolutionary Ecology*.
Harlow: Longman.

FOLEY, R. (1989). The ecology of speciation: comparative perspectives on the origins of modern humans. In *The Human Revolution: Behavioural and Biological Perspectives on the Origins of Modern Humans*, ed. P.A. Mellars & C.B. Stringer, pp. 298–320. Edinburgh: Edinburgh University Press.

FOLEY, R.A. (1991). How many hominid species should there be? *Journal of Human Evolution*, 20, 413–27.

FREEDMAN, L. (1976). South African fossil Cercopithecoidea: a reassessment including a description of new material from Makapansgat, Sterkfontein and Taung. *Journal of Human Evolution*, 5, 297–315.

GINGERICH, P.D. (1977). Patterns of evolution in the mammalian fossil record. In *Patterns of Evolution*, ed. A. Hallam, pp. 469–500. Amsterdam: Elsevier Science Publishers.

GROVES, C.P. (1989). *A Theory of Human and Primate Evolution*. Oxford: Clarendon Press.

HALL, C.M., WALTER, R.C., WESTGATE, J.A. & YORK, D. (1984). Geochonology, stratigraphy and geochemistry of Cindery Tuff in Pliocene hominid-bearing sediments in the Middle Awash, Ethiopia. *Nature*, 308, 26–31.

HARRIS, J.M., BROWN, F.H., LEAKEY, M.G., WALKER, A.C. & LEAKEY, R.E. (1988). Pliocene and Pleistocene hominid-bearing sites from west of Lake Turkana, Kenya. *Science*, 239, 27–33.

HILL, A. & WARD, S. (1988). Origin of the Hominidae: the record of African large hominoid evolution between 14 My and 4 My. *Yearbook of Physical Anthropology*, 16, 49–83.

ISAAC, G.L. (1977). *Olorgesailie: Archaeological Studies of a Middle Pleistocene Lake Basin in Kenya*. Chicago: University of Chicago Press.

JABLONSKI, N.G. (1986). The hand of *Theropithecus brumpti*. In *Primate Evolution*, ed. J.G. Else & P.C. Lee, pp. 173–82. Cambridge: Cambridge University Press.

JOHANSON, D.C., TAIEB, M. & COPPENS, Y. (1982). Pliocene hominids from the Hadar Formation, Ethiopia (1973–77): stratigraphic, chronologic and palaeoenvironmental contexts, with notes on hominid morphology and systematics. *American Journal of Physical Anthropology*, 57, 373–402.

JOLLY, C.J. (1970). The seed eaters. A new model of hominid differentiation based on a baboon analogy. *Man*, 5, 5–26.

JOLLY, C.J. (1972). The classification and natural history of *Theropithecus* (*Simopithecus*, Andrews 1916) baboons of the African Plio-Pleistocene. *Bulletin of the British Museum (Natural History), Geology*, 22.

KALB, J.E., JOLLY, C.J., MEBRATE, A., TEBEDGE, S., SMART, C., OSWALD, E.B., CRAMER, D., WHITEHEAD, P., WOOD, C.B., CONROY, G.C., ADEFRIS, T., SPERLING, L. & KANA, B. (1982a). Fossil mammals and artefacts from the Middle Awash Valley, Ethiopia. *Nature*, 298, 25–9.

KALB, J.E., OSWALD, E.B., TEBEDGE, S., MEBRATE, A., TOLA, E. & PEAK, D. (1982b). Geology and stratigraphy of Neogene deposits, Middle Awash Valley, Ethiopia. *Nature*, 298, 17–25.

LEAKEY, M.G. (1976). Cercopithecoidea from the East Rudolf succession. In *Earliest Man and Environments in the Lake Rudolf Basin*, ed. Y. Coppens, F.C. Howell, G. Isaac & R.E. Leakey, pp. 345–50. Chicago: Chicago University Press.

LEAKEY, M.G. & DELSON, E. (1987). Fossil Cercopithecidae from the Laetoli Beds. In *Laetoli: A Pliocene Site in Northern Tanzania*, ed. M.D. Leakey & J.M. Harris, pp. 91–107. Oxford: Clarendon Press.

NAKAMURA, S., TAKENAKA, O. & TAKAHASHI, K. (1983). Fibrinopeptides A and B of baboons (*Papio anubis, Papio hamadryas* and *Theropithecus gelada*): their amino acid sequences and evolutionary rates and a molecular phylogeny for the baboons. *Journal of Biochemistry*, 94, 1973–8.

OLSON, T.R. (1981). Basicranial morphology of the extant hominoids and Pliocene hominids: the new material from the Hadar Formation, Ethiopia, and its significance in early hominid evolution and taxonomy. In *Aspects of Human Evolution*, ed. C.B. Stringer, pp. 99–128. London: Taylor and Francis.

PATTERSON, B., BEHRENSMEYER, A.K. & SILL, W.D. (1970). Geology and fauna of a new Pliocene locality in northwestern Kenya. *Nature*, 226, 918–21.

PICKFORD, M. (1986). The chronology of the

Cercopithecoidea of East Africa. *Human Evolution*, 2, 1–17.

POTTS, R. (1989). Olorgesailie: new excavations and findings in Early and Middle Pleistocene contexts, southern Kenya. *Journal of Human Evolution*, 18, 477–84.

PRENTICE, M.L. & DENTON, G.H. (1988). The deep-sea oxygen isotope record, the global ice sheet record and hominid evolution. In *Evolutionary History of the 'Robust' Australopithecines*, ed. F.E. Grine, pp. 283–404. New York: Adine de Gruyter.

SENUT, B. & TARDIEU, C. (1985). Functional aspects of Plio-Pleistocene hominid limb-bones: implications for taxonomy and phylogeny. In *Ancestors: the Hard Evidence*, ed. E. Delson, pp. 193–201. New York: Alan R. Liss.

SHIPMAN, P., BOSLER, W. & DAVIS, K.L. (1981). Butchering of giant geladas at an Acheulean site. *Current Anthropology*, 22, 257–68.

STANLEY, S.M. (1979). *Macroevolution: Pattern and Process*. San Francisco: W.H. Freeman.

STANLEY, S.M. (1978). Chronospecies' longevity, the origin of genera, and the punctuational model of evolution, *Paleobiology*, 4, 26–40.

SZALAY, F.S. & DELSON, E. (1979). *Evolutionary History of the Primates*. New York: Academic Press.

TATTERSALL, I. (1986). Species recognition in palaeontology. *Journal of Human Evolution*, 15, 165–75.

TURNER, A. & CHAMBERLAIN, A. (1989). Speciation, morphological change and the status of African *Homo erectus*. *Journal of Human Evolution*, 18, 115–30.

VRBA, E.S. (1985). Ecological and adaptive changes associated with early hominid evolution. In *Ancestors: the Hard Evidence*, ed. E. Delson, pp. 63–71. New York: Alan R. Liss.

WALKER, A.C., LEAKEY, R.E., HARRIS, J.M. & BROWN, F.H. (1986). 2.5 Myr *Australopithecus boisei* from west of Lake Turkana, Kenya. *Nature*, 322, 517–22.

PART III

Anatomy of the fossil and living species of *Theropithecus*

10 Allometric aspects of skull morphology in *Theropithecus*

ROBERT D. MARTIN

Summary

1. Reliable quantitative comparisons of living and fossil *Theropithecus* with other primates require attention to non-linear scaling effects of body size. With special reference to Old World monkeys, allometric scaling analyses were conducted on (a) length of the upper cheek toothrow, (b) basal area of the upper canine, (c) sexual dimorphism in body weight, and (d) brain size.

2. *Papio* and *Theropithecus* both have relatively long cheek toothrows in comparison to other Old World monkeys, whereas colobine monkeys have relatively short toothrows. Because *Papio* and *Theropithecus* have comparatively large cheek teeth, associated with relatively long jaw length and overall skull length, the use of craniodental dimensions for the inference of body weight in fossil *Theropithecus* is problematic. It is hence preferable to apply scaling formulae determined exclusively for members of the baboon group (*Mandrillus, Papio, Theropithecus*) for this purpose (e.g. Dechow, 1983).

3. Relative to the general scaling relationship for Old World monkeys, there is no evidence of reduction of overall canine size in extant *Theropithecus*. Appropriate scaling studies are required to determine whether there has been reduction in relative size of the canines in fossil *Theropithecus*, as has been claimed. Modern *Theropithecus* show an interesting sexual difference: females have strikingly slender canines, relative to body size, in comparison to males and in comparison to other Old World monkeys. But canine length has not been reduced in female *Theropithecus* in comparison to other Old World monkeys. It remains to be determined whether this marked sexual difference in relative calibre of the canines also applies to fossil *Theropithecus*.

4. It is confirmed that the scaling of sexual dimorphism shows positive allometric scaling both in primates generally and in Old World monkeys specifically. In all comparisons, *Theropithecus* exhibits a somewhat greater degree of sexual dimorphism than expected from female body size.

Following the general scaling trend, it is likely that large-bodied fossil *Theropithecus* species showed even more extreme sexual dimorphism in body size.

5. Various analyses of the scaling of the brain in Old World monkeys consistently reveal a marked difference between *Papio* species (which have somewhat larger-than-expected brains on average) and *Theropithecus gelada* (which has a relatively small brain). The relative brain size of *Theropithecus* lies approximately on the boundary between cercopithecine and colobine monkeys. Certain dietary habits (e.g. leaf eating) are known to be associated both with relatively small brains and with relatively low basal metabolic rates. The hypothesis that low maternal metabolic levels constrain foetal brain development, and hence limit completed adult brain size, leads to the testable prediction that the special dietary habits of *T. gelada* are associated with a relatively low basal metabolic rate in comparison to *Papio* species. Preliminary inferences of brain size and body size for *T. oswaldi* indicate that in this fossil species relative brain sizes were at least as small as in modern geladas.

6. Several results of these scaling analyses conflict with the 'seed-eating' hypothesis, according to which geladas provide a parallel for early hominid evolution. Extant geladas are characterized by relatively large cheek teeth with relatively thin enamel, by canine teeth that show no reduction in overall size and by relatively small brain sizes, possibly associated with a reduction in basal metabolic rate in connection with their special dietary habits. Allometric scaling analyses of various features of fossil *Theropithecus* are required in order to test whether they differ from modern geladas and are more similar to early hominids in certain respects. Preliminary evidence indicating that fossil *Theropithecus* had relatively small brains is, however, in direct conflict with the proposal that they represent a suitable model for early hominid evolution. Given that geladas generally seem to have small brains associated with an unusual diet for primates, there is no basis for suggesting a dietary parallel between them and early hominids, which were characterized by expansion of the brain, probably in association with a shift to a high-energy, readily digestible diet.

Introduction

From a dietary point of view, living geladas (*Theropithecus gelada*) are of special interest because they are the only primates that may justifiably be called grazers (Dunbar, 1984). The existence of well-represented fossil species dating back to the mid-Pliocene increases their importance in this respect (Jolly, 1972; Szalay & Delson, 1979). Two distinct lineages can be recognized – lineage 1: *T. darti* + *T. oswaldi*; lineage 2: *T. baringensis* + *T. quadratirostris* +

T. brumpti – both terminating in forms of conspicuously large body size (Jablonski, 1986). Geladas have attracted additional interest because of the seed-eating hypothesis originally proposed by Jolly, which suggested parallels between the evolution of geladas (particularly large-bodied extinct species) and the early phases of human evolution (Jolly, 1970a, b, 1972).

For various reasons, it is vital to take into account the scaling effects of body size when comparing geladas with other primates. In the first place, this is necessary because numerous statements based on comparisons between species, for example with respect to 'reduction' in the size of various teeth, have quantitative implications that can only be tested effectively if scaling effects of body size are taken into account (Martin, 1980, 1989b). The relationships between individual parameters and body weight must be investigated through allometric analysis in order to provide a baseline for interpretation. As an extension of this requirement, if fossil species are to be included in comparisons, it is necessary to estimate their body sizes. This is particularly important for *Theropithecus* because some of the later fossil forms were notably large-bodied. With fossil forms, however, it is important to note that individual scaling relationships established for living species must be interpreted in the reverse direction, with body weight being predicted from individual measurements taken on available cranial and skeletal material (see Martin, 1990).

Allometric analysis

Scaling analysis (allometric analysis) explicitly confronts the problem of interspecific differences in body size in comparative studies and can yield valuable additional information for the reconstruction of phylogenetic relationships (Martin, 1989b, 1990). In the basic allometric approach, paired average values for body size (X) and a given biological parameter (Y) are analysed for a series of species, in order to separate the general influence of body size from more fundamental organizational differences between species. The approach is based on the widely accepted allometric formula:

$$Y = b \cdot X^{\alpha}$$

In this equation, the allometric exponent (α) represents the overall scaling principle involved, while the allometric coefficient (b) specifies the particular scaling relationship in any given case. For scaling analysis, the data are converted to logarithmic form to linearize the allometric formula and hence permit the use of simple line-fitting techniques:

$$\log Y = \alpha \cdot \log X + \log b$$

An empirical best-fit line for the logarithmically-converted data set directly yields values of α and $\log b$ for the scaling relationship concerned. There is usually an obvious trend in the relationship between a given biological parameter and body size (reflected by a relatively high value for the correlation coefficient, r), but points for individual species are typically scattered to varying degrees about the best-fit line. The best-fit line indicates the overall trend, or *scaling principle*, while *special adaptations* of individual species are indicated by their residual values (vertical distances above or below the line). A point above the line indicates a greater-than-expected value of the biological parameter for the body size of the species concerned; a point below the line a smaller-than-expected value. Bivariate allometric analysis, therefore, provides a practical solution to the theoretical problem of distinguishing simple scaling effects from more fundamental biological contrasts between species.

There are two persistent, fundamental problems involved in allometric analysis that are commonly given insufficient attention. In the first place, there is disagreement over the choice of an appropriate best-fit line (e.g. least squares regression, reduced major axis or major axis). Many authors have used least squares regressions almost as a matter of course, but there are good reasons for believing that for biological data the major axis might be a more appropriate line of best fit, at least in the majority of cases (Harvey & Mace, 1982). Recently, it has been demonstrated that the major axis may generally be the most suitable line of best-fit in *interspecific* comparisons, although in certain *intraspecific* comparisons least squares regressions may be more suitable in certain contexts (Martin & Barbour, 1989). There are compelling reasons for believing that the major axis is to be preferred over the least squares regression for determining the underlying *relationship* between two variables in any biological data set and a clear example of this is provided in the discussion of sexual

dimorphism below. For technical reasons, however, the least squares regression may provide the correct solution in the particular case where an empirical allometric formula is to be used for the purposes of *prediction*, especially in an intraspecific context. For interspecific data, by contrast, the major axis may be the appropriate line for prediction if the data conform to an 'extruded normal distribution' (Martin & Barbour, 1989). As a general rule, therefore, the major axis may be the most suitable best-fit line for all purposes wherever interspecific comparisons are involved. In any case, it is important to remember that the major axis (like the reduced major axis) provides a symmetrical approach in which no distinction between dependent and independent variables is implied. The least squares regression is an asymmetrical approach in which the X variable is taken to be independent and the Y variable is assumed to be dependent upon X. This asymmetry is a source of potential confusion. In a study of the scaling of the length of the cheek tooth row (see next section), for instance, if a least squares regression is applied, body size is taken as the X variable and cheek tooth length as the Y variable. By contrast, if body weight of a given fossil species is to be predicted from its cheek tooth length, a least squares regression used to make such a prediction must take body size as the Y variable and cheek tooth length as the X variable.

The second problem arises when the data set contains either extreme outliers or distinct subsets, in the latter case yielding separate 'grades' (which can be represented by parallel lines) in an allometric plot (Martin, 1980, 1989b, 1990). In either case, a single best-fit line is not appropriate. Successful allometric analysis, therefore, depends upon careful attention both to line-fitting procedures and to possible heterogeneity in the data set. As a precaution, it is advisable to examine the results obtained with different line-fitting techniques and to consider carefully any differences between them. It is also necessary to examine bivariate plots for the presence of extreme outliers or grade shifts. Practical examples of both problems are provided in the following sections.

Attention must also be given to the question of bias within the data set taken for allometric analysis. As a general rule, it is necessary to ensure that the data set is properly representative of the group studied. But even if the data set includes all species of a group, difficulties can arise because of over-representation. If, for example, a particular sample includes species-rich genera alongside species-poor genera, the former will exert a dominant influence on the best-fit line determined and may hence bias the results of the analysis (Harvey, Martin & Clutton-Brock, 1987). The essential problem is that species may not be, strictly speaking, statistically independent because of their relationship within a phylogenetic tree, and large groups of closely-related species, in particular, may bias an analysis. A pragmatic approach to counter this potential bias is to take average values for genera, rather than for individual species. When a relatively small taxonomic group (e.g. Old World monkeys) is examined, however, reduction of the data set to generic aver-

ages may reduce the size of the sample to an unsatisfactory level and may result in loss of valuable information. Further, because the correlation coefficient is commonly reduced with smaller data sets, the best-fit line determined is more likely to yield aberrant results. Although it is theoretically possible that species-rich genera may bias the results of analysis, this will not necessarily be the case in practice. Once again, it is best to try a number of approaches (e.g. both species-level analysis and generic-level analysis) in order to check for bias.

Scaling of the cheek teeth

The length of the cheek tooth row (the sum of the lengths of the premolars and molars in a given jaw) provides a simple indication of the overall size of the dental array involved in mastication. Plots of upper cheek tooth length against body weight for Old World monkey species, taking males and females separately, show that *Papio* and *Theropithecus* both have relatively long tooth rows (Fig. 10.1). Indeed, the residual values reveal that for both sexes these two genera rank very highly among Old World monkeys, with *Theropithecus* showing the maximum relative length of the upper cheek tooth row in both sexes (Fig. 10.2). This finding is doubtless related to the long jaws and skulls that typify baboons as a group and it suggests that the striking difference in skull shape between *Papio* and *Theropithecus* and the dietary differences between modern representatives of these two genera have not greatly affected the basic scaling relationship between length

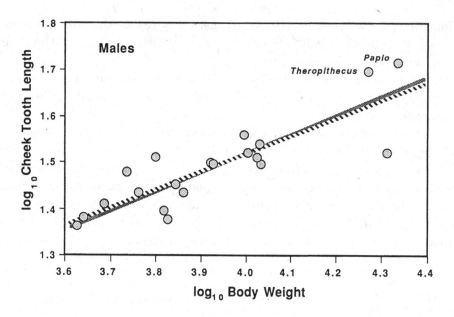

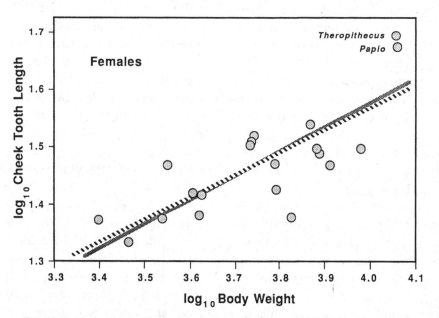

Fig. 10.1. Logarithmic plots of upper cheek tooth row length (mm) against body weight (g) for a sample of Old World monkey species (*n*=20). Data derived from Swindler (1976), with minor additions. The hatched lines are least squares regressions and the shaded lines are major axes.

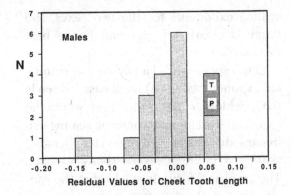

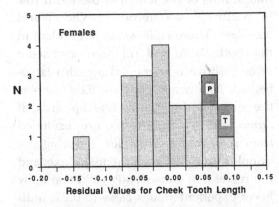

Fig. 10.2. Residual values for upper cheek tooth row length (mm) relative to body weight (g) for Old World monkey species (see also Table 10.1). *Theropithecus* (T) and *Papio* (P) both rank very highly in terms of relative length of the cheek teeth.

Table 10.1 *Residual values for scaling of cheek tooth length in Old World monkeys (major axis analysis).*

Species	Females	Males
Cercopithecinae		
Cercopithecus nictitans	− 0.035	− 0.050
C. mona	+ 0.052	+ 0.005
C. mitis	− 0.003	+ 0.010
C. neglectus	+ 0.010	− 0.003
C. ascanius	− 0.015	− 0.007
C. aethiops	− 0.005	+ 0.017
Cercocebus albigena	− 0.018	+ 0.005
C. torquatus	+ 0.021	+ 0.004
C. galeritus	+ 0.045	− 0.003
Macaca mulatta	+ 0.040	+ 0.070
M. nemestrina	+ 0.053	+ 0.040
M. fascicularis	+ 0.083	+ 0.065
Papio cynocephalus	+ 0.074	+ 0.060
Theropithecus gelada	+ 0.095	+ 0.067
Colobinae		
Colobus badius	− 0.040	+ 0.010
C. polykomos	− 0.029	− 0.019
Nasalis larvatus	− 0.070	− 0.127
Pygathrix nemaeus	− 0.069	− 0.039
Presbytis phayrei	− 0.062	− 0.032
P. aygula	− 0.126	− 0.073

of the cheek tooth row and body size. One might be tempted to link the long cheek tooth row of *Theropithecus* to its specific dietary habits. It is noteworthy, however, that colobine monkeys, which are generally more folivorous than cercopithecines, uniformly show quite low residual values. With only one exception (male *Colobus badius*), the residual values obtained are all negative (Table 10.1). Hence, there is no simple link between the relative length of the cheek tooth row and

the proportion of dietary matter requiring extensive mastication, though it might be argued that terrestrial feeding and especially grazing (as opposed to arboreal leaf eating) require special dental adaptations, for example to offset the abrasive effects of ingested silica.

As far as determination of a best-fit line for the entire sample of species is concerned, it can be seen that the major axis and the least squares regression yield quite similar results (Fig. 10.1). For females, the empirical allometric formula obtained with the major axis for the relationship between cheek tooth length (L in mm) and body weight (W in g) is:

$$\log_{10} L = 0.42 \cdot \log_{10} W - 0.12 \quad [r = 0.78]$$

The allometric formula obtained with the least squares regression is:

$$\log_{10} L = 0.39 \cdot \log_{10} W + 0.01$$

For males, the empirical allometric formula obtained with the major axis is:

$$\log_{10} L = 0.41 \cdot \log_{10} W - 0.01 \quad [r = 0.85]$$

while the allometric formula obtained with the least squares regression is:

$$\log_{10} L = 0.39 \cdot \log_{10} W - 0.00$$

With respect to the scaling of cheek tooth length, it is interesting to note that apparently spurious results are obtained when generic averages are used rather than the raw species values, reducing the sample size from 20 to 9. For females, the exponent value is markedly increased (to 0.55 for the major axis and to 0.48 for the least squares regression), whereas the exponent value is even more strikingly decreased for males (to 0.12 for both the major axis and the least squares regression). For males, reduction of the sample to generic averages leads to a major decrease in the correlation coefficient (from r = 0.85 to 0.40), whereas for females the correlation coefficient remains essentially unchanged (0.77 vs. 0.78). Given that the species-level analysis indicates similar scaling exponents for males and females, which is what would be expected, whereas the generic-level analysis indicates drastically different

scaling exponents for the two sexes, the latter is clearly suspect and hence best disregarded.

In the species-level analyses, the empirical exponent value (α) in all cases exceeds 0.33, which is the value that would be expected with simple isometric scaling of a linear dimension (tooth row length) against a cubic measure (body weight). The empirical formulae yielded at the species level might therefore seem to indicate that scaling of the length of the tooth row is positively allometric in Old World monkeys. There is, however, a problem in that both *Papio* and *Theropithecus* seem to be positive outliers and are also large-bodied. As a result, they are likely to bias the slope of the best-fit line upwards. If *Papio* and *Theropithecus* are excluded from the sample of species, reducing the sample size to 18, both the major axis and the regression consistently yield markedly lower exponent values close to 0.27, indicating that the relationship between cheek tooth length and body weight is in fact negatively allometric. Further, taking mean values for a broad sample of 47 non-human primate species including both prosimians and simians (male and female values averaged for each species), a major axis (Fig. 10.3) yields the following empirical scaling relationship between cheek tooth length and body weight:

$$\log_{10} L = 0.27 \cdot \log_{10} W + 0.47 \quad [r = 0.92]$$

In this case, the empirically determined allometric formula is indistinguishable from that yielded by a least squares regression (exponent value: 0.27). This exponent value is closely similar to that obtained with the sample of Old World

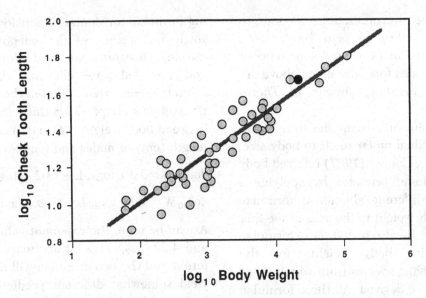

Fig. 10.3. Logarithmic plot of upper cheek tooth row length (mm) against body weight (g) for a broad sample of nonhuman primate species ($n=47$). The black point indicates *Theropithecus*, which shows a position deviation from the best-fit line (major axis). In this case, the major axis and the least squares regression yield almost identical results and only the major axis is shown.

monkey species after excluding *Papio* and *Theropithecus*, and the results overall are in concordance with the general conclusion that linear tooth dimensions show a negative allometric relationship to body weight in mammals. It is therefore safe to conclude that *Papio* and *Theropithecus* are indeed positive outliers in comparison to other Old World monkeys. The direct practical implication of this is that the upward deviations of *Papio* and *Theropithecus* from the general scaling relationship between cheek tooth length and body weight must in fact be even more extreme than indicated by the overall best-fit lines in Fig. 10.1.

Because *Papio* and *Theropithecus* are obvious outliers relative to most other Old World monkeys with respect to scaling of cheek tooth length, caution is obviously required when predicting body weights for fossil relatives. If modern *Papio* and *Theropithecus* have larger-than-expected cheek tooth rows relative to body size, the same is likely to apply to their fossil relatives. As a result, there will be a tendency to overpredict body weights for fossil *Papio* and *Theropithecus* when using general scaling formulae based on living primates. This can be illustrated by using the major axes for Old World monkeys illustrated in Fig. 10.1 to 'predict' body weights for male and female *Theropithecus gelada* from their cheek tooth lengths. When this is done, a body weight of 27.2 kg (actual value 18.7 kg) is predicted for males and a body weight of 18.9 kg (actual value 11.4 kg) is predicted for females. The overpredictions amount to 45 per cent and 65 per cent, respectively. Similar overpredictions are likely to occur if the major axis formulae given above are applied to fossil specimens of *Theropithecus*.

This conclusion also applies to the scaling of individual molar teeth to body size. For instance, Conroy (1987) inferred body weight in fossil primates by applying a variety of different allometric formulae relating body weight to the area of the first lower molar. He also tested these formulae by predicting body weights for the individual living species from which these formulae were derived. All three formulae that he applied to male *Papio cynocephalus* led to marked overprediction of body weight, yielding values of 35.1 kg, 35.6 kg and 40.1 kg, as compared to his actual starting value of 19.5 kg. Conroy did not provide data for *Theropithecus*, but if his formula for monkeys ($\log_e$ W = 1.561 · $\log_e$ A + 3.41, where A = area of M_1) is applied to male *Theropithecus*, it yields a predicted body weight of 29.3 kg, in contrast to the real value of 18.7 kg. The relatively greater length of the cheek teeth in baboons and geladas accounts, at least in part, for the tendency to overpredict body weights for these simian primates when using general relationships for scaling of craniodental variables derived from a broad sample of primate species (Martin, 1980, 1990).

Given that *Papio* and *Theropithecus* have relatively larger cheek teeth than most other Old World monkeys, an alternative strategy must be found for the inference of body weights for fossil *Theropithecus*. Dechow (1983) determined scaling formulae for dental dimensions based solely on species of the baboon group (*Papio*, *Theropithecus*, and *Mandrillus*) and provided the following allometric formulae (converted to match the conventions of this chapter) for the relationship between body weight (kg) and cheek tooth length (cm) in males and females:

$$\log_{10} W = 3.36 \cdot \log_{10} L - 1.02 \quad \text{[regression]}$$

$$\log_{10} W = 4.23 \cdot \log_{10} L - 1.60 \quad \text{[major axis]}$$

As can be seen, the exponent values (3.36 and 4.23, respectively) are markedly different and the two formulae will therefore yield somewhat different predictions for body weights of fossil species. Nevertheless, these formulae should yield more reliable results than a general scaling relationship for Old World monkeys that does not take into account the oversized cheek teeth of baboons and geladas. Interestingly, inversion of the major axis formula yields the following relationship between cheek tooth length and body weight for Dechow's baboon sample:

$$\log_{10} L = 0.24 \cdot \log_{10} W + 0.38$$

The exponent value in this equation (0.24) is quite close to that obtained for Old World monkeys excluding *Papio* and *Theropithecus* and to that for non-human primates generally (i.e. 0.27) and again indicates negative scaling of cheek tooth length relative to body weight.

Estimations of body weights of fossil geladas are plagued by two additional problems. In the first place, it is virtually certain that all species showed marked sexual dimorphism in body weight (to be discussed later). In addition, there is convincing evidence that average body weights increased quite markedly over time within certain fossil gelada species. Estimated body weights for the single species *T. oswaldi* based on dimensions of the femur head, have been determined by Leakey (chapter 3) and yield the following temporal sequence of average values over a period of just over one million years (2.4 to 1.3 Ma): Upper Burgi = 26.8 kg, KBS = 31.7 kg, Okote = 35.2 kg, Chari = 37.0 kg, Olduvai Upper Bed II = 52.3 kg. This approximate doubling of average body weights within a single species lineage over time presents a special difficulty for estimating typical body weights of fossil geladas for scaling comparisons with modern primates. In view of this, and given the additional difficulties posed by sexual dimorphism, it is clear that any predictions of body weight should ideally be made for individual specimens rather than for entire species.

As a final note on the scaling of dental dimensions, it should be noted that a scaling study of enamel thickness in primates (Kay, 1981) demonstrated that *T. gelada* has slightly thinner enamel than expected for its body size. This represents a major difference between modern geladas and early hominids and, if this finding also applies to fossil geladas, it raises doubt regarding Jolly's suggestion of specific dietary resemblances between fossil geladas and early hominids (Martin, 1990).

Scaling of the canines

Another aspect of the dentition that has received special attention is the size of the canine teeth. The canines are of interest because of the marked sexual dimorphism in canine size that is present in cercopithecine monkeys generally and because any realistic parallel with early hominids requires that the canines of *Theropithecus* should show reduction in size relative to those of *Papio*. Indeed, Jolly (1970a) specifically stated that the canines are relatively small in *Theropithecus*, especially in the larger-bodied fossil forms.

It is somewhat difficult to conduct a reliable comparison of canine length in adult Old World monkeys because the length of the canines may be secondarily modified during life. As a first approach, it is instructive to consider the scaling of the basal area of the upper canines (mesiodistal length × buccolingual breadth) in *Theropithecus* and other Old World monkey species. An allometric analysis of the basal area of the upper canines (Figs 10.4 and 10.5) shows that male *Theropithecus*, like male *Papio*, have slightly stouter canine bases than expected for their body sizes. Rather different scaling relationships are indicated for males by the major axis and regression, as is indicated in Fig. 10.4. The empirical allometric formulae relating basal area of the upper canine (C in mm^2) to body weight (W in g) are as follows (r = 0.71):

$$\log_{10} C = 0.70 \cdot \log_{10} W - 1.01 \text{ [regression]}$$

$$\log_{10} C = 0.97 \cdot \log_{10} W - 2.08 \text{ [major axis]}$$

Despite this discrepancy, however, *Theropithecus* and *Papio* are in both cases positive outliers and the difference between the regression and the major axis is simply

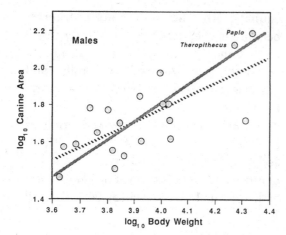

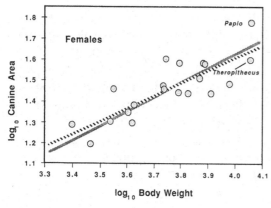

Fig. 10.4. Logarithmic plot of the basal area of the upper canine (mm²) against body weight (g) for a sample of Old World monkey species ($n=20$). Data derived from Swindler (1976). The hatched lines are least squares regressions and the shaded lines are major axes.

one of degree. With females, there is less of a discrepancy between the regression and the major axis, as is indicated by the empirical allometric formulae ($r = 0.81$):

$$\log_{10} C = 0.60 \cdot \log_{10} W - 0.77 \text{ [regression]}$$

$$\log_{10} C = 0.68 \cdot \log_{10} W - 1.11 \text{ [major axis]}$$

Both best-fit lines clearly reveal that female *Theropithecus* differ markedly

from female *Papio* in terms of relative basal canine area. The former have less robust canines than expected for their body size, while the latter have very stout canines relative to body size.

Restriction of the sample of Old World monkeys to generic averages (sample size reduced from 20 species to nine genera, as in the case of cheek tooth length) once again leads to confusing results. For males, the correlation coefficient is drastically reduced (from $r = 0.71$ to $r = 0.45$) and the exponent values in the empirical allometric formulae are also greatly reduced:

$$\log_{10} C = 0.28 \cdot \log_{10} W + 0.62 \text{ [regression]}$$

$$\log_{10} C = 0.39 \cdot \log_{10} W + 0.16 \text{ [major axis]}$$

For females, by contrast, the correlation coefficient is only slightly reduced (from $r = 0.81$ to $r = 0.75$) and the exponent values in the empirical allometric formulae remain virtually unchanged:

$$\log_{10} C = 0.57 \cdot \log_{10} W - 0.67 \text{ [regression]}$$

$$\log_{10} C = 0.69 \cdot \log_{10} W - 1.16 \text{ [major axis]}$$

Analysis at the species level indicates that in males basal area of the canines scales with a higher exponent than in females, whereas analysis at the generic level leads to exactly the opposite conclusion. Given that the degree of sexual dimorphism in canine size, like the degree of dimorphism in body size (see later), generally tends to increase with increasing absolute size in primates (see Leutenegger, 1978), the conclusion indicated by the generic-level analysis is obviously suspect.

Residual values derived from the species-level analysis (Fig. 10.5) show that female *Theropithecus* rank among the

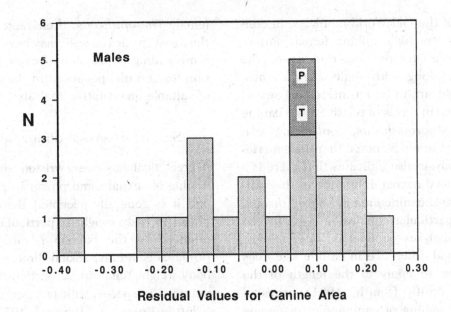

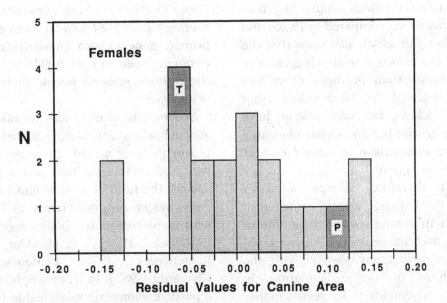

Fig. 10.5. Residual values for basal area of the upper canine (mm²) relative to body weight (g) for Old World monkey species. For males, *Papio* (P) and *Theropithecus* (T) both rank quite highly in terms of the relative area of the upper canine. For females, by contrast, there is a striking difference between *Papio* and *Theropithecus*. Female *Papio* rank very highly among Old World monkeys, whereas female *Theropithecus* occupy one of the lowest-ranking positions.

lowest of the Old World monkeys, in contrast to the high-ranking female *Papio*, while male *Theropithecus* rank among the highest, along with male *Papio*. Thus, there appears to be an interesting sexual difference in terms of relative basal canine area in *Theropithecus*, contrasting with *Papio*. It should be noted that the generic-level analysis also indicates that there is a pronounced sexual difference in the scaling of basal canine area in *Theropithecus*, so this particular conclusion is confirmed by all analyses conducted. Interestingly, this sexual difference in canine size does not seem to apply to the length of the canine. Smith (Smith, 1981) provided data on scaling of canine size in female primates which confirm that the basal area of the canine is relatively smaller in female *Theropithecus*, as compared to *Papio* and *Mandrillus*, but which also show that the length of the canine is relatively greater in *Theropithecus* than in these other two genera. Hence, it may be concluded that female *T. gelada*, for some reason, have relatively gracile but somewhat elongated canines in comparison to other members of the baboon group.

Overall, therefore, as was noted by Martin (1990) there is little evidence of reduction in canine size in living *Theropithecus*, and this presents a further problem in proposing any parallel with the early phase of dental evolution in hominids. It remains to be seen whether appropriate allometric analysis will reveal any evidence of reduction in canine size in fossil *Theropithecus*, though simple visual inspection of available specimens indicates that this is unlikely. It is possible that the relatively slender canines of female *Theropithecus*, if characteristic of the fossil species as well, may have created a misleading impression of canine reduction for certain specimens in the absence of suitable quantitative analysis.

Scaling of sexual dimorphism

A great deal has been written about the scaling of sexual dimorphism in primates and it is generally accepted that dimorphism in body weight is particularly pronounced in the baboon group. Among primates generally, sexual dimorphism in body weight tends to be particularly pronounced in species with terrestrial habits (Clutton-Brock & Harvey, 1977) and members of the baboon group provide a prime illustration of this generalization. If average values are taken for a sample of 49 primate genera, both *Theropithecus* and *Papio* are seen to rank highly in terms of the ratio of male to female body weight (Fig. 10.6).

Scaling effects must also be taken into account when considering sexual dimorphism in body weight, however, as it is now well established that across primate species the ratio of male weight to female body weight generally tends to increase with increasing female body weight (Clutton-Brock, Harvey & Rudder, 1977; Leutenegger, 1978; Leutenegger & Cheverud, 1982). In other words, there is a positive allometric relationship between male weight and female weight. This is confirmed by an allometric analysis of average male body weight in relation to average female body weight for 49 primate genera (Fig. 10.7). The major axis for this sample of primate genera yields the follow-

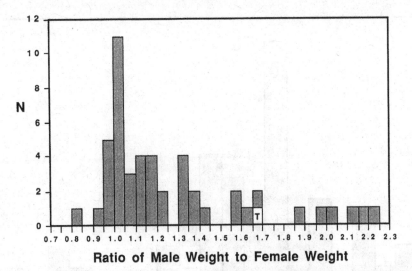

Ratio of Male Weight to Female Weight

Fig. 10.6. Frequency distribution of ratios of male weight to female weight based on average values for 49 modern primate genera. *Theropithecus* (T) ranks quite highly in comparison to other genera.

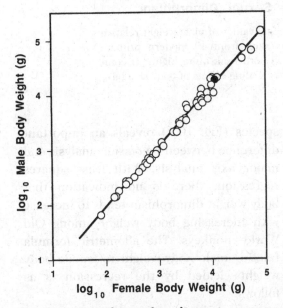

Fig. 10.7. Logarithmic plot of generic average values for male body weight (g) against female body weight (g) for a sample of 49 primate genera. The black point indicates *Theropithecus*, which shows a slight positive deviation from the best-fit line (major axis). The major axis and the least squares regression yield almost identical results, so only the major axis is shown.

ing empirical allometric formula relating male body weight (W_m in g) to female body weight (W_f in g):

$$\log_{10} W_m = 1.11 \cdot \log_{10} W_f - 0.27 \ [r = 0.995]$$

In this case, the correlation coefficient is very high and the least squares regression yields an almost identical formula:

$$\log_{10} W_m = 1.10 \cdot \log_{10} W_f - 0.25$$

In both cases, the scaling exponent is significantly greater than unity, which would indicate isometric scaling, and it can be safely concluded that sexual dimorphism in body weight among primates is positively allometric. Even after taking account of this scaling effect of body size, both *Papio* and *Theropithecus* are found to rank quite highly in relation to other primates in this generic-level allometric analysis (Fig. 10.8).

It is worth noting that an allometric analysis of sexual dimorphism in body

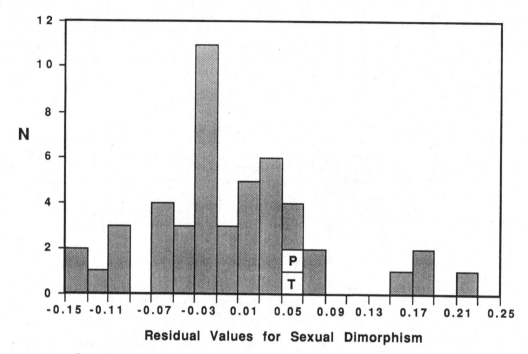

Fig. 10.8. Frequency distribution of residual values of male weight relative to female weight for generic averages taken from 49 modern primate genera. *Papio* (P) and *Theropithecus* (T) both rank quite highly in comparison to other primate genera even after scaling effects of body size have been taken into account.

weight using the raw values for 142 individual primate species, rather than the average values for the 49 genera represented, yields closely similar results, with a similarly high correlation (r = 0.992):

$$\log_{10} W_m = 1.10 \cdot \log_{10} W_f - 0.24 \text{ [major axis]}$$

$$\log_{10} W_m = 1.09 \cdot \log_{10} W_f - 0.20 \text{ [regression]}$$

Hence, although it is theoretically possible that species-rich genera might bias the results of allometric analysis, in this particular case any such bias is demonstrably negligible.

In a narrower taxonomic analysis, an allometric plot of male weight against female weight for 59 Old World monkey species (Fig. 10.9) reveals an important difference between regression analysis and major axis analysis. With least squares regression, there is no indication that body weight dimorphism tends to increase with increasing body weight among Old World monkeys. The allometric formula relating male body weight to female body weight yielded by the regression is as follows:

$$\log_{10} W_m = 1.00 \cdot \log_{10} W_f + 0.15 \text{ [r = 0.92]}$$

The slope of the best-fit line indicates an isometric relationship ($\alpha = 1.00$). With major axis analysis, by contrast, the generally accepted occurrence of positive

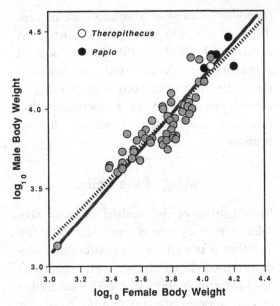

Fig. 10.9. Logarithmic plot of generic average values for male body weight (g) against female body weight (g) for a sample of Old World monkey species ($n=59$). The hatched line is the least squares regression and the shaded line is the major axis. Points for *Papio* (black) and *Theropithecus* (white) typically lie above both best-fit lines. The single point for *Papio* that lies below the lines is based on very limited data of questionable reliability for *P. papio*.

allometry between male weight and female weight is confirmed by the empirical allometric formula:

$$\log_{10} W_m = 1.11 \cdot \log_{10} W_f - 0.24$$

The scaling exponent ($\alpha = 1.11$) is in close agreement with that obtained from an allometric analysis of average values for 49 primate genera. Thus, an analysis based on Old World monkeys alone yields virtually the same result as that based on primates generally, provided that the major axis is taken as the appropriate best-fit line. By contrast, least squares regression yields very different answers for primates generally and for Old World monkeys alone. This example confirms the general rule noted in the introduction that the major axis is the appropriate best-fit line to take when the underlying *relationship* between two variables is at issue.

Once again, it is instructive to analyse the data for Old World monkeys at the generic level, following the rationale that this should exclude any bias due to species-rich genera. Calculation of generic averages in this case reduces the sample size from 59 to 16, but there is a slight improvement in the correlation coefficient (from $r = 0.92$ to $r = 0.95$). The exponent values in the empirical allometric formulae are somewhat increased, with the effect that both the least squares regression and the major axis now agree in confirming the existence of positive allometry in the scaling of sexual dimorphism in body weight:

$$\log_{10} W_m = 1.10 \cdot \log_{10} W_f - 0.20$$
[regression]

$$\log_{10} W_m = 1.17 \cdot \log_{10} W_f - 0.43$$
[major axis]

In this instance, the generic-level analysis seems to yield results that are in better accord with the pattern of scaling of sexual dimorphism in primates generally, although the exponent value obtained with the major axis (1.17) is rather higher than indicated by studies of primates generally.

Overall, it can be safely concluded that both *Theropithecus* and *Papio* show relatively pronounced sexual dimorphism in body weight even after the scaling effects of body size have been taken into account and it is therefore reasonable to expect that fairly extreme sexual dimorphism would also have characterized fossil species of *Theropithecus*. (The fact that *P. papio* is an apparent exception to this rule, as indicated in Fig. 10.9, may simply be an artefact attributable to the limited and questionable data so far available for this species, which are derived from a small number of captive specimens.) Sexual dimorphism in fossil *Theropithecus* has been specifically discussed by Jablonski (1986). Jablonski suggested that the relatively dense fossil record for *Theropithecus* may permit us to throw light on the development of sexual dimorphism among living forms, rather than depending in the usual way upon exclusive interpretation of fossil forms in the light of extant forms. If this suggestion is to be applied in practice, appropriate scaling analyses will be necessary.

It is also worth noting that there is a questionable assumption behind most discussions of body weight dimorphism in primates. Many authors have simply assumed that sexual dimorphism in size results from selection pressures acting on the male, leading to an increase in body size relative to the female. Males may increase in size either because of selection for fighting ability in interspecific conflicts with other males or because of the need to engage in active defence against predators. In fact, it is at least equally likely as a starting-point for any interpretation that selection pressures have acted to reduce body size in females relative to males (Willner & Martin, 1985). A considerable advantage of smaller female body size is that all reproductive variables are scaled down, such that higher reproductive

turnover at smaller absolute cost of production per offspring is permitted. A recent study (Willner, 1989) has produced substantial evidence that evolutionary reduction in female body size, leading to sexual dimorphism as a secondary by-product, has been common among primates.

Scaling of the brain

Investigation of the scaling of brain size relative to body size is complicated by the fact that it is well known that empirically-determined slopes change with taxonomic level (Martin & Harvey, 1985). The choice of an appropriate taxonomic level for analysis is, therefore, particularly important in this case. It could be argued that a suitable comparison for estimating relative brain size in geladas would be one confined to the taxonomic group containing Old World simians (monkeys and apes). In fact, however, one recent investigation has indicated that the smaller exponent values found at lower taxonomic levels may be, at least in part, a statistical artifact and that the true scaling exponent for the scaling of brain size in mammals generally is close to 0.75 (Pagel & Harvey, 1988). It is also important to give attention to bias (e.g. arising from statistical over-representation of species-rich genera or through the effects exerted by extreme outliers) when determining an empirical best-fit line. Analyses for the present study were therefore conducted in a number of ways, using both empirical best-fit lines and a line with a fixed slope value of 0.75, representing the overall scaling relationship for placental mammals generally (Martin, 1981).

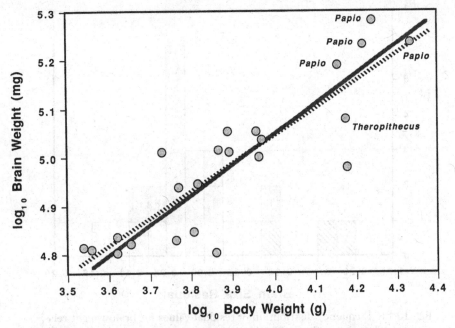

Fig. 10.10. Logarithmic plot of average values for brain weight (mg) against adult body weight (g) for a combined-sex sample of Old World monkey species ($n=23$). The hatched line is the least squares regression and the shaded line is the major axis. Points for *Papio* tend to lie above the best-fit lines, whereas the point for *Theropithecus* lies well below both lines.

A major axis determined using average values for a combined-sex sample of 17 Old World simian genera, excluding *Homo* as an extreme outlier, yielded the following allometric formula (Martin, 1983) relating brain weight (E, in mg) to body weight (W, in g):

$$\log_{10} E = 0.60 \cdot \log_{10} P + 2.68 \quad [r = 0.97]$$

This is only marginally different from the formula indicated by the best-fit line (major axis) obtained with the 23 individual species points for Old World monkey species (sexes combined) shown in Fig. 10.10:

$$\log_{10} E = 0.61 \cdot \log_{10} P + 2.61 \quad [r = 0.87]$$

The least squares regression in this case indicates a somewhat lower scaling exponent:

$$\log_{10} E = 0.57 \cdot \log_{10} P + 2.78$$

Residual values determined relative to the major axis for Old World monkey species shown in Fig. 10.10 are hence virtually identical with those that would be determined using the formula based on generic averages for Old World simians and would accordingly seem to be unaffected by bias through species-rich genera. Examination of the residual values (Fig. 10.11) yields the interesting conclusion that *T. gelada* has a relatively small brain in relation to body size. In fact, this is

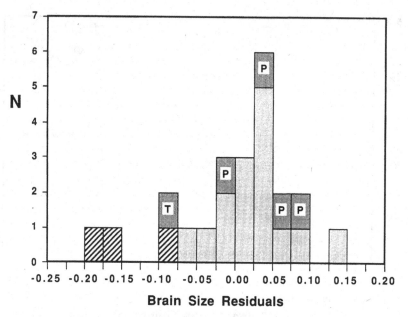

Fig. 10.11. Frequency distribution of residual values for brain weight relative to adult body weight for 23 Old World monkey species modern primate genera. Among cercopithecines (shaded boxes) *Papio* species (P) tend to have moderate to large relative brain sizes, whereas *Theropithecus* (T) lies on the boundary between cercopithecine and colobine monkeys (hatched boxes).

clearly apparent from Fig. 10.10, as the point for *Theropithecus* lies well below all four points for *Papio* species in the graph.

It has already been shown (Martin, 1983) that leaf-monkeys (Colobinae) generally have smaller brains that other Old World monkeys (Cercopithecinae). From Fig. 10.11, it can be seen that in terms of relative brain size *Theropithecus* lies at the boundary between the colobines (hatched boxes) and the other cercopithecines (shaded boxes). This does not apply to *Papio* species and the data on relative brain size from the empirical comparison therefore indicate an intriguing difference between extant geladas and baboons.

This analysis was repeated by fitting a line with a fixed slope of 0.75, which is the exponent value determined for placental mammals generally (Martin, 1981), a procedure that meets the objection that the low exponent value for Old World simians alone may be a statistical artifact (Pagel & Harvey, 1988). The allometric formula determined for the 23 Old World monkey species represented in Fig. 10.10 is as follows:

$$\log_{10} E = 0.75 \cdot \log_{10} P + 2.06$$

An analysis using this formula to calculate residual values does not alter the relative difference between *Theropithecus* and *Papio*, although it does alter the relationship between these genera and smaller-bodied colobines (Table 10.2). In this general mammalian comparison, both

Table 10.2 *Residual values for scaling of brain weight in Old World monkeys.*

Species	Major axis	Regression	Fixed slope
Cercocebus albigena	+ 0.048	+ 0.047	+ 0.052
C. galeritus	+ 0.031	+ 0.031	+ 0.031
C. torquatus	+ 0.030	+ 0.033	+ 0.022
Cercopithecus aethiops	− 0.010	− 0.022	+ 0.029
C. ascanius	+ 0.035	+ 0.020	+ 0.083
C. l'hoesti	+ 0.009	+ 0.005	+ 0.020
C. mitis	− 0.074	− 0.080	− 0.056
C. mona	+ 0.050	+ 0.034	+ 0.100
C. pygerythrus	+ 0.019	+ 0.007	+ 0.058
Erythrocebus patas	+ 0.003	+ 0.006	− 0.009
Macaca arctoides	− 0.027	− 0.023	− 0.037
M. fascicularis	− 0.017	− 0.028	+ 0.017
M. mulatta	+ 0.031	+ 0.025	+ 0.049
M. nemestrina	+ 0.075	+ 0.075	+ 0.076
M. niger	+ 0.126	+ 0.118	+ 0.149
Nasalis larvatus	− 0.178	− 0.166	− 0.218
Papio cynocephalus	+ 0.052	+ 0.067	+ 0.007
P. hamadryas	+ 0.046	+ 0.058	+ 0.010
P. papio	+ 0.086	+ 0.101	+ 0.037
P. ursinus	− 0.017	+ 0.003	− 0.079
Presbytis cristata	− 0.082	− 0.086	− 0.070
P. obscura	− 0.158	− 0.160	− 0.153
Theropithecus gelada	− 0.078	− 0.065	− 0.117

Theropithecus and *Papio* end up with lower values for relative brain size. The gelada emerges as particularly small-brained in comparison to *Papio*, but both *P. papio* and *T. gelada* are found to have lower values than the colobine *Presbytis cristata*.

The difference between colobines and cercopithecines in terms of relative brain size has been linked by several authors to the special adaptations of the former for folivory. Given that modern geladas, by any comparison, have smaller relative brain sizes than *Papio* species, this suggests the interesting possibility that *T. gelada* may have a relatively small brain in association with its unusual diet. It should be noted, however, that there are two competing hypotheses that have been advanced to account for the relatively smaller brain sizes of folivorous primates in comparison to frugivorous species. The first hypothesis proposes that foraging for leaves requires less complex central nervous processing than foraging for fruits and that folivorous primates have therefore been subject to less intense selection for increase in relative brain size (Clutton-Brock & Harvey, 1980). The alternative hypothesis proposes that the link between diet and relative brain size is indirect. Folivorous mammals generally have lower basal metabolic rates, relative to body size, and the alternative hypothesis suggests that the more severe metabolic constraints of folivorous species limit the level of maternal investment in foetal brain development, which in turn limits the ultimate size attained by the adult brain (Martin, 1981, 1983, 1984, 1989a). This second hypothesis, which is supported by several lines of evidence, generates the testable prediction that *Theropithecus* will be found to have lower basal metabolic rates than *Papio* species. Although *T. gelada* is not folivorous in the classical sense, it is possible that the grass-eating habits of geladas impose energetic constraints that are similar to those imposed by leaf-eating habits.

Clearly, it is of particular interest to study relative brain size in fossil species of *Theropithecus*. There is, however, a fundamental problem involved in the analysis of relative brain size for fossil primates in that reliable methods must be found for the estimation of body weight. In some cases it is easier to determine brain size through direct measurement of cranial capacity than to obtain a reliable estimate of body size, particularly when only the skull is available for measurement. In the case of species belonging to the baboon group, the problem is exacerbated by the fact that the skull and jaws have been secondarily elongated relative to body size, thus rendering dental dimensions and skull length unreliable as indicators of body size in comparison to other primates.

Using various available formulae, however, it is possible to obtain reasonable estimates of both brain size and body size for the following skulls of the fossil species *T. oswaldi*:

1. *Theropithecus oswaldi* [Peninj DAT 600/82, identified as female]: the cranial capacity of this new specimen was measured directly as 200 cm³ by packing the braincase with seeds. Body weight for this specimen can be estimated by applying Dechow's formulae for scaling of body size to the length of the upper cheek tooth row in species of the baboon group (Dechow, 1983). The length of the upper cheek tooth row in the Peninj skull (56.7 mm) indicates a body weight of 32.5 kg with the regression formula and a weight of 38.7 kg with the major axis formula. Taking the empirical major axis formula determined for the scaling of brain weight relative to body weight in Old World monkey species (Fig. 10.11), residual values of −0.061 and −0.107, respectively, are obtained with these two body weight estimates, indicating that relative brain size in this specimen of *Theropithecus oswaldi* was at least as small as that of extant *T. gelada* (see Table 10.2). The formula determined for the same sample of Old World monkeys with a fixed slope of 0.75 yields residual values of −0.143 and −0.200, respectively. Both of these

values are markedly lower than the residual value determined for extant *T. gelada* (Table 10.2). Thus, it would seem overall that this specimen of *T. oswaldi*, if anything, had a smaller relative brain size than modern *T. gelada*.

2. *Theropithecus oswaldi* [Kanjera BM 32102, identified as male]: the minimal cranial capacity of this specimen was measured directly as 155 cm³ by packing the braincase with seeds. This could be an underestimate because some residual matrix may be present in the braincase. Any underestimation, however, must be fairly limited because the external dimensions of the braincase are notably smaller than in the Peninj skull. Dechow's formulae applied to the length of the upper cheek tooth row in this skull (65.3 mm) indicate a body weight of 52.3 kg with the regression formula and 70.3 kg with the major axis formula. Taking the empirical major axis formula for Old World monkey species, residual values of −0.298 and −0.376, respectively, are obtained with these two body weight estimates, indicating that relative brain size in this specimen was considerably smaller than that of extant *T. gelada* or of any extant colobine monkey species (see Table 10.2). The formula determined with a fixed slope of 0.75 yields residual values of −0.408 and −0.505, respectively, and both of these values are well below those recorded for any extant Old World monkey species (Table

10.2). This specimen of *T. oswaldi*, therefore, seems to have had a far smaller relative brain size than modern *T. gelada*, though there may be some degree of underestimation because of the presence of residual matrix in the braincase.

3. *Theropithecus oswaldi* [Kanjera BM 14936, identified as female]: the minimal cranial capacity of this specimen was measured directly as 154 cm³ by packing the braincase with seeds. As with the previous skull, this may be an underestimate because of the presence of residual matrix in the braincase. Any underestimation, however, is likely to be slight because the external dimensions of the braincase are even smaller in comparison to the Peninj skull. Dechow's formulae applied to the length of the upper cheek tooth row in this skull (53.3 mm) yield a body weight of 26.4 kg with the regression formula and a weight of 29.8 kg with the major axis formula. Taking the empirical major axis formula determined for Old World monkey species, residual values of −0.120 and −0.152, respectively, are obtained with these two body weight estimates, indicating that relative brain size in this specimen of *T. oswaldi* was markedly smaller than in extant *T. gelada* and overlapped with that of extant colobine monkey species (see Table 10.2). The formula determined with a fixed slope of 0.75 yields residual values of −0.189 and −0.228, respectively. Both values are well below those

recorded for living *Theropithecus* and are close to the minimum established for living colobine monkeys (Table 10.2). Accordingly, this specimen of *T. oswaldi* also seems to have had a notably smaller relative brain size than modern *T. gelada*, though again there may be a slight degree of underestimation due to the presence of residual matrix in the braincase.

Although these calculations of relative brain size in fossil geladas are preliminary, they probably provide a fairly good guide. It is encouraging to note, for instance, that the overall average body weight inferred for the three specimens discussed above (41.7 kg) is quite close to the overall average body weight determined for the species *T. oswaldi* by Leakey (this volume) on the basis of independent postcranial measures (36.6 kg). It should also be noted that the use of cranial capacity, instead of brain weight, for fossil *Theropithecus* entails only a minor error amounting to an overestimation of about four per cent in the three cases examined. (For the formula relating cranial capacity to brain weight in primates, see Martin, 1990.) Thus, actual brain weight in the fossil *Theropithecus* specimens was possibly slightly smaller than indicated by the cranial capacity measurements cited.

In conclusion, it is very likely that *T. oswaldi*, like extant *T. gelada*, was characterized by relatively small brain size in comparison to *Papio*. The relatively smaller brain case of *T. oswaldi*, in comparison with many other Old World monkeys, accounts for the fact that the postorbital constriction is particularly marked in this species. Indeed, the combined evidence for the three skulls discussed above suggests that relative brain size was even smaller in *T. oswaldi* than in *T. gelada*. This presents another serious difficulty with respect to the hypothesis that geladas provide a parallel to the evolution of early hominids. Even the early phase of hominid evolution was in fact accompanied by an increase in relative brain size in comparison to great apes (Martin, 1983, 1989a). If brain size is linked to maternal metabolic turnover as suggested in the hypothesis outlined above, this indicates that the dietary adaptations of early hominids must have been markedly different from those of geladas, both fossil and extant. In fact, it would seem that *T. oswaldi* may have become even more specialized with respect to its diet than modern geladas, leading to an even more pronounced reduction in basal metabolic rate and an even tighter constraint on brain size.

Acknowledgements

Valuable assistance in compilation and analysis of some of the data for this chapter was provided by my research assistant Ms. Debbie Curtis. Most of the dental data were derived from Swindler (1976) supplemented in a few cases with additional measurements to fill in gaps. Data on brain size and body size (the latter in most cases based on measurements taken on wild animals) were derived from the author's extensive data base, compiled in collaboration with Dr Ann MacLarnon. The approach presented in this chapter owes a great deal to discussions with the following people: Dr Leslie Aiello, Dr Andrew Barbour, Mr Fred Brett, Dr Paul Harvey, Dr Michael Hills and Dr Ann MacLarnon. Particular thanks are due to Dr Glenn Conroy for providing the initial stimulus to examine relative brain size in geladas. Thanks are also due to Dr Caroline Ross and Dr Paul Harvey for making available the computer programmes used in line-fitting, and to Dr Prosper Ndessokia and Dr Peter Andrews for facilitating measurements on the Peninj skull and the Kanjera skulls, respectively.

References

CLUTTON-BROCK, T.H. & HARVEY, P.H. (1977). Primate ecology and social organization. *Journal of Zoology, London*, 183, 1–39.

CLUTTON-BROCK, T.H. & HARVEY, P.H. (1980). Primates, brains and ecology. *Journal of Zoology, London*, 190, 303–23.

CLUTTON-BROCK, T.H., HARVEY, P.H. & RUDDER, B. (1977). Sexual dimorphism, socionomic sex ratio and body weight in primates. *Nature*, 269, 797–800.

CONROY, G.C. (1987). Problems of body-weight estimation in fossil primates. *International Journal of Primatology*, 8, 115–37.

DECHOW, P. (1983). Estimation of body weights from craniometric variables in baboons. *American Journal of Physical Anthropology*, 60, 113–23.

DUNBAR, R.I.M. (1984). *Reproductive Decisions: An Economic Analysis of Gelada Baboon Social Strategies*. Princeton: Princeton University Press.

HARVEY, P.H. & MACE, G.M. (1982). Comparison between taxa and adaptive trends: Problems of methodology. In *Current Problems in Sociobiology*, ed. King's College Sociobiology Group, pp. 343–61. Cambridge: Cambridge University Press.

HARVEY, P.H., MARTIN, R.D. & CLUTTON-BROCK, T.H. (1987). Life histories in comparative perspective. In *Primate Societies*, ed. B.B. Smuts, D. Cheney, R.M. Seyfarth, R. Wrangham & T. Struhsaker, pp. 181–96. Chicago: Chicago University Press.

JABLONSKI, N.G. (1986). Patterns of sexual dimorphism in *Theropithecus*. In *Sexual Dimorphism in Living and Fossil Primates*, ed. M. Pickford & B. Chiarelli, pp. 171–82. Florence: II Sedicesimo.

JOLLY, C.J. (1970a). The seed eaters: A new model of hominid differentiation based on a baboon analogy. *Man*, new series, 5, 5–26.

JOLLY, C.J. (1970b). *Hadropithecus*, a lemuroid small-object feeder. *Man*, new series, 5, 619–26.

JOLLY, C.J. (1972). The classification and natural history of *Theropithecus* (*Simopithecus*) (Andrews, 1916), baboons of the African Plio-Pleistocene. *Bulletin of the British Museum (Natural History), Geology*, 22, 1–122.

KAY, R.F. (1981). The nut-crackers: A new theory of the adaptations of the Ramapithecinae. *American Journal of Physical Anthropology*, 55, 141–51.

LEUTENEGGER, W. (1978). Scaling of sexual dimorphism in body size and breeding system in primates. *Nature*, 272, 610–11.

LEUTENEGGER, W. & CHEVERUD, J. (1982). Correlates of sexual dimorphism in primates: ecological and size variables. *International Journal of Primatology*, 3, 387.

MARTIN, R.D. (1980). Adaptation and body size in primates. *Zeitschrift für Morphologie und Anthropologie*, 71, 115–24.

MARTIN, R.D. (1981). Relative brain size and metabolic rate in terrestrial vertebrates. *Nature*, 293, 57–60.

MARTIN, R.D. (1983). *Human Brain Evolution in an Ecological Context. (52nd James Arthur Lecture on the Evolution of the*

Human Brain). New York: American Museum of Natural History.

MARTIN, R.D. (1984). Body size, brain size and feeding strategies in primates. In *Food Acquisiton and Processing in Primates*, ed. D.J. Chivers, pp. 73–103. London: Plenum Press.

MARTIN, R.D. (1989a). Evolution of the brain in early hominids. *Ossa*, 14, 49–62.

MARTIN, R.D. (1989b). Size, shape and evolution. In *Evolutionary Studies – A Centenary Celebration of the Life of Julian Huxley*, ed. M. Keynes, pp. 96–141. London: Eugenics Society.

MARTIN, R.D. (1990). *Primate Origins and Evolution: A Phylogenetic Reconstruction*. London: Chapman & Hall.

MARTIN, R.D. & BARBOUR, A.D. (1989). Aspects of line-fitting in bivariate allometric analyses. *Folia primatologica*, 53, 65–81.

MARTIN, R.D. & HARVEY, P.H. (1985). Brain size allometry: ontogeny and phylogeny. In *Size and Scaling in Primate Biology*, ed. W.L.

Jungers, pp. 147–73. New York: Plenum Press.

PAGEL, M.D. & HARVEY, P.H. (1988). The taxon-level problem in the evolution of mammalian brain size: Facts and artifacts. *American Naturalist*, 132, 344–59.

SMITH, R.J. (1981). Interspecific scaling of maxillary canine size and shape in female primates: Relationships to social structure and diet. *Journal of Human Evolution*, 10, 165–73.

SWINDLER, D.R. (1976). *Dentition of Living Primates*. New York: Academic Press.

SZALAY, F.S., & DELSON, E. (1979). *Evolutionary History of the Primates*. New York: Academic Press.

WILLNER, L.A. (1989). *Sexual Dimorphism in Primates*. Ph.D. Thesis, University of London.

WILLNER, L.A. & MARTIN, R.D. (1985). Some basic principles of mammalian sexual dimorphism. *Symposia of the Society for the Study of Human Biology*, 24, 1–42.

11 Evolution of the masticatory apparatus in *Theropithecus*

NINA G. JABLONSKI

Summary

1. The nature of the adaptation of the earliest theropiths is not known, but there is now considerable evidence to support the hypothesis that the emergence and early diversification of *Theropithecus* was linked to the evolution of a feeding apparatus specialized for the eating of grasses. This made possible the invasion of grassland environments previously not colonized by primates. Specializations for grazing in *Theropithecus* included those of the hand – important for the harvesting of grass parts – and those of the masticatory apparatus.

2. In this study, simple scaling and biomechanical analyses are used to describe the functional changes of the masticatory apparatus that accompanied structural changes in the skull and teeth of Plio-Pleistocene and later Pleistocene species of *Theropithecus*.

3. In *T. brumpti* and its putative forebears *T. baringensis* and *T. quadratirostris*, evolution of longer muzzles was correlated with the anterior and lateral expansion of the origin of the masseter muscle. The masticatory apparatus of *T. brumpti* was designed to meet the requirements of a large gape during feeding as well as during canine displays. The muscles of mastication in this species, in particular the masseter and the temporalis, were very large. It is suggested that these muscles were mostly fleshy, with relatively long parallel fibres and long sarcomeres, as opposed to being multipennate in structure. Such an arrangement would have permitted the muscles to be stretched during the wide jaw opening (gape) necessary for canine displays yet – because of the large cross-sectional areas of the muscles – capable of generating high occlusal pressures between the molar teeth even when the jaws were widely separated. This situation differs from that seen in species of *Papio*, where long muzzles have been combined with a masticatory musculature comparable in proportions to those of macaques. This arrangement permits

a large gape, but the smaller cross-sectional areas of the major jaw elevators precludes production of occlusal pressures over the molar teeth as high as those possible in *T. brumpti* when the jaws are widely separated during gape. Delson and Dean's (chapter 4) interpretation of craniofacial morphology in *T. brumpti* is flawed because it fails to take into account these important considerations relating to the functional ramifications of internal muscular architecture.

4. In *T. oswaldi*, increasing body size and the demands of eating increasingly large quantities of low quality, high fibre vegetation led to the evolution of a 'chewing machine' featuring a relatively enlarged masseter muscle of complex internal architecture. The anatomy of the temporomandibular joint in later Pleistocene forms of this species further suggests that wide gapes were sacrificed to the demands of heavy molar chewing. The muscles of mastication, in particular the masseter, of *T. oswaldi* were more probably more pennate in structure than those of *T. brumpti*, an arrangement which permitted more efficient generation of high occlusal pressures between the cheek teeth, but which precluded wide gapes. This invites the speculation that open-mouthed canine displays were not a part of the males' behavioural repertoire.

5. With respect to the diet of *T. brumpti* and *T. oswaldi*, these results corroborate the findings of previous investigators who suggested that

T. oswaldi concentrated its feeding on objects of small diameter such as grass blades, seeds, and rhizomes, much as the extant gelada does. In contrast, the masticatory apparatus of *T. brumpti* was suited to large-object feeding, and the species may have shown a preference for objects of larger diameter such as large tubers, fruits, stem pith, or shoots.

Introduction

Among the earliest scientific comments on *Theropithecus* were those pertaining to the distinctiveness of the anatomy of its masticatory apparatus. In his anatomical study of *T. gelada*, Vram (1923) noted that the gelada was distinguished from other monkeys by several features of its skull and teeth, including the angle of the mandible and the disposition of the cusps of the molar teeth. Nearly 50 years later, Jolly (1972) recognized that the great similarities between the living gelada and species of the Plio-Pleistocene fossil monkey *Simopithecus* in the anatomy of the masticatory apparatus were a product of common ancestry. These similarities included features of molar morphology and shared possession of a relatively short face and an anteriorly placed temporalis muscle that together formed a functional complex distinct from that seen in other cercopithecines. Drawing upon these similarities, he placed *Simopithecus* as a subgenus within *Theropithecus*. Later still, Eck (1977) and Eck & Howell (1982) argued that the curious long-muzzled monkey from the Omo in Ethiopia, originally referred to as *Dinopithecus brumpti*, was in fact a member of *Theropithecus* by virtue of the several distinctive features of the masticatory apparatus it shared with the living gelada and *T. (Simopithecus) oswaldi*. A similar argument was made with regard to the morphology of *Papio baringensis* and *P. quadratirostris* by Eck & Jablonski (1984). These species, together with *T. brumpti*, were said to be united to the living gelada and other fossil species of the genus by a suite of specialized features of the masticatory apparatus, the most significant of which were the distinctively shaped molar teeth and the relationship of the cranial base and temporomandibular joint to the occlusal plane. It was recognized, however, that *T. baringensis*, *T. quadratirostris*, and *T. brumpti* were united by further specializations related to exaggerated facial prognathism, and that the shared possession of these features warranted their placement in a putative 'long muzzled' clade within the genus.

The list of shared derived characteristics of the skull and dentition that distinguish species of *Theropithecus* from other cercopithecines include: (1) a large and anteriorly placed temporalis musculature associated with a large temporal foramen, an enlarged infratemporal fossa, and a pronounced postorbital constriction; (2) a temporomandibular joint placed well above the level of the occlusal plane; (3) a deeply excavated posterior maxilla; (4) a high and upright mandibular ramus; (5) a superoinferiorly elongated mandibular symphysis; (6) high-crowned bilophodont molars with columnar cusps that, when worn, present a pattern of complexly curved enamel ridges; and (7) a mandibular cheek tooth row aligned in a reversed curve of Spee (Jolly, 1972; Meikle, 1977; Szalay & Delson, 1979; Jablonski, 1981; Eck & Jablonski, 1984). Most, if not all, of these characteristics are functionally related to the demands of mastication and are, therefore, closely interrelated.

In addition to these cranial specializations of the feeding apparatus, species of *Theropithecus* also appear to have shared a unique specialization of the postcranial skeleton related to feeding. The hand of the gelada has an elongated thumb and a somewhat abbreviated index finger that work together during feeding to make a precise, pincer-like grip that greatly facilitates the picking of specific parts of grasses (Maier, 1972; Etter, 1973). The relationship between thumb and index finger gives the gelada the highest opposability index of any non-human primate (Napier & Napier, 1967). A very similar relationship between the thumb and index finger was demonstrated for a fossil hand of *T. brumpti* (Jablonski, 1986). This finding suggested that specializations of the hand were also among those that marked the emergence of *Theropithecus*.

As noted above, further specializations characterized *T. brumpti* and its putative forebears, *T. baringensis* and *T. quadratirostris*. These included elongation of the muzzle and a tendency toward increasing size and robustness of the zygomatic arch, the latter characteristic reaching its most exaggerated expression in *T. brumpti* with its large, anterolaterally flaring zygomatic bones.

Still other specializations characterize *T. oswaldi*, the species of *Theropithecus* with the most chronologically and geographically extensive fossil record. The most noticeable of these are enlarged cheek teeth, reduced height of the male upper canine and the complementary honing surface of the canine on the lower third premolar, and reduced size of the incisors (Jolly, 1972).

In recent years, the specialized features of the masticatory apparatus of the species of *Theropithecus* have been convincingly related to extreme dietary specialization. Several reports on the feeding behaviour and ecology of the gelada have shown that grasses constitute the major fraction of the species's diet throughout the year (Dunbar, 1977; Iwamoto, 1979; and see Iwamoto, chapter 16). Similarities in dental morphology and patterns of dental wear between the living and fossil species would suggest similar or perhaps even more extreme dietary specializations in Plio-Pleistocene species of *Theropithecus*. The anatomical and behavioural similarities between living and fossil species of *Theropithecus* formed the basis of Jolly's (1970) now famous 'seed eater' or 'small-object feeding' hypothesis of early hominid dietary specialization.

There is now a reasonable body of evidence to suggest that the initial success and subsequent diversification of *Theropithecus* were linked to its development of a grass-eating niche. The earliest *T. darti* recognized in the fossil record, dated at about 3.4 Ma from Hadar in Ethiopia (see Eck, chapter 2), showed the complete suite of specialized features of the masticatory apparatus that distinguish the genus. Independent evidence of graminivory in early *Theropithecus* has

come from a study of the ratio of stable carbon isotopes in the enamel carbonate of the teeth of *T. darti* at Swartkrans (Lee-Thorp, van der Merwe & Brain, 1989). Perhaps more significantly, however, these earliest *Theropithecus* appear to have established themselves in relatively open, woodland grassland habitats at the same time as many species of grazing ungulates. The competitive advantage of early *Theropithecus* over ungulate species may be traced to its ability to utilize several different strata of the herb layer with its method of manual feeding. Grazing ungulates that rely on the harvesting of vegetation with the incisors tend to be specialized for the harvesting of only one stratum of the herb layer (Bell, 1970). Early *Theropithecus* thus occupied an environment similar to that of grazing ungulates, but had created a niche that was quite different (Dunbar, 1983).

Our understanding of the basic features of the masticatory apparatus of *Theropithecus* is now reasonably good, but gaps in our knowledge of some of the more extreme specializations of the genus remain. The purpose of this chapter is to trace the evolution of the masticatory apparatus of *Theropithecus* over time and through successive speciation events to understand better the functional transformations that accompanied some of the now well documented structural transformations of the skull and dentition that occurred during the evolution of the genus. The basis of this study is a simple scaling and biomechanical analysis designed to elucidate some of the more important functional changes that accompanied the evolution of species of

Theropithecus, particularly *T. oswaldi* and *T. brumpti*.

Materials and methods

The specimens examined and measured for this study represented all of the recognized species of *Theropithecus* except for the one known specimen of *T. quadratirostris*. These included specimens of *T. gelada* (n = 7); *T. darti* from Hadar (n = 2) and Makapansgat (n = 2); *T. oswaldi* from Koobi Fora (n = 34), Olduvai (n = 12), Kanjera (n = 4), and Olorgesailie (n = 3); *T. baringensis* from Chemeron (n = 2); and *T. brumpti* from the Omo (n = 3), East Turkana (n = 6), and West Turkana (n = 1). In the light of the relatively large size of the samples of *T. oswaldi*, especially from Koobi Fora and Olduvai, specimens of this species from different sites were treated as separate samples. It was recognized that the specimens within the larger samples, especially those from Koobi Fora and Olduvai, were not contemporaneous and that each of these samples was likely to contain populations of different sizes and characteristics. (An analysis of contemporaneous samples from different sites would have been ideal were it not for excessively small sample sizes and uncertainties remaining over the dating of some key sites.) For comparative purposes, specimens of *Macaca fascicularis* (n = 10), *M. radiata* (n = 10), *M. arctoides* (n = 8), *Papio hamadryas cynocephalus* (n = 6), and *P. h. anubis* (n = 10) were also examined. Macaques were taken to approximate the primitive condition of the papionine masticatory

apparatus (Szalay & Delson, 1979), whereas *Papio* baboons were chosen as the archetypal long-muzzled forms that differed in apparent proportions from both macaques and theropithecines. A complete list of specimens studied is given in Appendix 11.1 of this chapter.

Nineteen measurements of the skull were taken using digital calipers accurate to 0.01 mm (Digit-Cal, Tesa). These measurements included standard length and breadth measurements of the skull as well as several representing significant aspects of skull function such as moment arms of the major jaw adductors and estimates of jaw strength. A further two measurements for the estimated values of the second moment of area of the mandible (MANDXSA$^{1/4}$) and the area of the mandibular condyle (CONAREA$^{1/2}$) were calculated from various of the original measurements, as described in Appendix 11.2. Several of these measurements were based on those originally used by Emerson & Radinsky (1980) and Radinsky (1981, 1982, 1985). A complete list of abbreviations and definitions of variables is provided in Appendix 11.2.

For each species, means and standard deviations for each variable were calculated for the combined sex sample and for each sex separately. For selected variables, the mean value of the total sample (both sexes) was converted to natural logarithms. Natural log–log plots of selected variables against mandibular length (MANDL) were then constructed. These variables were: (1) the distance between the mandibular condyle and the first lower molar (CONM1); (2) the length of the glenoid fossa (GLENOL); (3) the width of the glenoid fossa (GLENOW); (4) the estimated second moment of area of the mandible (MANDXSA$^{1/4}$); (5) the length of the mandibular symphysis (MANDSYM); and (6) the estimated area of the mandibular condyle (CONAREA$^{1/2}$).

In order that the relative size of elements of the theropith masticatory apparatus could be accurately assessed, the variables were size-adjusted as follows. The first step of this process was to determine what species or group of species was to form the standard against which the size of the theropith masticatory apparatus was to be assessed. It was decided that the macaques were most appropriate for this purpose because they are thought to most closely approach the primitive condition for papionines in craniodental morphology (Szalay & Delson, 1979). The next step in the process was to generate regression values for the combined macaque species means so that master equations for transformation of variables for Theropithecus and Papio could be derived. This was done by performing linear least-squares regressions on macaque species means for selected measurements against MANDL, a measurement that is widely recognized as being a good estimator of overall body size. Because two of the fossil theropith samples (T. darti and T. oswaldi from Olorgesailie) were known to lack mandibles, however, regressions were performed against the GLENOW or the GLENOL. Thus, the master equations for transformation of the Theropithecus and Papio variables were derived from the combined means of the three macaque species only. The final step in the size-adjustment process involved a

calculation aimed at minimizing the effects of facial length on the analysis. Variables for Theropithecus and Papio were adjusted according to macaque mandibular length (or, in a few cases, glenoid width or glenoid length) using the method of Smith, Petersen & Gipe (1983). The formula used for calculating these size-adjusted values was:

Mandibular length adjusted value =
antilog (observed − predicted)^2

Values that had been adjusted for size using macaque mandibular length, glenoid width, or glenoid length were then used as the basis for subsequent computations and comparisons. The t-test was used to detect statistically significant deviations from the macaque mean of 1.00.

For Papio and Theropithecus species, an estimate of the size of the muzzle (MUZL), and of the size of the moment arms of the masseter (MAM) and temporalis (MAT) muscles relative to the moment arm of the resistance when biting at the first molar was calculated by dividing the size-adjusted values for these variables by CONM1.

Rough estimates of the differences in angulation of the masseter and temporalis muscles between Theropithecus gelada, T. baringensis, and T. brumpti were investigated by drawing the average lines of action of these muscles on line drawings of male skulls. The drawing for T. baringensis is based upon the Chemeron specimen KNM BC2. The drawing of T. brumpti is based on a composite of two Omo specimens, the male cranium

L345-287 and the mandible L576-8. The mandible has been reduced to 93 per cent of natural size in order to produce a 'best fit' with the cranium for purposes of illustration.

Results

In Tables 11.1 and 11.2 are presented the natural logarithmic transformations of the species means for selected cranial and mandibular variables. Natural log–log plots for CONM1, GLENOL, GLENOW, MANDXSA$^{1/4}$, MANDSYM and CONAREA$^{1/2}$ against MANDL are presented in Fig. 11.1 (a–f). For each species, a complete listing of the means and standard deviations for the combined sex and individual sex sam-

ples for all variables is presented · in Appendix 11.3.

The regression values for the macaque cranial and mandibular variables against MANDL are presented in Table 11.3. The correlation coefficients for most variables with mandibular length were very high, and for selected variables with glenoid width and length only slightly lower. For linear dimensions, geometric similarity (isometry) was indicated by a 1:1 relationship, that is, by a slope or regression coefficient of 1.00. These results showed that in macaques CRANL, PGA, MUZL, SMSL, BIZYGW, TEMFOSL, GLENOW, MAM, MAT, and CONM1 scaled nearly isometrically with mandibular length or glenoid width. Several variables, notably TEM-

Table 11.1. *Natural logarithmic transformations of species means for selected cranial variables in* Papio *and* Theropithecus *species*

Species	Variable									
	CRANL	PGA	MUZL	SMSL	BIZYGW	TEMFOSL	TEMFOSW	ZYGH	GLENOL	GLENOW
M.fas.	4.70	4.35	3.68	3.60	4.29	3.28	3.03	1.79	2.58	2.63
M.rad.	4.68	4.35	3.68	3.80	4.29	3.27	3.10	1.54	2.59	2.70
M.arc.	4.88	4.52	3.87	3.80	4.51	3.50	3.35	1.98	2.71	2.86
P.h.c.	5.15	4.84	4.46	3.93	4.60	3.59	3.36	2.28	2.91	2.92
P.h.a.	5.21	4.89	4.56	4.00	4.67	3.60	3.52	2.38	2.98	2.99
T.g.	5.09	4.79	4.34	4.00	4.63	3.59	3.57	2.16	3.04	3.11
T.d.	5.12	4.77	4.34	4.08	4.75	3.70	3.44	2.55	2.83	3.23
T.o. (KF)	5.30	4.29	4.49	4.16	4.89	3.73	3.58	2.69	3.13	3.38
T.o. (Old)	—	—	4.66	—	—	—	—	—	3.34	3.46
T.o. (Kan)	5.12	4.93	4.31	4.21	4.76	3.89	3.67	2.56	3.10	3.37
T.o. (Olo)	—	—	—	—	—	—	—	—	3.53	3.77
T.bar.	—	5.05	4.67	4.19	4.70	3.79	3.64	2.58	3.12	3.35
T.bru.	5.44	5.12	4.81	4.43	4.91	3.91	3.85	3.08	3.00	3.43

Note: The following abbreviations have been used for species names: M.fas. = *Macaca fascicularis*; M.rad. = *Macaca radiata*; M.arc. = *Macaca arctoides*; P.h.c. = *Papio hamadryas cynocephalus*; P.h.a. = *Papio cynocephalus anubis*; T.g. = *Theropithecus gelada*; T.d. = *Theropithecus darti*; T.o.(KF) = *T. oswaldi* from Koobi Fora; T.o.(Old) = *Theropithecus oswaldi* from Olduvai; T.o.(Kan) = *Theropithecus oswaldi* from Kanjera; T.o.(Olo) = *Theropithecus oswaldi* from Olorgesailie; T. bar. = *Theropithecus baringensis*; and T. bru. = *Theropithecus brumpti*. See Appendix 11.2 for an explanation of variables.

Table 11.2. *Natural logarithmic transformations of species means for selected mandibular variables in* Papio *and* Theropithecus *species*

Species	M'L	M'H	M'CORW	M'XSA	M'SYM	MAM	MAT	C'M1	C'DYL	C'DLW	C'AREA
M.fas.	4.31	2.84	2.00	2.17	3.20	3.63	2.91	4.00	1.77	2.36	3.62
M.rad.	4.35	2.85	2.04	2.21	3.22	3.68	2.87	3.94	1.54	2.44	3.44
M.arc.	4.54	3.18	2.33	2.65	3.48	3.82	3.09	4.18	1.78	2.68	3.87
P.h.c.	4.84	3.31	2.41	2.65	3.72	4.07	3.11	4.49	2.22	2.73	4.36
P.h.a.	4.88	3.38	2.41	2.84	3.74	4.16	3.16	4.58	2.18	2.85	4.42
T.g.	4.79	3.51	2.44	2.95	3.80	4.31	3.33	4.49	2.16	2.90	4.44
T.d.	—	—	—	—	3.77	—	—	—	—	—	—
T.o. (KF)	5.01	3.57	3.75	3.27	3.83	4.47	3.44	4.71	2.13	3.18	4.61
T.o. (Old)	5.23	3.60	2.88	3.41	3.99	4.76	3.73	4.93	2.20	3.58	4.98
T.o. (Kan)	4.85	3.52	2.69	3.19	3.84	4.38	3.31	4.60	2.08	3.10	4.50
T.o. (Olo)	—	3.64	2.95	3.50	3.96	—	—	—	—	—	—
T.bar.	5.06	3.58	2.56	3.11	4.00	4.46	3.38	4.74	2.48	3.16	4.96
T.bru.	5.14	3.77	2.67	3.33	4.04	4.62	3.74	4.83	2.46	3.20	4.97

Note: Abbreviations for species names are the same as those used in Table 11.1. In order to save space, 'MAN' or 'MAND' in the following variable names has been substituted by 'M'; 'CON' has been substituted by 'C'. 'M'XSA' stands for MANDXSA$^{1/4}$, and C'AREA stands for CONAREa$^{1/2}$. See Appendix 11.2 for an explanation of variables.

FOSW, ZYGH, MANDH, MANCORW, MANDXSA$^{1/4}$, MANDSYM, CONDYLW, and CONAREA$^{1/2}$ scaled positively allometric with mandibular length or glenoid width, while two others, GLENOL and condyle length (CONDYLL) scaled in a negatively allometric fashion.

Size-adjusted values for cranial and mandibular variables of *Theropithecus* and *Papio* are presented in Tables 11.4 and 11.5. Those significantly different from the macaque mean of 1.00 have been indicated. Values for MUZL/CONM1, MAM/CONM1, and MAT/CONM1 for *Papio* and *Theropithecus* species calculated from size-adjusted values are presented in Table 11.6. These figures present a most interesting picture. The values for MUZL/CONM1 confirm the visual impression that relative muzzle length is greatest in the two *Papio* species and in the long-muzzled species of *Theropithecus*, *T. baringensis* and *T. brumpti*. Significant from a functional point of view is the fact that while relative muzzle length in *Papio*, *T. baringensis*, and *T. brumpti* is very similar, the relative sizes of the moment arms of the masseter in these species are not. The relative size of the moment arm of the masseter in *Papio* is identical to that seen in macaques, whereas it is significantly larger in all the species of *Theropithecus*, especially in *T. brumpti* and the Olduvai sample of *T. oswaldi*. The relative size of the moment arm of the temporalis in all species of *Theropithecus* except *T. baringensis* is not significantly different from that

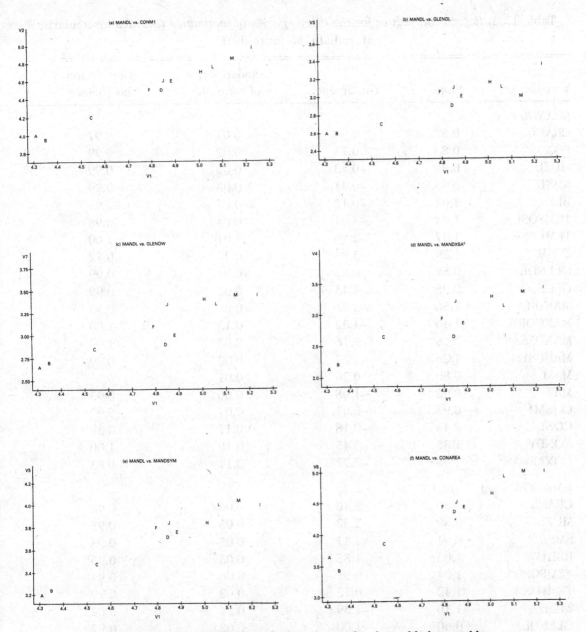

Fig. 11.1. Natural log–log plots of selected cranial and mandibular variables against mandibular length (MANDL, designated V1 on the x-axis). On all plots, individual species are designated by letters, as follows: A = *Macaca fascicularis*; B = *M. radiata*; C = *M. arctoides*; D = *Papio hamadryas cynocephalus*; E = *P. hamadryas anubis*; F = *Theropithecus gelada*; G = *T. darti*; H = *T. oswaldi* (Koobi Fora); I = *T. oswaldi* (Olduvai); J = *T. oswaldi* (Kanjera); K = *T. oswaldi* (Olorgesailie); L = *T. baringensis*; M = *T. brumpti*. Note: on each fig. two observations had missing values.

Table 11.3. *Regression values for the three species of macaques* (Macaca fascicularis, M. radiata, M. arctoides)

Variable	Slope	Y-intercept	Standard error of estimate	Correlation coefficient
(MANDL = x)				
CRANL	0.87	0.94	0.04	0.97
PGA	0.83	0.74	0.02	0.99
MUZL	0.88	−0.13	0.02	0.99
SMSL	0.93	−0.41	0.03	0.99
BIZ	1.02	−0.12	0.03	0.99
TEMFOSL	1.04	−1.21	0.04	0.98
TEMFOSW	1.37	−2.86	1.01	1.00
ZYGH	1.29	−3.93	0.22	0.72
GLENOL	0.59	0.05	0.01	0.99
GLENOW	0.95	−1.45	0.02	0.99
MANDH	1.56	−3.91	0.04	0.99
MANCORW	1.46	−4.32	0.13	1.00
MANDXSA$^{1/4}$	2.16	−7.16	0.03	1.00
MANDSYM	1.26	−2.27	0.02	0.99
MAM	0.80	0.20	0.01	0.99
MAT	0.92	−1.08	0.06	0.93
CONM1	0.93	−0.07	0.07	0.92
CONL	0.43	−0.18	0.17	0.39
CONDYLW	1.35	−3.45	0.19	1.00
CONAREA$^{1/2}$	1.46	−2.77	0.17	0.83
(GLENOW = x)				
CRANL	0.86	2.40	0.06	0.92
MUZL	0.89	1.32	0.05	0.95
SMSL	0.94	1.11	0.05	0.95
BIZYGW	1.03	1.55	0.05	0.95
TEMFOSL	1.04	0.51	0.06	0.94
TEMFOSW	1.42	−0.72	0.02	0.99
ZYGH	1.16	−1.39	0.24	0.62
GLENOL	0.60	1.00	0.02	0.97
MANDL	1.03	1.58	0.02	0.99
MANDH	1.58	−1.35	0.07	0.96
MANCORW	1.50	−1.97	0.05	0.98
MANDXSA$^{1/4}$	2.20	−3.67	0.08	0.97
(GLENOL = x)				
GLENOW	1.59	−1.44	0.04	0.97

Table 11.4. *Size-adjusted values for cranial measurements based on the results of the regression analysis on natural logarithmic transformations of combined macaque species means*

Species	Variable									
	CRANL	PGA	MUZL	SMSL	BIZYGW	TEMFOSL	TEMFOSW	ZYGH	GLENOL	GLENOW
P.h.c.	1.00	1.00	1.12	0.98	0.96	0.95	0.85	1.00	1.00	0.95
P.h.a.	1.00	1.00	1.16	0.99	0.97	0.94	0.91	1.00	1.00	0.96
T.g.	1.00	1.00	1.06	1.00	0.99	0.97	0.98	0.99	1.03	1.00
T.d.	1.00	1.00	1.02	1.00	0.98	0.97	0.83	1.04	0.99	1.03
T.o. (KF)	1.00	—	1.04	1.00	0.99	0.94	0.96	1.02	1.02	1.00
T.o. (Old)	—	—	1.04	—	—	—	—	—	1.05	1.00
T.o. (Kan)	1.00	1.02	1.03	1.02	0.95	1.00	0.99	1.04	1.04	1.04
T.o. (Olo)	—	—	—	—	—	—	—	—	1.08	0.86
T.bar.	—	1.01	1.13	0.99	0.89	0.94	0.84	1.00	1.01	1.00
T.bru.	1.00	1.01	1.18	1.01	0.96	0.96	0.90	1.13	1.00	1.00

Note: Abbreviations for species names are the same as those used in Table 11.1. Values significantly different from the macaque mean of 1.00 are underlined.

Table 11.5. *Size-adjusted values for mandibular measurements based on the results of the regression analysis on natural logarithmic transformations of combined macaque species means*

Species	Variable									
	M'L	M'H	M'CORW	M'XSA	M'SYM	MAM	MAT	C'M1	C'DYL	C'DLW	C'AREA
P.h.c.	1.00	0.90	0.88	0.66	0.98	1.00	0.94	1.00	1.12	0.89	1.01
P.h.a.	1.00	0.90	0.84	0.75	0.97	1.00	0.95	1.00	1.08	0.92	1.01
T.g.	1.00	1.00	0.94	0.94	1.00	1.09	1.00	1.01	1.09	0.98	1.05
T.d.	—	—	—	—	0.97	—	—	—	—	—	—
T.o. (KF)	1.00	0.89	0.93	0.86	0.94	1.08	1.00	1.01	1.03	0.98	1.01
T.o. (Old)	1.00	0.66	0.81	0.60	0.88	1.16	1.00	1.01	1.02	1.00	1.02
T.o. (Kan)	1.00	0.98	0.99	0.99	1.00	1.06	1.00	1.02	1.04	1.00	1.04
T.o. (Olo)	—	0.40	0.59	0.28	0.63	—	—	—	—	—	—
T.bar.	1.00	0.84	0.71	0.65	0.98	1.05	0.96	1.01	1.28	0.95	1.14
T.bru.	1.00	0.89	0.75	0.69	0.96	1.11	1.01	1.01	1.22	0.92	1.06

Note: Abbreviations for species names are the same as those used in Table 11.1. Values significantly different from the macaque mean of 1.00 are underlined.

of macaques, whereas it is significantly smaller in the two species of *Papio* and in *T. baringensis*.

In Fig. 11.2, the estimated lines of action of the masseter and temporalis muscles in *T. gelada*, *T. baringensis*, and *T. brumpti* have been indicated on line drawings of male skulls. This is acknow-

Table 11.6. *Relative size of the muzzle and the moment arms of the masseter and temporalis in species of* Papio *and* Theropithecus, *calculated by dividing muzzle length (MUZL) and values for the moment arms of the masseter (MAM) and temporalis (MAT) by the estimated moment arm of the resistance when biting at the first molar (CONM1). (All calculations based on size-adjusted values.)*

Species	MUZL/CONM1	MAM/CONM1	MAT/CONM1
P.h.c.	<u>1.12</u>	1.00	<u>0.94</u>
P.h.a.	<u>1.16</u>	1.00	<u>0.95</u>
T.g.	<u>1.05</u>	<u>1.08</u>	0.99
T.d.	1.02	—	—
T.o. (KF)	1.03	<u>1.07</u>	0.99
T.o. (Old)	1.03	<u>1.15</u>	0.99
T.o. (Kan)	1.01	<u>1.04</u>	0.98
T.o. (Olo)	—	—	—
T.bar.	<u>1.12</u>	<u>1.04</u>	<u>0.95</u>
T.bru.	<u>1.17</u>	<u>1.10</u>	1.00

Note: Abbreviations for species names are the same as those used in Table 11.1. Values significantly different from the macaque mean of 1.00 are underlined.

ledged to be a simplistic, two-dimensional representation of what is in nature a complex, three-dimensional arrangement. In the gelada, the line of action of the masseter is nearly perpendicular to the occlusal plane and the estimated average line of action of the temporalis is inclined only slightly away from the perpendicular. In *T. baringensis*, the line of action of the masseter also ran nearly perpendicular to the occlusal plane, while that of the temporalis (as judged in part from the inclination of the coronoid process) was inclined more obliquely relative to the occlusal plane. In *T. brumpti* the average estimated line of action of the masseter was not perpendicular to the occlusal plane, but was oblique to it in the direction opposite to the temporalis. The temporalis

in this species was extremely large both anteriorly and posteriorly and was oriented on average slightly more obliquely relative to the occlusal plane than is the muscle in the gelada. At first glance one would surmise that the posterior temporalis in this species acted mainly to retrude the mandible or stabilize the mandibular condyle in the glenoid fossa during wide jaw opening, as it does in many mammalian species. This impression is not, however, completely correct. The fibres of the posterior temporalis in *T. brumpti* turned nearly a right angle over the posterior aspect of the zygomatic arch before inserting on the coronoid process and appear to have strongly contributed to pulling the mandible upward as well as to maintaining

(a)

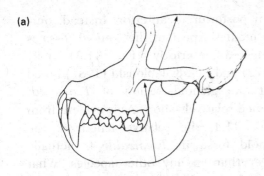

(b)

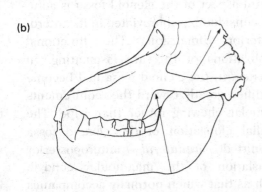

(c)

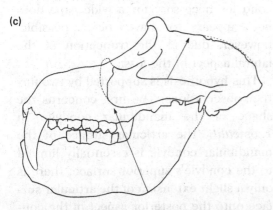

Fig. 11.2. Estimated lines of actions of the masseter and temporalis muscles in *Theropithecus gelada* (a), *T. baringensis* (b), and (c) *T. brumpti*.

the integrity of the temporomandibular joint.

Discussion

Theropithecus gelada

Of all the known species of *Theropithecus*, *T. gelada* most closely approximates the macaque condition in cranial proportions and may thus be considered to be the most primitive of the theropith species in this respect. In the gelada, the anatomical requirements for prolonged chewing have been balanced against those for a wide gape. The masticatory apparatus of the gelada is highly specialized for the chewing of grass parts (Jablonski, 1981), but the results of this study would indicate that it was not nearly as highly specialized as *T. oswaldi* for the trituration of large quantities of fibrous vegetation. Studies of the diet of the gelada (Iwamoto, 1979; chapters 16 and 17) have indicated that these animals favour tender grass parts, which contain the most protein and the least crude fibre.

Theropithecus darti

In the light of the scarcity of cranial remains of *T. darti*, conclusions about the functional anatomy of the masticatory apparatus of the species must be considered tentative. In some functional respects, this species can be considered intermediate between the gelada and *T. oswaldi*. The morphology of the glenoid fossa of the temporomandibular joint in *T. darti* (specimen MP222) would indicate that the anatomical compromise between the requirements of prolonged molar chewing and gape had been retained, and that, in contrast to later forms of *T. oswaldi* (see below), relatively wide gapes were possible.

Theropithecus oswaldi

The masticatory apparatus of *T. oswaldi* shows several interrelated specializations for the comminution of large quantities of fibrous vegetation. From the results of the mechanical analysis in this study it is clear that one of these specializations was the relative enlargement of the moment arm of the masseter muscle. Equally important, however, were specializations of the temporomandibular joint away from the primitive theropith condition.

As mentioned above, in the gelada the anatomical specializations for molar chewing of grasses appear to have been balanced against those that permit a wide gape. The glenoid fossa of the temporomandibular joint in the gelada is quite flat and is shaped roughly like an equilateral triangle with the postglenoid process centered at its base. This surface permits considerable anteroposterior translation of the mandibular condyle during symmetrical jaw opening (gape) as well as unilateral anteroposterior translation of the condyle during the asymmetrical jaw movements of molar chewing.

In many specimens of *T. oswaldi*, and in particular in later specimens from Kanjera, Olduvai, and the Turkana Basin (Koobi Fora) the glenoid fossa has quite a different shape. It is broad posteriorly near the postglenoid process, but it does not taper evenly into an equilateral triangle

from posterior to anterior. Instead, only the medial aspect of the glenoid fossa is elongated anteriorly (Fig. 11.3). Jolly (1972) and Eck & Jablonski (1987) noted that the glenoid region of *T. oswaldi* seemed relatively short. As is evident from Table 11.4, the relative length of the glenoid fossa in *T. oswaldi* is actually larger than in any other species. What these figures do not indicate is that the elongation of the glenoid fossa is restricted to the medial aspect of the fossa; the lateral aspect of the glenoid fossa is actually considerably abbreviated in its anteroposterior dimension. The functional implications of this are fascinating. In *T. oswaldi*, the glenoid fossa had become modified greatly toward the requirements of molar chewing rather than gape. The medial elongation of the glenoid fossa permitted *unilateral* anteroposterior translation of the mandibular condyle such as that which normally accompanies side-to-side movements of the jaw during molar chewing. Extensive bilateral anteroposterior translation of the mandibular condyles necessary for a wide gape does *not* appear to have been possible, however, due to the truncation of the lateral aspect of the fossa.

This hypothesis is supported by two further observations. The first concerns the shape of the mandibular condyle. In *T. oswaldi*, the articular surface of the mandibular condyle is essentially limited to the condyle's superior surface; there is only a slight extension of the articular surface onto the posterior aspect of the condyle where the condyle abuts the postglenoid process. In the gelada, the articular surface of the mandibular con-

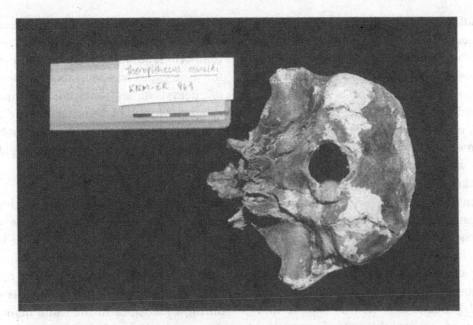

Fig. 11.3. View of the cranial base of *Theropithecus oswaldi* KNM-ER 969 from East Turkana to show the distinctive morphology of the glenoid fossa in the species. Note the antero–posterior shortening of the lateral aspect of the fossa indicated at the tip of the arrow.

dyle extends smoothly from the condyle's superior surface to its posteromedial surface in an arrangement that permits simultaneous rotation and forward sliding of the condyle. In *T. oswaldi*, lesser amounts of condylar rotation appear to have been possible, and it seems likely that the potential gape was not as wide as that possible in the gelada. The second observation concerns the size of the postglenoid process in *T. oswaldi*. In most specimens of this species, the postglenoid process is broad, thick, and tall, as opposed to thin, short, and parabolic in shape as it is in the gelada. The function of the large postglenoid processes in *T. oswaldi* was almost certainly to prevent posterior dislocation of the mandibular condyle and damage to the external acoustic meatus during molar chewing when the balancing side condyle was being forced backward. The process thus worked as a back-stop and a buttress during the exaggerated lateral movements of the mandible that characterized molar chewing in this species. It is also possible that the temporomandibular joint capsule of *T. oswaldi* itself restricted gape to some extent. Through its attachment to a relatively broad postglenoid process, the capsule may have permitted unilateral translation and pivoting of the mandibular condyles, but may have been pulled taut during attempted bilateral anteroposterior translations, thus preventing anterior dislocations and precluding wide gapes.

Detailed reconstruction of the architecture of the muscles of mastication in

T. oswaldi is impossible, but general inferences about the structure and function of these muscles in this species can be made using the gelada as a standard of comparison. The muscles of mastication in *T. oswaldi* appear to have been similar in their relative overall size, relative size of attachment areas, lengths of moment arms, and orientation to those of *T. gelada*. The only significant difference between *T. oswaldi* and *T. gelada* was in the greater length of the moment arm of the masseter (the height of the ascending ramus) in the Olduvai *T. oswaldi* sample, which contained a preponderance of specimens of large body size of Middle and Late Pleistocene age. In this sample and samples of comparable age from other sites, we also find the greatest relative enlargement of cheek tooth size (Eck, 1987) in the genus. It is interesting to speculate that evolution of the highly derived masticatory apparatus of later forms of *T. oswaldi* involved some reorganization of the internal architecture of the muscles of mastication, in particular the masseter. Reconstructions of the ecology of the Late Pleistocene theropiths (see chapters 18 and 19) suggest that, for these animals, the ability to process very large amounts of fibrous vegetation was of critical importance. The transformation of the masticatory apparatus of *T. oswaldi* into that of powerful chewing machine involved modifications of the temporomandibular joint as described above and modifications of the muscles of mastication that probably involved relatively economical changes in the muscles' internal architecture, toward a more highly pennate arrangement of muscle fibres. This transformation of the musculature occurred *pari passu* with that of the temporomandibular joint, as the capacity for wide gapes was sacrificed to the needs of heavy molar chewing. Comparable reorganizations of internal muscle architecture in response to similar selective pressures have been described for pigs and other mammals (Herring, 1980).

The relative shortness of the canine teeth in *T. oswaldi* has been a subject of considerable speculation over the years. The results here do not point clearly to any cause and effect relationships, but they do suggest that the stubbier canines of *T. oswaldi* were not being displayed during wide gapes in the same manner as those of the gelada or other large cercopithecines. The role of the canines in limiting the lateral movements of the mandible has been discounted for living species of monkeys (Kay, Scapino & Kay, 1986), but it cannot be assumed that large canines would not have hindered the more exaggerated lateral movements of the mandible in a species such as *T. oswaldi* that was much larger than any extant monkey.

Theropithecus baringensis

Theropithecus baringensis and its putative sister species *T. quadratirostris* present anatomical modifications of the zygomatic arch that presage those seen *T. brumpti* (Eck & Jablonski, 1984). In the masticatory apparatus of *T. baringensis*, we see a form intermediate between that of the gelada and that of *T. brumpti*. The muzzle of *T. baringensis* is as elongated as it is in *Papio*, and the moment arm of the temporalis is similarly shortened. Unlike *Papio*, however, the moment arm of the masseter in *T. barinensis* was relatively enlarged, possibly to compensate for the more posterior inclination of the temporalis (Fig. 11.2(b)). The anatomy of the temporomandibular joint indicates that the mandibular condyle could have moved forward during symmetrical jaw opening (gape) in a manner similar to that seen in modern large cercopithecines.

Theropithecus brumpti

The anatomy of the masticatory apparatus of *T. brumpti* is unique among mammals, and the functioning of the jaws of this species has been a subject of considerable speculation since the first fossils of the species were discovered. Early suggestions that the large, anterolaterally expanded zygomatic arches were decorative appendages promoted by sexual selection were swiftly countered by the discovery of smaller but equally flaring zygomatic arches in females of the species. Further study by Jablonski (1981, 1983) and Eck & Jablonski (1987) showed that the large zygomatic arches of *T. brumpti* afforded attachment for extremely large masseter muscles that were similar in their basic pattern of lamination to the masseters of *T. gelada*. The argument was put forth (Jablonski, 1983; Eck & Jablonski, 1987) that the very large masseter muscles of *T. brumpti* helped to maintain high occlusal pressures over the molar row. This was said to be necessary in order to compensate for the loss of mechanical advantage of the temporalis as the muscle had become more posteriorly inclined

(Fig. 2(c)). The results of the present study support this hypothesis and stimulate still others that warrant further investigation.

The results of the mechanical analysis indicate a significant relative enlargement of the moment arm of the masseter and a smaller relative enlargement of the moment arm of the temporalis in *T. brumpti*. The figures alone do not tell the whole story, however. Examination of the orientation of the masseter and temporalis muscles in Fig. 11.2(c) suggests that the muscles worked together to produce high occlusal pressures over the molar row, the anterior inclination of the masseter (and, presumably the medial pterygoid mirroring the masseter on the medial aspect of the mandible) countering the posterior inclination of the masseter. The question remains as to why this elaborate morphology was developed to maintain a functional relationship equivalent to that seen in the gelada. The answer may come from looking at the overall size of the muscles. The jaw adductors of *T. brumpti* were massive. Reconstruction of the pattern of lamination in the masseter of *T. brumpti* (Jablonski, 1983; Eck & Jablonski, 1987) indicated that the basic pattern of lamination was similar to that seen in the gelada, but that the laminae composed of fleshy fibres were significantly thicker than those of the living species. The structure of the jaw adductors of mammals represents a compromise between the demands of diet and display. The anatomical compromise can be struck in any number of ways (Herring & Herring, 1974; Herring, 1975). These include alteration of the origin-insertion ratios of muscles by relative enlargement and diminution of muscle attachment sites, alteration of the sizes of muscles, and alteration of the internal architecture of muscles to allow varying amounts of stretch. The jaw adductors of *T. brumpti* appear to have been extraordinarily fleshy, long-fibred muscles capable of both stretch and powerful contraction in different positions of the jaw. This capability made possible a fine open-mouthed gape for the display of large canines, while also making possible the biting of large diameter objects at full gape at high pressure. In *Papio* and *Mandrillus*, molar biting at full gape is weak because both the masseter and temporalis are nearly completely stretched, at which point – according to the length-tension curve – they cannot develop adequate tension against resistance. In *T. brumpti*, enlargement of the fleshy bellies of both the masseter and temporalis muscles by the addition of large numbers of long muscle fibres (and/or longer sarcomeres) permitted a very wide gape together with a powerful molar occlusion. Addition of a large number of muscle fibres to the muscle body of the superficial masseter in particular meant that the muscle could still assist in the production of high occlusal pressures even when the fibres were stretched past the point of developing peak tension. This arrangement was not economical in that it involved a great increase in relative muscle size, of the superficial masseter in particular, to accommodate the needs of a wide gape and the need to exert high occlusal pressures over the molar rows during partial or wide jaw opening. One can only speculate that the evolution of such a modification was influenced by strong selective pressures, the nature of which remain unknown.

Variation in the expression of 'zygomatic flare' in *T. brumpti* was great. In the West Turkana male *T. brumpti* (KNM-WT 16828), the zygomatic bones actually curl upwards at nearly right angles to the occlusal plane. The masseter muscle attaching to this curling extension was no doubt extremely massive. A wide gape would have stretched the muscle only slightly, leaving it in an advantageous mechanical position to contract powerfully with the mouth wide open.

Theropithecus brumpti was not a dweller of the open country, and accounts of its habitat preference (Eck & Jablonski, 1987) and locomotion (Jablonski, 1986; and see Krentz, chapter 14) suggest that it occupied a riverine forest niche and was at least partly arboreal. The reconstruction of the masticatory apparatus of the species offered here would suggest that *T. brumpti* was also very different in its diet from any other species of its genus. Its powerful masticatory apparatus would have facilitated the eating of large diameter objects including large fruits, tubers, and shoots that presumably would have been plentiful in riverine forest or back-swamp habitats. Just the opposite of *T. oswaldi*, the small-object feeding 'milling machine' of the savannah (Jolly, 1970), *T. brumpti* was the large object feeder of the forest that did not sacrifice a fearsome gape display to the exigencies of diet.

Conclusions

When the anatomy of the masticatory apparatus of the known species of *Theropithecus* is surveyed, we find one major theme with two divergent sets of variations. The major theme – an approximation of the primitive condition for the genus – is represented by the living gelada; the variations – or derived conditions – by *T. darti* and *T. oswaldi* on the one hand and *T. baringensis*, *T. quadratirostris*, and *T. brumpti* on the other. The modifications of the masticatory apparatus in *T. darti* and, in particular, in *T. oswaldi* were accompanied by more or less exclusive graminivory and increasing large body size. This occurred at the expense of canine gape displays, one of the more important 'traditional' methods among monkeys for maintaining social control and defending against predators. The modifications seen in *T. brumpti* and its putative forebears were far different, and involved the evolution of a masticatory apparatus in which gape displays were not sacrificed at the expense of powerful molar chewing. In both the *T. oswaldi* and *T. brumpti* lineages, transformations of the masticatory apparatus involved modifications of all of its elements, the teeth, jaws, jaw musculature, and temporomandibular articulation. These included important modifications of internal musculature architecture, the significance of which has not yet been realized by some students (e.g. Delson & Dean, chapter 4). Thus, within one genus we find evidence of two quite different and extreme evolutionary experiments with the masticatory apparatus, neither of which contributed to the long-term survival of the species involved.

Acknowledgements

I wish to thank Meave Leakey from the National Museums of Kenya and Peter Andrews and Paula Jenkins from the British Museum (Natural History) for allowing me study the collections of fossil and living *Theropithecus* in their care. I am also grateful to Eric Delson for permitting me to examine and measure specimens in his cast collection. I am particularly grateful to Peter Lisowski for allowing me to study his large collection of gelada cadavers at the University of Hong Kong for my dissertation research in 1979–80. This chapter benefitted greatly from discussions with George Chaplin, and from the comments of Marianne Bouvier and Chris Dean. Last, but certainly not least, I thank Gerald Eck for introducing me to *Theropithecus* and for encouraging my work on the evolution of monkeys over the years.

References

BELL, R.H.V. (1970). The use of the herb layer by grazing ungulates in the Serengeti. In *Animal Populations in Relation to Their Food Resources*, ed. A. Watson, pp. 111–24. Oxford: Blackwell Scientific Publications.

BOUVIER, M. (1986). A biomechanical analysis of mandibular scaling in Old World monkeys. *American Journal of Physical Anthropology*, 69, 473–82.

DUNBAR, R.I.M. (1977). Feeding ecology of gelada baboons: a preliminary report. In *Primate Ecology: Studies of Feeding and Ranging in Lemurs, Monkeys and Apes*, ed. T.H. Clutton- Brock, pp. 251–73. London: Academic Press.

DUNBAR, R.I.M. (1983). Theropithecines and hominids: contrasting solutions to the same ecological problem. *Journal of Human Evolution*, 12, 647–58.

ECK, G.G. (1977). Diversity and frequency distribution of Omo Group Cercopithecoidea. *Journal of Human Evolution*, 6, 55–63.

ECK, G.G. (1987). *Theropithecus oswaldi* from the Shungura Formation, Lower Omo Basin, Southwestern Ethiopia. *Les Faunes Plio-Pléistocènes de la Basse Vallée de l'Omo*

(Éthiopie). Tome 3. Cercopithecidae de la Formation de Shungura, ed. Y. Coppens & F.C. Howell, pp. 123–139, Cahiers de Paléontologie, Travaux de Paléontologie Est-Africaine. Paris: Editions du Centre National de la Recherche Scientifique.

ECK, G.G. & HOWELL, F.C. (1982). Un primate des formations plio-pléistocènes d'Afrique orientale: *Theropithecus brumpti* (Arambourg), (Primates, Cercopithecidae). *Comptes rendus des séances de l'Academie des Sciences*, 295 (Série II), 397–400.

ECK, G.G. & JABLONSKI, N.G. (1984). A reassessment of the taxonomic status and phyletic relationships of *Papio baringensis* and *Papio quadratirostris* (Primates: Cercopithecidae). *American Journal of Physical Anthropolology*, 65, 109–34.

ECK, G.G. & JABLONSKI, N.G. (1987). The skull of *Theropithecus brumpti* as compared to those of other species of the genus *Theropithecus. Les Faunes Plio-Pléistocènes de la Basse Vallée de l'Omo (Éthiopie).* Tome 3. Cercopithecidae de la Formation de Shungura, ed. Y. Coppens & F.C. Howell, pp. 11–122. Cahiers de Paléontologie, Travaux de Paléontologie Est-Africaine. Paris: Editions du Centre National de la Recherche Scientifique.

EMERSON, S.B. & RADINSKY, L. (1980). Functional analysis of sabertooth cranial morphology. *Paleobiology*, 6, 295–312.

ETTER, H.F. (1973). Terrestrial adaptations in the hands of Cercopithecinae. *Folia Primatologica*, 20, 331–50.

HERRING, S.W. (1975). Adaptations for gape in the hippopotamus and its relatives. *Forma et Functio*, 8, 85–100.

HERRING, S.W. (1980). Functional design of cranial muscles: comparative and physiological studies in pigs. *American Zoologist*, 20, 283–93.

HERRING, S.W. & HERRING, S.E. (1974). The superficial masseter and gape in mammals. *American Naturalist*, 108, 561–8.

IWAMOTO, T. (1979). Feeding ecology. *Ecological and Sociological Studies of Gelada Baboons*, ed. M. Kawai, *Contributions to Primatology*, 16, 279–330.

JABLONSKI, N.G. (1981). *Functional analysis of the masticatory apparatus of* Theropithecus gelada (*Primates: Cercopithecoidae*). Ph.D. Dissertation, University of Washington.

JABLONSKI, N.G. (1983). Evolution of a novel masticatory apparatus in a lineage of cercopithecoid primates. *American Zoologist*, 23, 1009 (Abstract).

JABLONSKI, N.G. (1986). The hand of *Theropithecus brumpti*. In *Primate Evolution*, Proceedings of the 10th Congress of the International Primatological Society, vol. 1, ed. P. Lee & J. Else, pp. 173–82. Cambridge: Cambridge University Press.

JOLLY, C.J. (1970). The seed-eaters: a new model of hominid differentiation based on a baboon analogy. *Man*, 5, 5–26.

JOLLY, C.J. (1972). The classification and natural history of *Theropithecus (Simopithecus)* (Andrews, 1916), baboons of the African Plio-Pleistocene. *Bulletin of the British Museum (Natural History), Geology*, 22, 1–123.

KAY, C.N., SCAPINO, R.P. & KAY, E.D. (1986). A cinephotographic study of the role of the canine in limiting lateral jaw movement in *Macaca fascicularis. Journal of Dental Research*, 65, 1300–2.

LEE-THORP, J.A., VAN DER MERWE, N.J. & BRAIN, C.K. (1989). Isotopic evidence for dietary differences between two extinct baboon species from Swartkrans. *Journal of Human Evolution*, 18, 183–90.

MAIER, W. (1972) Anpassungstyp und systematische Stellung von *Theropithecus gelada* Rüppell, 1835. *Zeitschrift fur Morphologische Anthropologie*, 63, 370–84.

MEIKLE, W.E. (1977). Molar wear states of *Theropithecus gelada. The Kroeber Anthropological Society Papers*, 50, 21–6.

NAPIER, J.R. & NAPIER, P.H. (1967). *A Handbook of Living Primates*. New York: Academic Press.

RADINSKY, L. (1981). Evolution of skull shape in carnivores. 1. Representative modern carnivores. *Biological Journal of the Linnean Society*, 16, 337–55.

RADINSKY, L. (1982). Evolution of skull shape in carnivores. 3. The origin and early radiation of the modern carnivore families. *Paleobiology*, 8, 177–95.

RADINSKY, L. (1985). Patterns in the evolution of ungulate jaw shape. *American Zoologist*, 25, 303–14.

SMITH, R.J., PETERSEN, C.E. & GIPE, D.P., (1983). Size and shape of the mandibular condyle in primates. *Journal of Morphology*, 177, 59–68.

SZALAY, F.S. & DELSON, E. (1979). *Evolutionary History of the Primates*. New York: Academic Press.

VRAM, U.G. (1923). Sul genere *Theropythecus*. *Archivio Zoologico*, 10, 169–214.

Appendix 11.1 Specimens of living and fossil species used in this Study

Under the heading of 'Nature of specimen' in the following table, 'Comp.' stands for complete, 'Inc.' for incomplete, 'Frag.' for fragmentary, and 'Mand.' for mandibular. The following abbreviations have been used under the heading of 'Repository of specimen': BM (NH) = British Museum (Natural History), NMK = National Museums of Kenya, HKU = University of Hong Kong, ED = cast collection of Eric Delson, and NGJ = personal cast collection of author.

Species	Specimen number	Male (n)	Female (n)	Nature of specimen	Repository of specimen
Macaca fascicularis	–	5	5	Comp. skulls	BM (NH)
M. radiata	–	5	5	Comp. skulls	BM (NH)
M. arctoides	–	4	4	Comp. skulls	BM (NH)
Papio hamadryas cynocephalus	–	3	3	Comp. skulls	NMK
P. h. anubis	–	6	4	Comp. skulls	NMK
Theropithecus gelada	–	3	1	Comp. skulls	BM (NH)
		2	1	Comp. skulls	HKU
T. darti (Hadar)	AL–205–1	1		Cranium with inc. mandible	ED
	AL–321–12		1	Inc. cranium	ED
T. darti (Makapansgat)	M.3073		1	Comp. skull	ED
	MP 222		1	Cranium	ED
T. oswaldi (Koobi Fora)	KNM–ER 581		1 (?)	Inc. cranium	NMK
	KNM–ER 881		1	Mand. corpus	NMK
	KNM–ER 4968A	1		Mand. corpus	NMK
	KNM–ER 2093	?		Mand. corpus	NMK
	KNM–ER 4964A	?		Mand. corpus	NMK
	KNM–ER 4977		1	Mand. corpus	NMK
	KNM–ER 119A & B		1	Mand. corpus	NMK
	KNM–ER 1525		1	Mand. corpus	NMK
	KNM–ER 1526		1	Inc. skull	NMK
	KNM–ER 613		1	Inc. mand.	NMK
	KNM–ER 4417		1	Inc. mand.	NMK
	KNM–ER 2003	1		Mand. corpus	NMK
	KNM–ER 1528		1	Mand. corpus	NMK
	KNM–ER 567A		1	Mandible	NMK
	KNM–ER 566	1		Frag. skull	NMK
	KNM–ER 969		1 ?	Inc. cranium	NMK
	KNM–ER 864	1		Mandible	NMK
	KNM–ER 557		1	Mandible	NMK
	KNM–ER 3847	1		Inc. mandible	NMK
	KNM–ER 3779		1	Mandible	NMK
	KNM–ER 560	1		Inc. mandible	NMK

Appendix 1 – *Contd.*

Species	Specimen number	Male (n)	Female (n)	Nature of specimen	Repository of specimen
	KNM–ER 865	1		Mandible	NMK
	KNM–ER 180		1	Inc. cranium	NMK
	KNM–ER 5490	1 ?		Frag. skull	NMK
	KNM–ER 2027		1	Mandible	NMK
	KNM–ER 1531	1		Cranium	NMK
	KNM–ER 151	1		Frag. skull	NMK
	KNM–ER 1572		1	Frag. skull	NMK
	KNM–ER 971		1	Cranium	NMK
	KNM–ER 1545	1		Mandible	NMK
	KNM–ER 3091A & B		?	Frag. Mandible	NMK
	KNM–ER 3077		1	Mand. corpus	NMK
	KNM–ER 136A–C	1		Frag. skull	NMK
T. oswaldi (Olduvai)	OLD BK II EX 1953/117		?	Mand. corpus	NMK
	OLD/63 DKI II/ L/2 3050	1		Mand. corpus	NMK
	OLD/69 S.89 SHK E	1		Inc. cranium	NMK
	OLD/63 BKII 3366	1		Inc. mandible	NMK
	OLD/60 SHK II 067/2771		1	Inc. mandible	NMK
	OLD/55 068/6511		1 ?	Inc. cranium	NMK
	OLD/69 S.13 DCII	1		Maxilla	NMK
	OLD BKII juv.		1 ?	Inc. mandible	NMK
	OLD/62 067/ 5603, 5608	1		Inc. skull	NMK
	OLD 068/6516		?	Mandible	NMK
	BM (NH) A M. 14953		?	Mand. corpus	BM(NH)
	BM (NH) A M. 14938		1	Mand. corpus	BM(NH)
T. oswaldi (Kanjera)	BM (NH) A M. 11539		1	Mandible	BM(NH)
	BM (NH) A M. 14936		1	Cranium	BM(NH)
	BM (NH) A M. 32104		1	Cranium	BM(NH)
	BM (NH) A M. 32102	1		Cranium	BM(NH)

Appendix 1 – Contd.

Species	Specimen number	Male (n)	Female (n)	Nature of specimen	Repository of specimen
T. oswaldi (Olorgesailie)	KNM–OG2		1	Mand. corpus	NMK
	KNM–OG4	1		Inc. mandible	NMK
	KNM–OG 1421		?	Frag. cranium	NMK
T. baringensis (Chemeron)	KNM–BC2 A–B	1		Skull	NMK
	KNM–BC 1647 A–B		?	Frag. mandible	NMK
T. brumpti (Koobi Fora)	KNM–ER 3780	1		Frag. mandible	NMK
	KNM–ER 2016	1		Mand. corpus	NMK
	KNM–ER 4985		?	Mand. corpus	NMK
	KNM–ER 2015	1		Inc. mandible	NMK
	KNM–ER 1564		1	Cranium	NMK
	KNM–ER 3038	1		Inc. mandible	NMK
T. brumpti (West Turkana)	KNM–ER 16828	1		Cranium	NMK
T. brumpti (Omo)	L338Y 2257	1		Cranium	ED
	L567–8	1		Mandible	ED
	L345–287	1		Cranium	NGJ

Appendix 11.2 Abbreviations and definitions of variables

CRANL: Cranial length. The distance between alveolare and inion.

PGA: The distance between the base of the anterior surface of the post-glenoid process and alveolare.

MUZL: Muzzle length. The distance between orbitale and alveolare.

SMSL: Superficial masseter scar length. The distance between the base of the anterior surface of the post-glenoid process to the most anterior point of the insertion of the superficial masseter on the zygomatic arch.

BIZYGW: Bizygomatic width. The distance between the two zygomatic arches measured at the midpoint of the zygomaticotemporal suture.

TEMFOSL: Temporal fossa length. Distance between the postorbital plate and the most posterior point of the infratemporal fossa.

TEMFOSW: Temporal fossa width. Distance between the point of the minimum width of the postorbital constriction to the medial surface of the zygomatic arch at the midpoint of the zygomaticotemporal suture.

ZYGH: Height of the zygomatic arch at the zygomaticotemporal suture.

GLENOL: Glenoid fossa length. Maximum anterior–posterior length of the glenoid fossa, measured in the sagittal plane.

GLENOW: Maximum width of the glenoid fossa.

MANDL: Mandible length. Distance between the back of the mandibular condyle and infradentale.

MANDH: Height of the mandible between the first and second permanent molars.

MANCORW: Width of the mandibular corpus between the first and second permanent molars.

MANDSYM: Length of the mandibular symphysis between gnathion and infradentale.

MAM: Moment arm of the masseter. Distance between the top of the mandibular condyle to the anterior border of the angle of the mandible (normally marked by a scar or indentation).

MAT: Moment arm of the temporalis. Distance between the midpoint of the curvature of the mandibular condyle to the apex of the coronoid process.

CONM1: Distance between the back of the mandibular condyle and the mesial border of the lower first permanent molar.

CONDYLL: Maximum anterior–posterior length (thickness) of the mandibular condyle.

CONDYLW: Maximum mediolateral length of the mandibular condyle.

MANDXSA: Mandibular cross-sectional area, calculated as the second moment of area (I), an estimator of resistance to bending. $I = \pi ab \; ^{3}\!/_{4}$, where $a = \frac{1}{2}$ MANCORW and $b = \frac{1}{2}$ MANDH. Converted to MANDXSA $^{1}\!/_{4}$ for comparison with linear measurements.

CONAREA: Condylar area, calculated using the formula for the area of an ellipse $= \text{CONDYLL} \times \text{CONDYLW} \times \pi$. Converted to CONAREA $\frac{1}{2}$ for comparison with linear dimensions.

Appendix 11.3 The means and standard deviations for the combined sex and individual sex samples for all variables

Total sample basic statistics

Name	N	Mean	SD	N (M)	Mean (M)	SD	N (F)	Mean (F)	SD
Macaca fascicularis									
CRANL	10	111.09	14.37	5	122.09	6.79	5	100.10	10.80
PGA	10	78.48	12.37	5	88.17	5.57	5	68.78	8.86
MUZL	10	40.43	8.71	5	46.84	4.95	5	34.03	6.60
SMSL	10	36.75	4.76	5	40.59	1.54	5	32.90	3.41
BIZYGW	9	73.70	10.17	5	81.14	3.08	4	64.41	7.50
TEMFOSL	10	26.78	3.71	5	29.28	2.13	5	24.29	3.31
TEMFOSW	10	20.93	3.39	5	23.72	1.49	5	18.14	2.05
ZYGH	10	6.23	1.72	5	7.54	1.22	5	4.92	0.92
GLENOL	10	13.46	2.45	5	15.23	0.60	5	11.69	2.32
GLENOW	10	14.11	2.32	5	15.81	1.41	5	12.41	1.71
MANDL	10	77.19	18.46	5	91.29	4.86	5	63.08	15.68
MANDH	10	17.23	2.01	5	18.88	1.09	5	15.58	1.03
MANCORW	10	7.45	0.79	5	7.73	0.22	5	7.16	1.08
MANDSYM	10	24.88	3.91	5	28.12	2.24	5	21.63	1.76
MAM	10	38.13	5.16	5	42.09	2.94	5	34.18	3.48
MAT	10	18.77	4.11	5	21.62	3.08	5	15.93	2.87
CONM1	10	55.49	9.06	5	63.12	3.16	5	47.86	5.39
CONDYLL	10	6.07	1.70	5	7.11	0.85	5	5.03	1.75
CONDYLW	10	10.68	1.58	5	11.93	0.50	5	9.43	1.20
Macaca radiata									
CRANL	10	109.30	14.66	5	120.19	4.25	5	98.41	12.99
PGA	10	77.51	7.68	5	84.52	2.87	5	70.49	1.29
MUZL	10	39.87	5.41	5	44.88	1.04	5	34.85	1.40
SMSL	10	36.64	2.49	5	38.32	1.64	5	34.95	2.05
BIZYGW	10	73.48	8.21	5	80.65	4.37	5	66.32	2.11
TEMFOSL	10	26.55	2.69	5	28.62	2.18	5	24.47	0.85
TEMFOSW	10	22.41	2.46	5	24.44	1.53	5	20.38	1.02
ZYGH	10	6.23	1.72	5	7.54	1.22	5	4.92	0.92
GLENOL	10	13.36	1.27	5	14.01	1.39	5	12.71	0.80
GLENOW	10	14.93	1.77	5	16.00	1.27	5	13.86	1.60
MANDL	10	77.92	7.62	5	84.84	3.26	5	71.00	0.79
MANDH	10	17.39	2.15	5	19.19	1.00	5	15.59	1.14
MANCORW	10	7.72	0.84	5	8.07	0.98	5	7.37	0.57
MANDSYM	10	25.23	3.48	5	28.15	1.84	5	22.31	1.62
MAM	9	39.93	3.85	5	42.16	3.80	4	37.15	1.28
MAT	10	17.66	2.05	5	18.51	2.40	5	16.81	1.37
CONM1	10	51.84	6.10	5	55.01	7.53	5	48.67	1.37
CONDYLL	10	4.77	1.00	5	5.39	1.00	5	4.15	0.53
CONDYLW	10	11.59	1.56	5	12.94	0.73	5	10.25	0.69

Appendix 11.3. *Continued*

Name	N	Mean	SD	N (M)	Mean (M)	SD	N (F)	Mean (F)	SD
Macaca arctoides									
CRANL	8	132.05	12.34	4	142.89	5.96	4	121.21	2.54
PGA	8	92.94	8.43	4	100.00	5.38	4	85.89	2.01
MUZL	8	48.36	6.18	4	53.19	3.70	4	43.53	3.63
SMSL	8	44.69	2.94	4	47.05	1.78	4	42.33	1.47
BIZYGW	8	91.58	9.53	4	99.78	4.43	4	83.38	3.62
TEMFOSL	8	33.14	2.69	4	34.75	2.17	4	31.54	2.30
TEMFOSW	8	28.67	3.56	4	31.62	1.17	4	25.73	2.26
ZYGH	8	7.38	1.54	4	8.52	1.13	4	6.24	0.90
GLENOL	8	15.14	1.80	4	16.27	1.90	4	14.00	0.72
GLENOW	8	17.58	2.05	4	19.06	1.54	4	16.11	1.28
MANDL	8	94.49	8.53	4	102.07	3.14	4	86.90	2.59
MANDH	8	24.25	3.10	4	26.47	0.95	4	22.04	2.90
MANCORW	8	10.34	0.74	4	10.87	0.46	4	9.81	0.59
MANDSYM	8	32.64	3.81	4	35.50	2.01	4	29.78	2.85
MAM	8	45.77	4.31	4	48.68	3.52	4	42.86	2.88
MAT	8	22.05	2.72	4	24.24	1.94	4	19.86	0.85
CONM1	8	65.71	5.38	4	70.25	2.50	4	61.18	2.54
CONDYLL	8	6.04	1.12	4	6.93	0.77	4	5.15	0.44
CONDYLW	8	14.73	2.05	4	16.48	1.22	4	12.99	0.46
Papio cynocephalus									
CRANL	6	174.07	19.70	3	191.66	5.50	3	156.48	3.52
PGA	6	127.39	16.79	3	142.44	4.54	3	112.34	2.26
MUZL	6	87.48	13.80	3	99.24	5.91	3	75.71	5.13
SMSL	6	51.37	5.32	3	56.15	1.50	3	46.59	0.47
BIZYGW	6	99.38	8.27	3	106.17	1.85	3	92.59	5.40
TEMFOSL	6	36.38	4.04	3	39.60	1.56	3	33.16	2.70
TEMFOSW	6	29.00	2.75	3	31.03	1.17	3	26.98	2.29
ZYGH	6	9.86	1.57	3	11.18	0.83	3	8.55	0.50
GLENOL	6	18.66	3.19	3	21.49	1.04	3	15.83	0.68
GLENOW	6	18.66	1.27	3	19.54	1.13	3	17.78	0.70
MANDL	6	127.09	16.86	3	141.85	6.71	3	112.34	3.57
MANDH	6	27.52	3.66	3	29.92	3.77	3	25.11	1.41
MANCORW	6	11.14	0.92	3	11.62	0.87	3	10.66	0.83
MANDSYM	6	41.56	6.68	3	46.92	4.75	3	36.20	1.66
MAM	6	58.76	6.47	3	63.70	4.09	3	53.82	3.83
MAT	6	22.62	3.99	3	25.92	1.86	3	19.31	1.90
CONM1	6	89.54	11.45	3	99.70	3.14	3	79.38	2.90
CONDYLL	6	9.33	1.65	3	10.66	1.14	3	79.38	2.90
CONDYLW	6	15.39	1.40	3	16.31	0.62	3	14.47	1.41
Papio anubis									
CRANL	9	184.77	23.33	5	204.20	4.65	4	160.50	2.88
PGA	10	134.46	18.86	6	148.80	3.89	4	112.94	3.72
MUZL	10	96.16	15.84	6	107.70	6.08	4	78.84	5.00
SMSL	10	54.98	6.92	6	60.02	2.12	4	47.43	3.06

Appendix 11.3. *Continued*

Name	N	Mean	SD	N (M)	Mean (M)	SD	N (F)	Mean (F)	SD
BIZYGW	10	106.69	9.65	6	114.05	1.75	4	95.65	1.83
TEMFOSL	10	36.90	4.06	6	39.87	1.45	4	32.45	1.44
TEMFOSW	10	33.90	3.92	6	36.51	1.94	4	30.00	2.42
ZYGH	10	11.02	2.05	6	12.14	1.16	4	8.88	0.49
GLENOL	10	19.87	2.72	6	21.54	1.96	4	17.38	1.38
GLENOW	10	20.00	2.87	6	21.88	1.06	4	17.18	2.28
MANDL	10	133.20	18.34	6	146.86	5.58	4	112.71	4.98
MANDH	10	29.83	4.38	6	32.93	1.68	4	25.17	2.18
MANCORW	10	11.21	1.06	6	11.82	0.44	4	10.29	1.09
MANDSYM	10	42.86	7.87	6	48.48	2.89	4	34.43	3.78
MAM	10	64.55	8.93	6	71.19	3.04	4	54.58	1.80
MAT	10	24.03	4.54	6	25.91	4.82	4	21.23	2.37
CONM1	10	92.40	11.99	6	101.12	3.98	4	79.33	5.00
CONDYLL	10	9.09	1.92	6	10.20	0.87	4	7.43	1.92
CONDYLW	10	17.48	2.53	6	19.17	0.77	4	14.94	1.99
Theropithecus gelada									
CRANL	7	163.51	16.85	5	172.60	8.01	2	140.79	1.11
PGA	7	120.76	12.72	5	127.68	5.73	2	103.45	1.08
MUZL	7	77.47	11.42	5	83.63	5.43	2	62.06	1.03
SMSL	7	54.93	3.74	5	56.82	2.28	2	50.18	0.04
BIZYGW	7	104.95	9.50	5	110.07	4.56	2	92.18	0.97
TEMFOSL	7	36.31	4.00	5	38.55	1.39	2	30.70	0.66
TEMFOSW	7	35.81	3.72	5	37.64	2.28	2	31.24	1.99
ZYGH	7	8.77	1.09	5	9.08	1.08	2	8.00	0.90
GLENOL	7	21.07	2.51	5	22.31	1.42	2	17.96	1.66
GLENOW	7	22.36	1.26	5	22.98	0.86	2	20.83	0.13
MANDL	6	121.62	14.62	4	130.32	7.29	2	104.21	0.21
MANDH	7	33.91	5.30	5	36.53	3.38	2	27.34	1.40
MANCORW	7	11.49	0.85	5	11.63	1.00	2	11.16	0.02
MANDSYM	6	45.59	9.86	4	51.13	5.98	2	34.50	3.20
MAM	7	74.66	5.17	5	77.11	3.67	2	68.54	1.35
MAT	7	28.20	3.17	5	29.65	2.43	2	24.58	0.43
CONM1	6	89.76	9.61	4	95.42	4.89	2	78.45	2.44
CONDYLL	7	8.88	2.07	5	9.71	1.82	2	6.82	0.57
CONDYLW	7	18.20	1.85	5	19.01	1.31	2	16.17	1.49
Theropithecus darti									
CRANL	4	167.12	8.69	1	177.00	0	3	163.83	6.95
PGA	3	120.00	5.46	1	124.10	0	2	117.95	5.86
MUZL	4	77.30	7.24	1	84.40	0	3	74.93	6.71
SMSL	3	59.36	3.47	1	61.90	0	2	58.10	3.81
BIZYGW	2	116.35	13.93	1	126.20	0	1	106.50	0
TEMFOSL	3	40.60	3.76	1	44.70	0	2	38.55	1.76
TEMFOSW	2	32.25	11.80	1	40.60	0	1	23.90	0
ZYGH	3	12.86	1.55	1	11.10	0	2	13.75	0.35

Appendix 11.3. *Continued*

Name	N	Mean	SD	N (M)	Mean (M)	SD	N (F)	Mean (F)	SD
GLENOL	3	17.03	2.19	1	18.40	0	2	16.35	2.61
GLENOW	2	25.30	1.41	1	26.30	0	1	24.30	0
MANDL	0	(All values missing)		0	(All values missing)		0	(All values missing)	
MANDH	0	(All values missing)		0	(All values missing)		0	(All values missing)	
MANCORW	0	(All values missing)		0	(All values missing)		0	(All values missing)	
MANDSYM	1	43.40	0	1	43.40	0	0	(All values missing)	
MAM	0	(All values missing)		0	(All values missing)		0	(All values missing)	
MAT	0	(All values missing)		0	(All values missing)		0	(All values missing)	
CONM1	0	(All values missing)		0	(All values missing)		0	(All values missing)	
CONDYLL	0	(All values missing)		0	(All values missing)		0	(All values missing)	
CONDYLW	0	(All values missing)		0	(All values missing)		0	(All values missing)	
Theropithecus oswaldi (Koobi Fora)									
CRANL	2	200.00	14.14	1	210.00	0	1	190.00	0
PGA	3	139.56	11.66	1	149.60	0	2	134.54	10.99
MUZL	3	88.94	4.48	1	93.85	0	2	86.49	2.02
SMSL	3	64.45	9.76	1	65.16	0	2	64.10	13.78
BIZYGW	1	132.65	0	0	(All values missing)		1	132.65	0
TEMFOSL	2	41.78	3.93	1	44.56	0	1	39.00	0
TEMFOSW	1	36.00	0	0	(All values missing)		1	36.00	0
ZYGH	1	14.68	0	0	(All values missing)		1	14.68	0
GLENOL	6	22.94	2.54	4	23.09	2.40	2	22.65	3.86
GLENOW	6	29.31	1.80	4	29.83	1.85	2	28.28	1.64
MANDL	4	150.61	15.80	2	157.93	22.71	2	143.28	4.36
MANDH	25	35.69	4.91	10	37.61	4.47	13	33.34	4.31
MANCORW	27	15.77	1.44	10	16.29	1.66	14	15.19	0.94
MANDSYM	14	47.15	10.04	6	54.90	10.42	8	41.34	4.44
MAM	5	87.49	5.70	2	82.71	7.33	2	90.77	0.24
MAT	4	31.73	5.66	2	36.05	0.21	1	24.13	0
CONM1	7	111.45	7.87	3	112.51	12.31	3	109.05	3.99
CONDYLL	7	8.60	1.77	4	9.19	1.87	2	6.95	0.77
CONDYLW	7	24.21	3.26	4	25.34	3.51	2	21.66	2.86
Theropithecus oswaldi (Olduvai)									
CRANL	0	(All values missing)		0	(All values missing)		0	(All values missing)	
PGA	0	(All values missing)		0	(All values missing)		0	(All values missing)	
MUZL	1	105.86	0	1	105.86	0	0	(All values missing)	
SMSL	0	(All values missing)		0	(All values missing)		0	(All values missing)	
BIZYGW	0	(All values missing)		0	(All values missing)		0	(All values missing)	
TEMFOSL	0	(All values missing)		0	(All values missing)		0	(All values missing)	
TEMFOSW	0	(All values missing)		0	(All values missing)		0	(All values missing)	
ZYGH	0	(All values missing)		0	(All values missing)		0	(All values missing)	
GLENOL	2	28.29	0.36	1	28.55	0	1	28.03	0
GLENOW	2	32.09	3.14	1	34.32	0	1	29.87	0
MANDL	2	187.00	4.24	2	187.00	4.24	0	(All values missing)	
MANDH	9	36.93	5.05	5	39.92	5.22	3	34.27	3.83

Appendix 11.3. *Continued*

Name	N	Mean	SD	N (M)	Mean (M)	SD	N (F)	Mean (F)	SD
MANCORW	9	17.94	2.49	5	18.99	2.75	3	16.79	1.81
MANDSYM	5	54.97	11.55	3	62.60	6.94	2	45.53	1.14
MAM	2	116.60	11.46	2	116.60	11.46	0	(All values missing)	
MAT	2	41.90	2.12	2	41.90	2.12	0	(All values missing)	
CONM1	2	138.12	3.41	2	138.12	3.41	0	(All values missing)	
CONDYLL	2	9.10	0.34	2	9.10	0.34	0	(All values missing)	
CONDYLW	2	36.00	3.39	2	36.00	3.39	0	(All values missing)	
Theropithecus oswaldi (Olorgesailie)									
CRANL	0	(All values missing)		0	(All values missing)		0	(All values missing)	
PGA	0	(All values missing)		0	(All values missing)		0	(All values missing)	
MUZL	0	(All values missing)		0	(All values missing)		0	(All values missing)	
SMSL	0	(All values missing)		0	(All values missing)		0	(All values missing)	
BIZYGW	0	(All values missing)		0	(All values missing)		0	(All values missing)	
TEMFOSL	0	(All values missing)		0	(All values missing)		0	(All values missing)	
TEMFOSW	0	(All values missing)		0	(All values missing)		0	(All values missing)	
ZYGH	0	(All values missing)		0	(All values missing)		0	(All values missing)	
GLENOL	1	34.00	0	0	(All values missing)		0	(All values missing)	
GLENOW	1	43.70	0	0	(All values missing)		0	(All values missing)	
MANDL	0	(All values missing)		0	(All values missing)		0	(All values missing)	
MANDH	2	38.28	4.82	1	41.70	0	1	34.87	0
MANCORW	1	19.20	0	0	(All values missing)		1	19.20	0
MANDSYM	2	53.90	16.12	1	63.30	0	1	42.50	0
MAM	0	(All values missing)		0	(All values missing)		0	(All values missing)	
MAT	0	(All values missing)		0	(All values missing)		0	(All values missing)	
CONM1	0	(All values missing)		0	(All values missing)		0	(All values missing)	
CONDYLL	0	(All values missing)		0	(All values missing)		0	(All values missing)	
CONDYLW	0	(All values missing)		0	(All values missing)		0	(All values missing)	
Theropithecus oswaldi (Kanjera)									
CRANL	3	182.00	22.11	1	207.00	0	2	169.50	6.36
PGA	2	139.01	19.45	1	152.77	0	1	125.25	0
MUZL	3	75.97	16.53	1	93.78	0	2	67.07	8.45
SMSL	2	67.60	9.17	1	74.09	0	1	61.11	0
BIZYGW	3	117.72	13.20	1	132.37	0	2	110.39	5.18
TEMFOSL	3	48.74	2.91	1	48.36	0	2	48.93	4.09
TEMFOSW	3	39.64	7.56	1	48.32	0	2	35.31	1.24
ZYGH	3	12.98	0.41	1	12.94	0	2	13.00	0.57
GLENOL	2	22.26	1.10	1	21.48	0	1	23.04	0
GLENOW	2	29.15	0.11	1	29.23	0	1	29.07	0
MANDL	1	127.47	0	0	(All values missing)		1	127.47	0
MANDH	1	33.70	0	0	(All values missing)		1	33.70	0
MANCORW	1	14.70	0	0	(All values missing)		1	14.70	0
MANDSYM	1	46.30	0	0	(All values missing)		1	46.30	0
MAM	1	75.40	0	0	(All values missing)		1	75.40	0
MAT	1	27.33	0	0	(All values missing)		1	27.33	0

Appendix 11.3. *Continued*

Name	N	Mean	SD	N (M)	Mean (M)	SD	N (F)	Mean (F)	SD
CONM1	1	99.34	0	0	(All values missing)		1	99.34	0
CONDYLL	1	7.99	0	0	(All values missing)		1	7.99	0
CONDYLW	1	22.14	0	0	(All values missing)		1	22.14	0

Theropithecus baringensis (only males)

Name	N	Mean	SD
CRANL	0	(All values missing)	
PGA	1	156.00	0
MUZL	1	105.27	0
SMSL	1	66.01	0
BIZYGW	1	110.20	0
TEMFOSL	1	44.39	0
TEMFOSW	1	38.02	0
ZYGH	1	13.26	0
GLENOL	1	22.73	0
GLENOW	1	28.50	0
MANDL	1	157.00	0
MANDH	2	36.06	4.48
MANCORW	2	13.03	1.81
MANDSYM	1	54.67	0
MAM	1	86.31	0
MAT	1	29.48	0
CONM1	1	114.58	0
CONDYLL	1	11.90	0
CONDYLW	1	23.63	0

Theropithecus brumpti

Name	N	Mean	SD	N (M)	Mean (M)	SD	N (F)	Mean (F)	SD
CRANL	2	229.50	4.94	2	229.50	4.94	0	(All values missing)	
PGA	2	166.50	2.12	2	166.50	2.12	0	(All values missing)	
MUZL	2	123.15	2.47	2	123.15	2.47	0	(All values missing)	
SMSL	3	83.92	6.78	3	83.92	6.78	0	(All values missing)	
BIZYGW	1	136.27	0	1	136.27	0	0	(All values missing)	
TEMFOSL	3	49.85	4.19	3	49.85	4.19	0	(All values missing)	
TEMFOSW	3	47.10	1.24	3	47.10	1.24	0	(All values missing)	
ZYGH	3	22.06	4.02	3	22.06	4.02	0	(All values missing)	
GLENOL	4	20.36	3.16	3	21.51	2.67	1	16.94	0
GLENOW	4	30.80	2.01	3	30.82	2.46	1	30.74	0
MANDL	1	171.00	0	1	171.00	0	0	(All values missing)	
MANDH	6	43.75	7.12	5	44.34	7.79	0	(All values missing)	
UM1M3L	1	41.30	0	5	14.58	1.61	0	(All values missing)	
MANDSYM	5	57.24	5.33	5	57.24	5.33	0	(All values missing)	
MAM	1	101.70	0	1	101.70	0	0	(All values missing)	
MAT	1	42.00	0	1	42.00	0	0	(All values missing)	
CONM1	1	124.90	0	1	124.90	0	0	(All values missing)	
CONDYLL	2	11.76	1.92	2	11.76	1.92	0	(All values missing)	
CONDYLW	2	24.75	4.31	2	24.75	4.31	0	(All values missing)	

12 Dental microwear and diet in extant and extinct *Theropithecus*: preliminary analyses

MARK F. TEAFORD

Summary

1. Recent work has shown that microscopic wear patterns on teeth can yield insights into diet and dental function in a variety of mammals. The purpose of this study was to see if (a) distinctive dental microwear patterns could be documented for museum samples of modern *Theropithecus*, and (b) similar patterns might be discernible on the teeth of fossil baboons from the Omo deposits in Ethiopia.

2. Epoxy casts were prepared from dental impressions as described by Rose (1983) and Teaford & Oyen (1989a). Scanning electron micrographs were taken of the second mandibular molars at a magnification of 500 times. Computations and analyses were the same as those described by Teaford & Robinson (1989).

3. Results indicate that there are significant differences in dental microwear between *Theropithecus gelada* and *T. brumpti*. There are also significant differences between *T. brumpti* and *T. oswaldi*, whereas the latter is generally more similar to *T. gelada* in terms of its dental microwear patterns.

4. The molars of *T. gelada* are characterized by the presence of enamel prism relief and relatively little microwear. Microwear features that are present tend to be fine scratches and occasional large pits. The molars of *T. brumpti* are characterized by the presence of more microwear and more pits than in either *T. oswaldi* or *T. gelada*. *Theropithecus oswaldi* differs from *T. gelada* in the higher percentage of pits and smaller size of the pits on its molars.

5. These data suggest that *T. brumpti* and *T. oswaldi* had different diets. They also add support to the idea that *T. brumpti* had a different diet from that of *T. gelada*. Finally, they suggest that *T. oswaldi* had a slightly different diet from that of *T. gelada*.

Introduction

One of the most intriguing changes during the last 20 million years of primate evolution has been the dramatic increase in the diversity of Old World monkeys. Baboons form a particularly interesting part of this adaptive radiation, as they now inhabit a wide range of environments, and fossil baboons have been recovered in large numbers from most of the early hominid sites of East Africa. The genus *Theropithecus* forms a puzzling part of this adaptive radiation because, over the past two million years, it has gone from the most common large monkey in certain habitats to one of the most ecologically-restricted Old World monkeys. As a result, two questions continue to perplex investigators: if there were indeed different species of *Theropithecus* 1–3 Ma ago, what were those species doing differently from each other, and what, if anything, were they doing differently from modern *Theropithecus*?

Two lineages of extinct *Theropithecus* are now generally recognized. The *T. oswaldi* lineage began more than 3.5 Ma ago (Eck, 1987; and see chapter 2), and ultimately included the largest monkeys (Jolly, 1972; Ruff, 1988). Morphological and paleoecological evidence have consistently pointed to similarities between *T. oswaldi* and modern gelada baboons (Eck 1987; Jolly, 1970, 1972). This has led to a general consensus that *T. oswaldi* was the quintessential primate grazer.

The second lineage of extinct *Theropithecus*, the *T. brumpti* lineage, has proven to be more problematic. In short, there are just enough differences between *T. brumpti* and modern geladas to suggest that *T. brumpti* had a different habitat preference, and perhaps a different diet (Eck, 1977a; Eck & Jablonski, 1984, 1987; Ciochon, 1986; Jablonski, 1986; Benefit, 1987). The problem here is that, in trying to distinguish between *T. brumpti* and *T. gelada*, or between *T. brumpti* and *T. oswaldi* for that matter, we are comparing two morphologically-similar species. In such cases, standard morphological analyses may be at the limits of their resolution. Even if we can document significant differences in certain comparisons (e.g. tooth shape), we are still left with inferences of *capabilities* rather than evidence of *use*. If *T. brumpti* is different from *T. gelada*, then what do we do? We might be left with an adaptive surprise that is indecipherable by standard morphological means: an 'ancient' gelada baboon that is doing something different from modern geladas. If we rely solely on standard morphological analyses for our interpretations, then we must assume that certain morphological differences are more useful for certain activities and thus confer selective advantages that might be reflected in the fossil record. The end result of such analyses is that we can never be sure of the behaviour of specific individuals found in the fossil record. To counter this, we must hope that a variety of analyses will paint a consistent picture for us to interpret. Clearly, in the face of such difficulties, paleobiological interpretations of prehistoric *Theropithecus* must use every available piece of evidence.

Analyses of microscopic wear patterns on teeth, or dental microwear analyses, have the potential to yield direct evidence of the effects of different behaviour patterns on teeth. As a result, they could provide new evidence for paleobiological interpretations. Admittedly, dental microwear analyses are still in their infancy. Thus, they are not without their methodological problems – most notably, a lack of standardization of techniques and an over-reliance upon analyses of museum specimens (Gordon, 1984b, 1988; Grine, 1986; Peters, 1987; Teaford, 1988a, 1991; Walker & Teaford, 1989; Robson & Young, 1990). They have only begun to deal with such potential complications as intraspecific differences in diet (Teaford & Robinson, 1989; Teaford & Glander, 1991) and variations in dental microstructure (Maas, 1988). Despite these shortcomings, however, certain patterns are beginning to emerge. First, primate hard-object feeders show consistent molar microwear patterns characterized by the presence of large pits (10–100 μm in diameter) on the enamel (Teaford & Walker, 1984; Harmon & Rose, 1988; Teaford, 1988a, Teaford & Oyen, 1989c). Secondly, small microscopic pits (< 5 μm in diameter) are probably formed differently than large microscopic pits – the former perhaps by adhesion and the latter by compressive fracture (Teaford & Oyen, 1989b; Teaford & Runestad, 1992). As a result, interpretations of the incidence of pitting on teeth must also take into account the size of the pits on the teeth. Thirdly, primates that eat more foliage and less fruit tend to have relatively more scratches and fewer pits on their molars (Teaford & Walker, 1984; Teaford, 1985, 1988a; Teaford & Robinson, 1989). Fourthly, the width of scratches may be a relatively poor indicator of dietary dif-

ferences (Maas, 1988; Solounias, Teaford & Walker, 1988; Teaford & Robinson, 1989). Fifthly, individual microwear features are generally created and obliterated relatively quickly (Walker, Hoeck & Perez, 1978; Teaford & Oyen, 1989b; Teaford & Glander, 1991; Teaford & Tylenda, 1991), thus premortem microwear patterns on fossil teeth are probably only a record of the last week or two of tooth use in an animal's life. Sixthly, premortem wear and postmortem damage are fairly easy to distinguish on most fossil teeth. Premortem wear is normally found in regular patterns on the chewing surfaces of the teeth while postmortem damage can occur anywhere on the tooth and in irregular or unusual patterns (Grine, 1977, 1986; Teaford, 1988b).

The purpose of this chapter is to use dental microwear analyses to take another look at the paleobiology of *Theropithecus* from the Omo deposits in Ethiopia.

Materials and methods

Materials

Hundreds of fossil *Theropithecus* specimens have been recovered from the Plio-Pleistocene sites of East Africa (Eck, 1976, 1977a, and see chapter 2 in this volume; Leakey, 1976, and see chapter 3), and dental microwear analyses are only just beginning on them. For preliminary comparisons, I chose to examine some of the specimens collected during the 1967 to 1975 field seasons of the International Omo Research Expedition – specifically those that could be confidently assigned to either *T. brumpti* or *T. oswaldi* (Eck, 1987; Eck & Jablonski, 1987). This immediately reduced the potential sample sizes for this study to 135 specimens of *T. brumpti* and 27 of *T. oswaldi*, since it is extremely difficult (if not impossible) to assign isolated teeth to either of these species (Eck & Jablonski, 1987). One of the morphological traits which best distinguishes between partial specimens of *T. brumpti* and *T. oswaldi* is the development of the mandibular fossa (Eck & Jablonski, 1987). Thus, only mandibular specimens were used in this study. As previous molar microwear analyses of modern primates had focused on M2 (Teaford & Walker, 1984; Teaford, 1985, 1986, 1991; Teaford & Robinson, 1989; Teaford & Runestad 1992; Teaford & Glander, 1991), only specimens preserving M_2 were used in this study. This further reduced the potential sample size to 59 specimens of *T. brumpti* and 18 of *T. oswaldi*.

Wild-trapped specimens of *T. gelada* are notoriously rare in museum collections (Eck, 1977b). The present study involved comparisons with samples from the Department of Anthropology, University of California, Berkeley and from the Field Museum of Natural History in the USA. In all 17 specimens of *T. gelada* were used.

Data collection

Each specimen was cleaned and replicated using the techniques described by Rose (1983) and Teaford & Oyen (1989a). Thus, a polysiloxane (or addition-curing) dental impression material was used to take the impression (either 'Express, Light Body, Regular Set, Hydrophobic', 3M, or 'President Jet Regular', Coltene)[1], and each impression was then poured with epoxy ('Araldite', Ciba-Geigy).

The epoxy casts were sputter-coated with approximately 200 Å of gold and examined in an AMRAY 1810 scanning electron microscope (SEM) at 20 kV in secondary emissions mode. Methods of SEM data collection were essentially the same as those described by Teaford & Walker (1984). However, in this study, micrographs were taken only of crushing-/grinding areas (i.e. facets 9, 10n, or x) (Kay, 1977) on M_2 of each specimen. Since microwear patterns may vary between leading and trailing edges of facets (Teaford, 1988a; Robson & Young, 1990), most micrographs were taken near the middle of the facets. Since the teeth of *Theropithecus* tend to wear-down rather quickly, micrographs were also taken of worn specimens – in the middle of enamel infoldings, either along the buccal notch or along the buccal side of the central basin (see Fig. 12.1). For each individual, two representative micrographs were taken at a magnification of 500 times.

All microwear features within each micrograph were measured in microns by using a digitizer controlled by a pc-compatible computer. For each micrograph, the total number of features was recorded

---

1 It should be noted that 3M has stopped making the hydrophobic version of 'Express'. The current, hydrophilic version of 'Express', has yielded significant pitting artifacts (Gordon, 1984a) and poor resolution of detail when poured with various industrial epoxies (e.g. 'Araldite', Ciba-Geigy).

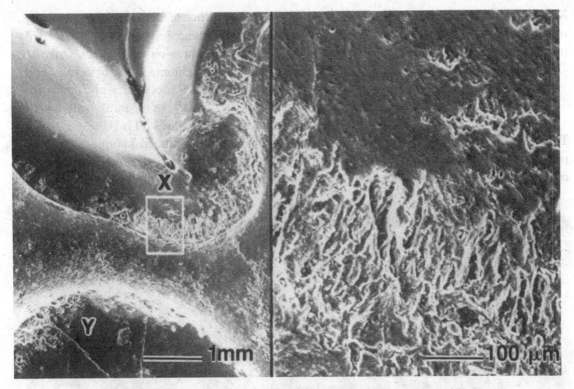

Fig.12.1. SEM micrograph of worn left mandibular M_2 of *Theropithecus brumpti* (L70–4) showing postmortem erosion of enamel. X = enamel infolding on buccal side of remaining portion of central basin. Y = enamel infolding along remaining portion of buccal cleft. Left = mesial, Right = distal. Band of material stretching mesiodistally between X and Y on left half of micrograph = dentin exposure.

and the maximum length and width of each feature were digitized.

Data Analysis

As in previous studies (Teaford & Walker, 1984; Teaford, 1985, 1986, 1988a; Teaford & Robinson, 1989), the ratio of the maximum length to the maximum width of each microwear feature was used to categorize that feature as either a pit or a scratch. A 4:1 length:width ratio was used as the cut-off between pits and scratches

(Teaford, 1988a; Teaford & Robinson, 1989).

Mean values for the following microwear measurements were computed for each individual: number of features per micrograph, proportion of pits, pit length, pit width, scratch length, and scratch width. A variance stabilizing transformation (the arcsine transformation) was performed on all proportions before the computation of any statistics. Single-factor analysis of variance and a multiple comparison test (the Student-Newman-

Keuls test) were used to test for differences in the various dental microwear measurements between the samples of *T. brumpti*, *T. oswaldi*, and *T. gelada*.

Results

The Omo fossil material showed more postmortem damage than in any museum collection examined to date. The most common forms of damage were probably due to chemical erosion (Fig. 12.1) and postmortem abrasion (Fig. 12.2). Only 21 of the 59 specimens of *T. brumpti*, and seven of the 18 specimens of *T. oswaldi* could be used in dental microwear analyses (see Table 12.1).

Despite the reduced sample sizes, the M_2s of *T. brumpti* showed significantly more features, more pits, smaller pits, and

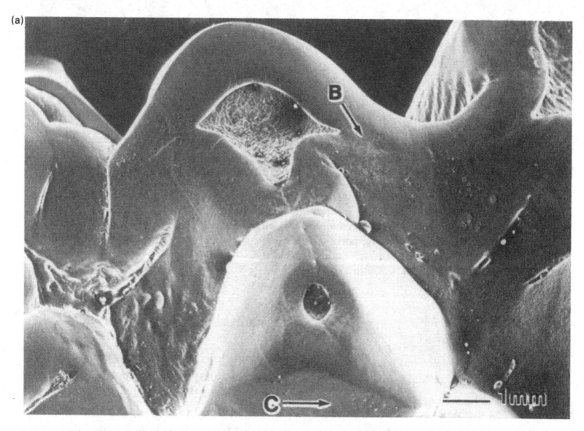

Fig. 12.2. (a) Low magnification SEM micrograph of left mandibular M_3 of *Theropithecus brumpti* (Omo 75N-'71-C1). Right = mesial, left = distal. B = site of Fig. 12.2 (b). C = site of Fig. 12.2 (c); (b) Higher magnification SEM micrograph taken on occlusal surface; (c) Higher magnification SEM micrograph taken on lingual (non-chewing) side of tooth.

Note that, despite the fact that one micrograph was taken on the occlusal surface and one was taken on the lingual side of the tooth, both 12.2 (b) and 12.2 (c) show similar microwear patterns characterized by a preponderance of small-to-medium-sized pits.

(b)

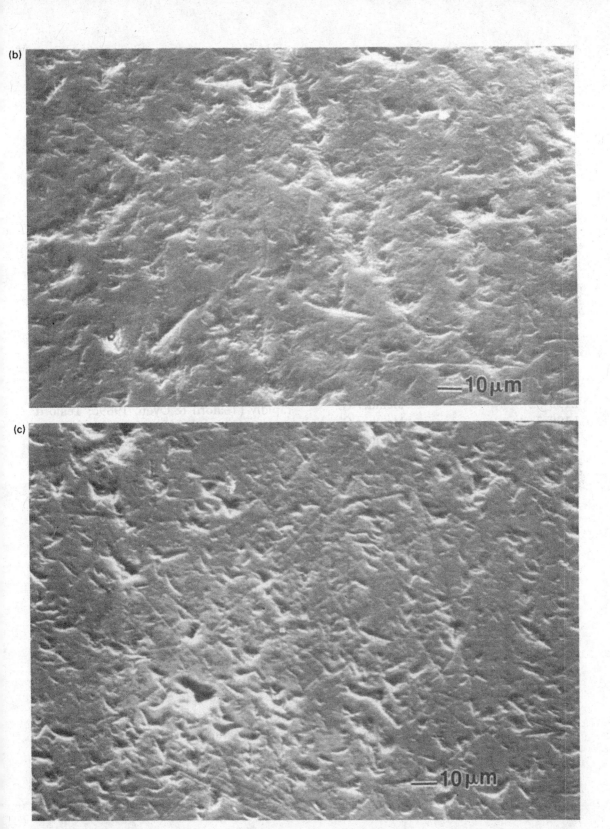

(c)

Fig. 12.2 – *cont*.

Table 12.1 *Omo fossil* Theropithecus
used in this study*.

Specimen number	Stratigraphic provenience
Theropithecus brumpti	
Omo 28–'68–2196	Member B
Omo 40–'69–436	Member C
Omo 75–C40	Member G
Omo 75I–'70–1055	Member G
Omo 75N–'71–C1	Member G
Omo 76–'72–24	Member F
Omo 175.2–'73–1577	Member D
Omo 207–'73–1781	Member E
L17–72	Member C
L28–29x	Member F
L37–16	Member C
L70–4	Member C
L193–32	Member C
L199–3	Member C
L199–5	Member C
L227–6	Member D
L345–4 & 25	Member C
L345–31	Member C
L576–8	Member C
L790–1	Member F
L865–1	Member E
Theropithecus oswaldi	
Omo 25–'67–5	Member G
Omo 33–'70–2534	Member F
Omo 33–'73–3092	Member F
Omo 75–C51	Member G
Omo 75I–'70–1003	Member G
Omo 233–'73–4553	Member G
L465–82a	Member F

* Specimen numbers and provenience from
Eck (1987) and Eck & Jablonski (1987).

shorter scratches than did the M_2s of
T. gelada (Table 12.2 and Figs 12.3, 12.4
and 12.5). *Theropithecus brumpti* also

showed significantly more features and
more pits on its M_2s than did *T. oswaldi*
(Table 12.2 and Figs 12.3 and 12.4). The
only significant differences between
T. oswaldi and *T. gelada* involved the percentage of pits and the size of the pits on
M_2 (smaller in *T. oswaldi*) (Table 12.2 and
Fig. 12.5). There were no significant differences between any of the samples in the
width of scratches (Table 12.2).

Discussion

The amount of postmortem wear on the
teeth of the Omo fossils is probably to be
expected since most of the material comes
from river deposits. As individual
microwear features can change quite
rapidly (Teaford & Oyen, 1989c; Teaford
& Glander, 1991; Teaford & Tylenda,
1991), the small sample sizes make interpretations of the fossils difficult. Still, the
results are both encouraging and
intriguing.

Molar microwear of modern geladas

Molar microwear patterns in modern
Theropithecus are apparently quite distinctive. The molars of all individuals in
this sample showed a microwear pattern
that is best characterized as fine scratches
overlaying enamel prism relief. The few
pits that were present tended to be large.
The most notable variations within the
sample involved the number of features
per micrograph – with individual averages
ranging from 56 to 154. Still, the average
number of features per micrograph was
relatively low. This information corresponds well with the published dietary

Table 12.2 *Descriptive statistics for microwear measurements in* Theropithecus *spp.*

Measurement	T. gelada (mean ± SD)	T. brumpti (mean ± SD)	T. oswaldi (mean ± SD)
Percentage of pits	9.7 ± 5.0	29.5 ± 7.8*	14.0 ± 3.0[†]
Number of features/microg.	94.2 ± 28.8	119.1 ± 27.9**	84.6 ± 19.6
Pit width	4.3 ± 1.8[††]	3.3 ± 0.8	2.8 ± 0.7
Pit length	9.0 ± 3.1[†††]	6.9 ± 1.5	5.5 ± 1.2
Scratch width	0.8 ± 0.07	0.8 ± 0.11	0.7 ± 0.12
Scratch length	36.7 ± 6.2	30.5 ± 7.3***	34.0 ± 3.6

* = significantly greater than values for *T. gelada* and *T. oswaldi* ($p < 0.001$).

† = significantly greater than value for *T. gelada* ($p < 0.05$).

** = significantly greater than values for *T. gelada* ($p < 0.01$) and *T. oswaldi* ($p < 0.025$).

†† = significantly greater than values for *T. brumpti* ($p < 0.025$) and *T. oswaldi* ($p < 0.05$).

††† = significantly greater than values for *T. brumpti* ($p < 0.01$) and *T. oswaldi* ($p < 0.005$).

*** = significantly less than value for *T. gelada* ($p < 0.025$).

(a)

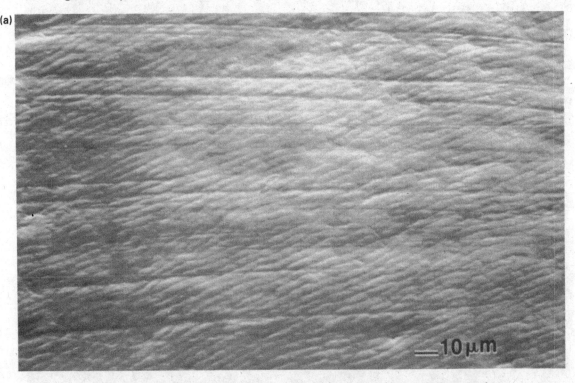

Fig. 12.3. Representative SEM micrographs from the left mandibular M_2s of modern and fossil *Theropithecus*. (a) *Theropithecus gelada* (FMNH 27039); (b) *Theropithecus oswaldi* (Omo 751-'70–1003); (c) *Theropithecus brumpti* (L790–1).

(b)

——10μm

(c)

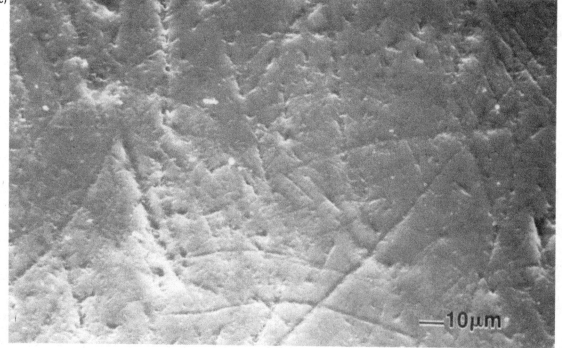

——10μm

Fig. 12.3 – *cont*.

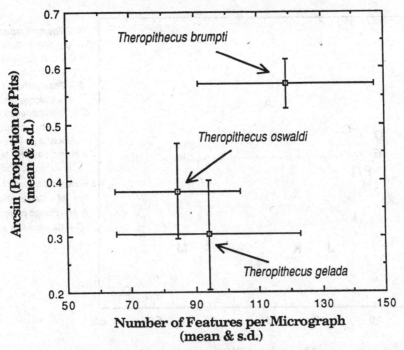

Fig. 12.4. Bivariate plots of molar microwear measurements (means and standard deviations) for modern and fossil *Theropithecus*.

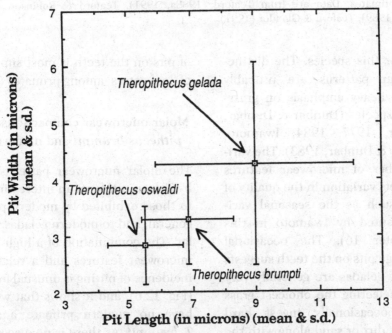

Fig. 12.5. Bivariate plots of molar microwear measurements (means and standard deviations) for modern and fossil *Theropithecus*.

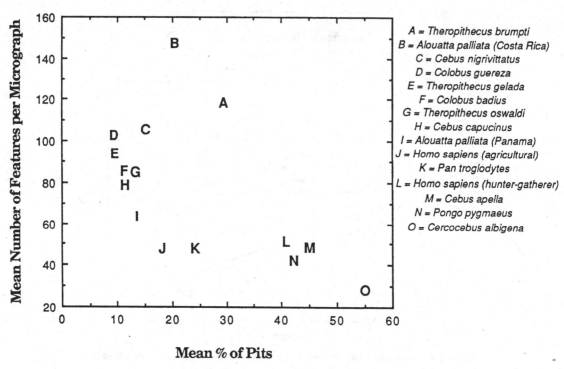

Fig. 12.6. Bivariate plots of molar microwear measurements for different primates. Data are from Teaford (1988a, 1991), Teaford & Robinson (1989), Teaford & Glander (1991).

information for this species. The distinctive microwear patterns are probably related to the heavy emphasis on grass-eating in this species (Dunbar & Dunbar, 1974; Dunbar, 1977, 1984; Iwamoto, 1979; Iwamoto & Dunbar, 1983). The variation in number of microwear features probably reflects variation in the quality of grass eaten (such as the seasonal variations documented by Iwamoto in this volume, chapter 16). The occasional presence of large pits on the teeth suggests that, while the geladas are generally very successful at selecting the choicest grass leaves, they occasionally ingest hard objects such as dirt or sand along with the grass. Not surprisingly, the low proportion

of pits on the teeth is most similar to that of 'leaf-eaters' among primates (Fig. 12.6).

Molar microwear comparisons of Theropithecus brumpti and modern geladas

The molar microwear patterns exhibited by T. brumpti form an interesting contrast to those exhibited by modern primates in general, and to modern geladas in particular. The combination of a high number of microwear features and a relatively high incidence of pitting is unusual for primates (Fig. 12.6), and it shows that we probably have no modern primate analogue for T. brumpti. As there is no discernible correlation between the number of microwear features and the biological age of individual specimens of *T. brumpti* (cf. Gordon, 1984b), the larger number of features on the teeth of this species suggests that *T. brumpti* had more abrasives in, or associated with, its diet than does *T. gelada*. The shorter scratches on the teeth of *T. brumpti* probably just reflect the increased number of microwear features on the teeth – that is, with more features per micrograph, there is a better chance that individual scratches might be truncated by other wear features. The fact that the pits on the teeth of *T. brumpti* are more frequent, yet smaller, than those on the teeth of modern *Theropithecus* is more difficult to interpret. As noted earlier, microscopic pits on teeth are probably not all formed in the same fashion. Unfortunately, most dental microwear studies of modern primates (e.g. Teaford & Walker, 1984; Teaford, 1985; Teaford & Robinson, 1989) have not looked at the incidence of pits of different sizes. Since small pits may be formed by adhesive wear – that is, in repeated tooth–tooth contacts in the mastication of soft food (Teaford & Oyen, 1989c; Teaford & Runestad, 1992), the easiest interpretation of the present results is that *T. brumpti* had a more variable diet than that of *T. gelada* and included significant amounts of soft fruit in its diet. The sheer number of microwear features on the teeth of *T. brumpti* would argue against significant amounts of acidic fruit in the diet, as acid erosion might also be expected to remove microwear features from the tooth surface (Teaford, 1988a; Teaford & Glander, 1991). At the present time, it is impossible to rule out the possibility that leaves formed a significant part

of the diet of *T. brumpti*. It should be noted, however, that no modern primate 'leaf-eater' has shown an incidence of pitting as high as that documented for *T. brumpti* (Fig. 12.6). Only *Alouatta palliata* (from the dry tropical forest of Costa Rica) comes close to *T. brumpti* in this respect (Teaford & Glander, 1991), and we know that those populations of *A. palliata* include significant amounts of fruit in their diet (Glander, 1981).

Molar microwear comparisons of *Theropithecus oswaldi* and modern geladas

As for *T. oswaldi*, its molar microwear pattern is more similar to that of the modern gelada than to that of *T. brumpti*. The consistently low proportion of pits on its teeth suggests that, like modern geladas, *T. oswaldi* had a fairly restricted diet. However, the smaller pits and higher percentage of pits on its teeth, as compared with the teeth of *T. gelada*, suggest three possibilities for subtle differences in diet between these species. *Theropithecus oswaldi* could have been: (1) consuming more leaves – thus avoiding the occasional sand and dirt that might be adhering to grasses, (2) consuming younger, less fibrous leaves or grasses – thus allowing more tooth–tooth contacts in chewing, or (3) occasionally supplementing its diet with soft fruit. The relatively high proportion of scratches on the teeth (as compared with other primates) suggests that variations in leaf- or grass-eating might be the more reasonable interpretations, although, given the small samples of *T. oswaldi* used in this study, more work is necessary before we can sort through these alternatives.

Molar microwear comparisons of *Theropithecus brumpti* and *Theropithecus oswaldi*

Finally, while it is tempting to throw caution to the wind and merely say that the differences in molar microwear between *T. brumpti* and *T. oswaldi* show that there were dietary differences between these species, the microwear differences raise another possibility that must be addressed before interpretations can proceed: what about paleoecological changes within the Omo sequence? It is no secret that, within the Omo sequence, *T. brumpti* is found primarily in the lower levels, whereas *T. oswaldi* is found primarily in the upper levels (Eck, 1976; Eck & Jablonski, 1987). If there was a climatic shift during the deposition of the Omo sequence, could *T. oswaldi* show less microwear and a lower incidence of pitting on its teeth solely because it lived in a wetter environment? Recent dental microwear studies of modern primates (e.g. Teaford & Robinson, 1989; Teaford & Glander, 1991) might support such an interpretation. However, two other sources of information argue strongly against it. First, all available paleoecological data (e.g. Bonnefille, 1976; Eck, 1976; de Heinzelin, 1983; Wesselman, 1984; Feibel, 1988) point to a progressive increase in aridity as one moves upwards through the Omo sequence. Thus, if anything, *T. oswaldi* lived in a drier habitat than *T. brumpti*. Second, these two species coexisted at certain levels within the Omo sequence. If their

dental microwear differences were solely tied to ecological changes within the sequence, one might expect those microwear differences to vanish if microwear comparisons could be made between specimens collected at the same stratigraphic levels. Fortunately, such comparisons are possible for a total of 13 specimens collected from Members F and G of the Omo sequence (see Table 12.1). Even in this small sample of specimens, dental microwear differences were maintained between the species, as *T. brumpti* from Members F and G showed significantly more features and more pits than did *T. oswaldi* from the same Members (Table 12.3). This strongly suggests that, as in comparisons with modern geladas, *T. brumpti* had a more variable, abrasive diet than did *T. oswaldi*.

Broader implications of the results of this study

Two additional points deserve mentioning as a result of this work. First, not only does this study add further support to the idea that *T. brumpti* had a different diet than that of modern geladas, it also adds support to the idea that dental microwear analyses can aid in paleobiological interpretations. Admittedly, some of the interpretations presented here are speculative, and some are slightly different than those derived from other morphological analyses. For instance, Benefit (1987) has suggested that *T. brumpti* had a less abrasive diet than did *T. oswaldi* – based on analyses of dental morphology. Still, the morphological analyses and the dental microwear analyses are yielding the same

Table 12.3 *Microwear measurements for specimens of* Theropithecus brumpti *and* T. oswaldi *collected from Members F & G of the Omo Sequence.*

	T. brumpti	T. oswaldi
Percentage of pits	26.9 ± 5.4*	14.0 ± 3.0
Number of features/microg.	122.2 ± 21.9**	84.6 ± 19.6
Pit width	3.0 ± 0.2	2.8 ± 0.7
Pit length	6.5 ± 0.5	5.5 ± 1.2
Scratch width	0.8 ± 0.1	0.7 ± 0.1
Scratch length	31.3 ± 5.6	34.0 ± 3.6

* = significantly greater than value for *T. oswaldi* ($p < 0.002$).
** = significantly greater than value for *T. oswaldi* ($p < 0.02$).
(Both determined by the Mann-Whitney test.)

basic conclusion: *T. brumpti* was different than *T. gelada*. Thus, even though there will be discussions and debates over subtleties of interpretation, the fact remains that those discussions will centre around new dietary subtleties that have traditionally been unrecognizable in interpretations of fossil material (e.g. variable vs. restricted diets and relative amounts of soft vs. hard fruits in the diet). In essence, we are a step closer to the true intricacies of diet.

The second implication of these results is that not all *Theropithecus* species are the same. Ever since Jolly's pioneering work in the early 1970s (1970, 1972), investigators have emphasized the specialized nature of gelada molar morphology, and the similarities in molar morphology between various species of *Theropithecus* (Delson, 1973, 1975; Eck, 1977a; Meikle, 1977; Szalay & Delson, 1979; Eck & Jablonski, 1987). This has naturally led to the assumption that all gelada molars were designed, or used, for grass-eating. The results of the present study, together with Benefit's (1987) analyses of molar morphology and Jablonski's (chapter 11) research, cast this into doubt. At the very least, these studies raise another question that now needs to be answered: if *T. brumpti* had a different diet than either modern *Theropithecus* or *T. oswaldi*, and if *T. oswaldi* had a slightly different diet than modern *Theropithecus*, when did these dietary changes occur in relation to evolutionary changes in morphology? In an attempt to answer this question, future molar microwear analyses will include (1) comparisons of isolated teeth collected throughout the Omo sequence, and (2) examination of samples from earlier and later sites (e.g. Hadar, Kanjera) in Africa.

Conclusion

Preliminary molar microwear analyses were performed on samples of modern and fossil *Theropithecus*. Results indicate that the lower second molars of modern *Theropithecus* show a fairly distinctive microwear pattern characterized by relatively little microwear – fine scratches overlaying enamel prism relief, with a few large pits. The molar microwear patterns of *T. oswaldi* are more similar to those of modern *Theropithecus* than to those of *T. brumpti* suggesting that *T. oswaldi* had a diet that was more similar to that of modern geladas. By contrast, the lower second molars of *T. brumpti* show more microwear, and a higher incidence of pitting, than do those of either modern *Theropithecus* or *T. oswaldi*. This suggests that *T. brumpti* had a more variable, abrasive diet as compared with either *T. oswaldi* or modern geladas. The smaller pits on the teeth of *T. brumpti* and *T. oswaldi* (as compared with modern *Theropithecus*) are more difficult to interpret, but they suggest additional, more subtle differences in diet between prehistoric and modern geladas. The safest interpretation for *T. brumpti* is that it consumed more soft fruit than does modern *Theropithecus*. High proportions of scratches on the teeth of *T. oswaldi* lend more support to the idea that it consumed more leaves, or younger grasses or leaves, than does modern *Theropithecus*. Future analyses of larger samples, and samples from different sites, should help to sort through these alternatives and, in the process, shed new light on the paleobiology of *Theropithecus*.

Acknowledgements

I wish to thank Nina Jablonski and Robert Foley for inviting me to participate in the symposium 'Theropithecus as a case study in primate evolutionary biology'. I also wish to thank them for their patience, understanding, and encouragement when I could not attend the symposium at the last minute. I also wish to thank Rose Keller for her help in preparing and cataloguing casts and for taking many of the SEM micrographs used in this study. Special thanks go to Bruce Patterson for allowing access to specimens in his care at the Field Museum and to F. Clark Howell for allowing access to specimens in his care at Berkeley. Dr Howell's interest and enthusiasm, together with his willingness to allow easy access to his lab, really made this project feasible. As is so often the case in a project like this, numerous people deserve additional thanks for their help and hospitality. This would include Susan Anton, Brenda Benefit, Walter Hartwig, Don Johanson, Monte McCrossin, Yoel Rak, Gary Richards, and Tim White. Bill Kimbel deserves special thanks for his enthusiastic discussions – not to mention his untiring tours of the Bay Area. Nina Jablonski and two anonymous reviewers provided useful and tactful comments on a very rough draft of this paper. Gerald Eck and Meave Leakey provided stimulating discussions and encouragement throughout the evolution of this project. This work was supported by a grant from the L.S.B. Leakey Foundation and NSF grants 8904327, 8803570, 8605172.

References

BENEFIT, B. (1987). *The Molar Morphology, Natural History, and Phylogenetic Position of the Middle Miocene Monkey* Victoriapithecus. Ph.D. Dissertation, New York University.

BONNEFILLE, R. (1976). Palynological evidence for an important change in the vegetation of the Omo Basin between 2.5 and 2 million years. In *Earliest Man and Environments in the Lake Rudolf Basin*, ed. Y. Coppens, F.C. Howell, G.L. Isaac & R.E.F. Leakey, pp. 421–31. Chicago: University of Chicago Press.

CIOCHON, R.L. (1986). *The Cercopithecoid Forelimb: Anatomical Implications for the Evolution of African Plio- Pleistocene Species*. Ph.D. Dissertation, University of California, Berkeley.

DELSON, E. (1973). *Fossil Colobine Monkeys of the Circum-Mediterranean Region and the Evolutionary History of the Cercopithecidae (Primates, Mammalia)*. Ph.D. Dissertation, Columbia University.

DELSON, E. (1975). Evolutionary history of the Cercopithecidae. *Contributions to Primatology*, 5, 167–217.

DUNBAR, R.I.M. (1977). Feeding ecology of gelada baboons: a preliminary report. In *Primate Ecology*, ed. T.H. Clutton-Brock, pp. 251–73. New York: Academic Press.

DUNBAR, R.I.M. (1984). *Reproductive Decisions: An Economic Analysis of Gelada Baboon Social Strategies*. Princeton, New Jersey: Princeton University Press.

DUNBAR, R.I.M. & DUNBAR, E.P. (1974). Ecological relations and niche separation between sympatric terrestrial primates in Ethiopia. *Folia Primatologica*, 21, 36–60.

ECK, G.G. (1976). Cercopithecoidea from Omo Group deposits. In *Earliest Man and Environments in the Lake Rudolf Basin*, ed. Y. Coppens, F.C. Howell, G.L. Isaac & R.E.F. Leakey, pp. 332–44. Chicago: University of Chicago Press.

ECK, G.G. (1977a). Diversity and frequency distribution of Omo Group Cercopithecoidea. *Journal of Human Evolution*, 6, 55–63.

ECK, G.G. (1977b). *Morphometric Variability in the Dentitions and Teeth of Theropithecus and Papio*. Ph.D. Dissertation, University of California, Berkeley.

ECK, G.G. (1987). *Theropithecus oswaldi* from the Shungura Formation, Lower Omo Basin, southwestern Ethiopia. In *Les Faunes Plio-Pléistocènes de la Basse Vallée de l'Omo (Éthiopie)*, Tome 3. Cercopithecidae de la Formation de Shungura, ed. Y. Coppens & F.C Howell, pp. 123–40. Cahiers de Paléontologie, Travaux de

Paléontologie Est-Africaine. Paris: Éditions du Centre National de la Recherche Scientifique.

ECK, G.G. & JABLONSKI, N.G. (1984). A reassessment of the taxonomic status and phyletic relationships of *Papio baringensis* and *Papio quadratirostris* (Primates, Cercopithecidae). *American Journal of Physical Anthropology*, 65, 109–34.

ECK, G.G. & JABLONSKI, N.G. (1987). The skull of *Theropithecus brumpti* compared with those of other species of the genus *Theropithecus*. In *Les Faunes Plio-Pléistocènes de la Basse Vallée de l'Omo (Éthiopie)*. Tome 3. Cercopithecidae de la Formation de Shungura, ed. Y. Coppens & F.C. Howell, pp. 11–122. Cahiers de Paléontologie, Travaux de Paléontologie Est-Africaine. Paris: Éditions du Centre National de la Recherche Scientifique.

FEIBEL, C.S. (1988). *Paleoenvironments of the Koobi Fora Formation, Turkana Basin, Northern Kenya*. Ph.D. Dissertation, University of Utah.

GLANDER, K.E. (1981). Feeding patterns in mantled howling monkeys. In *Foraging Behavior: Ecological, Ethological, and Psychological Approaches*, ed. A. Kamil & T.D. Sargent, pp. 231–59. New York: Garland Press.

GORDON, K.D. (1984a). Pitting and bubbling artefacts in surface replicas made with silicone elastomers. *Journal of Microscopy*, 134, 183–88.

GORDON, K.D. (1984b). Hominoid dental microwear: complications in the use of microwear analysis to detect diet. *Journal of Dental Research*, 63, 1043–6.

GORDON, K.D. (1988). A review of methodology and quantification in dental microwear analysis. *Scanning Microscopy*, 2, 1139–47.

GRINE, F.E. (1977). Postcanine tooth function and jaw movement in the gomphodont cynodont *Diademodon* (Reptilla; Therapsida). *Palaeontologica Africana*, 20, 123–35.

GRINE, F.E. (1986). Dental evidence for dietary differences in *Australopithecus* and *Paranthropus*: a quantitative analysis of permanent molar microwear. *Journal of Human Evolution*, 15, 783–822.

HARMON, A.M. & ROSE, J.C. (1988). The role of dental microwear analysis in the reconstruction of prehistoric diet. In *Diet and Subsistence: Current Archaeological Perspectives*, ed. B.V. Kennedy & G.M. LeMoine, pp. 267–72. The Archaeological Association of the University of Calgary, Calgary, Alberta.

DE HEINZELIN, J. (Ed.) (1983). *The Omo Group: Archives of the International Omo Research Expedition*. Annales, S. 8, Sciences Géologiques, (85), Musée Royal de l'Afrique Centrale, Tervuren.

IWAMOTO, T. (1979). Feeding ecology. In *Ecological and Sociological Studies of Gelada Baboons*, ed. M. Kawai, pp. 279–330. Tokyo: Kodansha.

IWAMOTO, T. & DUNBAR, R.I.M. (1983). Thermoregulation, habitat quality and the behavioral ecology of gelada baboons. *Journal of Animal Ecology*, 52, 357–66.

JABLONSKI, N.G. (1986). The hand of *Theropithecus brumpti*. In *Primate Evolution*, ed. J.G. Else & P.C. Lee, pp. 173–82. Cambridge: Cambridge University Press.

JOLLY, C.J. (1970). The seed-eaters: A new model of hominid differentiation based on a baboon analogy. *Man*, 5, 5–26.

JOLLY, C.J. (1972). The classification and natural history of *Theropithecus* (*Simopithecus*) (Andrews, 1916), baboons of the African Plio-Pleistocene. *Bulletin of the British Museum (Natural History), Geology*, 22, 1–123.

KAY, R.F. (1977). The evolution of molar occlusion in the Cercopithecidae and early catarrhines. *American Journal of Physical Anthropology*, 46, 327–52.

LEAKEY, M.G. (1976). Cercopithecoidea of the East Rudolf succession. In *Earliest Man and Environments of the Lake Rudolf Basin*, ed. Y. Coppens, F.C. Howell, G.L. Isaac & R.E.F. Leakey, pp. 345–50. Chicago: University of Chicago Press.

MAAS, M.C. (1988). *The Relationship of Enamel Microstructure and Microwear*. Ph.D. Dissertation, State University of New York at Stony Brook.

MEIKLE, W.E. (1977). Molar wear states of *Theropithecus gelada*. *The Kroeber Anthropological Society Papers*, 50, 21–6.

PETERS, C.R. (1987). Nut-like oil seeds: food for

monkeys, chimpanzees, humans, and probably ape-men. *American Journal of Physical Anthropology*, 73, 333–63.

ROBSON, S.K. & YOUNG, W.G. (1990). A comparison of tooth microwear between an extinct marsupial predator, the Tasmanian tiger *Thylacinus cynocephalus* (Thylacinidae) and an extant scavenger, the Tasmanian devil *Sarcophilus harrisii* (Dasyuridae: Marsupialia). *Australian Journal of Zoology*, 37, 575–89.

ROSE, J.J. (1983). A replication technique for scanning electron microscopy: Applications for anthropologists. *American Journal of Physical Anthropology*, 62, 255–61.

RUFF, C. (1988). Hindlimb articular surface allometry in Hominoidea and *Macaca*, with comparisons to diaphyseal scaling. *Journal of Human Evolution*, 17, 687–714.

SOLOUNIAS, N., TEAFORD, M.F. & WALKER, A. (1988). Interpreting the diet of extinct ruminants: the case of a non-browsing giraffid. *Paleobiology*, 14, 287–300.

SZALAY, F.S. & DELSON, E. (1979). *Evolutionary History of the Primates*. New York: Academic Press.

TEAFORD, M.F. (1985). Molar microwear and diet in the genus *Cebus*. *American Journal of Physical Anthropology*, 66, 363–70.

TEAFORD, M.F. (1986). Dental microwear and diet in two species of *Colobus*. In *Primate Ecology and Conservation*, ed. J. Else & P. Lee, pp. 63–6. Cambridge: Cambridge University Press.

TEAFORD, M.F. (1988a). A review of dental microwear and diet in modern mammals. *Scanning Microscopy*, 2(2), 1149–66.

TEAFORD, M.F. (1988b). Scanning electron microscope diagnosis of wear patterns versus artifacts on fossil teeth. *Scanning Microscopy*, 2(2), 1167–75.

TEAFORD, M.F. (1991). Dental microwear: what can it tell us about diet and dental function? In *Advances in Dental Anthropology*, ed. M.A. Kelley & C.S. Larsen, pp. 341–56. New York: Alan R. Liss.

TEAFORD, M.F. & GLANDER, K.E. (1991). Dental microwear in live, wild-trapped *Alouatta* from Costa Rica. *American Journal of Physical Anthropology*, 85, 313–19.

TEAFORD, M.F. & OYEN, O.J. (1989a). Live primates and dental replication: new problems and new techniques. *American Journal of Physical Anthropology*, 80, 73–81.

TEAFORD, M.F. & OYEN, O.J. (1989b). Differences in the rate of molar wear between monkeys raised on different diets. *Journal of Dental Research*, 68, 1513–18.

TEAFORD, M.F. & OYEN, O.J. (1989c). *In vivo* and *in vitro* turnover in dental microwear. *American Journal of Physical Anthropology*, 80, 447–60.

TEAFORD, M.F. & ROBINSON, J.G. (1989). Seasonal or ecological differences in diet and molar microwear in *Cebus nigrivittatus*. *American Journal of Physical Anthropology*, 80, 391–401.

TEAFORD, M.F. & RUNESTAD, J.A. (1992). Dental microwear and diet in Venezuelan primates. *American Journal of Physical Anthropology*, 88, 347–364.

TEAFORD, M.F. & TYLENDA, C.A. (1991). A new technique for assessing rates of human tooth wear. *Journal of Dental Research*, 70, 204–7.

TEAFORD, M.F. & WALKER, A. (1984). Quantitative differences in dental microwear between primate species with different diets and a comment on the presumed diet of *Sivapithecus*. *American Journal of Physical Anthropology*, 64, 191–200.

WALKER, A.C., HOECK, H.N. & PEREZ, L. (1978). Microwear of mammalian teeth as an indicator of diet. *Science*, 201, 908–10.

WALKER, A. & TEAFORD, M.F. (1989). Inferences from quantitative analysis of dental microwear. *Folia Primatologica*, 53, 177–89.

WESSELMAN, H.B. (1984). The Omo micromammals: systematics and paleoecology of early man sites from Ethiopia. *Contributions to Vertebrate Evolution*, vol. 7, ed. M.K. Hecht & F.S. Szalay.

13 The development and microstructure of the dentition of *Theropithecus*

DARIS R. SWINDLER AND A. DAVID BEYNON

Summary

Radiographs were made of 11 specimens of *Theropithecus gelada* of unknown chronological age ranging from newborn to young adult. Tooth buds were removed from the younger specimens and stained with alizarin red S. The tooth buds were then studied using a dissecting microscope. In addition, teeth were removed from four juvenile mandibles and embedded in polyester using a prolonged impregnation and curing schedule. These teeth were then studied employing various histological techniques.

2. At birth, the deciduous teeth of *T. gelada* were similar in development to those of *Papio hamadryas cynocephalus*, *Macaca mulatta*, *M. nemestrina* and *Alouatta caraya*. The first permanent molars were also calcified at birth in all species examined and were always more mature in the monkey species than in *Pan troglodytes* and *Homo sapiens*.

3. The sequence of permanent tooth eruption in *T. gelada* was M1, I1, I2, M2, [P3, P4, C], M3, a pattern not unlike that reported for other cercopithecid monkeys as well as for the fossils, *T. brumpti* and *T. oswaldi*. The estimated age (in years) of eruption of the permanent teeth using histological data was: M1 = 1.7; I1, I2 = 3.0–3.2; M2 = 3.5–4.0; P3, P4, C = 4.2–5.3; M3 = 6.0–6.5.

4. It was suggested that the distal accessory cuspules on M_{1-2} in *T. gelada* are serially homologous with the M_3 hypoconulid and that they represent an adaptation for maintaining mesiodistal contact between adjacent molars thereby limiting early interstitial attrition. Likewise, the hyposodont molars and late eruption of M_3 appear to be dental adaptations to prolong or delay tooth loss.

5. The uniform rate of seven daily enamel increments between striae was the most consistent 'near weekly' rhythmicity encountered to date in anthropoid primates. *Theropithecus gelada* appears to be unique in the absence of hypoplastic bands or accentuated striae indicating that they grew up in a stable environment lacking environmental and dietary stresses or illness during the dental growth period.

6. The comparative odontogenetic information presented here suggests that when tooth formation data are available for more comprehensive comparisons among the taxa of Papionini, the major events in sequence and timing will be similar, that is, of the same general order of magnitude.

Materials and methods

The sample consisted of 11 specimens of *Theropithecus gelada*. The ages were unknown but they were all young animals according to their state of dental emergence (Table 13.1). The animals were collected in 1966 in Ethiopia, by Dr F.P. Lisowski, from a single region west of Debra Berhan (Jablonski, 1981). They were embalmed in the field with a 4 per cent formaldehyde solution. Since the early 1970s, they have been at the University of Hong Kong stored in 5 per cent formalin. Deciduous and permanent tooth-buds were removed intact from the younger specimens under a dissecting microscope. After extraction, the tooth-germs in their follicles were stained in a

Table 13.1. *Theropithecus gelada* specimens

Specimen	Sex	Dental emergence	Age category
Newborn	M	No emergence	Newborn
HKU 0233	M	ii c dm^1 dm^2 M^1 ii c dm_1 dm_2 M_1	Young
HKU 0228	F	ii c dm^1 ii c dm_1	Infant
HKU 0248	M	ii c dm^1 dm^2 ii c dm_1 dm_2	Infant
HKU 0252	F	II C P^3 dm^2 M^1 M^2 II C P_3 dm_2 M_1 M_2	Young
HKU 0237	M	II c P^3 dm^2 M^1 M^2 II c dm_1 dm_2 M_1 M_2	Young
HKU 0242	M	II c P^3 dm^2 M^1 M^2 II C dm_1 dm_2 M_1 M_2	Young
HKU 0235	F	II C P^3 P^4 M^1 M^2 II C P_3 P_4 M_1 M_2 M_3	Adult
HKU 0251	M	ii c dm^1 dm^2 M^1 ii c dm_1 dm_2 M_1	Young
HKU 0243	F	II C P^3 P^4 M^1 M^2 II C P_3 P_4 M_1 M_2	Young
HKU 0232	M	II C P^3 P^4 M^1 M^2 II C P_3 P_4 M_1 M_2 M_3	Young

weak aqueous solution of alizarin red S which differentiates calcified material. Subsequent storage, handling, and examination was in glycerin. In order to reveal the shape of the developing tooth, the follicle was cut away and the stellate reticulum was removed. The tooth-buds were studied under a dissecting microscope, photographed and drawn from the occlusal view.

In addition to *T. gelada* material, stained tooth-buds from several species of anthropoids (*Alouatta caraya*, *Papio anubis*, *Macaca mulatta*, *M. nemestrina*, *M. fascicularis*, *Pan troglodytes*, and *Pongo pygmaeus*) were used for the comparative discussion. These tooth-germs are part of the senior author's collection of primate tooth-germs and dental casts housed in the Department of Anthropology, University of Washington, USA.

Lateral radiographs of the Hong Kong specimens were available for study as well as cephalograms of *Papio hamadryas cynocephalus*, the yellow baboon, and *M. nemestrina*, the pigtailed macaque. The latter material is from the collection of longitudinal cephalograms taken of these animals between 1968–76 at the Regional Primate Research Center (RPRC), University of Washington.

Although there is a small difference in the chronology of pre- and postnatal tooth formation between the maxilla and mandible, the mandibular events are usually considered to reflect the development of the entire dentition (Demirjian, Goldstein & Tanner, 1973). Moreover, radio-

graphs of the lower teeth are clearer and easier to examine; therefore, unless otherwise noted, the mandibular teeth are discussed in this chapter.

Four juvenile mandibles were used for the histological study (HKU 0237, HKU 0252, HKU 0243, and HKU 0232). Their sexes and dental emergence schedules are presented in Table 13.1 along with all specimens used in this study. The specimens were radiographed and teeth were dissected out on one side of the jaw to allow crown height and root length to be measured (Tables 13.2 and 13.3).

The teeth were embedded in polyester resin using a prolonged impregnation and curing schedule. In anterior teeth a single axial section was prepared from each tooth. In the molar teeth buccolingual sections were prepared through the mesial and distal cusp pairs. Care was taken to orientate the specimen so that the final section included the tips of the underlying dentine horns. Initially 'thick' sections were prepared with an initial thickness of around 500 μm, and sections were then

Table 13.2. *Table of buccal (B) and lingual (L) crown heights (mm) for* Theropithecus gelada

Specimen		I_1	I_2	C	P_3	P_4	M_1	M_2	M_3
HKU 0237	B	13.0	10.5	28.0	21.5	—	> 7.0	11.5	—
	L							8.5	
HKU 0252	B	11.8	9.0	12.0	12.5	9.2	> 7.0	10.0	> 8.5
	L						> 5.5	7.5	
HKU 0243	B	> 11.1	> 10.3	> 12.6	11.8	> 10.0	> 4.7	10.2	11.0
	L						> 6.2	> 7.0	7.5
HKU 0232	B	> 9.3	> 9.4	> 12.6	11.4	> 7.5	> 4.6	8.5	10.0
	L								6.5

Table 13.3. *Table of root lengths for* Theropithecus gelada

Specimen		I_1	I_2	C	P_3	P_4	M_1	M_2	M_3
HKU 0237	B L	11.5	10.5	4.5	0.0 9.6	9.5	14.0*	10.0	—
HKU 0252	B L	11.0	10.0	14.0	7.5	7.5	13.5*	8.0	—
HKU 0243	B L	11.5*	10.5*	16.5*	11.5*	12.0*	15.0*	12.0	< 1.0
HKU 0232	B L	10.0*		14.5*			12.0*	13.0	< 8.0

*roots complete

B = buccal

L = lingual

lapped down on one side to the desired level, bonded to a microscope slide, and thinned from the other surface to final working thickness of between 80 and 120 µm. The surface was etched for 20 seconds in a 0.5 per cent solution of orthophosphoric acid to remove the surface smear layer. The P_3 from HKU 0237 was attached to a stump, briefly etched with a 0.5 per cent phosphoric acid, sputter coated with gold and examined in an S600 Cambridge scanning electron microscope (SEM).

Perikymata were frequently prominent on labial, buccal, and lingual surfaces of both anterior and posterior teeth. Two stage impressions were made of the tooth surfaces using Coltene President impression material after the method of Beynon (1987) prior to embedding impressions. Replicas were made using Spurr resin and were subsequently sputter coated with gold to facilitate the counting of surface perikymata. Counts of perikymata per mm were made from the cervix towards the occlusal plane (Tables 13.4–13.7).

Ground sections were viewed using transmitted (including polarized) light on a Zeiss Universal microscope. Specimens were studied using ×10, ×16, and ×40 objectives. Measurements were made on cross-striation spacing using ×16 and ×40 objectives; and stria spacing measurements were made along the direction of the prism axis using ×10 objectives, with a Filar micrometer eyepiece. Measurements were made along the enamel surface, counting the numbers of striae reaching that surface, expressed per mm of surface length, starting from the tooth cervix. These values are presented with perikymata counts where these were possible in Tables 13.4–13.7.

Numbers of cross-striations between striae in individual teeth were estimated using two different methods. In favourable sections direct counts were made of cross-striations between striae in individual

Table 13.4. *Counts of striae per mm, and perikymata per mm, along the enamel surface from cervical to occlusal levels in* Theropithecus gelada

HKU 0237 Male		Cervical → Occlusal	SS	BI	Total	Years
Striae	R I_1	8* 9 10 10 9 9 7 8 8 5	101	28	129+	2.5
Striae	R I_2	6 11 11 10 9 11 8 8 7 7	103	27	130	2.5
Striae	R C	5 10 10 10 11 11 9 10 11 9 8 9 8 8 8 9 8 6 5 4	260	10	270	5.2
perikymata	R	9* 10 9 8 8 9 10 9 10 11 9 11 10 11 12 10 12 11 2 —			249	5.0
Striae	R P_3	14 13 13 12 13 13 14 11 13 15 13 13 13 15	241	20	261	5.0
Striae	R M_1 Mesiolingual	12 10 6 5 5 3 Attrition	41	25†	66	1.3
perikymata	R	11 11 — — —				
Striae	R M_1 Mesiobuccal	12 15 9 6 5 6 2 Attrition	55	23†	78	1.5
Striae	R M_1 Distolingual	10 10 6 4 4 4 Attrition	38	25†	63	1.2
perikymata	R	13 11 9 5! 3!				
Striae	R M_1 Distobuccal	13 13 8 7 7 4 2 Attrition	54	23†	77	1.5
Striae	R M_2 Mesiolingual	12 10 7 5 4 5 4 4 2 Attrition	53	20	73	1.4
perikymata	R	12 11 9† 7† 5† 5† 4†			58	
Striae	R M_2 Mesiobuccal	7* 12! 9 6 4 4 4 3 4 2	59	20	79+	1.5
Striae	R M_2 Distolingual	• 9 11 7 4 5 5 3	44	20	64+	1.2
perikymata	R	10 12 10 6† 5† 5 2			60	
Striae	R M_2 Distobuccal	• 17 14 12 10 6 5 4 4 4 3 4 4	87	20	107+	2.1

SS = total numbers of surface striae. BI = buried increments in occlusal appositional enamel (buried increments include estimates in teeth showing slight occlusal wear). Total striae counts are given in the penultimate column and converted into age in years by multiplying by the seven-day cross-striation periodicity. * = fractured cervical enamel. † = estimated perikymata counts. Estimates are given for the labial or buccal surfaces of anterior teeth. In molars values are given for each of the four principal cusps on buccal and lingual surfaces. The greatest value is generally found on the distobuccal cusp and is used as the maximum estimate of crown formation time. Total estimates of crown formation time for individual teeth in each animal are given in Table 13.8. These data were used to construct a provisional bar chart (Fig. 13.13).

Table 13.5. *Counts of striae per mm along the enamel surface from cervical to occlusal levels in* Theropithecus gelada

HKU 0252 Female		Cervical →												Occlusal	SS BI	Total	Years
Striae	R I₁	8	9	10	11	10	10	9	10	9	8	9	5!		108+20	128	2.5
Striae	R C̄	*	9	9	12	12	11	10	9	9	8	8			119+20	139	2.7
Striae	R P₃	4	8	13	12	14	13	13	12	14	9	8	5		138+20	158	3.0
Striae	R P₄ Lingual	11	12	8	7	4	4								46+23	69	1.3
Striae	R P₄ Buccal	12	16	9	5	4	6	5							62+21	71	1.4
Striae	R M₂ Distolingual	11	12	8	6	4	5	4†	1						51+20†	71	1.4
Striae	R M₂ Distobuccal	14	14	11	9	7	5	4	3	3	4	4			78+25†	103	2.0

SS = total numbers of surface striae. BI = buried increments. BI = buried increments in occlusal appositional enamel (buried increments include estimates in teeth showing slight occlusal wear). Total striae counts are given in the penultimate column and converted into age in years by multiplying by the seven-day cross-striation periodicity. * = fractured cervical enamel. † = estimated perikymata counts. Estimates are given for the labial or buccal surfaces of anterior teeth. In molars values are given for each of the four principal cusps on buccal and lingual surfaces. The greatest value is generally found on the distobuccal cusp and is used as the maximum estimate of crown formation time. Total estimates of crown formation time for individual teeth in each animal are given in Table 13.8. These data were used to construct a provisional bar chart (Fig. 13.13).

Table 13.6. Counts of striae per mm along the enamel surface from cervical to occlusal levels in Theropithecus gelada

HKU 0243	Female	Cervical										Occlusal	SS + BI	Total	Years
Striae	R I₁	*	10	11	10	10	8	8	5	6	5	5 7	85+20	105+	2.0
Striae	R I₂	11	13	13	11	8	7	8	9	9			100+20	120	2.3
Striae	R C̄	8	11	14	10	11	10	9	8	8	7	8 3 #	128+30	158+	3.0
Striae	R M₁	Distolingual 10†	10†	6†	6†	Attrition							32+28	60+	1.2
Striae	R M₁	Distobuccal 10	15	11	9	8	6	2	Attrition				61+20†	81	1.6
Striae	R M₃	Mesiolingual 16	12	9	5	5	5	6	5				63+20†	83	1.6
Striae	R M₃	Mesiobuccal 13	12	12	12	10	8	†	†				67+20	87	1.7
Striae	R M₃	Distolingual 10	10	7	5	5	4	3					44+20†	64	1.2
Striae	R M₃	Distobuccal 10	14	14	11	5	5	5	4	4			77+20†	97	1.9

SS = total numbers of surface striae. BI = buried increments. SS = total numbers of surface striae. BI = buried increments in occlusal appositional enamel (buried increments include estimates in teeth showing slight occlusal wear). Total striae counts are given in the penultimate column and converted into age in years by multiplying by the seven-day cross-striation periodicity. * = fractured cervical enamel. † = estimated perikymata counts. Estimates are given for the labial or buccal surfaces of anterior teeth. In molars values are given for each of the four principal cusps on buccal and lingual surfaces. The greatest value is generally found on the distobuccal cusp and is used as the maximum estimate of crown formation time. Total estimates of crown formation time for individual teeth in each animal are given in Table 13.8. These data were used to construct a provisional bar chart (Fig. 13.13).

Table 13.7. *Counts of striae per mm along the enamel surface from cervical to occlusal levels in* Theropithecus gelada

HKU 0232			Cervical													Occlusal	SS	BI	Total	Years
Striae	R	I_1	8	11	11	11	11	11	11	8	8						90+20		110	2.1
Striae	R	I_2	7†	12	13·	13	11	10	8	7	5						86+20		116	2.2
Striae	R	Č	7+	11	12	13	12	11	12	12	11	9	7	5+			122+30		152+	2.9
Striae	R	P_3	†	8	14	16	15	16	14	14	15	13	14	12			151+20		177+	3.4
	R	M_1	Mesiolingual																	
Striae			11	9	5	4	Attrition										29+28†		57	1.1
	R	M_1	Mesiobuccal																	
Striae			13	14	11	9	4	Attrition									51+20†		71	1.4
	R	M_1	Distolingual																	
Striae			10	9	7	6	Attrition										32+28†		60	1.2
	R	M_1	Distobuccal																	
Striae			12	10	7	6	Attrition										35+30		65	1.3
	R	M_2	Mesiolingual																	
Striae			9	12	8	5	5	6									45+30		75	1.4
	R	M_2	Mesiobuccal																	
Striae			16	14	12	9	7	6	3								67+32		99	1.9
	R	M_2	Distolingual																	
Striae			9	8	7	6	5	2									37+30		67	1.3
	R	M_2	Distobuccal																	
Striae			17	18	†	†	†	†	†											
	R	M_3	Mesiolingual																	
Striae			9	10	8	7	6	8	7	1							56+25		81	1.6
	R	M_3	Mesiobuccal																	
Striae			15	17	16	9	8	7	5	5	5	3					90+25		115	2.2

SS = total numbers of surface striae. BI = buried increments in occlusal appositional enamel (buried increments include estimates in teeth showing slight occlusal wear). Total striae counts are given in the penultimate column and converted into age in years by multiplying by the seven-day cross-striation periodicity. * = fractured cervical enamel. † = estimated perikymata counts. Estimates are given for the labial or buccal surfaces of anterior teeth. In molars values are given for each of the four principal cusps on buccal and lingual surfaces. The greatest value is generally found on the distobuccal cusp and is used as the maximum estimate of crown formation time. Total estimates of crown formation time for individual teeth in each animal are given in Table 13.8. These data were used to construct a provisional bar chart (Fig. 13.13).

areas of enamel. The second indirect method was to measure an average size of cross-striations in a particular area and divide that value into the spacing between striae measured along the prism axis. Planimetric measurements were made on enamel thickness in sections of molar teeth, after the method of Martin (1985).

Results

Odontogenesis

The youngest gelada specimen is a newborn male (Fig. 13.1). The crown of the deciduous central incisor is formed but there is, as yet, no root formation. The incisal border is slightly higher mesially

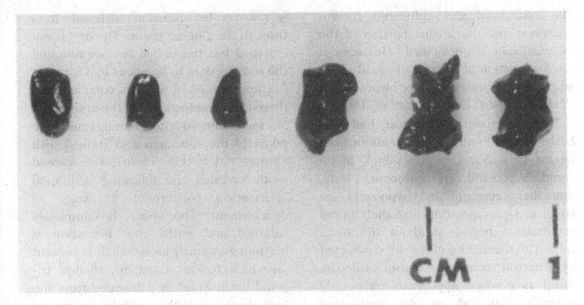

Fig. 13.1. The stage of dental development of the deciduous teeth and M_1 at birth in *Theropithecus gelada*. The teeth are stained with alizarin red S.

than its sloping distal margin. The lingual surface is slightly concave and the mesial and distal marginal ridges converge cervically to form a slight lingual tubercle. The buccal surface is convex mesiodistally.

The deciduous lateral incisor is at the same stage of development as the central incisor. Neither central nor lateral incisor show any evidence of mammelon formation.

The deciduous canine crown is present and, as in the incisors, there is no root formation.

The dm_1 crown is complete and the bilophodont cuspal configuration is apparent at this stage of morphodifferentiation. The four cusps are high, columnar projections with wide developmental grooves separating them. There is a mesiobuccal projection extending forward from the protoconid which terminates as a low,

rounded elevation, or cusplet. There is a shallow, narrow basin in the centre of the extension. The distal marginal ridge protrudes slightly distally as a rounded, narrow ledge and is separated from the buccal surface by a narrow distobuccal groove while the lingual side blends smoothly onto the lingual surface of the tooth.

The dm_2 crown displays the following morphodifferentiation. The two mesial cusps are connected by a transverse enamel lophid (metalophid) that is wider and thicker than the distal lophid (hypolophid) uniting the distal two cusps. This difference in size indicates the earlier coalescence of the anterior moiety. The bilophodont pattern of the mature tooth is obvious at this stage of development. In addition, the protoconid and hypoconid are united by the coalescence of their mesial and distal marginal cusp ridges.

The metaconid and entoconid remain separated and the lingual portion of the talonid basin is uncalcified. This area is the last portion of the crown to calcify in all primate species that have been studied (Swindler, 1961; Kraus & Jordan, 1965).

The permanent first molar has four cusps calcified and the size gradient from largest to smallest is protoconid, meta-conid, hypoconid, and entoconid. Note, that the metaconid and hypoconid are equal in size suggesting that their initial calcification begins at about the same time. The mesial two cusps are connected by a narrow, transverse lophid while the distal two cusps appear to have only recently coalesced, as the connecting lophid is thin and narrow. There are no buccal or lingual marginal connections between the mesial and distal portions of the crown and the talonid basin is uncalci-fied. There is a delicate extension of enamel from the protoconid that will ultimately become the mesial shelf of the mature tooth. Also, two short enamel extensions from the hypoconid and ento-conid protrude distally anticipating the future distal accessory cusp. The four cusps are high, columnar projections with wide developmental grooves between them. The median buccal cleft, a con-sistent feature of gelada lower molars, is present.

HKU 0228 (Figs 13.2a and b). This is an infant female with only its deciduous teeth erupted. The lower incisors have their roots formed and appear to be about full length; however, as yet, there is no apical closure. The canine root is present and open. The mesial half of dm_1 has emerged into the oral cavity while the distal portion is not yet in occlusion, although it is through the gingival tissue. The dm_2 is not emerged but the crown is complete and the root appears to be about half formed.

The crown of M_1 appears complete and there is no emergence into the oral cavity. M_1 was removed after radiographs were taken of the specimen and stained with alizarin red S (Fig. 13.2b). The stained tooth revealed the following additional information concerning its stage of development. The crown is completely calcified and initial root formation is beginning. A small mesial shelf is present on which resides a narrow, shallow tri-gonid basin which is delineated from the buccal surface of the tooth by the mesial buccal cleft. The median buccal cleft separates the protoconid from the hypo-conid. A well developed distal accessory cuspule (hypoconulid) is present project-ing from the distal surface of the tooth (Fig. 13.2b). The distal buccal cleft passes vertically between the cuspule and the hypoconid.

HKU 0248 (Figs 13.3a and b). This is a female with a complete deciduous denti-tion in occlusion. The roots appear closed (where visible on the radiographs) thus indicating a mature deciduous dentition. The crowns of the permanent lower incisors are present without roots. A small permanent canine crown lies slightly distal and below the incisors. A crypt is present for P_3 but not for P_4. M_1 is present at a crown complete stage in the radiograph and appears about to emerge into the oral cavity. When removed and stained, M_1 revealed a completely calcified crown, morphologically identical to the M_1 of HKU 0228, except that it had a root 3 mm long

(a)

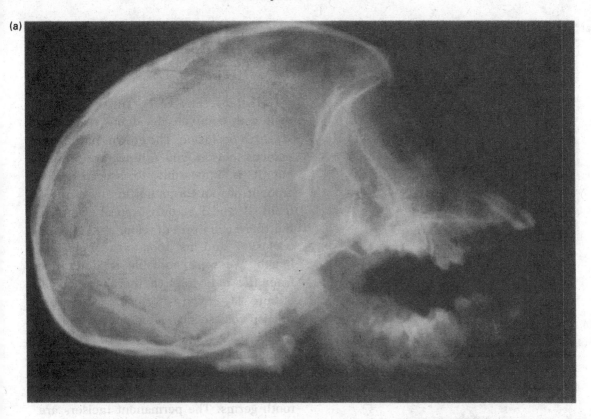

(b)

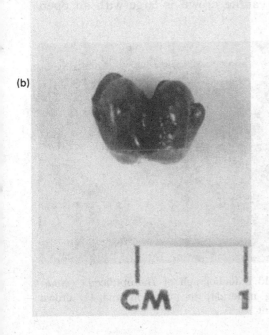

Fig. 13.2. (a) Radiograph of *Theropithecus gelada* infant female. Deciduous teeth erupted and M_1 crown complete. (b) Alizarin red S stained M_1 tooth bud of specimen in Fig. 13.2 (a) showing complete crown formation and beginning root formation.

(a)

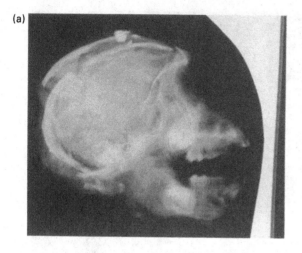

(b)

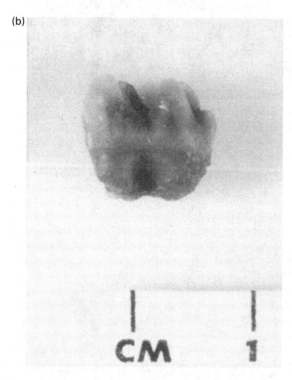

CM   1

Fig. 13.3. (a) Radiograph of *Theropithecus gelada* infant female. Complete deciduous dentition in occlusion. (b) Alizarin red S stained M_1 tooth bud of specimen in Fig. 13.3 (a). Crown complete with 3 mm root formed.

(Fig. 13.3b). Note, that the root was not visible in the radiograph (Fig. 13.3a). M_2 is not visible in the radiograph and could not be found by dissection.

HKU 0251 (Fig. 13.4). This is a young male with complete deciduous teeth plus M_1, all in occlusion. The crowns of the permanent incisors and canines are present but it is impossible to ascertain the amount of root formation in any of these teeth. P_3 and P_4 crowns appear complete but there is no root development. M_1 is in occlusion and its roots are about full length. We were not able to determine from the radiograph if there was apical closure. M_2 crown is formed with respect to cusp coalescence. M_3 is not present.

HKU 0233 (Figs 13.5a and b). This is a young male with a complete deciduous dentition. The roots of the deciduous teeth are partially resorbed by the permanent tooth germs. The permanent incisors are erupted and the roots are nearly complete. The canine crown is large with an open

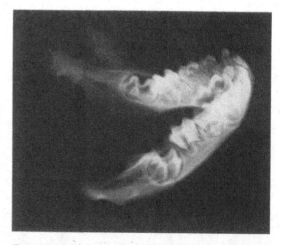

Fig. 13.4. Radiograph of *Theropithecus gelada* young male. M_1 in occlusion and M_2 crown present.

(a)

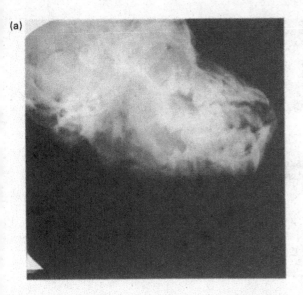

(b)

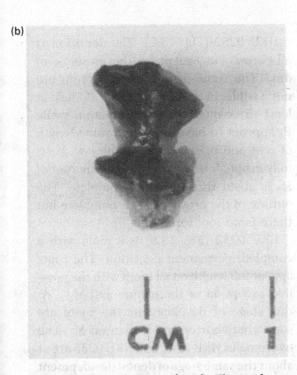

CM 1

Fig. 13.5. (a) Radiograph of *Theropithecus gelada* young male. All permanent teeth in various stages of calcification. (b) Alizarin red S stained M_3 showing development of hypoconulid. See text for discussion.

root and it appears to have moved orally. P_3 and P_4 possess nearly complete crowns and each show some degree of root formation. P_3 is slightly more orally in position than P_4. M_1 is fully emerged and in occlusion and the roots are closed. The M_2 crown is completely calcified. There appears to be some root development and it is about ready to emerge into the oral cavity (note, the roots were measured after removal and are 6 mm long, i.e. about one half their mature length). M_3 is present as a crown consisting of four cusps connected by narrow marginal ridges; the hypoconulid is not visible in the radiograph. M_3 was removed and stained (Fig. 13.5b). The mesial two cusps are connected transversely by the metalophid while the distal pair are joined by the hypolophid. The buccal marginal ridge connects the protoconid and hypoconid, although the lingual marginal ridge is uncalcified as is the lingual part of the talonid basin. The hypoconulid is uncalcified but calcified distal marginal ridges are extending distally from the hypoconid and entoconid. The question as to whether the M_3 hypoconulid is a separate calcification centre or not, unfortunately, cannot be answered from the present data. This region of the tooth will be discussed later.

HKU 0237 (Fig. 13.6). This is a male with mixed dentition. The permanent incisors are fully erupted and their roots appear to be full length. The permanent canines are about to emerge into the oral cavity and the lower premolars have full crown development. Unfortunately, it is impossible to see the amount of root development from the radiograph. M_1 and M_2 are fully emerged. The roots are closed

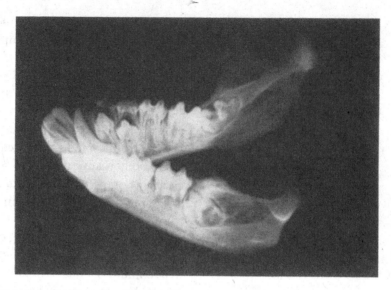

Fig. 13.6. Radiograph of *Theropithecus gelada* male with mixed dentition. M_3 in crypt.

on M_1 but only about three- quarters developed on M_2. The crown of M_3 is complete
but we cannot determine the degree of
root formation from the radiograph.

In the male specimen HKU 0242, dental
development is about the same as that in
HKU 0237 discussed above. The permanent incisors are present and their roots
appear closed. The deciduous canines are
still present in the upper jaw while in the
mandible, the left permanent canine is
present. The permanent canines possess
open roots. Both deciduous molars are
present in the mandible. The crowns of
P_{3-4} are complete with well developed
roots and lie just below dm_{1-2}. P_3 is present
in the upper jaw while the dm_2 is still in
place. M_1 is fully emerged and its roots are
closed. M_2 is present in the oral cavity and
its roots are long but still open. The crown
of M_3 is completely calcified but there
appears to be no root formation at this
time.

HKU 0252 (Fig. 13.7). The dentition is
all permanent except for the presence of
dm_2. The roots of the anterior teeth are
not visible in the radiograph. P_3 has at
least three-quarters root formation while
P_4 appears to have about the same length
of root and it is still within the jaw. M_1 is
fully erupted, its root is closed. The root of
M_2 is about three-quarters complete. The
outline of the crown of M_3 is complete but
there is no root formation.

HKU 0232 (Fig. 13.8) is a male with a
complete permanent dentition. The roots
appear full length on all teeth with the possible exception of the canine and M_{2-3}. At
this stage of development the roots are
somewhat shorter on M_3 than on M_2. The
two females HKU 0243 and HKU 0235 are at
about the same stage of dental development
as HKU 0232 except that in HKU 0243 the
M_3 has not erupted into the oral cavity. The
crown of M_3 is fully formed but there is little
if any root visible on the radiograph.

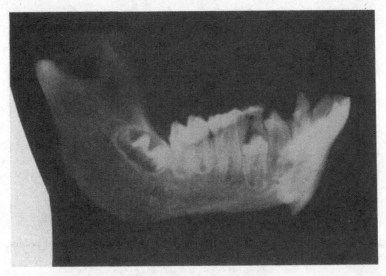

Fig. 13.7. Radiograph of *Theropithecus gelada* female. P_4 not erupted and M_3 crown present, no root formation.

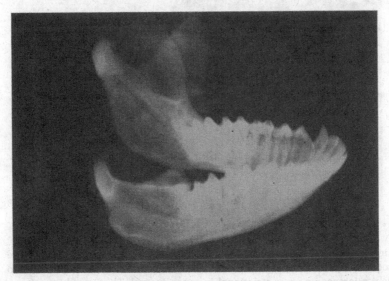

Fig. 13.8. Radiograph of *Theropithecus gelada* male. All permanent teeth erupted, M_3 just beginning root formation.

Histology

The four specimens used in this part of the study show regular prominent striae in ground sections (Figs 13.9–13.12). There is little evidence of enamel hypoplasia or developmental disturbance during tooth formation. Indeed, the four animals show an unusual uniformity of striae with an absence of accentuated incremental

Fig. 13.9. Ground section of I_2 [HKU 0237] viewed under crossed polars. Regular striae reach the surface giving rise to surface perikymata.

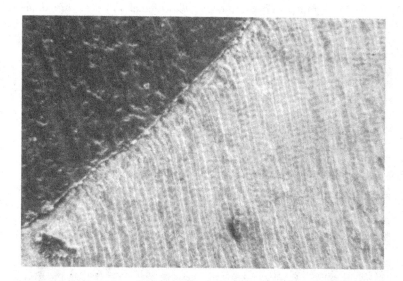

Fig. 13.10. SEM micrograph of P_3 [HKU 0237]. Outer enamel near to the occlusal tip showing regular varicosities and constrictions [equivalent to light microscopic cross-striations] in outer enamel. There are hints of surface striae and perikymata. Scale bar = 40 μm.

growth lines. Cross-striation spacing intervals are relatively large, ranging between 4 and 6 μm, with smaller values dominating in cervical levels and larger values in occlusal levels. In the SEM preparation of P_3 (HKU 0237), prism varicosities are prominent after light etching. In this specimen, average varicosity spacing is approximately 6 in outer enamel (Figs 13.10 and 13.11), which is consistent with optical microscopy measurements of cross-striation values at the same site. In

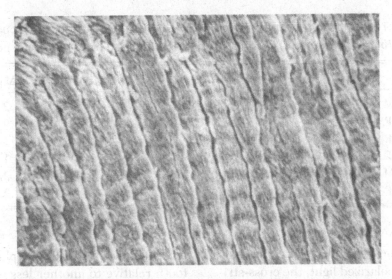

Fig. 13.11. Detail of above specimen showing varicosities and constrictions [cross-striations] ranging around 6 μm. There is loss of prism structure in the outer 15 μm of aprismatic surface enamel (top left). Scale bar = 10 μm.

Fig. 13.12. Photomicrograph of section of I₂ [HKU 0243], ×40 objective viewed with crossed polars. Enamel etched for 20 seconds in 0.5% orthophosphoric acid. Objective focussed on section surface showing prism boundaries with parazones and diazones. Regular striae run obliquely from ADJ into enamel (solid white arrows). Open arrow indicates parazone with prominent cross-striations. The labelled marker on the upper left part of the figure shows the same cross-striation spacings transposed over a prism along its axis between two prominent striae. Seven cross-striation intervals separate two adjacent striae crossed polars. Scale bar = 20 μm.

Table 13.8. *Crown formation times in years based on buccal surface striae counts (x7), and estimates on buried increment formation times in* Theropithecus gelada

Specimen		I_1	I_2	C	P_3	P_4	M_1	M_2	M_3
HKU 0237	M	2.5	2.5	5.2	5.0	—	1.5	2.0	—
HKU 0252	F	2.5	—	> 2.7	3.0	1.6	—	2.0	—
HKU 0243	F	> 2.0	> 2.3	> 3.0	—	—	1.6	—	1.9
HKU 0232	F	> 2.1	> 2.2	> 2.9	3.4	—	> 1.4	> 1.9	2.2
$\bar{X}$	M	2.5	2.5	5.2	5.0	—	1.5	2.0	—
	F	2.5	2.3	≥ 3.0	3.4	1.6	1.6	2.0	2.2

ground sections from all four animals viewed using polarized light, the cross-striation count between striae is seven, using both direct counts and by striae spacing measurements (Fig. 13.12). This seven-day periodicity is used with counts on striae (and perikymata) to make estimates of crown formation times. Tooth surface striae (and perikymata) counts per mm are presented in Tables 13.4–13.7. These values are multiplied by seven to obtain an estimate of imbricational enamel formation times. Estimates of buried increments in appositional enamel over the cusps are obtained either directly from measurements of enamel thickness, or from reconstructions of enamel outlines in worn teeth. This enamel thickness measurement is divided by a median cross-striation value of 5 μm to obtain an estimate of appositional enamel formation times, which are summed with the imbricational values to give a total crown formation time. Values are expressed in years, rounded to the nearest 0.1, in Table 13.8. These data are used to construct a provisional bar chart of dental chronology (Fig. 13.13). The absence of accentuated

incremental lines made the placing of one tooth relative to another less certain. The overall relationships are based on estimates of crown formation times combined with the pattern of dental development observed in the series of juvenile specimens discussed below (Fig. 13.14).

Planimetric measurements on enamel thickness (Martin, 1985) in unworn molar teeth were made using sections in the 'ideal' plane through the cusp and dentine horn tips. Sections in other planes were rejected. The area of the enamel cap (c), the length of the enamel-dentine junction (e), and the combined area of dentine and pulp (b) were measured using a digitizing tablet. Average enamel thickness (c/e) was calculated and values are presented in Table 13.9.

Discussion

Since the gelada specimens are wild captured animals their ages are unknown and the data cannot be analyzed by age categories. The data can, however, be compared with odontogenetic data from other primates in order to assess the similarities

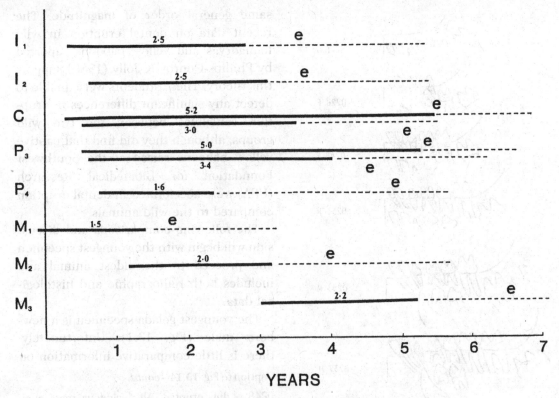

Fig. 13.13. Bar chart showing chronology of mandibular tooth development in *Theropithecus gelada*. Solid line (with number above) represents crown formation time (in years), the dotted line indicates approximate root formation period. Eruption times (e) are estimates. Crown formation times are based on average values for each tooth. Placing of the teeth relative to one another is based on sequences of dental development. Two values are given for Č and P_3, corresponding to male (upper) and female (lower).

and differences in sequence and pattern of tooth formation that may occur between the gelada baboon and other primates, particularly other Old World monkeys. When possible, radiographic comparisons of the gelada material are made with radiographs of comparable dental formation (pattern) in *Papio hamadryas cynocephalus*. Since these taxa are closely related members of the Papionini, it seems likely that the relative sequence and perhaps timing of odontogenetic events are fairly similar (Meikle & Day, 1986; Ohsawa, 1979). Indeed, in a previous paper (Swindler, 1991), it was shown that there were few significant differences between the stages of molar formation and chronological age in *P. hamadryas cynocephalus* and *Macaca nemestrina*. We suspect that when tooth formation data are available for more comprehensive comparisons among the species of Papionini, the major events in sequence and timing will be rather similar, that is, of the

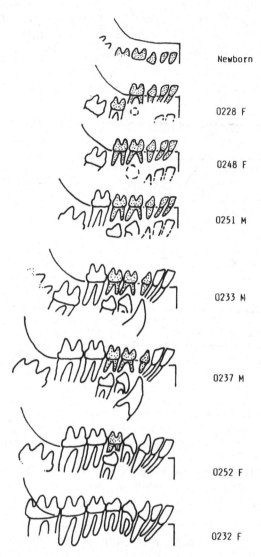

Newborn

0228 F

0248 F

0251 M

0233 M

0237 M

0252 F

0232 F

Fig. 13.14. Summary chart showing stages of dental development in mandibles of *Theropithecus gelada*. Deciduous teeth are stippled. Incomplete teeth are indicated by incomplete outlines. Stages of incomplete crown and root formation are proportional representations. Teeth with uncertain stages of development are indicated in interrupted outline.

Newborn – Unerupted deciduous crowns complete except dm_2. Cusps M_1 calcified.

0228 – Deciduous teeth erupted except dm_2, all with incomplete roots. M_1 crown complete. Incisor crowns are probably forming.

same general order of magnitude. The recent data on dental eruption in wild hamadryas and yellow baboons reported by Phillips-Conroy & Jolly (1988) support this theory. These students were unable to detect any significant differences in eruption schedules between the two wild groups, although they did find that captive yellow baboons raised at the Southwest Foundation for Biomedical Research (USA) were accelerated in dental eruption compared to the wild animals.

The following comparisons and discussion will begin with the youngest specimen and proceed to the oldest animal and includes both radiographic and histological data.

The youngest gelada specimen is a newborn male (Fig. 13.1). Unfortunately, there is little comparative information on

Caption to Fig. 13.14 – *contd.*

0248 – dm_2 erupted. All deciduous roots complete. M_1 crown complete, with 3.0 mm of root. No evidence M_2.

0251 – M_1 erupted, roots near complete. Unerupted premolars are advanced (P_4 crown complete, elongated P_3 crown incomplete). M_2 incomplete with fused cusps.

0233 – M_1 roots are just complete. Premolars are slightly further advanced, although P_3 crown still incomplete. Large $\check{C}$ crown probably incomplete. I_1 and I_2 erupted with near complete roots. Unerupted M_2 crown complete with some root. M_3 cusps coalesced.

0237 – M_2 erupted with incomplete roots. $\check{C}$ crown complete, start of root formation. Premolars unerupted but P_3 and P_4 crowns complete with root formation. M_3 crown complete, no root.

0252 – $\check{C}$ and P_3 erupted with incomplete roots. Incisors and canines near complete roots. M_3 crown near complete.

0232 – All permanent teeth erupted with complete roots, except M_3 with only mesial cusps exposed in the mouth.

Table 13.9. *Enamel thickness data for* Theropithecus gelada

Specimen (HKU)	Tooth	Section	Area of dentine cap	Average enamel thickness
0237	M_2	M	64.01	1.09
		D	58.15	1.09
0252	M_2	M	54.83	1.11
0243	M_3	M	54.54	1.26
		D	49.10	1.19
0232	M_3	M	46.69	1.36
		D	44.64	1.03

the status of deciduous tooth development at birth in non-human primates (Anthropoidea). One exception is the recent paper by Swindler & Emel (1990), reporting on dental and skeletal development at birth in *M. nemestrina*. They found both central and lateral incisor crowns completely formed with initial root formation. The canine crown was complete but there was no root. The first and second deciduous molar crowns were completely calcified but neither tooth showed any evidence of root formation. The only differences in dental development between *Macaca* and *Theropithecus* at birth are the absence of incisor root formation and the slightly less mature dm_2 of gelada (Fig. 13.1). The teeth of several newborn non-human primates have been removed and stained for comparisons with gelada. The degree of dental maturation at birth is similar among all of these primates in that the crowns of all deciduous teeth are calcified as well as the beginning of calcification of the permanent first molar. There are, however, quantitative dif-

ferences in the amount of crown calcified in individual teeth and the extent of root formation in the incisors. For example, root formation has commenced in both incisors in *Alouatta* and *Macaca* but not in the three hominoids. Also, dm_2 is less calcified in the apes than in the monkeys, particularly in the gibbon. In the monkeys, M_1 has all four cusps calcified and the mesial two are connected by a transverse lophid, whereas the distal cusps are still separate. In hominoids, the mature M_1 usually consists of five cusps of which four or five are calcified at birth (see below).

The degree of deciduous tooth formation at birth in humans has been described by Schour & Massler (1941) and Kraus & Jordan (1965). In both studies, the human deciduous teeth were less mature than those of *Macaca* discussed here but were similar in maturity to those of the apes. We need more information on the development of the deciduous teeth at birth in all primates; however, it appears from the evidence presented here that anthropoids have rather similar developmental status of their deciduous teeth at birth and that macaques are slightly more precocious than other taxa. Indeed, macaques are also more mature in skeletal development at birth than other primates (Newell-Morris *et al.*, 1980; Swindler & Emel, 1990).

The first permanent molar is calcified at birth in higher primates (Swindler, 1961; Kraus & Jordan, 1965; Tarrant & Swindler, 1973; Swindler & Emel, 1990). The M_1 of gelada is calcified at birth with four cusps recently connected by transverse lophids (Fig. 13.1). The largest cusp is the protoconid followed by the metaconid, hypoconid, and entoconid and, although not direct evidence of the order of calcification, this is the usual sequence in primates (Swindler & McCoy, 1965). The maturity of the gelada M_1 is similar to that of *Macaca*, *Papio*, and *Alouatta* at birth, whereas it lacks any evidence of cusp coalesence in hominoids (Fig. 13.1). In humans, Kraus & Jordan (1965) reported a mean age of 32 weeks gestation for initial calcification of M_1 and they never found a, 'completely uncalcified first permanent molar' (p. 117) in a full-term foetus. Incidently, they also reported finding anywhere from one to five calcified cusps but never any coalesence of cusps during the circumnatal period.

HKU 0228 (Fig. 13.14), has complete deciduous eruption except for dm_2. When its radiograph was compared with RPRC baboons for pattern of development, it was found to be most comparable to females between the ages of 0.43 and 0.66 years. The length of the incisor roots is almost identical and the degree of dm_2 eruption is similar. The roots of dm_1 are the same length and both are still open at their ter-

minal ends. The M_1 crown is formed in both animals, an event which occurs between 0.55 and 1.10 years in the RPRC animals (Swindler, 1991). There is just a hint of initial root formation in the gelada radiograph which was substantiated when the tooth was removed for further study. Initial root formation takes place between 0.89 and 1.30 years in the RPRC female baboons. There is no indication of M_2 development in either specimen. M_2 appears between 0.89 and 1.30 in the RPRC sample (Swindler, 1991).

The next specimen, HKU 0248 (Fig. 13.14), is developmently similar to the male baboons that are just over one-year-old. The actual age of the colony baboon is 1.11 years. Both animals have a complete deciduous dentition in occlusion. The crowns of the permanent incisors and canines are formed. The incisors show a small amount of root formation while the canines lack any root formation. In the lower jaw, the crypts for P_3 are present in both radiographs but there is no indication of P_4 development. M_1 is present in both jaws and the crown appears to be complete with the beginning of root formation. The tooth was removed for further study and the root was found to be 3 mm long.

HKU 0251 (Fig. 13.14), with its complete deciduous dentition plus M_1 in occlusion, matches a 2.24-year-old male baboon. The crowns of all permanent teeth are present except M_3 which appears between 2.50 and 3.43 years in RPRC male baboons (Swindler, 1991). It is impossible to discern the amount of root development of the incisors and canines from the radiograph of gelada. The P_4 crown appears complete while P_3 is still incomplete in both animals with no root formation in either. M_1 is fully erupted and its roots are long and still open in *Papio* as well as in *Theropithecus*. The M_2 crown is formed and there is coalesence between the cusps and around the occlusal margins of the tooth. In the RPRC males, M_2 crown completion can be expected between 2.01 and 2.63 years (Swindler, 1991). Histologically, the specimen appears to be about 2.4 years old.

HKU 0233 (Fig. 13.14), is about three-years-old from the appearance of M_3, since in male *P. hamadryas cynocephalus*, M_3 commences calcification between 2.59 and 3.43 years (Swindler, 1991). The radiographs also reveal similar developmental patterns regarding the maturation of the canine, P_{3-4}, M_1 and M_2. These teeth all have complete crown formation and are in various stages of root formation. There is complete crown formation of P_{3-4} and there is root extension in both teeth; P_3 is in a slightly more oral position than P_4 and will usually erupt before P_4. The M_1s are in full occlusal contact and the roots are terminally converged and closed, a condition which occurs between 2.3 and 2.9 years of age in the colony reared baboons. M_2 crowns are complete with root formation. As noted above, the crowns of M_3 are present in both animals but they appear somewhat more mature in the radiographs of gelada. A detailed description of the gelada M_3 was presented in the previous section. The permanent incisors are erupted in both species and their roots are nearly complete. The specimen is between 3.0 and 3.5-years-old.

When HKU 0252 (Fig. 13.14), is compared with the colony reared baboons, an age of 3.8 to 4.4 years is the best approximation. The major points of similarity are the following: the roots of the permanent molars show the same stage of development; P_4 is unerupted; M_2 is erupted and the crown of M_3 is complete with no root formation. HKU 0237 is a male at about the same stage of dental maturation as HKU 0252, the major differences being the unerupted permanent canine and the presence of dm_{1-2}. In fact, the most comparable baboon radiograph is of a 3.7-year-old male. HKU 0242 is at the same developmental level except it has just erupted the left lower canine. The right deciduous canine is still present.

The remaining gelada specimen, HKU 0232 (Fig. 13.14), 0235 and 0243, are all young adults according to their dental eruption. HKU 0232 and HKU 0235 have erupted M_3 but not M^3, while all M_3s are unerupted in HKU 0243. Also, in the former two animals, the roots of M_2 are converged with apical closure while, in the latter animal, the roots are not converged and are open. In our laboratory baboons, M_2 root closure takes place between 4.0 and 5.1 years (Swindler, 1991). Since the lower M_3s have just erupted in two of these animals they can also be used for an approximate age estimate of the specimens. According to Phillips-Conroy & Jolly (1988), lower M_3s erupt in wild male yellow baboons at about 6.4 years while in captive pigtailed macaques this tooth erupts at 5.7 years (Sirianni & Swindler, 1985). Indeed, lower M_3 appears to erupt between 5.5 and 6.4 years in all male Papionini studied: for example, *M. mulatta* 5.5 years; *M. nemestrina* 5.7 years; *M. fascicularis* 6.2 years and

P. hamadryas cynocephalus 6.4 years (Swindler, 1985). Thus, there is less variability in the range of M_3 eruption among these species than there is in the single species of *H. sapiens*.

Dental eruption has been used for many years for estimating the age of primates and for making evaluations of their stage of maturation. These data can be collected either by repeated examinations of the teeth of living primates or by making observations on the teeth of a large series of dead specimens of differing ages. To date, neither of these methods have been applied to *T. gelada*, although there is one study of dental eruption in *T. brumpti* and *T. oswaldi* from the lower Omo Valley of Ethiopia by Meikle & Day (1986). They proposed the following eruption sequence for the permanent teeth of these fossils based on 60 specimens: M1, [1, I2], M2, [P3, P4, C], M3. The present sample, though small, is still useful for looking at the sequence of eruption in *T. gelada*. The order of eruption is, M1, I1, I2, M2, [P3, P4, C], M3 the same as that found by Meikle & Day (1986) for *T. brumpti* and *T. oswaldi*. The teeth in brackets indicate that they may erupt differently from the order presented, suggesting a sequence polymorphism in the premolar, canine region in both groups. For example, in HKU 0252, P_3 and the canine are erupted while P_4 is unerupted. Also, in HKU 0242, the canine is erupted before the premolars. Unfortunately, the sample is too small to ascertain if there is a sex difference in the timing or sequence of dental eruption in *T. gelada*.

The bar chart (Fig. 13.13), shows the chronology of mandibular tooth development using the histological techniques discussed earlier. The following eruption times in years are estimates based on these methods: M1 = 1.7; I1, I2 = 3.0–3.2; M2 = 3.5–4.0; P3, P4, C = 4.2–5.3; M3 = 6.0–6.5. It should be noted that the estimates based on histology in the developmental series are broadly similar to the radiographic ages in the younger age range, but they tend to become greater in the older animals. This may reflect differences in radiological and histological methods of estimating dental age. It is possible, however, that the period of tooth development in *T. gelada* is prolonged. *Theropithecus* is approximately twice as heavy as *M. nemestrina* and is half as large again as *P. cynocephalus* judged on female body weights. Teeth in *Theropithecus* are absolutely larger than those of *M. nemestrina* and crown height is approximately twice as high. It is crown length which finally determines the overall period of crown formation, and crowns are hypsodont in *T. gelada*. The most extreme case is the male canine, which takes 5.2 years to form, only slightly longer than P_3 at 5.0 years. Martin (chapter 10) has reported a relatively high sexual dimorphism ratio in *T. gelada*. The canine in the male could begin to erupt before root formation commences but this tooth cannot function at the level of the occlusal plane in the absence of root attachment tissues. The sectorial P_3 in the male could erupt and function in occlusion before the buccal extension is complete since the lingual root is initiated after half of the buccal crown height is formed. These observations impose a lower limit of around 5.0 to 5.5 years on the eruption of the lower canine and P_3 in males.

It would appear that there has been no major change in the sequence of eruption from the fossil forms to the modern gelada baboon. In fact, according to Schultz (1935), this is the usual order of tooth eruption in the majority of Old World monkeys and pongids if we allow some shifting of the premolar order. For example, Phillips-Conroy & Jolly (1988) reported the lower P_4 erupting slightly earlier than P_3 in males than females in their sample of Tanzanian yellow baboons. A similar finding was reported for *M. nemestrina* (Sirianni & Swindler, 1985). It is interesting to note that the upper P^3 came in earlier than P^4 in both of these species. There is obviously a great deal of heterochrony during the development of these teeth and their eruption into the oral cavity. If we include the human sequence of tooth eruption there is a further reordering of events in that M2 now appears after the premolars and canines resulting in an eruption pattern that is more specialized than that of any other primate, M1, I1, I2, [C, P3, P4, M2], M3 (Schultz, 1935). Although this is the usual pattern of tooth eruption in hominids, Garn & Lewis (1957) and Garn (1963) have indicated the taxonomic importance of such sequence polymorphisms, particularly the [P4, M2] sequence, when studying fossil and recent hominids.

Several authors have reported the late eruption of M_3, or at least a tendency to delay full occlusal contact, relative to M_1 and to the anterior teeth in *T. gelada* (Jolly, 1972; Delson, 1975; Jablonski, 1979). The delayed eruption of M_3, relative to the other molars, is seen as an adaptation against heavy molar attrition due to a diet consisting mainly of grasses as it affords an unworn surface as the anterior molars are worn down; and as pointed out by Jolly (1972), there are parallels to other grass-eating mammals such as the horse and elephant. The possible late eruption of M_3 is just one of several factors contributing to the significantly prolonged availability of efficient masticating surfaces in *T. gelada* (Jablonski, 1981). As mentioned earlier, there is no information on the chronological age of tooth eruption or the time interval between molar eruption in *T. gelada*. The present radiographic data offer little support for or against since the information is relative to the stage of maturation of other teeth when M_3 erupts. There are two animals in the present sample with erupted M3s, HKU 0232 and 0235, in the former animal, only the lower M_3 has erupted and, quite recently, since the tooth has only the mesial portion erupted. In the latter animal, the lower M_3 is erupted as is the left upper M^3. In both monkeys, all teeth except M_2 and the canine in HKU 0232 have apical closure of their roots and the roots of M_3 appear about one half to three quarters grown. This is similar to the degree of dental maturation in *M. nemestrina* and *P. hamadryas cynocephalus* when M_3 erupts (Sirianni & Swindler, 1985; Swindler, 1991). Thus, the sequence of events as well as the general stage of overall dental maturation at the time of M_3 eruption is similar in these three groups of Papionini. Unfortunately, these data do not offer direct chronological information on the length of the interval between eruption of M1, M2 and M3, however the histological data do. If we look again at the bar chart (Fig. 13.13), the beginning of M_3 calcification appears to be at about three years, resulting in crown completion at approximately 5.2 years. A later eruption of M_3 estimated at 6.0 to 6.5 years following the emergence of C and P_3 into the oral cavity suggests a relative delay in M_3 emergence following crown completion compared with the M_1 and M_2, similar to, but of much shorter duration, the delay in the eruption of P_4 after crown completion.

There is a relatively steep wear gradient in the molars of *Theropithecus* compared to other cercopithecoids, including *Papio* (Jolly, 1970, 1972; Meikle, 1977). Teaford (chapter 12) has shown fine microwear pattern with relatively few pits indicating wear due to processing of grass rather than ingested grit particles. Despite the fine microwear pattern it is apparent that the diet is highly abrasive. Walker, Hoeck & Perez (1978), showed fine microwear patterns and rapid wear rates in grazing hyraxes, which was attributed to the ingestion of large quantities of phytoliths. Juvenile geladas exhibit a considerable gradient of microwear in the molars, which erupt at approximately 1.7, 3.8 and 6.5 years based on histological data. This is also apparent in the present sample, for example, as noted above in both HKU 0232 and 0235, M_3 is erupting or has just erupted and there is extreme wear on M_1 and moderate attrition on M_2. Such a steep molar wear gradient implies that, at any given time, the more mesial molars will show more attrition relative to the more distal ones, a position which supports Jolly

(1972) when he suggests that, 'Whether relative or absolute, the delayed eruption of M_3 would seem to represent an adaptation combating heavy molar wear, by replacing and crowding forward worn-out anterior molars, while retaining the mechanically advantageous short dental row (p. 113).

Several other authors (Clark, 1950, 1967; Jolly, 1970; Molnar, 1972; Simons, 1977; Smith, 1983) have reported that a marked wear gradient is characteristic of hominids. The finding of marked molar wear gradients in *T. gelada* illustrates the effects of ingesting an abrasive diet in animals lacking prolonged intervals between emergence in the molar series. Recent studies on dental development in early hominids suggests that they may not have had prolonged juvenile growth periods (Beynon & Dean, 1988; Smith, 1989). Marked molar wear gradients in early hominids may instead reflect a high rate of tooth wear, although microwear evidence for an abrasive diet is lacking (Walker, 1981), rather than long intervals between emergence in the molar series.

Molar teeth in the gelada baboon emerge into the mouth with high cusps which require some wear to produce an efficient self-sharpening grinding surface. In unworn teeth, the bilophodont mesial and distal cusp pairs are initially high and are partly separated by the deeply waisted buccal and lingual enamel folds. This arrangement, when worn, produces a series of incomplete transverse ridges with harder enamel alternating with softer dentine, resembling tooth form in herbivores (Jolly, 1972). Microwear patterns (see Teaford, chapter 12) suggest a predominately anteroposterior jaw movement which is consistent with this functional design.

Jolly (1970) reported that enamel in cheek teeth of *T. gelada* was thin compared with enamel in hominids. Kay (1981) presented linear measurements on enamel thickness on slightly worn teeth and he reported in *T. gelada* that enamel thickness was slightly less than expected (−4.8 per cent) relative to body size. Preliminary analyses of planimetric measurements on tooth sections in 31 anthropoid primate species (in preparation) suggests that enamel thickness in *T. gelada* is average for its body weight although enamel appears relatively thin over the cusp tip. Other dental dimensions, however, suggest that molar teeth are large relative to body size including the area of the dentine cap and the combined area of dentine and enamel. The unworn tooth height is also large relative to animal size.

The molar teeth are hyposodont with long buccal surfaces extending down into the root bifurcation, allowing the long anatomical crown to function until the root supporting system is compromised. The relatively late eruption of M_3 with its large hypoconulid and its pattern of gradual horizontal exposure of cusps from front to back is reminiscent of horizontal exposure in the elephant molar. The perikymata counts on tooth surfaces corresponded well with internal striae counts, showing that surface counts of perikymata can be used to estimate the formation period of imbricational enamel. The finding of a completely uniform rate of seven daily enamel increments between striae is noteworthy, being the most consistent 'near

weekly' rhythmicity that we have encountered in anthropoids. In modern humans, the number of striae range between six and ten, with a modal value of eight (Beynon & Reid, 1987). *Theropithecus* is unique in our experience in the absence of hypoplastic bands or accentuated striae. The absence of such lines made the relative placing of teeth in the bar chart provisional. The absence of accentuated growth lines during tooth development in wild shot animals suggests that they grew up in a stable environment, lacking environmental and dietary stresses or illness during the dental growth period (Rose, 1977).

In Old World monkeys, the hypoconulid, or 5th cusp, is present on the lower M_3 but only in some taxa (all members of the tribe Cercopithecini lack the hypoconulid on M_3). Some years ago the senior author published a paper on the primate hypoconulid with particular reference to its development in *T. gelada* (Swindler, 1983). This paper attempted to trace the phylogenetic and ontogenetic history of the hypoconulid in primates and came to the conclusion that the well-formed distal accessory cuspule on M_2 and sometimes on M_1 of *T. gelada* was a functional hypoconulid since it was similar in topographic position, morphology, occlusal relations, and ontogeny with the hypoconulid on M_3. It should be noted that Szalay & Delson (1979) stated that the ancestral cercopithecoid lower molar morphotype had five cusps and that the hypoconulid was probably slightly reduced on M_{1-2}, an observation in agreement with Tattersall, Delson & Couvering (1988). Thus, the primitive cercopithecoid morphotype for M_{1-3} included the hypoconulid which was subsequently reduced and finally lost on M_{1-2} in all cercopithecoids.

It was the belief at that time, and still is, that the distal accessory cuspules on M_{1-2} and the hypoconulid of M_3 in gelada are serially homologous and are structurally and functionally comparable to such structures in other species. Such corresponding features in different animals, as noted by Butler (1985) are given the same name (Gr. *homos*, same; *logos*, word). The use of homology in biology goes back to Owen's definition in 1843, 'Homologue: the same organ in different animals under every variety of form and function' and has been the source of much confusion and debate ever since (Boyden, 1943; Van Valen, 1982). In the present context, we follow Butler (1985) and use the term homology to imply morphologically comparable features whether or not their resemblance is due to common ancestry. Thus, similar animals under similar selective influences are likely to respond in the same way and the result of such processes (parallelisms) can produce species of cusps defined by topographical and functional relations which may not necessarily imply strict homology (Butler, 1978). It seems that the development of the hypoconulid in gelada is under the influence of a morphogenetic field which has its maximum effect on M_3 and decreases anteriorly from M_{3-1}.

There was then, and still is, little information available on the ontogenetic pattern of calcification of the lower permanent molars in Old World monkeys except for a few radiographic investigations of longitudinal tooth development and these do not reveal the details of M_3 hypoconulid formation (Sirianni & Swindler, 1985). In hominoids, the hypoconulid is generally present on all lower molars but displays a size reduction gradient from M_1 to M_3. The hypoconulid has a separate centre of calcification on the lower molars in hominoids (Siebert & Swindler, 1985). What little evidence there is for cercopithecids indicates that the hypoconulid appears as a separate calcification centre on the distal edge of the distal marginal ridge of M_3. Thus, in both cercopithecids and hominoids, the hypoconulid has a separate centre of calcification.

It was hoped that the Hong Kong material would help clarify the ontogeny of the hypoconulid with respect to it being a separate calcification centre or not on the lower molars of *T. gelada*. If there were independent centres of calcification for the accessory cuspules of M_{1-2} and the hypoconulid of M_3, then this would be strong ontogenetic support for all three structures being hypoconulids in *T. gelada* since this is the condition in hominoids. Unfortunately, the information from most of the specimens was either too old or too young. As discussed earlier, the newborn male has a developing M_1 (Fig. 13.1) which shows enamel ridges passing from the distal aspects of the hypoconid and entoconid which, when they coalesce, will form the distal accessory cuspule. There is no suggestion of a separate calcification centre at this time. The other specimen, HKU 0233, has M_3 in the beginning stages of calcification as discussed earlier in this paper (Fig. 13.5a and b). The talonid basin, distal fovea, and hypoconulid are uncalcified at this time. The distal marginal ridges extend distally from the hypoconid (this ridge was accidently broken and appears separate from the hypoconid, see Fig. 13.5b) and entoconid and are beginning to curve toward each other which when complete, will form the distal margin of the hypoconulid. Unless the calcification centre for the hypoconulid appears later, it would seem that the M_3 hypoconulid develops as enamel outgrowths from the distal part of the tooth and, therefore, does not have a separate calcification centre. Unfortunately, this is the only specimen at this critical stage of M_3 development. The distal accessory cuspules of M_{1-2} in the other specimens have already calcified. Thus, the question still remains as to the mode of formation of the hypoconulid on M_3 as well as the distal accessory cuspule on M_{1-2} in T. gelada. If it should prove that the M_3 hypoconulid does not have a separate calcification centre but develops as an extension from the distal margin of M_3 as it appears to do in HKU 0233, then this would imply parallel development of the hypoconulid among primate taxa. Such parallel evolution in dental structures is not uncommon in the history of mammalian radiations (Butler, 1978).

We believe the distal accessory cuspules on M_{1-2} are serially homologous with the M_3 hypoconulid, and that they function as such when the teeth are in occlusion, and further, that these structures as well as the enlargement of the mesial shelf on these same teeth are part of an adaptive complex that is, 'structurally and functionally unique among primates' (Jablonski, 1981, p. 157). Extending, even slightly, the mesial and distal length of the occlusal surface of lower molars would seem to help to maintain the contact of these teeth during interstitial wear and mesial drift as well as retaining the mechanically advantageous short dental row as noted by Jolly (1972).

Acknowledgements

We would like to thank Nina Jablonski for inviting us to participate in the Cambridge Theropithecus Conference. We thank Ian Bell for his care and skill in making the tooth sections and Donald Reid for measuring them. The senior author wishes to acknowledge the L.S.B. Leakey Foundation for a grant permitting him to go to Hong Kong to study the gelada specimens housed in the Department of Anatomy, University of Hong Kong.

References

BEYNON, A.D. (1987). Replication technique for studying microstructure in fossil enamel. *Scanning Microscopy*, 1, 663–9.

BEYNON, A.D. & DEAN, M.C. (1988). Distinct dental development patterns in early fossil hominids: a review. *Nature*, 335, 509–14.

BEYNON, A.D. & REID, D.J. (1987). Relationships between perikymata counts and crown formation times in the human permanent dentition. *Archives of Oral Biology*, 32, 773–80.

BOYDEN, A. (1943). Homology and analogy: a century after the definitions of 'homologue' and 'analogue' of Richard Owen. *The Quarterly Review of Biology*, 15, 228–41.

BUTLER, P.M. (1978). Molar cusp nomenclature and homology. In *Development, Function and Evolution of Teeth*, ed. P.M. Butler & J.D. Joysey, pp. 439–53. London: Academic Press.

BUTLER, P.M. (1985). Homologies of molar cusps and crests, and their bearing on assessments of rodent phylogeny. In *Evolutionary Relationships Among Rodents*, ed. W. Luckett & J.L. Hartenberger, pp. 381–401. London: Plenum Publishing Corporation.

CLARK, W.E. LE GROS (1950). Hominid characters of the australopithecine dentition. *Journal Royal Anthropological Institute*, 80, 37–54.

CLARK, W.E. LE GROS (1967). *Man-Apes or Ape-Men? The story of Discoveries in Africa*. New York: Holt, Rinehart and Winston.

DELSON, E. (1975). Evolutionary history of the Cercopithecidae. In *Approaches to Primate Paleontology. Contributions to Primatology*, 5, ed. F. Szalay, pp. 167–217. Basal: Karger.

DEMIRJIAN, A., GOLDSTEIN, H. & TANNER, J.M. (1973). A new system of dental age assessment. *Human Biology*, 45, 211–32.

ECK, G.G. (1977). *Morphometric Variability in the Dentitions and Teeth of* Theropithecus *and* Papio. Ph.D. Dissertation, University of California, Berkeley.

GARN, S.M. (1963). Phylogenetic and intra-specific variations in tooth sequence polymorphism. In *Dental Anthropology*, ed. D.R. Brothwell, pp. 53–73. New York: Macmillan Company.

GARN, S.M. & LEWIS, A.B. (1957). Relationship between the sequence of calcification and the sequence of eruption of the mandibular molar and premolar teeth. *Journal of Dental Research*, 36, 992–5.

JABLONSKI, N.G. (1979). Functional analysis of the masticatory apparatus of the gelada baboon. *American Journal of Physical Anthropology*, 50, 451.

JABLONSKI, N.G. (1981). *Functional Analysis of the Masticatory Apparatus of the Gelada Baboon*, Theropithecus gelada. Ph.D. Dissertation, University of Washington, Seattle.

JOLLY, C.J. (1970). The seed-eaters: a new model of hominid differentiation based on a baboon analogy. *Man*, 5, 5–26.

JOLLY, C.J. (1972). The classification and natural history of *Theropithecus* (*Simo-pithecus*) (Andrews, 1916), baboons of the African Plio-Pleistocene. *Bulletin of the British Museum (Natural History), Geology*, 22, 1–123.

KAY, R.F. (1981). The nut-crackers – a new theory of the adaptations of the Ramapithecinae. *American Journal of Physical Anthropology*, 55, 141–51.

KRAUS, B.S. & JORDAN, R.E. (1965). *The Human Dentition Before Birth*. Philadelphia: Lea and Febiger.

MARTIN, L. (1985). Significance of enamel thickness in hominoid evolution. *Nature*, 314, 260–3.

MEIKLE, W.E. (1977). Molar wear stages in *Theropithecus gelada*. In *The Kroeber Anthropological Society Papers*, No. 50, ed. N.T. Boaz & J.E. Cronin, 22–9. California: Berkeley.

MEIKLE, W.E. & DAY, M.B. (1986). Dental development in fossil *Theropithecus*. *American Journal of Physical Anthropology*, 69, 239.

MOLNAR, S. (1972). Tooth wear and culture: a survey of tooth functions among some prehistoric populations. *Current Anthropology*, 34, 175–90.

NEWELL-MORRIS, L., TARRANT, L.H., FAHRENBRUCH, C.E., BURBACHER, T.M. & SACKETT, G.P. (1980). Ossification in the hand and foot of the pigtail-macaque (*Macaca nemestrina*). *American Journal of Physical Anthropology*, 53, 423–39.

OHSAWA, H. (1979). The local gelada population and environment of the Gich area. In *Contributions to Primatology*, 16, ed. M. Kawai, pp. 3–45.

PHILLIPS-CONROY, J.E. & JOLLY, C.J. (1988). Dental eruption schedules of wild and captive baboons. *American Journal of Primatology*, 15, 17–29.

ROSE, J.C. (1977). Defective enamel histology of prehistoric teeth from Illinois. *American Journal of Physical Anthropology*, 46, 439–46.

SCHOUR, I. & MASSLER, M. (1941). The development of the human dentition. *Journal of American Dental Association*, 28, 1153–60.

SCHULTZ, A.H. (1935). Eruption and decay of the permanent teeth in primates. *American Journal of Physical Anthropology*, 19, 489–581.

SIEBERT, J.R. & SWINDLER, D.R. (1985). Ontogenetic changes in the chimpanzee dentition. *American Journal of Physical Anthropology*, 66, 228.

SIMONS, E.L. (1977). Ramapithecus. *Scientific American*, May, 28–35.

SIRIANNI, J.E. & SWINDLER, D.R. (1985). *Growth and Development of the Pigtailed Macaque*. Boca Raton: CRC Press, Inc.

SMITH, B.H. (1983). *Dental Attrition in Agriculturalists and Hunter–Gatherers*. Ph.D. Dissertation, University of Michigan.

SMITH, B.H. (1989). Growth and development and its significance for early hominid behavior. *Ossa*, 14, 63–96.

SWINDLER, D.R. (1961). Calcification of the permanent first mandibular molar in rhesus monkeys. *Science*, 134, 566–7.

SWINDLER, D.R. (1983). Variation and homology of the primate hypoconulid. *Folia primatologica*, 41, 112–23.

SWINDLER, D.R. (1985). Nonhuman primate dental development and its relationship to human dental development. In *Nonhuman Primate Models for Human Growth and Development*, ed. E. Watts, pp. 67–94. New York: Alan R. Liss, Inc.

SWINDLER, D.R. (1991). Tooth formation in the yellow baboon (*Papio hamadryas cynocephalus*). *American Journal of Human Biology*, 3, 371–80.

SWINDLER, D.R. & EMEL, L.M. (1990). Dental development, skeletal maturation, and bodyweight at birth in pig-tail macaques (*Macaca nemestrina*). *Archives or Oral Biology*, 35, 289–94.

SWINDLER, D.R. & McCOY, H.A. (1965). Primate odontogenesis. *Journal of Dental Research*, 44, 283–305.

SZALAY, F. & DELSON, E. (1979). *Evolutionary History of the Primates*. London: Academic Press.

TARRENT, L.H. & SWINDLER, D.R. (1973). Prenatal dental development in the Black Howler monkey (*Alouatta caraya*). *American Journal of Physical Anthropology*, 38, 255–60.

TATTERSALL, I., DELSON, E. & COUVERING, J.V. (1988). *Encyclopedia of Human Evolution and Prehistory*. New York: Garland Publishing.

VAN VALEN, L.M. (1982). Homology and causes. *Journal of Morphology*, 173, 305–12.

WALKER, A.C. (1981). Dietary hypothesis and human evolution. *Philosophical Transactions of the Royal Society of London*, 292, 57–64.

WALKER, A.C., HOECK, H.N. & PEREZ, L. (1978). Microwear of mammalian teeth as an indicator of diet. *Science*, 202, 908–10.

14 Postcranial anatomy of extant and extinct species of *Theropithecus*

HARTMUT B. KRENTZ

Summary

1. In this study, the postcrania of *Theropithecus gelada*, *T. oswaldi*, *T. brumpti*, and *T. darti* were examined and compared to extant Old World monkey postcrania in order to present a morphological description of these elements for the genus, and for each species.

2. Functional analysis of the postcrania was provided to explain morphological differences between species.

3. The locomotor behaviour of each species was discussed.

4. The phylogenetic implications of morphological variation within the genus was addressed.

5. Comparisons with extant cercopithecoids show that extant and extinct *Theropithecus* possess a unique suite of characters on their postcrania that are related to locomotor and feeding behaviours, and that these characters appear early in the evolution of the genus.

6. *Theropithecus gelada* manually graze grasses, seeds, and rhizomes sitting upright most of the day plucking food items with their thumb and index finger. They 'shuffle forward bipedally' to move to new sites. Although geladas are terrestrial quadrupeds, these behaviours produce detectable characters in the shoulder, elbow, hand, and hip regions.

7. *Theropithecus oswaldi*, although two or three times larger in size, is very similar in morphology to *T. gelada*, indicating similar feeding and locomotor patterns.

8. In contrast, *Theropithecus brumpti* possesses a suite of characters in its shoulder, elbow and hip joints more indicative of an arboreal quadruped, suggesting the species was better adapted to arboreal habitats.

9. *Theropithecus darti* postcrania, although known only from one site, are more similar in morphology to *T. gelada* and *T. oswaldi* than to *T. brumpti*, indicating closer phylogenetic ties to the former two species.

Introduction

Theropithecus gelada is among the most terrestrial of the Old World monkeys.

They rarely, if ever, climb trees preferring instead the open grasslands of the high plateaus of central Ethiopia (Dunbar, 1977). Geladas do climb but do so mainly on the steep cliff faces of the gorges that cut the plateaux. Their locomotor behaviour has been described simply as 'quadrupedal' (Napier & Napier, 1967) or more formally as 'terrestrial quadruped; ground stander and walker' (Rose, 1974). Their postcranial anatomy reflects this type of locomotion by converging on a body plan similar to that found in other non-primate quadrupeds: elongated fore- and hindlimbs roughly equal in length, stoutly built limb bones, and reduced digits and tail.

However, *T. gelada* also possess a series of postcranial characters that are associated, not with locomotion, but with feeding behaviours. Unlike most other primates that eat primarily leaves or fruit, 90–95 per cent of the gelada's diet consists of grass, seeds, or rhizomes (see Iwamoto, chapter 16). Geladas are bulk feeders but preference is given to the greenest or youngest parts of the grass. These are obtained by plucking the grass with their hands using their thumb and index fingers as pincers. Geladas sit upright for extended periods of time efficiently harvesting large quantities of food. They move to new feeding sites mainly by 'shuffling forward bipedally' with hips, knees, and ankles highly flexed (Wrangham, 1980). Because grass and seeds tend to be found in bunches, this type of locomotion is more efficient than returning to all four limbs and it allows the gelada to feed continuously throughout the day.

Schultz (1970) noted the basic similarities in postcranial morphology among the cercopithecoids as a general adaptation towards terrestrial quadrupedal locomotion and, as a result, distinquishing differences in the postcrania of extant and extinct monkeys is difficult. However, because of their unique feeding, postural, and locomotor behaviour, detectable differences are seen in the postcranial anatomy of *Theropithecus*. Compared to other terrestrial primates (e.g. *Papio*), the gelada forearm is elongated with greater flexibility at the shoulder, elbow, and wrist joints (Krentz, 1992). Digital flexor and extensor musculature is relatively larger to facilitate fine grasping of small objects (Jolly, 1972; Maier, 1972), and a high degree of opposability is produced by the elongation of the thumb and the shortening of the index finger, giving the species the highest opposability index of any non-human primate (Napier & Napier, 1967; Etter, 1973). On the hindlimb, continuous habitual squatting results in a femur with reduced greater trochanter heights, laterally extended articular surfaces on the femoral head, femoral shafts that angle laterally, and tibial articular surfaces that are uneven in size (Krentz, 1988).

When and why both 'manual grazing' and 'bipedal shuffling locomotion' first appear within the lineage is difficult to determine but it does appear that both are basic adaptations of the genus directly related to their terrestrial lifestyle, diet, and feeding behaviour (Szalay & Delson, 1979). These aspects are observed in the living *T. gelada* but some can be detected in fossil theropiths as well.

Craniodental remains of *Theropithecus* are abundant in many Plio-Pleistocene aged sites of Africa, but postcranial elements attributable to the genus are more limited in number and range. They have been recovered from Hadar (Taieb *et al.*, 1976), Hopefield (Singer, 1962), Kanjera (Andrews, 1916), Koobi Fora (Leakey, 1976), Olduvai (Leakey & Whitworth, 1958), Olorgesailie (Issac, 1977), Omo (Eck 1976, 1977), and West Lake Turkana (Harris, Brown & Leakey, 1988). Fossil postcrania have been identified from three of the five known extinct species – *T. darti*, *T. brumpti*, and *T. oswaldi*. The postcrania show substantial variation in size and morphology indicating that these species possessed differing locomotor adaptations.

Until recently, what was known of the locomotion of extinct theropiths was based on the morphometrical analysis of the most recent (i.e. 0.5 to 2.4 Ma) species – *T. oswaldi*. Jolly (1972) examined postcrania from Kanjera, Olduvai, and Olorgesailie. Sample sizes were small but most skeletal elements were represented. Based upon anatomical and metrical comparisons with extant species, Jolly concluded that *T. oswaldi* was a terrestrial quadruped much like modern *T. gelada*. Between 1967 and 1986, additional *T. oswaldi* postcrania were recovered from the Shungura Formation of the Omo (Eck, 1976) and from the Kubi Algi Formation at Koobi Fora (Harris, 1978). Preliminary analysis of these remains support the terrestrial nature of *T. oswaldi* locomotion (Birchette, 1982; Ciochon, 1977, 1986).

Postcranial bones from a second theropith – *T. brumpti* – were also recovered from the Shungura Formation

(Eck, 1976, 1977) and from the Kubi Algi Formation (Harris, 1978). Their postcranial anatomy varied in a number of significant ways from that of *T. oswaldi* and *T. gelada* (Ciochon, 1986; Krentz, 1992). Several of the traits found in the fore- and hindlimbs of *T. brumpti* are more similar in morphology to arboreal rather than terrestrial monkeys indicating significant differences in locomotor behaviour.

The postcrania of a third, and older, species of *Theropithecus* – *T. darti* – were recovered from deposits dated at 2.9 to 3.4 Ma at the Hadar (Taieb *et al.*, 1976). Although craniodental remains of *T. darti* are found throughout East and South Africa, the specimens from the Hadar are the only postcrania identified from this species. The collection has not yet been described in detail but initial observation indicated that *T. darti* was a small monkey similar in locomotion to the modern gelada (Birchette, 1982).

In this study, the postcrania of *T. gelada*, *T. oswaldi*, *T. brumpti*, and *T. darti* were examined and compared to extant Old World monkey postcrania in order to present a morphological description of these elements for the genus, and for each species. Functional analysis of the postcrania are provided when possible to explain the morphological differences between species, and to discuss the locomotor behaviour of each. The phylogenetic implications of these findings are also discussed.

The size, diversity, and time span of the theropith postcranial collection have several important consequences. First, the collection is by far the largest and most diverse yet recovered from a single genus of non-human primate. Although no complete skeletons have been found, almost all postcranial elements are represented for each of the three extinct species. Many elements are represented by complete bones but most are fragmentary or incomplete. Within each species, elements range in size indicating that male and female postcrania are represented. The large sample size permits the examination of variation for a character or a suite of characters for the genus, and for each species. Secondly, because the postcrania represent individuals from time-consecutive species, morphological and locomotor changes can be observed to determine how changes are effected or have affected locomotion, habitat, and/or speciation events. This is important for *T. brumpti* and *T. oswaldi* who overlap in time and space (Eck, 1977). Third, the collection represents postcrania from different East African sites with different environmental habitats. Questions of habitat preference can be addressed. And fourth, the extinct species can be compared to *T. gelada* with differences and similarities duly noted. These characters can be studied for their phylogenetic and locomotor importance.

Understanding how differences in skeletal morphology are related to locomotor abilities, and how these, in turn, are related to environmental conditions and speciation events, significantly contributes to our understanding of theropith evolution.

Materials

In this study the postcrania of extant and extinct species of *Theropithecus* were examined. Specimens of *T. gelada* are housed at the Department of Zoology, Smithsonian Institute, Washington, D.C., and at the Department of Anthropology, University of California, Berkeley, California, USA. All specimens were wild-caught adults free of skeletal pathologies. Fossil postcrania of *T. oswaldi* are housed at the British Museum (Natural History), London, England; Kenya National Museum, Nairobi, Kenya; and the Department of Anthropology, University of California, Berkeley, California. *Theropithecus brumpti* specimens are housed at the Department of Anthropology, University of California, Berkeley, California. Specimens of *T. darti* are housed at the Institute of Human Origins, Berkeley, California. A complete list of specimens including catalogue number, site, age, sex, and a brief description of each is provided elsewhere (Krentz, 1992).

The comparative collection includes male and female adults of *Papio anubis*, *Cercopithecus aethiops*, *Cercocebus albigena*, and *Colobus guereza*. The complete list of specimens is provided elsewhere (Krentz, 1992). Extant species were selected to provide a broad spectrum of phylogenetic relatedness, locomotor, and postural behaviour.

Classification of locomotor types is based on that of Rose (1974). Primate quadrupedalism is subdivided into arboreal and terrestrial with further subdivisions for 'Old World semibrachiators' (*Colobus*), 'branch sitters and walkers' (*Cercocebus* and *Cercopithecus*), and 'ground standers and walkers' (*Papio* and *Theropithecus*). The significance of these locomotor categories will be discussed in greater detail below.

The forelimb and hindlimb bones of

extant and extinct species were measured using 176 variables. The anatomical description of these along with the data set including sample size, mean, standard deviation, and standard error, are given in Krentz (1992). All measurements were taken with standard dial calipers or a millimetre-graduated osteometric board, and are accurate to 1.0 mm. In addition, a series of indices was developed to quantify various morphological traits. The description and importance of the indices is discussed below. Student's t-tests were performed between groups to determine their statistical significance. Additional discussion of the data is found in Krentz (1992).

Comparative anatomy

Variation in postcranial morphology of extant primates is closely related to differences in locomotor and postural behaviours (Fleagle, 1988). These, in turn, have been influenced by the ancestry of the group through selection and adaptation to environmental pressures. By tracing the modifications of these skeletal systems, postcranial bones provide information on phylogeny and locomotor adaptations. Tables 14.1(a) and (b) and 14.3(a) and (b) present characters on the humerus, ulna, radius, femur, and tibia that describe and differentiate the extant comparative collection including *Theropithecus*. Tables 14.2 and 14.4 present the same characters for the genus including extinct species. Figures 14.1 to 14.8 show the location of these characters. Because so few carpal, metacarpal, tarsal, metatarsal, and phalanges are identified as belong-

ing to extinct theropiths, these are not included in the Tables but will be discussed below.

For the ease of description the presentation of the data is depicted by a (+) or (−) system with the (+) indicating the presence of the named character while (−) denotes its absence; (++) represent stronger expression of the character while (+/−) indicates variable expression. The naming of the character is based upon its appearance in *T. gelada* in order to maintain consistency throughout the discussion. It does not imply a derived state for the character, although in some cases this may be true.

Theropith forelimb

Theropithecus gelada postcrania share a number of characters with other cercopithecoids (see Table 14.1). They share only 10 of 38 with *Colobus* but 21 and 22 with *Cercocebus* and *Cercopithecus* respectively. *Theropithecus* and *Papio* share a full 28 characters reflecting their close phylogenetic and locomotor ties (Jolly, 1967; Sarich, 1970). Interestingly, *T. gelada* and *Colobus* share five characters not found in the other cercopithecines – a narrow bicipital groove (char. 5), a concave lateral border of the bicipital groove (char. 6), an extended medial epicondyle (char. 11), a highly concave medial surface of the olecranon process (char. 21), and a smooth lateral border of the radial tuberosity (char. 33). As will be discussed in the next section, these characters are associated with maintaining flexibility in the forelimb.

Forelimb variation exists within *Thero-*

Table 14.1(a) *Morphological characters on the forelimb are described as to their appearance in* Theropithecus gelada.

Character	Description
Humerus	
1	Greater tubercle proximally extended above head.
2	Insertion for m. infraspinatus shallow.
3	Insertion for m. teres minor not pronounced.
4	Insertion for m. coracobrachialis not pronounced.
5	Narrow bicipital groove width.
6	Proximal medial border of deltopectoral crest concave.
7	Distinct groove on distal tuberosity.
8	Shaft robust.
9	Ridge of medial border of distal shaft absent.
10	Width of distal articular surface relatively wide.
11	Medial epicondyle relatively extended.
12	Medial epicondyle highly retroflexed.
13	Superior border of medial epicondyle horizontal.
14	Trochlear margin pronounced.
15	Olecranon fossa deep and narrow.
16	Proximal border of trochlea on the posterior humerus is angled.
17	Trochlear flange relatively short.
18	Sulcus for the ulnaris nerve not pronounced.
Ulna	
19	Olecranon process relatively short.
20	Olecranon process highly angulated.
21	Medial surface of olecranon process highly concave.
22	Tip of coronoid process pointed.
23	Deep trochlear notch.
24	Articular surface of the ulnar notch is complete.
25	Undulating superior trochlear margin.
26	Radial notch lies horizontal to shaft.
27	Shaft of ulna posteriorly concave.
28	Pronounced posterior ulnar border.
29	Long styloid.
Radius	
30	Radial head oblong.
31	Slope of radial head great.
32	Radial tuberosity not greatly raised.
33	Lateral border of radial tuberosity smooth.
34	Longitudinal groove on radial tuberosity absent.
35	Shaft robust.
36	Pronounced groove on medial surface of shaft.
37	Groove on distal posterior shaft present.
38	Distinct distal ulna articulation.

Table 14.1(b) *Morphological characters on the forelimb of extant cercopithecoids.*

Character	Theropithecus	Papio	Cercopithecus	Cercocebus	Colobus
1	++	++	+	+	–
2	+	+	+	–	–
3	+	+	–	+	–
4	+	+	+/–	+	–
5	+	–	–	–	+
6	+	–	–	–	+/–
7	+	–	–	–	–
8	+	+	+	–	–
9	+	–	–	–	–
10	++	++	+	+	–
11	+	+	–	–	–
12	+	+	+	+	–
13	+	–	–	–	–
14	+	+	+	+/–	–
15	+	+	+	+	–
16	+	+	+/–	–	–
17	+	–	+	+	+
18	+	+	–	+/–	–
19	+	+	+/–	+/–	–
20	+	+	+	+	–
21	++	–	–	–	+
22	+	+	–	–	–
23	+	++	+	+	–
24	+	+	+	+/–	–
25	+	+	+	+	–
26	+	+	+/–	+/–	–
27	++	+	+/–	+/–	–
28	+	+/–	–	–	+/–
29	++	+	+	+	–
30	+	+	+	+/–	–
31	+	+	–	–	–
32	+	+	+	+/–	–
33	+	–	+	+	+
34	+	–	–	–	+
35	+	+	+	+	–
36	+	++	+	+	–
37	++	+/–	+	+	–
38	+	+	+	+/–	–

Table 14.2 *Morphological characters within* Theropithecus.

Characters	T. gelada	T. oswaldi	T. brumpti	T. darti	Papio	Colobus
Humerus						
1	+	+	−	+/−	+	−
2	+	+′	−	+	+	−
3	+	+	−	+	+	−
4	+	+	−	+	+	−
5	+	−	+	+/−	−	+
6	+	−	+	+/−	−	−
7	+	+	+	+	−	−
8	+	+	+	+	+	−
9	+	+	−	+	+	−
10	+	+	+	+	+	−
11	+	+	+	+	−	+
12	+	+	+	+	+	−
13	+	+	+	+	−	−
14	+	+	−	+/−	+	−
15	+	+	+	+	+	−
16	+	+	−	−	+	−
17	+	+	+/−	+/−	−	+
18	+	−	+	+	−	+
Ulna						
19	+	+	−	−	+	−
20	+	+	+	+/−	+	−
21	+	++	++	+	−	+
22	+	+	−	−	+	−
23	+	+	−	+/−	+	−
24	+	+	+	+	+	−
25	+	+	+	+	+	−
26	+	+	+	+	+	−
27	+	+	+	?	+/−	−
28	+	+	++	?	+/−	−
29	+	+	+	?	+	−
Radius						
30	+	+	+/−	+/−	+	−
31	+	+	+/−	+	+	−
32	+	+	+	+	+	−
33	+	+	+	+	−	+
34	+	+	−	−	+	−
35	+	++	++	−	+	−
36	+	+	+	+	+	−
37	+	+	++	+	+	−
38	+	+	+	?	+	−

See Table 14.1(a) for an explanation of the characters.

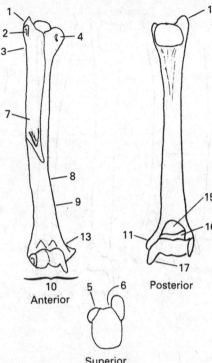

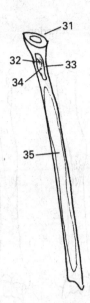

Fig. 14.3. Generalized radius.

Fig. 14.1. Generalized right humerus showing
characters described in text and listed in Tables
14.1 and 14.2.

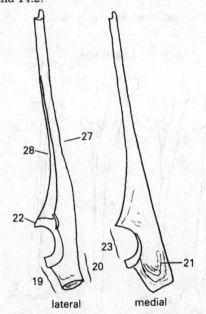

Fig. 14.2. Generalized left ulna.

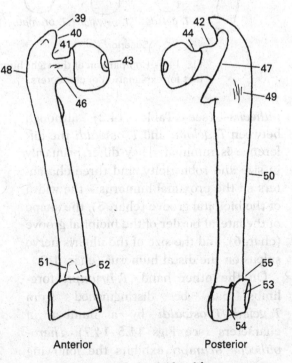

Fig. 14.4. Generalized femur.

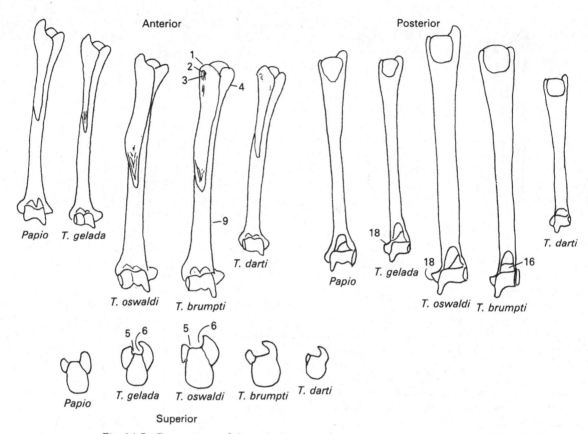

Fig. 14.5. Comparison of the right humeri of *Theropithecus* and *Papio*. See text for explanation of characters.

pithecus (see Table 14.2) although between *T. gelada* and *T. oswaldi* the difference is minimal. They differ primarily in size and robusticity, and three characters on the proximal humerus – the width of the bicipital groove (char. 5), the shape of the lateral border of the bicipital groove (char. 6), and the size of the ulnaris nerve sulcus on the distal humerus (char. 18).

On the other hand, *T. brumpti* forelimbs can be distinguished from *T. gelada*/*T. oswaldi* by a number of characters (see Figs 14.5–14.7). *Theropithecus brumpti* exhibits the following traits not seen on the other two species: greater tubercles of the humerus lie below

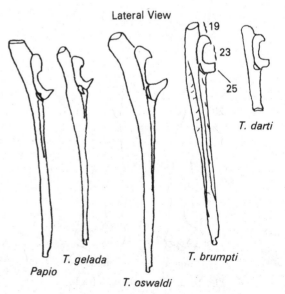

Fig.14.6. Comparison of right ulnae.

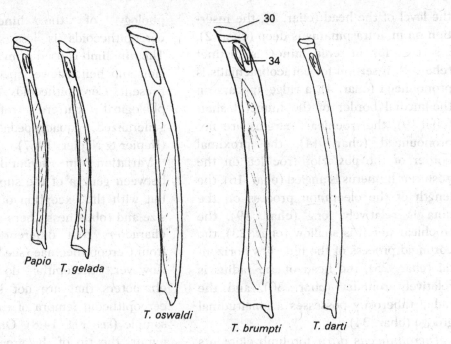

Fig. 14.7. Comparison of radii.

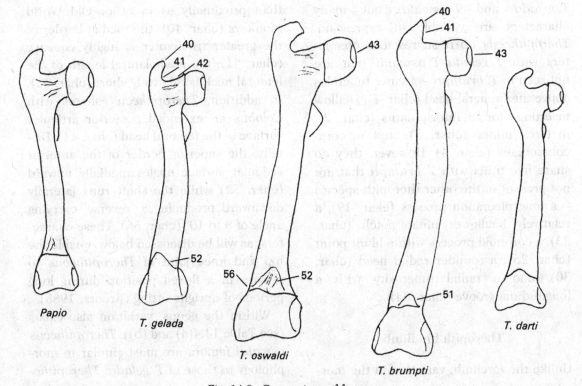

Fig. 14.8. Comparison of femur.

the level of the head (char. 1), the insertion on m. infraspinatus is deep (char. 2),
insertion for m. teres minor is distinct
(char. 3), insertion for coracobrachialis is
pronounced (char. 4), a ridge appears on
the medial border of the humeral shaft
(char. 9), the trochlear margins are not
pronounced (char. 14), the proximal
border of the posterior trochlea on the
posterior humerus is angled (char. 16), the
length of the olecranon process on the
ulna is relatively long (char. 19), the
trochlear notch is shallow (char. 23), the
coronoid process of the ulna lies horizontal (char. 25), the head of the radius is
relatively rounder (char. 30), and the
radial tuberosity possesses a longitudinal
groove (char. 34).

Theropithecus darti forelimb elements
are more similar in morphology to
T. gelada and *T. oswaldi* but many
characters are variable in expression.
Theropithecus darti shares four characters with *T. gelada–T. oswaldi* that are
not seen in *T. brumpti* – greater tubercles
above the humeral head (char. 1), shallow
insertions for m. infraspinatus (char. 2),
m. teres minor (char. 3), and m. coracobrachialis (char. 4). However, they do
share five traits with *T. brumpti* that are
not present in the other theropith species
– a long olecranon process (char. 19), a
relatively shallower ulnar notch (char.
23), a coronoid process with a blunt point
(char. 25), a rounder radial head (char.
30), and a radial tuberosity with a
longitudinal groove (char. 34).

Theropith hindlimb

Unlike the forelimb, variation in the mor-

phology of the hindlimb among
cercopithecoids is less straightforward.
The hindlimb is used primarily for propul-
sion and hence fewer specializations are
present. Cercopithecoids exhibit a morphological pattern reflecting their
generalized quadrupedal locomotion
(Napier & Napier, 1967).

Variation in femoral morphology
between genera of the superfamily exists
but with the exception of differences in
size and robustness, there are no absolute
characters that differentiate Colobinae
from Cercopithecinae (see Table 14.3(b)).
However, theropiths do possess five
characters that are not seen on other
cercopithcoid femora of the comparative
sample (see Fig. 14.8). On the proximal
femur, the tip of the greater trochanter
points medially in *Theropithecus* rather
than proximally as in other Old World
monkeys (char. 40); the medial border of
the greater trochanter is highly concave
(char. 41); the mediolateral length of the
femoral neck is relatively short (char. 42).
In addition, *Theropithecus* shares with
Colobus an extended posterior articular
surface of the femoral head (char. 44). Distally, the superior border of the anterior
articular surface angles medially upward
(char. 52) while the shaft runs laterally
downward producing a 'reverse' carrying
angle of 8 to 10° (char. 56). These characters, as will be discussed below, enable the
hip and knee joint of *Theropithecus* to
remain in a flexed position during long
periods of upright sitting (Krentz, 1988).

Within the genus, variation also exists
(see Table 14.4(a) and (b)). *Theropithecus
oswaldi* femora are most similar in morphology to those of *T. gelada*. *Theropithe-*

Table 14.3(a) *Morphological characters on the hindlimb are described for their appearance in* Theropithecus gelada.

Character	Description
Femur	
39	Proximal extension of the greater trochanter lies above the head of the femur.
40	Proximal end of greater trochanter extends medially.
41	Medial border of greater trochanter concave.
42	Femur neck relatively short.
43	Fovea capitis oval.
44	Posterior articular surface of the femoral head extends on to the neck.
45	Superior inclination of femoral head.
46	Ridge on anterior surface of femoral neck present.
47	Intertrochanteric ridge incomplete.
48	Groove on lateral edge of anterior shaft absent.
49	Gluteal tuberosity weakly marked.
50	Shaft robust.
51	Depression superior to the distal anterior articulation present.
52	Superior border of distal anterior articulation angled.
53	Lateral condyle larger than the medial condyle.
54	Intercondylar fossa narrow.
55	Notch on superior border of lateral epicondyle absent.
56	Angle of shaft on to condyle distolaterally.
Tibia	
57	Horizontal line on anterior surface of tibial tuberosity present.
58	Tibial tuberosity broad.
59	Ridge descending posteriorly from intercondylar notch absent.
60	Anterior surface of tibial tuberosity flat.
61	Shaft robust.
62	Notch for m. tibialis posterior on medial malleolus present.
63	Medial malleolus angled approximately 45 degrees.
64	Lateral edge of distal articular surface not notched.
65	Notch on lateral edge of the distal articular surface present.

cus oswaldi differs by possessing an elongated fovea capitis (char. 43), a horizontal angulation of the head (char. 45), larger muscular attachments for m. gluteus minimus (char. 48) and maximus (char. 49), a depression superior to the distal articular surface (char. 52), and a notch for the tendon of biceps femoris on the lateral epicondyle (char. 55).

Despite its smaller size, *T. darti* femora are more similar to those of *T. oswaldi* than *T. gelada* sharing a number (chars 45, 49, 51, 52) of characters not seen in *T. gelada*. *Theropithecus darti* femora dif-

Table 14.3(b) *Morphological characters on the hindlimb of Cercopithecoidea.*

Character	*Theropithecus*	*Papio*	*Cercopithecus*	*Cercocebus*	*Colobus*
Femur					
39	+	+	+/−	−	−
40	+	−	−	−	+/−
41	+	−	−	−	−
42	+	−	−	−	−
43	+	+	+	−	−
44	+	−	−	−	+
45	+	+	−	+	−
46	+	+	−	+	−
47	+	−	−	+/−	+
48	+	+	+	+	+
49	+	+	−	.+/−	+
50	+	+	+	+	+
51	+	+	+	−	−
52	+	−	−	−	−
53	+	+	+	−	−
54	+	+	+	+/−	+/−
55	+	+	+	+	+
56	+	−	−	−	+/−
Tibia					
57	+	−	−	−	+
58	+	+	+	−	−
59	+	+/−	+/−	−	+/−
60	+	+/−	−	−	+
61	+	+	+/−	+/−	−
62	+	−	+	−	+/−
63	+	−	−	−	+
64	+	+/−	−	−	+/−
65	+	+	−	+	+

See Table 14.3(a) for an explanation of the characters.

fer by the absence of a bony ridge on the anterior surface of the neck (char. 46), the lack of a depression for m. vastus lateralis on the shaft (char. 48), and the absence of the notch for m. biceps femoris (char. 55).

Theropithecus brumpti femora differ from the other theropiths by possessing a more proximally pointed tip to the greater trochanter (char. 40) with less concave medial borders (char. 41), an elongated

Table 14.4 *Morphological characters in the hindlimb of* Theropithecus.

Character	T. gelada	T. oswaldi	T. brumpti	T. darti
Femur				
39	+	+	+	+
40	+	+		+
41	+	+	–	+
42	+	+	+	+/–
43	+	+/–	–	+
44	+	+	+	+/–
45	+	–		
46	+	+		
47	+	+/–	+	+
48	+	–	+/–	+
49	+	–	–	+
50	+	+	+	+
51	+	–	++	–
52	+	–	+/–	–
53	+	+	+	+
54	+	+	+	+
55	+	–	–	+
56	+	+	+	+
Tibia				
57	+		+	–
58	+	+	+	+
59	+	+	–	+
60	+	+	+	+
61	+	+	+	+
62	+	+/–	+	+
63	+	+	+	+/–
64	+		+	+
65	+	+	–	–

See Table 14.3(a) for an explanation of the characters.

fovea capitis (char. 43) and a very deep depression superior to the distal articular surface (char. 51). They share with *T. oswaldi* and *T. darti* a superior inclination of the femoral head (char. 45), a ridge on the anterior surface of the femoral neck (char. 46), and a weakly marked gluteal tuberosity (char. 49).

Differences in the morphology of the tibia between cercopithecoids are less than those seen on the femur (see Table 14.3). There are no characters that clearly distinguish colobine from cercopithecine other than general robustness. *Theropithecus gelada* does not possess any distinct characters although it does share two only with *Colobus* – a horizontal line on the anterior surface of the tibial tuberosity (char. 57) and a medial malleolus angled at 45° (char. 63).

Within the extinct theropiths, there is little variation (see Table 14.4). *Theropithecus oswaldi* differs from *T. gelada* by only the absence of the horizontal line on the tibial tuberosity (char. 57). *Theropithecus brumpti* shares this character with *T. gelada* but differs by possessing a ridge that descends from the intercondylar notch (char. 59) and by the presence of a

notch on the lateral edge of the distal articular surface (char. 65). *Theropithecus darti* tibiae are similar to *T. gelada* but also lack the horizontal line of the tibial tuberosity while they do possess a notch on the distal articular surface.

Theropith species can be differentiated by a suite of postcranial characters that have phylogenetic and functional significance. The importance of these differences is discussed below.

Functional anatomy

Theropith postcrania vary on a number of characters that have functional importance. A more detailed examination of

Table 14.5(a) *Indices used.*

Index	Description
Humerus	
A	Greater Tubercle Projection
B	Bicipital Groove Width
C	Projection of Deltoid Tuberosity
D	Relative Width of Distal Articular Surface
E	Projection of Medial Epicondyle
F	Medial Epicondyle Retroflexion
G	Relative Depth of Olecranon Fossa
H	Relative Width of Olecranon Fossa
I	Projection of Distal Trochlear Flange
Ulna	
J	Relative Length of Olecranon Process
K	Olecranon Angulation
L	Styloid Length
Femur	
M	Lateral Extension of Articular Surface
N	Relative Length of Neck
O	Relative Height of Greater Trochanter

For complete description see Krentz, 1992.

Table 14.5(b) *Indices for Cercopithecoidea.*

	Theropithecus gelada	T. oswaldi	T. brumpti	T. darti	Papio	Cercopithecus	Cercocebus	Colobus
A	109.1	105.6	93.5	101.2	107.2	105.3	106.0	95.2
B	22.4	28.2	21.1	26.2	29.8	21.3	29.8	25.1
C	42.1	42.4	38.4	–	37.4	35.2	35.0	36.4
D	71.6	70.1	72.4	70.1	73.4	72.0	70.4	66.9
E	19.5	16.0	16.1	18.7	13.0	15.1	14.2	18.6
F	31.0	31.0	29.7	30.4	32.2	33.1	30.4	21.9
G	43.8	43.8	45.8	41.8	42.9	47.5	46.3	53.4
H	53.0	61.8	56.3	60.4	55.7	55.5	54.0	55.4
I	71.4	70.9	69.6	70.4	55.1	60.4	62.3	80.2
J	39.2	35.8	44.3	45.9	39.4	47.5	42.3	43.9
K	38.3	43.6	39.5	34.3	40.0	35.2	30.1	18.3
L	75.8	59.9	72.3	–	69.9	63.2	63.2	74.3
M	46.8	48.6	48.0	45.3	42.0	43.0	49.2	45.5
N	20.9	20.5	19.1	22.5	24.9	24.6	25.0	23.5
O	6.4	6.2	6.1	6.0	5.9	5.5	5.2	3.6

For complete data set see Krentz, 1992.

these characters provides an understanding of the postural and locomotor behaviour for each species, and addresses questions of habitat selection and speciation. Many of the characters listed in Tables 14.1 to 14.4 can be quantified in order to show the degree of difference between species for that trait. Table 14.5(a), (b) provides data on these quantified traits. A description of the variables and indices used is provided elsewhere (Krentz, 1992).

Forelimb

Proximal Humerus Differences in the size and shape of the greater and lesser tubercles have long been associated with the locomotor habits of extant primates (e.g. Clark & Thomas, 1951; Napier & Davis, 1959; Jolly, 1967; Fleagle & Simons, 1982). Functionally, they are the attachment sites for the rotator cuff muscles of the shoulder joint and for some of the protractors and retractors of the humerus. They also contribute to stability around the glenohumeral joint by providing skeletal support. Variation in tubercle morphology reflects the importance of each of these actions in locomotor and/or postural behaviour.

Proximal prolongation of the greater tubercle has been related to terrestrial quadrupedalism (Savage, 1957). The supraspinatus muscle inserts onto the proximal surface of the tubercle; this muscle stabilizes the shoulder joint by keeping the head of the humerus closely aligned to the glenoid fossa of the scapula, and functions as a flexor (or protractor) of the humerus. Jolly (1967) argued that the expansion of the tubercle proximally raises the insertion of the supraspinatus thereby lengthening its moment arm at the glenohumeral joint. A longer moment arm provides more power for fast quadrupedal gaits. It also allows for the stronger muscular forces that are needed to counteract ground reaction forces generated during standing and locomotion that could lead to passive extension in the shoulder during the fore-aft locomotion seen in terrestrial primates.

In contrast, the greater tubercles in arboreal quadrupeds lie below the humeral head thereby lowering the insertion of supraspinatus. This produces weaker but faster protraction of the forelimb (Jolly, 1972). Smaller tubercles also allow more movement at this joint.

Recently, Larson & Stern (1989) have argued that the primary role of the supraspinatus muscle was not one of protraction but as a joint stabilizer during the support phase of quadrupedalism. They suggest that differences in greater tubercle height are the result of the muscle's role as an elevator of the forearm against gravity. In the more arboreal primates, the 'lowering of the greater tubercle' is better viewed as raising the humeral head upward above the tubercle for the purpose of increasing overall mobility at the humeral joint. In terrestrial primates the trend is to reduce mobility and the overall size of the

humeral head resulting in a large greater tubercle.

Although the exact role of the supraspinatus is debatable, the association between a projecting greater tubercle (or a reduced humeral head) and terrestrial locomotion is well documented (Walker, 1974; Birchette, 1982; Fleagle & Simons, 1982; Ciochon, 1986). In this study, Index A (see Table 14.5) measures greater tubercle projection – indices above 100 indicate tubercle heights above the head, those below 100 indicate reduced tubercles. The data support previous studies in showing the extant terrestrial cercopithecoids having high indices reflecting their extended tubercles with the reverse – low indices and low tubercles – being the case for arboreal monkeys.[1]

Within *Theropithecus* (see Table 14.5(a), (b)), *T. gelada* has a very high index (109.1) clearly aligning it with the terrestrial quadrupeds. *Theropithecus oswaldi* have greater tubercles that rise above the head but they are not as extended as those seen in the gelada as indicated by their lower index of 105.6. On the other hand, *T. brumpti* has a very low index of 93.5 with the greater tubercles lying well below the humeral head (see Fig. 14.5). As discussed above, this trait is characteristically found in primates that have greater flexibility at the shoulder joints (i.e. arboreal quadrupeds). The difference in greater tubercle height between *T. brumpti* and *T. oswaldi* clearly distinguish the two extinct species. *Theropithecus darti* tubercles extend to

---

1. The differences reported in this paper are statistically significant to 0.05% using Student's t-test (see Krentz, 1992).

the height of the head or slightly higher (index 101.2).

In addition to greater tubercle height, two other traits associated with the greater tubercle and one with the lesser tubercle are of taxonomic and functional importance: (a) the insertion of m. infraspinatus on the lateral surface of the greater tubercle; (b) the insertion m. teres minor inferior to the greater tubercle; and (c) the insertion of m. coracobrachialis profundus inferior to the lesser tubercle.

The infraspinatus muscle is an important stabilizer of the shoulder and acts as a lateral rotator of the humerus (Maier, 1972). Arboreal primates have relatively large infraspinatus muscles, and rotatory cuff muscles in general, to enhance flexibility. This is reflected on the humerus by a deep concavity on the lateral surface of the greater tubercle. Terrestrial primates possess relatively smaller rotatory muscles and do not exhibit deep concavities on the humerus.

A similar trend is seen for m. teres minor insertion but here there is a longitudinal depression rather than a concavity. Arboreal monkeys generally show this depression while terrestrial ones do not. The teres minor muscle is a lateral rotator of the humerus and is larger in arboreal species (Maier, 1972).

A third indicator of large rotatory muscles is seen by the insertion of m. coracobrachialis profundus inferior to the lesser tubercle. This muscle is a medial rotator of the humerus (Maier, 1972). A depression can be seen just inferior to the lesser tubercle in arboreal forms but not in terrestrial ones.

The presence or absence of these characters are seen in Tables 14.1 and 14.2. Extant terrestrial monkeys (i.e. the ground sitters and walkers) are consistent in not exhibiting these characters. These traits are variable in the arboreal quadrupedal group with *Colobus* possessing all three traits. *Cercopithecus* shows a deep concavity for the infraspinatus but the elongated depression for the teres minor muscle is absent. The opposite trend is seen by *Cercocebus*.

Within *Theropithecus*, geladas lack these characters (see Fig. 14.5); they have relatively smaller rotatory muscles (Jolly, 1972; Maier, 1972). The humeri of younger specimens of *T. oswaldi* also do not show these characters although the expression of them is more variable in the older specimens. These characters are also not seen in *T. darti*. In contrast, all three traits are present in *T. brumpti* indicating the presence of relatively large rotatory muscles for this species.

The width and shape of the bicipital groove varies between locomotor groups. Groove width is determined by anterior extension of the tubercles, and by the muscles that insert within the groove. Terrestrial monkeys have greater tubercles that are much larger and more anteriorly placed than their lesser tubercles thus creating a broad groove (see char. 5). In arboreal primates, the tubercles are approximately equal in size, an anterior extension making their groove appear narrow (Fleagle & Simons, 1982).

Functionally, the more anteriorly extended lesser tubercles and narrow bicipital groove aid in keeping the tendon for the biceps brachii within its proper track through a wide range of movement about the shoulder joint (Ford, 1980). The narrow groove is advantageous for suspensory or brachiating primates that need a great deal of circumduction around the shoulder (Ford, 1980; Fleagle & Simons 1982). Conversely, a wide groove produced by a small lesser tubercle is characteristically seen in terrestrial species and indicates a lack of anatomical specialization for suspensory postures (Ford, 1980).

Index B presents bicipital groove width as a percentage of proximal end width. The highest indices – and widest grooves – are found in the terrestrial species such as *Papio* (29.8) and in the less terrestrial *Cercocebus* (29.8). *Colobus* has an index of 25.1 and does show a narrow groove on visual inspection. However, both *T. gelada* (22.4) and *Cercopithecus aethiops* (21.3) have low indices which is surprising given their terrestrial habits. Either the function of the biceps brachii muscle is different in these species or something else is contributing to bicipital groove width.

The width of the bicipital groove is also affected by the muscles that insert onto its borders. The lateral edge of the groove is defined by the deltopectoral crest. While the deltoid muscle inserts onto the surface of the tuberosity itself, the pectoralis major muscle inserts onto the medial edge of the tuberosity and the lateral edge of the groove. This muscle is a major retractor of the forelimb. In monkeys who use their forelimbs primarily for fore-aft movements during locomotion, this muscle is relatively large (Maier, 1972). Associated with the increase in the size of the deltoid muscle is a larger and straighter (i.e. more anteriorly projecting) deltopectoral crest. This contributes to a wider bicipital groove

by allowing room for larger pectoralis major muscles. Conversely, this muscle is smaller in arboreal species and, as a result, the medial edge of the deltopectoral crest is concave or angles medially. This restricts the size of the muscle but it acts to brace it through a wide range of movement around the shoulder (Ford, 1980).

Character 6 on Tables 14.1 and 14.2 shows the appearance or absence of a concave medial border of the deltopectoral crest (see Fig. 14.1). It can be seen that those monkeys that have wide bicipital grooves (char. 5) do not have concave borders but ones that are 'straight' or anteriorly projecting when viewed superiorly, do. The reverse is true as well – narrow grooves, concave borders, although this is not absolute. The correlation between these traits and locomotor type is not clear but it does appear that terrestrial monkeys with narrow sulci use their forelimbs in a somewhat different fashion from those with wide grooves.

Theropithecus gelada is different from the other terrestrial quadrupeds in possessing a narrow bicipital groove and a concave medial border (see Fig. 14.5). *Theropithecus oswaldi* shows the opposite trend of a wide groove with straight borders. *Theropithecus brumpti* humeri have narrow grooves and concave borders similar to those seen in the gelada. *Theropithecus darti* is variable in both traits.

Shaft

The deltoid tuberosity dominates the upper portion of the humeral shaft. It is the insertion site for the deltoid muscle that acts as a retractor and an abductor of the forelimb in monkeys (Howell & Strauss, 1933). The distal extent of the tuberosity has been used as an indicator of arboreal versus terrestrial locomotor behaviours.

Napier & Davis (1959) have argued that brachiating (i.e. arboreal) primates have lower insertions for the deltoid muscle, and hence more distally extended tuberosities to allow the anterior fibers of the muscle to insert inferior to the fibers of the pectoralis major muscle. This would alter the range over which efficient humeral abduction can occur as well as increasing the leverage of the deltoid (Campbell, 1974). In non-brachiating (i.e. terrestrial) primates, the deltoid does not extend past the insertion of the pectoralis major muscle and thus the deltoid tuberosity is much shorter.

However, the correlation between locomotor behaviour and deltoid tuberosity length within the cercopithecoids is not clear. Birchette (1982) did not find a correlation among a sample of extant and extinct colobines nor did Ciochon (1986) in his study of cercopithecines. Both caution that the locomotor constraints of a brachiating primate are substantially different from those of an arboreal quadruped like *Colobus*. In brachiators such as the apes, the deltoid functions as an abductor while in monkeys it acts as a flexor or protractor of the forelimb during quadrupedal walking or running. Furthermore, as Ciochon (1986) points out, the shape of the thorax differs between apes and monkeys, and this difference effects the function of the deltoid muscle. Index C depicts deltoid tuberosity extension as a percentage of total humeral

length. Among extant cercopithecoids, there is little difference in relative deltoid length (e.g. *Papio* has an index of 37.4 while *Colobus* has an index of 36.4) with the exception of *Theropithecus*. The gelada has a long deltoid tuberosity indicated by its high index of 42.1. Ciochon (1986) also found this to be the case in his study. He attributes this not to locomotor requirements of the deltoid but to the unique manual foraging behaviour of gelada and the need to use the forelimb during feeding.

Only two complete humeri are known for *T. oswaldi* but both have indices comparable to *T. gelada* (41.4 and 43.2 vs. 42.1). *T. brumpti* humeri (n = 2) show an index closer to modern *Colobus* than *Papio* at 38.4. Because no complete humeri of *T. darti* are known, the length of their tuberosity cannot be calculated but they appear long on visual inspection.

All four species of *Theropithecus* do have distinct depressions on the distal surface of the deltoid tuberosity. This depression is not seen in any other monkey. The functional significance of this trait is not clear but it may relate to the increased size of the deltoid muscle.

A final character on the shaft is only seen in *T. brumpti*. There is a distinct ridge on the medial border of the distal shaft in these theropiths, which is not seen in the other species. It is also unclear what function this trait may have other than to provide an enlarged surface area for the digital flexors.

Distal Humerus The distal end of the humerus has been used to differentiate arboreal and terrestrial primates. It articulates with the ulna and radius to form the elbow joint. It is the site of attachment for many of the forearm and digital flexors and extensors. Variation in morphology reflects the differing requirements of stability, forearm movement, and digit dexterity.

Jolly (1972) states that the relative breadth of the distal articular surface is a good indicator of terrestrial locomotion. He used the ratio of articular breadth to biepicondylar breadth to distinguish locomotor types with an index of 70 per cent or higher indicating terrestrial locomotion. An index this high implies that the articular surface is quite broad and that the medial epicondyle is relatively short: a shorter medial epicondyle implies reduced flexor musculature and somewhat more restricted movement around the elbow. The broader articular surface also provides more structural support to the upper limb during terrestrial quadrupedalism.

Index D used in this study is identical to Jolly's index and shows the similar results. *Papio* has a high index (73.4) while *Colobus* is much lower (66.9). All of the theropiths cross Jolly's terrestrial rubicon – *T. gelada* (71.6), *T. oswaldi* (70.1), *T. brumpti* (72.4), and *T. darti* (70.1).

A second indicator of locomotor habits in the distal humerus is the projection of the medial epicondyle (Jolly, 1967). More medial projection of the epicondyle is associated with the relatively larger carpal and digital flexors that originate there (Jolly, 1972; Maier, 1972; Birchette, 1982). In addition, the greater medial displacement increases the medial rotatory torques exerted by the digital and carpal

flexors around the radiohumeral and radioulnar joints (Jenkins, 1973), and increases the moment arm of the pronator teres muscle about the axis of the forearm thus increasing the mobility of the forearm (Jolly, 1972). These forces are most often utilized in climbing, reaching, or hanging behaviours associated with arboreal quadrupedalism.

Among most ground-living cercopithecines (and cursorial mammals in general) the epicondyle does not extend medially but projects more posteriorly in order to enhance the action of the pronators and flexor muscles (Fleagle & Simons, 1982). Smaller epicondyles also assist in reducing medial torques when the elbow is already pronated during pronograde quadrupedalism.

Index E shows the relative projection of the medial epicondyle as a percentage of total biepicondylar breadth. Among extant cercopithecoids, there is a clear separation between terrestrial (i.e. *Papio* 13.0) and arboreal (i.e. *Colobus* 18.6) monkeys. The exception to this is *T. gelada*. It has a high index (19.5) and more medially projecting epicondyles. This finding is unusual especially since Birchette (1982) found that larger forms decrease the size of the epicondyles allometrically. The large epicondyles in the gelada may be explained by the need to maintain larger carpal and distal flexors in order to grasp fine objects during feeding (Jolly, 1970). *Theropithecus gelada* has the most opposable pollex among the cercopithecoids and heavy reliance on digital dexterity (Jablonski, 1986a).

Extinct theropiths exhibit a similar pattern. *Theropithecus oswaldi* has an index of 16.0 indicating a smaller epicondyle than the gelada but still larger than seen in extant baboons. *Theropithecus brumpti* (16.1) and *T. darti* (18.7) show a similar extension in their epicondyle. In addition, theropiths possess a horizontal superior border of the medial epicondyle (char. 16) in contrast to other monkeys in which this border angles inferiorly. This trait further increases the surface area for the digital and carpal flexors and allows for larger muscles. Both traits appear early within the genus and may indicate a change in the function of the digits, from one associated with manual dexterity in an arboreal setting to one of fine manipulation of food objects on the ground.

Backward displacement (or retroflexion) of the medial epicondyle is part of the above complex. Retroflexion adds stability to the elbow joint in terrestrial species by providing a solid structure when the weight-bearing axis of the forearm habitually shifts toward the ulnar side (Jolly, 1965; Birchette, 1982). Retroflexion also involves displacing the origins of the flexor musculature posteriorly to accommodate a more substantial humeroulnar ligament (Birchette, 1982).

Index F reflects the degree of medial epicondyle retroflexion as a percentage of distal epicondylar breadth. Not unexpectedly there is a clear distinction between terrestrial and arboreal cercopithecines. *Papio* has a high index (32.2) with a highly retroflexed epicondyle while the opposite is true for *Colobus* (21.9).

Theropithecus gelada, with an index of 31.0, aligns with the terrestrial monkeys as does *T. oswaldi* (31.0). *Theropithecus brumpti* has slightly reduced retroflexion

(29.7) but still significant. The retroflexion of *T. darti* (30.4) follows the trend seen in the other species.

The relative breadth and depth of the olecranon fossa reflects the size and shape of the olecranon process of the ulna, and indirectly of the size of the triceps muscle. This muscle is the major extensor of the forelimb. The range of forearm flexion is relatively greater in arboreal primates while forearm extension is much more limited (Birchette, 1982). Movement along branches in a pronograde quadrupedal fashion is most frequently accomplished with pronounced forearm flexion for better support and in preparation for leaping between branches (Ziemer, 1978). Because the extension of the forearm is limited, the size of the triceps is reduced producing a smaller olecranon process, and a shallower broader fossa.

In terrestrial quadrupeds, extension of the forearm plays a greater role in locomotion requiring a larger triceps muscle, a larger olecranon process, and a larger olecranon fossa. Furthermore, fossae are deep and narrow to enhance stability at the elbow joint.

Indices G and H depict the relative depth and width of the olecranon fossa. Cercopithecines follow the trends given above. Arboreal monkeys have low indices while terrestrial monkeys have high ones. Modern geladas have deep and narrow fossae as indicated by their indices as do *T. oswaldi*, *T. brumpti* and *T. darti*.

A final trait on the distal humerus is the projection of the trochlear flange. A more distally projecting trochlear flange has been associated with terrestrial locomotion (Jolly, 1965). Jolly states that a trochlea that extends distally will abut against the articular surface of the ulnar coronoid process and thereby contribute to the stability of the elbow joint by counteracting the forces that would tend to either displace the ulna medially or the humerus laterally. Conversely, the moderately developed flange found in arboreal species reflects an adaptation for elbow stability in all positions of the forearm rather than just during pronation (Napier & Davis, 1959).

The relative distal projection of the trochlear flange is presented in Index I. Arboreal (*Colobus* 80.2) and terrestrial (*Papio* 55.1) forms are clearly separated, with the former having shorter and the latter longer flanges.

Theropithecus gelada has a much shorter (71.4) flange than seen in *Papio* as does *T. oswaldi* (70.9), *T. brumpti* (69.6), and *T. darti* (70.4). The shorter flange allows for greater extension of the forearm than seen in the other terrestrial monkeys. A shorter flange in these species indicates more flexibility during most forearm movements.

Ulna

Like the distal humerus, the proximal ulna contributes to elbow joint stability and sites of attachment for forelimb retractors, protractors, forearm and digital flexors, and extensors.

Three aspects of the olecranon process have functional importance – its length, angulation, and the configuration of its medial surface. The relative length of the olecranon process is related functionally to the action of the triceps muscle and to the position of the forearm (Gray, 1968). Process length is a measure of the lever arm of the triceps muscle. When the process is short relative to the length of the distal ulna, the arc traversed at the end of the limb is greater for a given muscle contraction, increasing movement and speed of the load arm (Rodman, 1979). A short process also allows for full extension of the elbow thus increasing the length of the whole forelimb during and particularly at the end of a stride. This, in turn, increases the length of the stride for a given set of muscle contractions. Conversely, a relatively longer process increases its leverage, and hence the mechanical efficiency of the triceps by transmitting more power or strength to the forearm when the forelimb is habitually flexed (Gray, 1968). Oxnard (1963), Ashton *et al.* (1976), Jolly (1967) and others have found that shorter olecranon processes are characteristic of terrestrial primates that normally have their forelimbs in an extended position during locomotion. Longer processes are found in the arboreal primates.

The relative length of the olecranon process is measured by Index J which gives olecranon length as a percentage of proximal ulnar length. *Papio* (39.4) and *Theropithecus* (39.2) have low indices and short processes while *Colobus* (43.9) and *Cercopithecus* (47.5) have much longer ones.

The extinct theropiths show a wide range of values. *Theropithecus oswaldi* has a low index (35.8) and a short process similar to *T. gelada* but *T. brumpti* has a much longer process (44.3) more characteristic of arboreal primates. Ciochon (1986) found this trait in his study of extinct cercopithecines. *Theropithecus darti* has a long olecranon process (45.9) as well. The long processes of *T. brumpti* and *T. darti* imply that these species had large tricep muscles and may further imply locomotor behaviours with forelimbs held in a more habitually flexed position similar to that seen in extant arboreal monkeys.

The significance of olecranon process length is closely linked to the degree of olecranon angulation. Angulation is functionally related to the different requirements of the triceps for maximum efficiency in climbing and in level surface locomotion (Jolly, 1967). A retroflexed olecranon provides maximum mechanical advantage to the triceps when the forearm is close to vertical and when the elbow angle is high because this pushes the insertion of the triceps further away from the pivotal axis of the elbow joint. Hence, a retroflexed olecranon is most advantageous for terrestrial locomotion. In arboreal quadrupeds, the elbow is seldom, if ever, at full extension and thus there is no significant mechanical advantage to retroflex the olecranon.

Index K gives the degree of retroflexion of the olecranon process. The results here confirm earlier studies with the terrestrial species displaying a greater degree of angulation than the arboreal ones. *Papio* (40.0) is clearly separated from *Colobus* (18.3) with the arboreal *Cercocebus* (30.1) and *Cercopithecus* (35.2) intermediate. *Theropithecus gelada* has a highly angulated olecranon process (38.3) as do the other theropiths – *T. oswaldi* (43.6), *T. brumpti* (39.5), and *T. darti* (34.3).

A third aspect of the olecranon process related to locomotion and manual dexterity is the configuration of its medial surface. This is the site of origin for m. flexor carpi ulnaris and m. flexor digitorum profundus (Hartman & Strauss, 1933). The head of the former lies more proximal than the latter but both extend distally to insert at the wrist and on the digits. M. flexor carpi ulnaris is a medial deviator of the wrist while m. flexor digitorum profundus is the primary flexor of the digits.

In most primates, these muscles are well developed and produce a concavity on the medial surface of the olecranon process. However, in those primates that use their digits in arboreal modes of locomotion, these muscles are relatively larger, producing greater concavities and proximally extended processes (Hartman & Strauss, 1933). This can be seen in Table 14.1, character 21 which shows the appearance of a deep medial concavity on arboreal monkeys but not terrestrial ones.

The exception to this trend is *Theropithecus*. Maier (1972) found that the origins of m. flexor carpi ulnaris and m. flexor digitorum profundus are more strongly marked on the ulna on *Theropithecus* than is seen in *Papio*. Jolly (1972) found that portions of the m. flexor digitorum profundus that flex the first digit were much larger in *Theropithecus*. He attributed this to the extreme manual dexterity of their digits as an adaptation to 'manual-grazing'. *T. oswaldi* and *T. brumpti* also exhibit a large concavity along the olecranon extending distally under the olecranon notch. In *T. brumpti* it is quite marked as it is in *T. darti*. This trait is present in all members of the genus and was an early adaptation to digital dexterity.

The height of the coronoid process and the depth of the trochlear notch have been associated with locomotor preferences. Knussman (1967) found a correlation between low coronoid height and a long trochlear notch; these traits are associated with arboreality. Conversely, a high coronoid process serves to brace the forelimb and prevent anterior dislocation of the humerus during extension (Conroy, 1974). A deeper and more closed trochlear notch further aids in stability of the elbow during hyperextension in terrestrial primates.

Characters 23 and 25 show these traits in the extant sample. Terrestrial monkeys have high coronoid processes and deep notches. *Theropithecus gelada* and *T. oswaldi* follow these monkeys but *T. brumpti* and *T. darti* have coronoid processes and notches more similar to that seen in the arboreal monkeys.

Theropithecus brumpti and *Theropithecus darti* also share a morphological trait of their coronoid processes not seen in the other theropithecines. These species have a 'flat' upper border (char. 22) of the process. *Theropithecus gelada* and *T. oswaldi*, and other terrestrial monkeys, have a lateral downward slope to this border. In primates that locomote with extended elbows, the coronoid process curves forward ending with a marked upturned tip. The angulation allows for parasagittal movements of the forelimb at the elbow joint. In contrast, arboreal primates have coronoid processes with anterior edges that are flat or horizontal. This arrangement allows straighter fore-aft

movements of the forearm.

The ulnar shaft provides a measure for arboreal or terrestrial locomotion among primates (Conroy, 1974). In arboreal genera, the shafts have an anterior concavity while terrestrial ones exhibit shafts that are straight or anteriorly convex. The anterior convexity of the shaft in terrestrial monkeys provides a stable union between the ulna and radius, and acts to limit the degree of pronation and supination (Conroy, 1974). It also limits the area for the interosseous membrane thereby reducing the corresponding area for the origin of m. flexor digitorum profundus and the extensors, flexors, and abductors of the thumb (Jolly, 1965).

An anteriorly concave shaft permits increased areas for the digital flexors and pollical extensors and abductors while allowing greater degrees of pronation and supination (Birchette, 1982). It also increases the bending strength of the ulna against the pull of the forearm flexors without additional deposition of bone (Schon-Ybarra & Conroy, 1978). This would be an effective method by which an arboreal monkey requiring strong powers of forearm flexion could withstand the resultant muscular forces without compromising an equally important requirement for a light, gracile skeleton (Birchette, 1982).

The shaft of *Theropithecus gelada* is highly convex anteriorly supporting its terrestrial nature. Some shafts do show variation in the degree of curvature. The shaft of *T. oswaldi* and *T. brumpti* exhibit a similar pattern.

The shaft of *T. brumpti* has very distinct medial and posterior borders which apparently bracket m. abductor pollicus longus and m. extensor pollicus longus. This arrangement is not seen in other theropiths or cercopithecoids and its function remains unclear. It may imply a larger or stronger pollex than seen in other species but this remains to be documented. The 'double-ridge' formation is quite distinct in this species.

Ulnar styloid length on the distal ulna varies between terrestrial and arboreal species. A long styloid in direct contact with the pisiform and triquetral bones provides stability to the wrist while limiting ulnar deviation (Ciochon, 1986). A shorter more gracile styloid is more advantageous in arboreal leaping where a mobile wrist joint with greater movement capabilities is needed. The shorter styloid provides a greater range of adduction in arboreal locomotion.

All cercopithecines, including all theropiths have relatively longer styloid processes than do colobines (see character 29).

Radius

Among the forelimb bones, the radius has been the most difficult to analyze in terms of correlating morphological traits and locomotor behaviours. Studies of the radii of Old World monkeys (Napier & Davis, 1959; Conroy, 1976; Ciochon, 1986; Jolly, 1965; Birchette, 1982) have produced mixed and inconsistent results. The lack of distinguishable traits identifying locomotor behaviour is surprising given the habitual pronation of the forearm required during terrestrial locomotion (Jolly, 1967). However, general tendencies

are evident and if these are not directly correlated with locomotion, they do indicate the amount of movement allowed at the elbow and wrist joints.

The shape of the radial head often indicates the range of movement of the forelimb. An oblong or 'elliptical' head may imply reduced ranges of pronation and supination while a more circular head is associated with greater ranges of forearm mobility and an increased capacity for supination (Howell, 1944). In habitually pronograde primates this trend holds true (e.g. Howell, 1944; Conroy, 1974) but the distinction between head shape and locomotion is less clear among cercopithecoids. Jolly (1965) and Birchette (1982) found no clear connection but Ciochon (1986) finds a distinct difference between the oblong heads of cercopithecines and the rounder heads of the colobines. He feels that this difference in radial head shape is one that has an underlying phylogenetic basis separating the two groups.

Character 30 shows radial head shape. Cercopithecines generally have oblong heads; colobines much rounder ones. Head shape in T. gelada and T. oswaldi is oblong but shape is more variable in T. brumpti and T. darti; some are quite round.

In addition to head shape, the angulation of the proximal surface of the head also restricts pronation and supination. Pronograde quadrupeds are characterized by distinct proximal projections of the medial surface of the head (Conroy, 1976). This projection fits into the deep trochlear groove of the humerus severely limiting the ability of the radius to rotate into the supinated position. Terrestrial cercopithecines move with hands in a fully or partially pronated position and distribute the weight of their bodies through both the ulna and radius. The obliquely slanting radial head maintains lateral elbow stability when the head is in the pronated position. A less angulated head enhances supination and pronation.

Arboreal quadrupeds in the comparative sample possess a less angulated head (char. 31) while head slope is great in terrestrial species. Head slope is also relatively great in *T. gelada* and *T. oswaldi* but less so in *T. brumpti* and *T. darti*; in both species the articular surface around the head is larger indicating more supination and pronation in this species.

The importance of the biceps brachii muscle is reflected in the projection of the radial tuberosity. This muscle is a flexor of the elbow. Conroy (1976) found that in many pronograde mammals, this tuberosity is only weakly developed whereas in arboreal animals it is much larger. However, Birchette (1982) found no discernible pattern in the projection of the radial tuberosity and preferred locomotion among cercopithecoids.

Character 32 shows that cercopithecines and colobines do differ in radial tuberosity projection. The terrestrial species show only mildly raised tuberosities whereas they are much higher in arboreal species. All theropith species follow the trend seen in the terrestrial quadrupeds by possessing a weakly developed tubercle.

A sharp lateral border of the radial tuberosity restricts movement of the radius and is seen in terrestrial quadrupeds (char. 33). The exception is

Theropithecus which has a smooth lateral border. A rounded, smooth surface lateral to the radial tuberosity helps the play of the biceps tendon during pronation and supination (Conroy, 1976) and is seen most often in arboreal primates. *Theropithecus oswaldi*, *T. brumpti*, and *T. darti* all exhibit this character.

Shaft robustness is functionally associated with terrestrial locomotion. Increased stoutness relative to general body size implies an overall increase in the amount of compressive stress to be withstood by the bone, which in turn, may imply more massive musculature associated with terrestrial locomotion. Ciochon (1986) showed that living colobines have more slender radii than cercopithecines, however, the separation in the two groups is not compelling. These findings were confirmed by this study (char. 35). All theropiths have robust shafts with *T. oswaldi* and *T. brumpti* possessing very robust shafts.

On the shaft itself, arboreal quadrupeds show a large flattened volar groove for the broad origin of the flexor pollicus longus muscle (char. 36). This is associated with a strongly grasping, opposable pollex. Terrestrial quadrupeds have smaller muscle and less extensive volar surfaces.

In the theropiths, *T. gelada* and *T. oswaldi* do not possess an extensive groove, which is surprising given their highly opposable hands. But this groove is quite large in *T. brumpti* and *T. darti* suggestive of a broad origin for flexor pollicus longus muscle.

The projection of the styloid process of the radius is related to the projection of the ulnar styloid, and follows the same trend. The radial styloid articulates with the scaphoid providing a brace and stability to the wrist (Lewis, 1965). Longer styloids are found in most cercopithecines while colobines have relatively shorter styloids. All theropiths have long styloids.

Hand

One of the unique features of the hand of *T. gelada* is the high degree of opposability of the first and second digits (Napier & Napier, 1967; Etter, 1973). This is produced by the relative elongation of the thumb (especially of the pollical metacarpal) and the relative shortening of the proximal and middle phalanges of the index finger (Jablonski, 1986a). Geladas spend considerable feeding time plucking grasses with these highly opposable digits.

Carpals, metacarpals, and phalanges of extinct theropiths have been recovered from various sites but they have not yet been analyzed in detail (Krentz, unpub. data). Most hand bones are found isolated and thus probably represent different individuals. The most complete collection of hand bones from a single individual was recovered in Member E, Locality 865 of the Shungura Formation (Eck, 1977). Because of their proximity to *T. brumpti*, these specimens have been assigned to that species. Jablonski's (1986a) analysis of these elements clearly demonstrates that the elongated thumb and reduced index finger was present within the genus at an earlier age (the hand skeleton is dated at 2.2 Ma).

Preliminary examination of metacarpals and phalanges loosely associated with *T. brumpti* and *T. oswaldi* from other East African sites reveal a similar complex in the hand (Krentz, unpub. data). Little variation, other than size, is noted. The hand bones from the Hadar assigned to *T. darti* hint that this complex was present in the earliest members of the genus although these bones await further analysis.

Femur

Theropith femora differ from other terrestrial cercopithecoids in five areas (see Fig. 14.8): the articular surface of the head, the length of the neck, the height and configuration of the greater trochanter, the angulation of the shaft onto the condyles, and the unequal size of the condyles. Although some variation exists, these characters are consistently found within the genus.

The extent of the articular surface of the femoral head is associated with movements around the hip joint (Fleagle, 1976; Rose, 1983). A spherical head with the articular surfaces extending anteriorly and especially posteriorly onto the neck allows for relatively free movement about all axes of the hip with abduction and lateral rotation favoured (Rose, 1983). Posteriorly extended articular surfaces further imply hindlimbs that are held laterally for parasagittal movements associated with climbing or leaping (Fleagle, 1976).

Index M depicts the amount of lateral excursion of the articular surface as a percentage of the mediolateral length of the proximal end. *Cercocebus* (49.2) shows the greatest extension but *Theropithecus* is second longest at 46.8, closely related to *Colobus* at 45.5. *Papio* has the shortest extension with an index of 42.0. Among

the extinct theropiths, all show high indices – *T. oswaldi* (48.6), *T. brumpti* (48.0),
T. darti (45.3).

The relative length of the neck is associated with the positioning of the hindlimb
beneath the hip joint in a quadrupedal
stance (Harrison, 1982). In fully terrestrial
animals, the hip, knee, and ankle joints
tend to align vertically to efficiently move
the limb in a fore-aft manner (Howell,
1944). Longer necks provide more surface
area for the muscles of protraction and
retraction, and assist in keeping the hip
joint aligned in much the same way as the
extended greater tubercle in the humerus
aids in stability. In less fully adapted
quadrupeds (i.e. climbers and leapers),
necks are shorter to allow more parasagittal movement about the hip.

Index N is a measure of the relative
length of the femoral neck. The terrestrial
Papio (24.9) and *Cercopithecus* (24.6)
have high indices and long necks. *Colobus*
has a shorter neck (23.5) but *T. gelada* has
the shortest of the sample (20.9). This is
also seen in the extinct species –
T. oswaldi (20.5), *T. brumpti* (19.1), and
T. darti (22.5).

The greater trochanter acts as the end of
the true lever arms that act around the
femoral head and extend the hindlimb
(Rodman, 1979). A higher (i.e. more
proximal) greater trochanter allows for a
longer lever arm producing more power
per stroke as needed by terrestrial
quadrupeds in much the same fashion
seen in an increased greater tubercle of
the humerus. A longer greater trochanter
also aids in stability about the hip.

Index O gives the relative height of the
greater trochanter relative to femoral

length. *Theropithecus gelada* has the
highest greater trochanter (6.4); much
high than *Papio* (5.9) and almost twice as
high as *Colobus* (3.6). Once again, extinct
theropiths follow this trend: *T. oswaldi*
(6.2), *T. brumpti* (6.7), *T. darti* (6.0).

Distally, the shaft of theropiths angle
laterally onto the condyles producing a
'reverse' carrying angle of eight to ten
degrees. This angulation is functionally
associated with the short neck, lateral
excursion of the articular surface discussed above. Unlike other terrestrial
quadrupeds that have their hindlimb
beneath the hip – and hence have no carrying angle – theropiths splay their hindlimb laterally, not for locomotion but for
sitting during long periods of food gathering. Femoral shafts deviate laterally from
3° in *T. gelada* to 12° in *T. oswaldi*
(Krentz, 1988). This also occurs in
T. darti, and is a trait characteristic for the
genus.

Tibia

Functionally, the tibiae of the cercopithecoids do not vary significantly (see Table
14.3). Proximally, both subfamilies are
similar in morphology with only slight
variation on the posterior surface of the
intercondylar notch (char. 59). Condyles
are unequal in size reflecting the influence
of the femoral condyles. The tibial
tuberosities (chars 57 and 58) are broad
and increase with size. Distally,
cercopithecines show more differences
from colobines primarily around the ankle
joint with the former stressing stability
and the later flexibility but even these differences are not striking. Colobines pos-

sess an angulated medial malleolus (char. 63) and a pronounced notch for m. tibialis posterior (char. 62). Both characters aid in inverting the foot during locomotion and postural behaviour. *Theropithecus gelada* shares the latter two characters with *Colobus* indicating slightly more flexibility around the ankle than seen in other cercopithecines. The gelada inverts its foot much of the time while feeding (Maier, 1972) making a more flexible ankle advantageous. Within the extinct theropiths, little variation exists in tibia morphology. The tibia of these species is similar to that of modern *T. gelada*.

Foot

Like the hand bones, foot bones attributed to the genus *Theropithecus* are fairly abundant in the fossil record but have not yet been described in detail. The most extensive analysis was made by Jolly (1972) on *T. oswaldi* remains from Olorgesailie, Kanjera, and Olduvai. Among living Cercopithecinae, Jolly found that the more ground-living species had shorter and stouter proximal and middle phalanges of digits II–V. This is an almost clear progression in relative shortness from arboreal genera to terrestrial ones. Jolly found that the phalanges of *T. oswaldi* were more stout than those of the more terrestrial extant monkeys.

A second complex discussed by Jolly was the relative size of the hallux and the muscles that power its movements and degree of abductability. These are larger in arboreal monkeys. The relative development of the hallux is dependent on the absolute size of the animal as well as its

degree of arboreality. Reduction of the size and importance of the hallux as seen in terrestrial monkeys is expressed by a small and weak hallucial metatarsal, a relatively small articular surface of the first cuneiform, and a small articular surface on the second cuneiform. Jolly found each of these characters on *T. oswaldi* fossils from Olorgesailie.

Preliminary investigation of the foot bones of *T. brumpti* and *T. darti* show similar morphological characters (Krentz, unpub. data). There is very little variation, other than size, in the morphology of the astragalus or calcaneus. These latter two elements are morphologically similar to those of terrestrial monkeys. Although the foot bones of extinct theropiths need further analysis, it does appear that their morphology – and function – has changed very little through time.

Size of *Theropithecus*

Theropithecus gelada is a medium-sized Old World monkey with a body weight of 18–22 kg for males and 12–14 kg for females (Napier & Napier, 1967). This is slightly less than that of *Papio anubis*. Determining body size or weight for extinct theropiths is difficult because no complete or associated limbs are known. Hence, estimates of size are based upon either single specimens or a combination of specimens from different individuals. In either case, given the variation seen in time, size, sex, and geographical location, only very crude body-size approximations can be attempted. However, these can contribute to an understanding of the locomotor behaviour of these extinct species.

Comparisons of the length of the limb bones gives relative size differences between the species (see Table 14.7). Modern *T. gelada* are much smaller than *T. oswaldi* or *T. brumpti*. For example, whereas the average length of the humerus for a male gelada is 190 mm, the length for *T. oswaldi* ranges from 235–310 mm (Krentz, 1992). *Theropithecus brumpti* humeri average 264.1 mm. *Theropithecus darti* are also much smaller than these latter theropiths averaging an estimated 181 mm for male humeri.

Body weight estimates show similar size differences. Jolly (1972), using cross-sectional length of the humerus, radius, tibia, and femur, estimated the body weight of male *T. oswaldi* from Kanjera at 34.7 kg; females 20.8 kg. A very large male from Olorgesailie was estimated to have a weight of 62.8 kg. In this study, body weights were estimated using Jolly's Robusticity Quotient and Aiello's (1981) correlations using transverse diameters of the femur. When both approaches were used for individuals of the same species at the same site, differences in weight estimates ranged from 4 to 18 per cent; this is not unreasonable given the great variation in the collection.

Table 14.6 presents body weight estimates for the theropiths. The smallest members of the genus are also the oldest – *T. darti*. Body weights based mainly on femur diameters are estimated at 15.2 kg and 10.6 kg for males and females respectively. They are slightly smaller than modern gelada.

Body weights increase two to three fold from *T. darti* to both *T. brumpti* and *T. oswaldi*. Body weight for male *T. brumpti* range from 39.4 to 44.9 kg while females are much smaller at 24.7 kg. Similarly, the oldest specimens of *T. oswaldi* exhibit a dramatic increase in estimated weight from that seen in *T. darti*. *Theropithecus oswaldi* males from Shungura Formation average 38.5 kg; females 17 to 23 kg; *T. oswaldi* males are approximately 40 to 50 per cent larger than females. Younger (in time) members of *T. oswaldi* (i.e. from Olduvai and Olorgesailie) increase in size by another 20 kg to average 55–60 kg for males and 37–41 kg for females. These latter body sizes support Jolly's (1972) estimates showing that these theropiths were indeed among the largest of all cercopithecoids.

Sexual dimorphism

Jablonski (1986b) found significant amounts of sexual dimorphism in craniodental remains of extinct theropiths extending back to the earliest representatives of the *T. darti* lineage. She stated that sexual dimorphism was a 'hallmark' of theropith evolution. Estimates of sexual dimorphism based on estimates of body size derived from the limb bones of extinct theropiths confirm Jablonski's statements.

Table 14.6 presents degrees of sexual dimorphism within the lineage. All species exhibit significant amounts of dimorphism. Sexual dimorphism in body weight of modern *T. gelada* is approximately 67 per cent – females are roughly two-thirds the size of males. Sexual dimorphism is greater in *T. oswaldi* ranging from 55 to 75 per cent. *Theropithecus brumpti* males

Table 14.6 *Estimated body size of* Theropithecus *based upon Jolly's (1972) Robusticity Quotient (a) and/or Aiello's (1981) Femur Correlation (b).*

| Species | | Body weight (kg) | | Sexual dimorphism |
		(a)	(b)	F/M
T. darti	M	–	15.2	70.1
	F	8.2	10.6	
T. brumpti	M	39.4	44.8	55.0
	F	–	24.7	
T. oswaldi				
Shungura F.	M	38.5	–	?
	F	–	8.2 (n=1)	
Kanjera	M	–	–	?
	F	20.0	17.6	
Koobi Fora	M	–	42.5	55.1
	F	27.2	23.4	
Olorgesailie	M	58.7	60.3	62.8
	F	–	37.8	
Olduvai	M	53.7	55.4	75.5
	F	37.1	41.8	
T. gelada	M	20.5	21.6	67.3
	F	13.8	14.0	

are much larger than females with a sexual dimorphism index of 55 per cent, although the sample sizes used are quite small. The estimated sexual dimorphism in *T. darti* is 70 per cent, similar to modern gelada.

Limb proportions

Jolly (1972) stated that *T. oswaldi* possessed a relatively long humerus and a short forearm. This was based primarily upon specimens from Kanjera and Olorgesailie. Specimens of *T. oswaldi* from Koobi Fora generally follow this trend (Krentz, 1992). Table 14.7 presents estimated limb proportions and indices of the extinct theropith species. Sample sizes are small but humeri and radii are approximately equal in length in the sample of *T. oswaldi* from Koobi Fora, giving them an intercrural index of 100.1 and 100 for males and females, respectively, as opposed to the estimates made by Jolly for

Table 14.7 *Limb proportions in* Theropithecus.

Species		Humerus	Radius	Femur	R/H	H/F
T. darti	M	c. 181.2	c. 186.7	196.5	103.3	92.3
	F	c. 173.0	c. 165.0	c. 180.0	104.4	96.0
T. brumpti	M	264.1	222.2	261.0	84.1	101.2
	F	–	–	–	–	–
T. oswaldi						
Kanjera	M	c. 235.0	224.0	234.0	95.3	100.4
	F	209.0	198.0	201.0	94.7	104.0
Koobi Fora	M	c. 253.2	253.8	249.0	100.1	101.6
	F	223.8	223.3	241.7	100.0	92.5
Olorgesailie	M	310.0	–	297.0	–	104.0
Olduvai	M	278.9	270.1	279.0	96.9	100.0
T. gelada	M	190.0	196.1	194.3	103.8	98.4

Kanjera at 95.3. In *T. brumpti* samples from the Shungura Formation, the radii are much shorter than the humerus producing an intercrural index of 84.1. The intercrural index estimates for *T. darti* from the Hadar are quite close (103.3) to those of modern geladas.

Estimates of brachial indices (see Table 14.7) are closely aligned around 100 with modern *T. gelada* slightly under at 98.4 and *T. oswaldi* slightly over at 101.5. The brachial index for *T. brumpti* is 101.2. Interestingly, *T. darti* possess the lowest indices at 92.3 for males and 96.0 for females. Given the small sample sizes and the fact the estimates here are based on fossil bones from different individuals, the significance of these indices awaits recovery of more complete individuals.

Locomotion

As Day (1979) points out, the main difficulty in analyzing fossil postcrania is being able to recognize those morphological features or combination of features that reflect locomotor behaviour in the fossil forms. Since most of the differences in morphology are directly related to the way primates move, hang, or sit, the best approach to understanding locomotor behaviour in fossil species is to first consider how living primates have dealt with these problems (Fleagle, 1988). Locomotion in extinct theropith species is constructed by comparisons with *T. gelada*, and other cercopithecoids as described in the previous section.

Theropithecus gelada The gelada has always been considered one of the most terrestrial of the Old World monkeys, and certainly the analysis of their postcranial skeleton confirms this. *Theropithecus gelada* forelimbs are very similar to those of *Papio* with differences related not to locomotion but to the unique feeding adaptations seen in the gelada. For example, *T. gelada* has a narrower bicipital groove with a concave medial border of the deltoid tuberosity; these are related to increased flexibility at the shoulder, compared to *Papio*. Geladas also show an increase in the surface area for the attachment of the flexors and extensors of the digits and pollex. In addition, *Theropithecus* and *Papio* differ in the proportions of the hand and various aspects of the hindlimb; these latter characters are related to 'squatting' during long periods of sitting. However, the overall morphology of *T. gelada* is that of a terrestrial quadruped.

Theropithecus oswaldi Is very similar to *T. gelada* in postcranial morphology differing on only a few traits. Locomotor behaviour was undoubtedly very similar as well. *Theropithecus oswaldi* also aligns closely with terrestrial *Papio* in most traits except those that are unique to *Theropithecus*. Stability is important in the shoulder, elbow, and wrist joints in this large monkey. Locomotion was that of a terrestrial quadruped. *Theropithecus oswaldi* does have some traits – for example, extension of the medial epicondyle, highly concave medial surface of olecranon process – indicative of opposability of the pollex, and shuffling types of locomotor behaviour.

Theropithecus brumpti As seen in the list of characters in Tables 14.2 and 14.4, *T. brumpti* differs most from the other theropiths in a number of important characters. Most of these relate to increased flexibility at the shoulder joint. The elbow joint is adapted to stability for the most part but does show the increased surface areas for the digital musculature that the other theropiths display. This mosaic of functional traits is unique among the cercopithecoids. Given *T. brumpti* fossils are most often associated with a canopy forest along rivers (Eck & Jablonski, 1987), flexibility about the shoulder would be an advantageous adaptation for an arboreal quadruped. Jablonski (1986a) has shown that *T. brumpti* possessed the long pollex and short second digit seen in the later theropith. Hindlimb characters are similar to modern *T. gelada* indicating a squatting type of behaviour. The mosaic of characters seen in *T. brumpti* indicates that it was a terrestrial quadruped but spent more time in the trees than modern gelada or *T. oswaldi*.

Theropithecus darti Determining the locomotion of *T. darti* is difficult. Sample sizes are small and they were recovered from only one site. Many of the characters are variable and are not expressed as fully as they are in later theropiths. *Theropithecus darti* postcrania are more similar to the terrestrial *T. oswaldi/T. gelada* than to *T. brumpti* although the morphological characters are not as distinct. The smaller size of *T. darti* may be contributing to this. Stability is greater at the shoulder, elbow, and wrist joints than in *T. brumpti* while adaptations in the hand bones hint at opposable first and second digits. Hindlimb characters, forelimb proportions, and degrees of sexual dimorphism are similar to that seen in modern *T. gelada*. *Theropithecus darti* can be characterized as a generalized terrestrial quadruped with some arboreal tendencies.

Conclusion

The variation in the postcrania of *Theropithecus* is the result of locomotor and postural differences within the genus. This has both phylogenetic and evolutionary significance.

Recently, questions have arisen as to the interrelatedness of the theropith species. Eck & Jablonski (1984, 1987) have stated that, based upon craniodental remains, three lineages exist within the genus: (1) *T. baringensis–T. quadratirostris–T. brumpti*, (2) *T. darti–T. oswaldi*, and (3) *T. gelada* and its unknown ancestors. According to Eck & Jablonski (1987), *T. brumpti* diverged first from the common ancestor prior to four million years with *T. gelada* diverging from the *T. darti–T. oswaldi* line at a later date but before the appearance of *T. darti* at the Hadar Formation at 3.4 Ma.

Evidence derived from the postcrania support the divergence of *T. brumpti*, or perhaps more aptly, the early divergence of the 'non-brumpti' group of *T. darti/ T. oswaldi/T. gelada*. *Theropithecus brumpti* shares many traits with the arboreal quadrupeds (e.g. *Colobus*) that may be retentions of an as yet unknown ancestral morphotype. The locomotion and preferred habit of the ancestral cercopithecine is still unknown but both an arboreal (Napier & Napier, 1970) and a terrestrial (Andrews, 1916) ancestor have been proposed. *Theropithecus brumpti* either retained its arboreal life-style and skeletal morphology from the ancestral condition or 'reacquired' a series of characters as it readapted to the arboreal habitat after a terrestrial existence. Since *T. brumpti* is much larger than most living arboreal quadrupedal primates, and in fact most terrestrial primates, the latter scenario would be highly unlikely. In either case, the postcrania of *T. brumpti* are unique for the genus.

Evidence for a *T. darti–T. oswaldi* linkage, or the separation of *T. gelada* from this group is less clear but there is no evidence to reject these propositions. *Theropithecus darti* exhibits characters not seen in *T. brumpti* but which are present in *T. oswaldi* and *T. gelada*. *Theropithecus darti* evolving into *T. oswaldi* is clearly possible. Furthermore, there are no characters to eliminate *T. gelada* from this group. The greatest difference within these two lineages is that of size – *T. oswaldi* increased dramatically while *T. gelada* increased by a much smaller amount from that of *T. darti*.

Differences in locomotor behaviours are also affected by the environment and habitat preferences of the species. *Theropithecus darti* remains are found associated with forested or more closed habitats than later species although open grasslands may have been present. *Theropithecus darti* locomotion may have been similar to that of the modern *Cercopithecus* with both arboreal and terrestrial components. *Theropithecus brumpti* also exhibits morphological characters associated with arboreal quadrupedalism. Their fossils are recovered in deposits indicative of a riverine forest. *Theropithecus oswaldi*, in turn, is found associated with open woodlands and grasslands even in areas where *T. brumpti* are present. It is not coincidental that in the Shungura Formation, the density of *T. brumpti* decreases while *T. oswaldi* increases in response to the changing environmental conditions towards a more open habitat. The adaptation of extant *T. gelada* appears to have been the result of a long evolutionary process towards terrestrial life-styles.

Theropithecus postcrania can be characterized by their adaptations towards terrestrial locomotion, the possession of a highly opposable hand, a specialized femur and hindlimb adapted for long periods of upright sitting, and a high degree of sexual dimorphism. These adaptations are present, to some extent, in all members of the lineage regardless of size. They are found in the oldest members of the genus indicating that these adaptations were basic to the lineage and contributed to their separation from a Pliocene ancestor. Theropiths adopted this life-style early in their history, and given the longevity of the genus, it has been a highly successful one.

Acknowledgements

This study was funded by a grant from the L.S.B. Leakey Foundation. Additional funding was provided by Dept. of Anthropology, University of Washington. Special thanks to Dr Nina Jablonski, Dr Robert Foley, Dr Gerry Eck, Catherine Krentz, Christopher Thornton, Madalene Krentz, Nancy Christensen, and Anna Krentz for their support and patience.

References

AIELLO, L. (1981). The allometry of primate body proportions. *Symposium of the Zoological Society, London*, 48, 331–53.

ANDREWS, C.W. (1916). Notes on a new baboon (*Simopithecus oswaldi* gen. et. sp. nov.) from the (?) Pliocene of British East Africa. *The Annals and Magazine of Natural History*, 18, 410–19.

ASHTON, E.H., FLINN, R.M., OXNARD, C.E. & SPENCE, T.F. (1976). The adaptive and classificatory significance of certain quantitative features of the forelimb in primates. *Journal of Zoology, London*, 179, 515–56.

BIRCHETTE, M.G. (1982). *The Postcranial Skeleton of* Paracolobus chemeroni. Ph.D. Dissertation, Harvard University.

CAMPBELL, B.G. (1974). *Human Evolution*. Chicago: Aldine Publishing Company.

CIOCHON, R.L. (1977). A methodological approach to the study of the Omo group cercopithecoid postcranial remains. *American Journal of Physical Anthropology*, 47, 123.

CIOCHON, R.L. (1986). *The Evolution of the Cercopithecoid Primate Forelimb with Special Reference to African Plio-Pleistocene species*. Ph.D. Dissertation, University of California, Berkeley.

CLARK, W.E. LeGROS & THOMAS, D. (1951). Associated jaws and limb bones of *Limnopithecus macinnesi*. Fossil Mammals of Africa, No. 3 (British Museum of Natural History, London), pp. 1–40.

CONROY, G.C. (1974). *Primate Postcranial Remains from the Fayum Province, Egypt, UAR*. Ph.D. Dissertation, Yale University.

CONROY, G.C. (1976). Primate postcranial remains from the Oligocene of Egypt. *Contributions to Primatology*, 8, 1–134.

DAY, M.H. (1979). The locomotor interpretation

of fossil primate postcranial bones. In *Environment, Behavior and Morphology: Dynamic Interactions in Primates*, ed. M.E. Morbeck, H. Preuschoft & N. Gomberg, pp. 245–58. New York: Gustav Fischer.

DUNBAR, R.I.M. (1977). Feeding ecology of gelada baboons: a preliminary study. In *Primate Ecology: Studies of Feeding and Ranging in Lemurs. Monkeys, and Apes*, ed. T.H. Clutten-Brock, pp. 251–73. London: Academic Press.

ECK, G.G. (1976). Cercopithecoidea from Omo Group Deposits. In *Earliest Man and Environments in the Lake Rudolf Basin*, ed. Y. Coppens, F.C. Howell, G.H. Isaac & R.E.F. Leakey, pp. 332–44. Chicago: University of Chicago Press.

ECK, G.G. (1977). Diversity and frequency distribution of Omo Group Cercopithecoidea. *Journal of Human Evolution*, 6, 55–63.

ECK, G.G. & JABLONSKI, N.G. (1984). A reassessment of the taxonomic status and phyletic relationship of *Papio baringensis* and *Papio quadratirostris* (Primates: Cercopithecidae). *American Journal of Physical Anthropology*, 65, 109–34.

ECK, G.G. & JABLONSKI, N.G. (1987). The skull of *Theropithecus brumpti* compared with those of other species of the genus *Theropithecus*. In *Les Faunes Plio-Pléistocènes de la Basse Vallée de l'Omo (Éthiopie)*. Tome 3. Cercopithecidae de la Formation de Shungura, ed. Y. Coppens & F.C. Howell, pp. 11–122. Cahiers de Paléontologie, Travaux de Paléontologie Est-Africaine. Paris: Éditions du Centre National de la Recherche Scientifique.

ETTER, H.F. (1973). Terrestrial adaptations in the hands of Cercopithecinae. *Folia Primatologica*, 20, 331–50.

FLEAGLE, J.G. (1976). Locomotor behavior and skeletal anatomy of sympatric Malaysian leaf-monkeys (*Presbytis obscura* and *Presbytis melalophos*). *Yearbook of Physical Anthropology*, 20, 440–63.

FLEAGLE, J.G. (1988). *Primate Adaptation and Evolution*. New York: Academic Press.

FLEAGLE, J.G. & SIMONS, E.L. (1982). The humerus of *Aeqyptopithecus zeuxis*, a primitive anthropoid. *American Journal of Physical Anthropology*, 59, 175–93.

FORD, S.M. (1980). Phylogenetic relationships of the Platyrrhini; The evidence of the femur. In *Evolutionary Biology of the New World Monkeys and Continental Drift*, ed. R. Ciochon, & B. Chiarelli, pp. 317–29. Plenum Press, New York.

GRAY, J. (1968). *Animal Locomotion*. London: William Clowes and Sons.

HARRIS, J.M. (1978). Paleontology. In *Koobi Fora Research Project*, vol. 1, ed. M.G. Leakey & R. Leakey, pp. 32–63. Oxford: Clarendon Press.

HARRIS, J.H., F.H. BROWN & M.G. LEAKEY (1988). Stratigraphy and paleontology of Pliocene and Pleistocene localities west of Lake Turkana, Kenya. *Contributions in Science* (399), pp. 1–125. Los Angeles: Natural History Museum of Los Angeles County.

HARTMAN, C.G. & STRAUSS, W.L. (1933). *Anatomy of the Rhesus Monkey*. New York: Hafner Publishing Co.

HARRISON, T. (1982). *Small Bodied Apes from the Miocene of East Africa*. Ph.D. Dissertation, University of London.

HOWELL, A.B. (1944). *Speed in Animals: Their Specialization for Running and Leaping*. Chicago: University of Chicago Press.

HOWELL, A.B. & STRAUSS, W.J., Jr. (1933). The muscular system. In *The Anatomy of the Rhesus Monkey* (Macaca malatta), ed. C. Hartman & W.L. Strauss, Jr., pp. 89–175. New York: Hafner Publishing Company.

ISSAC, G. (1977). *Olorgoresailie*. New York: Academic Press.

JABLONSKI, N.G. (1986a). The hand of *Theropithecus brumpti*. *Primate Evolution*. Proceedings of the 10th Congress of the International Primatological Society, vol. 1, pp. 173–82. Cambridge: Cambridge University Press.

JABLONSKI, N.G. (1986b). Patterns of sexual dimorphism in *Theropithecus*. In *Sexual Dimorphism in Living and Fossil Primates*, ed. M. Pickford & B. Chiarelli, pp. 171–82. Florence: Il Sedicesimo.

JENKINS, F.A. (1973). The functional anatomy and evolution of the mammalian humero-ulnar articulation. *American Journal of Anatomy*, 137, 281–98.

JOLLY, C.J. (1965). *Origins and Specializations of the Longfaced Cercopithecoidea*. Ph.D. thesis, University of London.

JOLLY, C.J. (1967). The evolution of the baboons. In *The Baboon in Medical Research*, vol. 2, ed. H. Vagtborg, pp. 427–57. Austin: University of Texas Press.

JOLLY, C.J. (1970). The seed eaters: a new model of hominid differentiation based on a baboon analogy. *Man*, 5, 5–26.

JOLLY, C.J. (1972). The classification and natural history of *Theropithecus (Simopithecus)* (Andrews, 1916), baboons of the African Plio-Pleistocene. *Bulletin of the British Museum (Natural History), Geology*, 22, 1–123.

KNUSSMAN, R. (1967). Humerus, Ulna und Radius der Simiae. *Bibliotheca primatologica*, Fasc. 5.

KRENTZ, H.B. (1988). The femur of *Theropithecus*: evidence for the appearance of shuffling behavior. *American Journal of Physical Anthropology*, 75(2), 234.

KRENTZ, H.B. (1992). *The Forelimb Anatomy of Theropithecus brumpti and Theropithecus oswaldi from the Shungura Formation, Ethiopia*. Ph.D. Dissertation, University of Washington.

LARSON, S.G. & STERN, J.T. (1989). Role of supraspinatus in the quadrupedal locomotion of vervets (*Cercopithecus aethiops*): implications for the interpretation of humeral morphology. *American Journal of Physical Anthropology*, 79, 369–77.

LEAKEY, L.S.B. & WHITWORTH, T. (1958). Notes on the genus *Simopithecus*, with a description of a species from Olduvai. *Corydon Memorial Museum Occasional Papers*, 6, 3–14.

LEAKEY, M.G. (1976). Cercopithecoidea of the East Rudolf succession. In *Earliest Man and Environments in Lake Rudolf Basin*, ed. Y. Coppens *et al.*, pp. 345–50. Chicago: University of Chicago Press.

LEAKEY, M.G. & LEAKEY, R.E.F. (1976). Further Cercopithecinae (Mammalia, Primates) from the Plio/Pleistocene of East Africa. *Fossil Vertebrae of Africa*, 4, 121–46.

LEWIS, O. (1965). Evolutionary changes in the primate wrist and inferior radio-ulnar joint. *Anatomical Record*, 151, 275–86.

MAIER, W. (1972). Anpassungstyp und systematische Stellung von *Theropithecus geleda* Rüppell, 1835. *Zietschrift von Morphologie Anthropologia*, 63, 370–84.

NAPIER, J.R. & DAVIS, P.R. (1959). The forelimb skeleton and associated remains of *Proconsul africanus*. *Fossil Mammals of Africa (British Museum, Natural History)*. 16, 1–69.

NAPIER, J.R. & NAPIER, P.H. (1967). *A Handbook of Living Primates*. London: Academic Press.

NAPIER, J.R. & NAPIER, P.H. (1970). *Old World Monkeys*. London: Academic Press.

OXNARD, C.E. (1963). Locomotor adaptations in the primate forelimb. *Symposium of the Zoological Society, London*, 10, 165–82.

RODMAN, P.S. (1979). Skeletal differentiation of *Macaca fascicularis* and *Macaca nemestrima* in relation to arboreal and terrestrial quadrupedalism. *American Journal of Physical Anthropology*, 51, 51–62.

ROSE, M.D. (1974). Postural adaptations in New and Old World monkeys. *Primate Locomotion*, ed. F. Jenkins, pp. 201–22. New York: Academic Press.

ROSE, M.D. (1983). Miocene hominid postcranial morphology. Monkey-like, ape-like, neither or both? In *New Interpretations of Ape and Human Ancestry*, ed. R. Ciochon & R. Corruccini. New York: Plenum Press.

SARICH, V.M. (1970). Primate systematics with special reference to Old World monkeys: a protein perspective. *Old World Monkeys*, ed. J.R. Napier & P.H. Napier, pp. 34–53. London: Academic Press.

SAVAGE, R. (1957). Quadrupedal locomotion. *Proceedings of the Zoological Society, London*, 129, 151–72.

SCHON-YBARRA, M. & CONROY, G.C. (1978). Nonmetric features in the ulna of *Aegyptopithecus, Alouatta, Ateles*, and *Lagothrix*. *Folia Primatolgica*, 29, 178–95.

SCHULTZ, A. (1970). The comparative uniformity of the Cercopithecoidea. *Old World Monkeys*, ed. J.R. Napier & P.H. Napier, pp. 3–17. London: Academic Press.

SINGER, R. (1962). *Simopithecus* from Hopefield, South Africa. *Bibliotheca primatologica*, 1, 43–70.

SZALAY, F.S. & E. DELSON (1979). *Evolutionary History of the Primates*. Academic Press, New York.

TAIEB, M., JOHANSON, D.C., COPPENS, Y. & ARONSON, J.L. (1976). Geological and palaeontological background of Hadar hominid site, Afar, Ethiopia. *Nature*, 260, 289–93.

WALKER, A. (1974). Postcranial remains of the Miocene Lorisidae of East Africa. *American Journal of Physical Anthropology*, 33, 249–61.

WRANGHAM, R.W. (1980). Bipedal locomotion as a feeding adaptation in gelada baboons, and its implication for hominid evolution. *Journal of Human Evolution*, 9, 329–31.

ZIEMER, L.K. (1978). Functional morphology of forelimb joints in the woolly monkey *Lagothrix lagothricha*. *Contributions to Primatology*, 14, 1–130.

PART IV

Behaviour and ecology of living and fossil species of *Theropithecus*

15 Social organization of the gelada

R. I. M. DUNBAR

Summary

1. Gelada live in a multi-level social system consisting of at least three increasingly inclusive groupings (coalitions, reproductive units and bands).
2. The basic social group is the one male reproductive unit, consisting of a single breeding male and up to 12 reproductive females, plus their dependent young. Some units contain additional adult males.
3. Reproductive units consist of a number of longterm alliances between two to three reproductive females and their dependent offspring, and these constitute the lowest grouping level in gelada society. Reproductive units that share a common ranging area are termed a *band*. The band is a relatively closed social unit.
4. Although gelada have low reproductive rates, mortality rates are so low that population growth rates are among the highest recorded for any primate population.
5. The cohesion of reproductive units through time is mainly a function of the relationships among the reproductive females. Males are relatively peripheral, being used as substitute females only by those individuals who lack close relatives with whom to form a more conventional alliance.
6. Females remain in their natal units throughout their lives. Males leave their natal units as subadults to join an all-male group; some two to four years later, they return to the reproductive units to acquire their own breeding females.
7. Males can pursue two main options for acquiring breeding females: they can either (a) take over an entire unit intact after challenging a haremholder or (b) join a unit as a submissive follower in order to build up a nuclear unit-within-a-unit with one or two of the socially more peripheral females.

Introduction

Behaviour acts as the interface between the environment and the processes of evolution. Evolution itself is neither more nor less than the consequence of successful reproduction, for the success with which animals reproduce themselves is what ultimately determines the course of evolution for the taxon to which they belong. But successful reproduction in this sense depends on the success with which animals are able to solve the immediate problems of survival, mating, and rearing in environments that are not always conducive to these activities. Primates are, above all, social animals and their solutions to these problems are primarily social.

Hence, if our aim is to understand the evolutionary history of a taxon, elucidating its social behaviour will be at least as important as understanding the ecological factors that challenge it. Social behaviour is not simply a consequence of environmental conditions, rather, it constitutes the animals' collective attempts to circumvent the problems posed by the environment. Although we can, at present, say little about the social behaviour of extinct species, our understanding of the forces that shape the social systems of contemporary species is growing exponentially with each passing decade. Perhaps the most important lesson of recent years has been the recognition that the behaviour of individual species cannot necessarily be inferred by analogy from the behaviour of other species, even when they are closely related (see Dunbar, 1988). If we are to achieve anything of consequence in this respect, it must be done by a process of extrapolation from general principles applied to the particular ecological and demographic contexts of the species concerned.

In this chapter, I briefly summarize the main characteristics of the gelada's social system and their demographic processes. This account draws on field studies carried out by myself and my wife at Sankaber in the Simen Mountains on the northern edge of the Ethiopian plateau and at Bole near the southern limit of the species' range during 1971–72 and 1974–75 (Dunbar & Dunbar, 1975; Dunbar, 1984) and by Kawai and his colleagues in the Gich area of the Simen Mountains during 1973 (Kawai, 1979). The relevant data and analyses are given in full in the following publications: Dunbar & Dunbar (1975), Kawai (1979), Dunbar (1978, 1979, 1980a, b, 1983a, b, c, d, 1984, 1986, 1989). But before doing so, let me reiterate once more the point that the social behaviour of the extant gelada cannot necessarily tell us anything about the behaviour of their extinct sister-species.

The gelada social system

The gelada social system consists of a tiered hierarchy of social groupings, each more inclusive than the one below (Kawai et al., 1983). The main groupings and their demographic characteristics are summarized in Table 15.1.

The basic social group is the reproductive unit which typically consists of a single breeding male and four or five (maximum 12) reproductive (i.e. post-pubertal) females and their dependent young. About one quarter of the units contain additional adult males. About half of these are former harem males who, despite having been displaced by younger rivals, continue to remain within their units for some years. The rest are young adult males who have joined a unit as a 'follower' – a subordinate male who gradually builds up the nucleus of a unit of his

Table 15.1 *Social units of the gelada.*

Unit	Demographic structure
Reproductive unit	1–12 reproductive females, plus dependent young, with one to four adult males.
All-male group	2–15 non-breeding males aged 3–10 years.
Band	2–27 reproductive units, plus one to three all-male groups, which share a common ranging area.
Herd	Temporary aggregation of 2–60 reproductive units, usually from the same band.
Community	One to four bands whose ranging areas overlap extensively and whose reproductive units can form mixed band herds relatively easily.

own by developing grooming (and later mating) relationships with one or two of the more peripheral females. Table 15.2 summarizes the typical compositions of these units in the three study areas. The reproductive units themselves are highly structured and consist of a number of alliances formed between two or three females and their dependent young.

A number of reproductive units share a common ranging area and therefore tend to forage together during the day more often than they do with other units in the vicinity. This level of grouping is referred to as a *band* (Kawai *et al.*, 1983). Bands typically consist of 2–27 reproductive units, plus one to three all-male bachelor groups – in all around 100 animals of all ages and both sexes (see Table 15.3). The largest bands recorded contained 325–50 animals (Main band at Sankaber in 1971 prior to fission; Emetgogo band at Gich in 1973: see Kawai, 1979; Dunbar & Dunbar, 1975). At any one time, the units of a band may be dispersed throughout their ranging area either on their own or in the company

of other units in herds of varying size. The herds themselves are very unstable in composition, with units leaving and joining more or less at will. Sometimes, herds may consist of reproductive units from several different bands in areas where their respective ranges overlap.

Kawai *et al.* (1983) identified at least one further level of grouping (the *community*), which consists of those bands whose ranging areas overlap extensively and who therefore tend to be found in mixed-band herds more often than is normally the case. The precise significance of this level of grouping remains obscure.

The composition of these various groupings changes gradually over time as a result of births and deaths, with only the occasional group fission resulting in a major change in size. Permanent migration of reproductive units between bands appears to be very rare, with no certain case of transfer being observed in 120 unit-years of observation. In contrast, emigration by reproductive units to new ranging areas is more common and occurs

Table 15.2 *Demographic structure of reproductive units in the three study populations.*

Study area	Mean number per unit:				Total size		Units with > 1 male	N
	Males		Reproductive	Immatures*	mean	range	· (%)	
	adult	subad	females					
Bole Valley	1.1	0.6	5.9	9.5	17.1	8–28	20.0	10
Sankaber (Simen)	1.3	0.1	4.1	6.5	12.0	3–26	38.5	48
Gich (Simen)	1.2	0.3	3.9	4.4	9.9	2–17	22.8	31

* Animals that have not yet undergone puberty (usually around 3.5 years of age).

Source: Dunbar (1984).

Table 15.3 *Demographic structure of the three study populations.*

Study area	Number of bands	Band size mean	Band size range	Mean number per band reproductive units	Mean number per band all-male groups
Bole Valley	3	60.3	48–78	3.3	1.3
Sankaber (Simen)	11	131.5	30–262	10.7	1.5
Gich (Simen)	6	107.2	27–170	9.7	1.5

Source: Dunbar (1984).

with a probability of 0.032 per unit per year (Ohsawa & Dunbar, 1984). Such emigrations, however, occur only as a consequence of the fission of a band and usually involve two to six reproductive units moving away to establish a new ranging area elsewhere.

Although birth rates are relatively low (typically in the order of one infant every 18–33 months, depending on the habitat), mortality rates are unusually low for a free-ranging primate population. Dunbar (1980a) estimated that nearly 90 per cent of infants born survive to puberty (four years of age). As a result, population growth rates are relatively high: Ohsawa & Dunbar (1984) reported a mean growth rate for six bands of 13.7 per cent per annum (with only one of the bands having a growth rate of less than 10 per cent). Such a high growth rate clearly could not be sustained indefinitely without resulting in serious overcrowding. In fact, bands undergo fission at intervals of eight to nine years, with a section of the band migrating out into a new ranging area in a less densely used habitat lower on the escarpment face. This habitat appears to act as a demographic sink in which mortality rates

exceed birth rates and recruitment to the population occurs mainly through immigration from above.

On the basis of protein polymorphism distributions, Shotake (1980) estimated a rate of gene flow between bands of only about five per cent per generation. This is commensurate with the observed migration rates which suggest an equivalent rate of gene exchange of about 6.4 per cent per generation. Most of this comes through the movement of adult males between bands during the process of acquiring reproductive units. From this, we may infer that the members of a given band are closely related to each other.

The reproductive units themselves are rather similar in their demographic processes. Females remain within their natal units throughout life. Transfers between units by females seem to be extremely rare: only one female was ever known to have transferred from one unit to another (and that within the same band) in either of the two Simen study areas (equivalent to a rate of 0.005 per female per year). Some confirmation of this is provided by Shotake's (1980) finding that reproductive units are significantly more homogenous in their protein polymorphisms than the average for the bands to which they belong.

One consequence of this is that, all other things being equal, reproductive units will grow in size with time. Dunbar (1984) estimated that, on average, the number of females in a unit of the Sankaber population doubles every 6.7 years. Sustained growth of this magnitude would clearly be impossible for any length of time. In fact, it is offset by the periodic fission of units into two daughter units. At Sankaber, fission occurred at a rate of 0.12 per unit per year. However, only large units with at least four females ever underwent fission, and the fission rate for large units (0.156 per year) exactly matches the harem doubling time. During her 10-year reproductive lifetime, then, the average female can expect her harem size (the number of reproductive females in her unit) to vary considerably and her unit to undergo fission approximately 1.5 times. Fission appears to occur when the number of females in the group is such as to yield two separate matrilineages (Dunbar, 1989). This probably happens when two grandmothers who are sisters die, leaving no direct familial links between their respective female lineages. With no residual social ties to hold them together, the females remaining in the unit may be more willing to separate into different units.

Structure of reproductive units

The reproductive units are highly structured, with a characteristic division into small sets of two to three post-pubertal females who groom almost exclusively with each other and who support each other both in altercations with members of their own unit and during agonistic encounters with other reproductive units. There is strong circumstantial evidence from the field to suggest that the females who form these coalitions are close relatives (mother–daughter or sisters), and this is supported by data from a captive group of known pedigree (Dunbar, 1982).

These alliances are serviced by grooming and can be shown to have important functional consequences for the females. A female is significantly more likely to support her grooming partner when she becomes involved in an agonistic interaction than she is to support a female whom she rarely grooms; females who are members of such a dyad are significantly less likely to be attacked by other members of the group; young and old females who are members of a coalition occupy significantly higher dominance ranks within the unit than those that are not; and, finally, females who are members of a coalition have a significantly higher birth rate than those of the same age and rank that are not (Dunbar, 1980b, 1989).

This latter point is crucial to an understanding of the social behaviour of the gelada. The reproductive units themselves seem designed to buffer the females against the stresses imposed on them when they are in large herds of several hundred animals (Dunbar, 1986). Under these conditions, reproductive units are spread out over a significantly smaller area and become involved in many more agonistic interactions with neighbouring units than when they are on their own or in small herds of only three or four units. Most encounters between units are initiated by the females, who seem unwilling to tolerate the adult members of other units encroaching into their own unit's social space. The harem males become involved only if the encounter escalates to include most of the females in his unit (Dunbar, 1983c). Such interactions rarely extend to physical contact, even when the males become involved; rather, in most cases they are settled by the two units moving apart after an exchange of sometimes intense threats. Within the units themselves, it seems that the coalitions among the females are mainly concerned with the same problem of reducing the stresses of crowding once the unit exceeds a certain size.

Females low in the hierarchy suffer an accumulation of low-level aggression and harassment from the females that rank above them, and this accumulation becomes increasingly serious as the female's rank declines. From extensive experimental studies of captive primates, as well as work in human infertility clinics, we now have a very detailed understanding of how stress disrupts the reproductive endocrinology of female primates, leading to high frequencies of anovulatory cycles and reduced conception rates. In the wild gelada, this is reflected in a decline in birth rate with decreasing rank (once the influence of age has been removed) and in the fact that low-ranking females take more menstrual cycles to conceive than high-ranking ones (Dunbar, 1980b).

In this context, coalitions function to reduce the stress on females by minimizing the frequency with which they are har-

assed by other members of their unit. In the gelada, an individual female's dominance rank within her unit depends on two factors, her intrinsic power due to her own physical capacities (largely an inverted-U-shaped function of age) and the extrinsic power that she gains through support from other more powerful individuals. Because an alliance with a higher ranking individual causes a female's rank within her unit to rise to a position immediately below that of her ally, females of low intrinsic power (i.e. younger and older females) will tend to want to establish alliances with the most dominant group-members available to them. In most cases, these will tend to be close female relatives, partly because long-term familiarity will generally make it easier to establish the requisite relationships with kin and partly because high-ranking females will prefer to form alliances with immediate kin rather than with less closely related females in order to benefit through kin selection (see Dunbar, 1988). In general among the gelada, time budget constraints limit the females to forming alliances with one (at most two) other reproductive females (Dunbar, 1984).

These coalitionary relationships among the females have important implications for the stability of reproductive units through time. In contrast to the one-male units of the hamadryas, the death or experimental removal of the male does not lead to the immediate dissolution of the gelada reproductive unit (compare experimental manipulations by Kummer, 1968 and Kawai, 1979). This is a consequence of the fact that the gelada unit is essentially a group of females with very strong coalitionary relationships with each other, with the male being socially peripheral. In contrast, the hamadryas one male unit exists only because the male forces the females to stay with him by punishing them if they stray too far away (Kummer, 1968; Dunbar, 1983d).

Detailed modelling of the females' reproductive strategies indicates that coalitions formed with very close female relatives (mother, daughters, sisters) are much more profitable than coalitions with less closely related individuals. From a female's point of view, the optimal alliance is with her first-born daughter, and the value of daughters as allies declines with their birth rank. This is mainly a consequence of the fact that, given the life history patterns typical of the Simen gelada, a mother is likely to die before a late-born daughter has reached an age where she can act as an effective coalition partner.

In this respect, the harem male is a less valuable ally than a female relative. The male has little or no effect on the female dominance hierarchy so that, unlike a female ally, he cannot raise his partner's rank in the hierarchy by supporting her against her rivals. However, because the male's partner spends a significant amount of time grooming with him, she can mate with the male with less interference from the more dominant females than other females can. As a result, the male's main female grooming partner gains an estimated eight per cent extra offspring over the course of her lifetime (roughly half that gained from an alliance with a female relative). As may be expected,

females who have no close female relatives compete for the male as a social partner and ally, and the most dominant among them is usually able to monopolize access to him. In contrast, females that have close female relatives with whom to form alliances show little interest in the male.

That females treat the male simply as a substitute female is clear from the fact that the detailed structure of the relationship between the male and his main partner female is identical to that between two female coalition partners. In contrast, females who are not the male's grooming partner (and especially those who have alliances with other female relatives) tend to respond in a most perfunctory manner to his social advances. Whereas the male's partner female grooms him about as often as he grooms her, the other females in the unit are less willing to reciprocate grooming bouts with the male; even when they do groom him, they tend to do so for less time than his partner female does. The male has to work harder to elicit grooming from non-partner females, tends to initiate most interactions with them and to find himself being abandoned by them more often than is the case when he interacts with his partner female. Moreover, the male's partner female continues to remain his grooming partner even after he has been displaced as harem-holder, despite the fact that she now mates with the new harem male.

The main reason why the male is of less interest as a social partner seems to be that he is rarely around for more than a few years. As a result, a female who establishes an alliance with her harem male early on in her reproductive career is likely to find herself without an ally at the crucial point later on in her life when her own physical powers start to wane. Hence, it is mostly older females who tend to show an interest in the harem male. In contrast, it is the younger low-ranking females who show most interest both in the new harem male after a takeover has occurred and in any young follower males that happen to join the unit.

Follower males are particularly valuable for two reasons. First, they are relatively young and are therefore likely to be around for much longer than the incumbent harem male. Second, a follower will precipitate fission of the unit about two years after joining it and in doing so, will take with him up to half the females in the unit. For a low-ranking female, a shift of allegiance to the follower at this point will automatically result in a dramatic increase in her absolute rank (and hence a corresponding reduction in the amount of reproductive suppression she suffers).

Male reproductive strategies

Gelada males usually leave their natal units soon after puberty and join an all-male group (Dunbar & Dunbar, 1975). The evidence suggests that they tend to join a group in which they already have relatives (Dunbar, 1984). A male spends two to four years in an all-male group and then sets about acquiring a group of females with whom he can breed. Although nearly full-grown by this stage, six-year-old males are much smaller than most harem-holders, and are thus at a significant disadvantage in terms of fighting power, most of whom are typically around ten years of age.

A male's ability to reproduce depends on whether or not he can monopolize a group of breeding females. (It should be noted that, in this context, 'monopolize' refers not to the male's ability to dictate how the females should behave – something he is virtually powerless to do – but rather to the extent to which he can prevent other males from gaining access to the females in his unit.) Essentially, a gelada male has two options open to him. He can either take over an existing unit intact by ousting the incumbent male or join a unit as a subordinate follower in order to acquire a few peripheral females with whom to form the nucleus of a unit of his own (Dunbar & Dunbar, 1975; Dunbar, 1984).

These two strategies turn out to be characterized by rather different features (Table 15.4). Compared to males that become followers, those that opt for a takeover strategy are: (1) significantly older, (2) acquire more females at the outset of their reproductive careers (because they take over an entire unit intact), (3) suffer higher failure rates, and, finally, (4) do not have as long a tenure as harem-holder. While takeovers invariably involve serious fighting and are not always successful, the follower strategy minimizes the risks of aggression. Prospective followers are invariably greeted with aggression by the incumbent male of the unit they seek to join, but they always respond submissively. Providing the male responds submissively to the incumbent male's attacks and makes no attempt to interact with any of the females, he will usually be allowed to join the unit within a day or so. Once there, he begins first to interact with the unit's juveniles, and then later with

Table 15.4 *Characteristic features of the two strategies whereby gelada males acquire their own reproductive units.*

	Takeover	Follower
Mean number of females at start	4.71	1.89
	(N=7)	(N=10)
Probability of wounding per attempt	0.50	0.00
	(N=16)	(N=7)
Probability of success per attempt	0.70	1.00
	(N=10)	(N=7)
Mean age at start of tenure (years)	7.7	6.2
	(N=15)	(N=16)
Expected tenure (years)*	3.46	6.1
Expected number of offspring**	4.94	4.72

* Estimated from observed harem growth rates and takeover rates.
** Total lifetime reproductive success predicted by a simulation model.
Source: Dunbar (1984).

one or two of the unit's lower ranking and socially more peripheral females. Over a period of about two years, he builds up what amounts to a nuclear harem within the unit. At this point, he then leaves, taking his females with him, to establish an independent unit of his own.

Although he generally remains submissive towards the harem-holder throughout this period, the follower will often support the females when they become involved in disputes with their harem male. In exceptional circumstances, this may even give a follower an opportunity to take over the unit if the harem male begins to show signs of weakness. However, this is likely to occur only rarely because other males from the all-male groups are more likely to be in a better position in terms of age and power to challenge the harem-holder

before a young follower reaches an age at which he can do so.

When a young male opts for a takeover, he necessarily commits himself to serious risks: takeover contests involve the most serious fighting ever observed among the gelada, and the protagonists often incur injuries from razor-sharp canines. Whether or not a male succeeds in taking over the unit once he has begun to challenge for it depends more on the females' collective willingness to desert in his favour than on the two males' respective fighting abilities. In fact, the fighting between the two males is less about control over the unit as such than about whether the intruder male is to be permitted to interact with the females. Most of the intruder's time is taken up with approaching the females and soliciting grooming from them, while the incumbent male alternates between chasing the intruder away and grooming frenetically with all his females (apparently in a desperate attempt to strengthen their loyalty to him). Only if the intruder refuses to be driven away will the two males engage in face-to-face combat during which they fence with exposed canines.

Whether the incumbent male succeeds in driving the intruder away or loses his unit to the intruder depends wholly on the reactions of the females. If enough of the females are sufficiently motivated to transfer allegiance to the new male, then the incumbent will lose his harem no matter how dominant he is to the intruder in physical terms. The actual process by which this decision is made involves what amounts to a casting of votes by the females. Females cast their votes in favour of the intruder by grooming with him when he solicits them. Once a majority of females have groomed the intruder, the incumbent male seems to give up fighting and accepts defeat as inevitable. During this process, the females themselves are commonly engaged in violent disputes in which those who do not wish to abandon their current male actively try to prevent the other females from interacting with the intruder.

The shorter tenure of takeover strategists is largely a consequence of the females' socio-reproductive strategies. In units with fewer than four reproductive females, the cumulative effects of stress on the lowest ranking females are insufficient to have a significant impact on the female's reproductive performance. However, as harem size increases beyond this, low-ranking females who lack female relatives with whom to form alliances begin to perform increasingly poorly and are therefore more willing to accept a new male entering the unit. Males are consequently attracted to large units since they seem to realize that their chances of taking over such units are significantly higher (see Dunbar, 1984).

However, the very fact that a male is able to take over a large unit successfully means that he is at once susceptible to the same fate from some other male. Harem takeovers arouse a great deal of interest among all-male group males, who commonly sit around observing the proceedings very carefully: if the intruder is finally driven off by the incumbent and gives up, another male from an all-male group may launch a challenge of his own. Harem-holders are seldom able to resist sequential challenges by several different males. If the unit is very large (10–12 females), then several males may be able to take it over in rapid succession and divide the females up among themselves.

In contrast, followers begin with a small number of females and, because their harem only grows slowly as the juvenile females mature into it, they seldom achieve a harem size at which they are at risk of takeover. They are thus able to hold onto their harems for most of their natural lives.

The life history implications of these alternative strategies have been modelled using the demographic parameters of the Sankaber gelada (Dunbar, 1984). These analyses suggest that the two strategies are equally profitable over a lifetime because the initial advantage incurred by takeover strategists in terms of a larger initial harem size is offset by the fact that they cannot survive as harem male for as long as a follower can. Furthermore, the evidence clearly suggests that the frequencies of the two strategies are held in balance by frequency-dependent processes: as the number of takeovers increases, males who pursue this strategy have shorter tenures, so that it becomes more profitable from a male's point of view to opt for a follower strategy, and vice versa (Dunbar, 1984).

Once a harem-holder has been defeated, he remains in the unit as an old follower and maintains his relationships with his former females, though he does not normally mate with them. During this period, he is extremely protective of the young infants born to him shortly before he lost control of the unit. In this way, he probably both helps to reinforce the new male's hegemony over the unit in the face of other rivals and prevents infanticide, thus reducing his own lifetime reproductive output. Old followers remain in their units for about two years (just long enough to ensure that their last batch of infants are old enough not to be at risk of infanticide) and then seem to drift off to rejoin the all-male groups.

Evolution of the gelada social system

In trying to explain the evolution of the gelada social system, we need to account for two main features, namely (1) the fact that the females live in small cohesive matrilineal groups and (2) the fact that these groups combine to form unstable herds. Most other aspects of the gelada social system can then be explained as inevitable consequences of these two key features.

Thus, the fact that the gelada reproductive units are one-male groups can be explained by the number of females in these groups. Dunbar (1988) has shown that, among primates at least, the number of males in a group is a simple function of (a) the number of females in the group and (b) the extent to which their reproductive cycles are synchronized. Thus, whether a group has a one male or a multimale structure depends essentially on its size (see also Andelman, 1986). Similarly, we can explain the existence of bands among the gelada as a consequence of the fact that the reproductive units prefer to forage in large herds, and therefore tend to remain near to those units with whom they can most easily form larger aggregations. Since reproductive units undergo fission from time to time and kinship is an important factor enabling animals to live together, it follows that closely related units will tend to remain in the same general area. This naturally creates a band-like structure with kinship (probably mediated by familiarity) being the main factor determining whether units remain in the same band.

Paradoxically, it is perhaps easier to begin by asking why gelada should live in large herds. Analysis of the observed variation in herd size in relation to a number of environmental variables suggests that the cause of this variation lies in predation risk: herd sizes are largest in areas where the animals are most at risk of being caught by predators (Dunbar, 1986). Although natural predators are relatively scarce on the Ethiopian highlands now, it is only within the last century or so that this has been the case. Leopard, caracal, jackal, and hyaena are still to be found (though only the last of these in significant numbers). Lions were present up to around the middle of the last century, but are not found there today. The most common predators now, however, are native dogs and it is undoubtedly these that the gelada are most afraid of. Whereas gelada will react with only mild watchfulness to jackals or caracal, they will run in precipitous flight for the cliff edge if dogs appear near them.

The gelada's problem appears to be that while the cliff faces are safe from predators, they offer relatively poor pickings as far as grass is concerned. In contrast, the flat plateau top has a rich carpet of grass, but it offers little or no protection from predators. Reproductive units are decidedly nervous about venturing onto the plateau top on their own, and will invariably remain on the escarpment face if they cannot join other units. Herd size, in fact, correlates inversely with the steepness of the terrain (Dunbar, 1986). Hence, in order to be able to exploit the rich food sources offered by the plateau top, the gelada need to be able to form large herds for mutual protection against predators. But these herds cannot be sustained on the cliff face, where the density of grasses is too low to allow so many animals to forage together. Large herds simply drift apart as they forage along the cliff face. The gelada's social system therefore offers them the flexibility to exploit the various kinds of habitat that they encounter on the Ethiopian highlands.

The problem created by these large herds, however, is that the compression of many animals into a relatively small space inevitably causes tensions: such an effect has been widely documented in many other primate species, and the stresses caused by it can in turn be expected to lead to reproductive suppression among the females (see Dunbar, 1988). The females' problem here is how to minimize the stresses caused by group-living without obviating the very benefits for which the groups were formed in the first place. The solution to this classic problem that is near-universal among the higher primates is the formation of coalitions. I argue, then, that the gelada reproductive units are female coalitions designed to reduce the stresses imposed on the females by living in large herds (Dunbar, 1986, 1988).

In effect, then, the whole social system of the gelada can be interpreted as a series of social solutions both to the problem of how to exploit their habitat in the most effective way and to the difficulties that these solutions themselves create. Thus, herds are formed to enable the animals to avoid the problems of predation in open habitats; this leads to stress effects that the females try to solve by forming kin-based coalitions. However, as these coalitions (the reproductive units) themselves increase in size over time with the maturation of daughters, so they begin to create on a small scale the very problem of stress that they were designed to overcome; consequently, closely related females within the units form coalitions among themselves in order to reduce these stresses without driving the other members of their group away altogether.

Acknowledgements

The field work on which this account is based was supported by grants from the Science & Engineering Research Council (UK) and the Wenner Gren Foundation for Anthropological Research. I am grateful to the Institute of Ethiopian Studies of the University of Addis Ababa for sponsoring me as a Visiting Scholar and to the Ethiopian Government Wildlife Conservation Organisation both for allowing us to work in the Simen Mountains National Park and for considerable logistic support. I am also grateful to Professor Masao Kawai and his colleagues (Toshitaka Iwamoto, Hideyuki Ohsawa, and Umeyo Mori) for their continued collaboration over the years.

References

ANDELMAN, S. (1986). Ecological and social determinants of cercopithecine mating patterns. In *Ecological Aspects of Social Evolution*, ed. D. Rubenstein & R. Wrangham, pp. 201–16. Princeton: Princeton University Press.

DUNBAR, R.I.M. (1978). Sexual behaviour and social relationships among gelada baboons. *Animal Behaviour*, 26, 167–78.

DUNBAR, R.I.M. (1979). Structure of gelada baboon reproductive units. I. Stability of social relationships. *Behaviour*, 69, 72–87.

DUNBAR, R.I.M. (1980a). Demographic and life history variables of a population of gelada baboons (*Theropithecus gelada*). *Journal of Animal Ecology*, 49, 485–506.

DUNBAR, R.I.M. (1980b). Determinants and evolutionary consequences of dominance among female gelada baboons. *Behavioural Ecology and Sociobiology*, 7, 253–65.

DUNBAR, R.I.M. (1982). Structure of social relationships in a captive group of gelada baboons: a test of some hypotheses derived from studies of a wild population. *Primates*, 23, 89–94.

DUNBAR, R.I.M. (1983a). Structure of gelada baboon reproductive units. II. Social relationships between reproductive females. *Animal Behaviour*, 31, 556–64.

DUNBAR, R.I.M. (1983b). Structure of gelada baboon reproductive units. III. The male's relationships with his females. *Animal Behaviour*, 31, 565–75.

DUNBAR, R.I.M. (1983c). Structure of gelada baboon reproductive units. IV. Integration at group level. *Zeitschrift für Tierpsychologie*, 63, 265–82.

DUNBAR, R.I.M. (1983d). Relationships and social structure in gelada and hamadryas baboons. In *Primate Social Relationships*, ed. R.A. Hinde, pp. 299–307. Oxford: Blackwell Scientific Publications.

DUNBAR, R.I.M. (1984). *Reproductive Decisions: An Economic Analysis of Gelada Baboon Social Strategies*. Princeton: Princeton University Press.

DUNBAR, R.I.M. (1986). The social ecology of gelada baboons. In *Ecological Aspects of Social Evolution*, ed. D. Rubinstein & R. Wrangham, pp. 332–51. Princeton: Princeton University Press.

DUNBAR, R.I.M. (1988). *Primate Social Systems*. London: Chapman & Hall.

DUNBAR, R.I.M. (1989). Reproductive strategies of female gelada baboons. In *The Sociobiology of Sexual and Reproductive Strategies*, ed. A. Rasa, C. Vogel & E. Voland, pp. 74–92. London: Chapman & Hall.

DUNBAR, R.I.M. & DUNBAR, P. (1975). *Social Dynamics of Gelada Baboons*. Basel: Karger.

KAWAI, M. (Ed.) (1979). *Ecological and Sociological Studies of Gelada Baboons*. Basel: Karger.

KAWAI, M., DUNBAR, R.I.M., OHSAWA, H. & MORI, U. (1983). Social organisation of gelada baboons: social units and definitions. *Primates*, 24, 1–13.

KUMMER, H. (1968). *The Social Organisation of Hamadryas Baboons*. Basel: Karger.

OHSAWA, H. & DUNBAR, R.I.M. (1984). Variations in the demographic structure and dynamics of gelada baboon populations. *Behavioural Ecology and Sociobiology*, 15, 231–40.

SHOTAKE, T. (1980). Genetic variability within and between herds of gelada baboons in central Ethiopian highland. *Anthropologica Contemporanea*, 3, 270.

16 The ecology of *Theropithecus gelada*

TOSHITAKA IWAMOTO

Summary

1. Gelada baboons (*Theropithecus gelada*) are found mainly on the Amhara highlands in Ethiopia at altitudes ranging between 1500–4500 m. Recently, a new population was found in the Arussi region (Mori & Belay, 1990).

2. The gelada habitat is characterized as wet and cool compared to that of lowland baboons. The higher the altitude, the longer the vegetation is kept green, because of the high rainfall and low temperature.

3. Gelada baboons can achieve higher population densities and biomass compared to other sympatric primates and ungulates; this is made possible by their efficient food processing techniques and anti-predator strategies, including group living and the use of cliffs as refuges.

4. They are highly graminivorous, with grasses forming more than 90 per cent of their diet in most habitats and seasons. However, when the availability and the nutritional content of grasses declines, they shift to eating herbs.

5. At present, we know little about the digestive processes that enable gelada to achieve a high biomass.

6. The time spent feeding is the highest level among herbivorous primates, ranging from 67 per cent in the high Simen to 36 per cent at Bole.

History of field studies on gelada baboons

All of the ecological field data on gelada baboons (*Theropithecus gelada* Rüppell) were accumulated during a single decade (1964–75). John H. Crook carried out his pioneer study on this species in Ethiopia in 1964–5. He was able to provide a basic description of the species' social system as well as the species' unusual ecology (Crook, 1966; Crook & Aldrich-Blake, 1968). He mainly observed baboon populations living along the cliff edge at Debra Libanos 80 km north of Addis Ababa, the capital city of Ethiopia and in the Amba Ras area of the Simen Mountains (approximately 120 km north-east of the old capital, Gondar).

His study was succeeded by that of R.I.M. Dunbar and P. Dunbar. They concentrated on the Sankaber area of the Simen Mountains National Park, and carried out two sessions (1971–2 and 1974–5) of intensive field work. At Sankaber, their observations were focused on the adaptive significance of gelada social organization and ecology (Dunbar, 1978; Dunbar, 1984; Dunbar & Dunbar, 1974). During their stay in Ethiopia, they made three short studies on the gelada population in Bole Valley, about 50 km west of Debra Libanos, where they studied niche separation between three sympatric primates including the gelada (Dunbar & Dunbar, 1974).

During the same period, M. Kawai and his team carried out observations on gelada baboons in the Gich area of the Simen Mountains National Park. They carried out two seasons of field work in 1973–4 and 1975–6. Based on the complete individual identification of herd members, they made detailed analyses on inter-unit and inter-individual relationships in the society and on bioeconomical aspects of the population (Kawai, 1979).

The last field study on gelada was carried out by R.W. Wrangham at Sankaber in the Simen. He mainly studied the feeding strategy of this species (Wrangham, 1976).

Thus, the basic features of ecology and sociology of this species have been made clear by these intensive field studies and by excellent observation conditions in the field. However, knowledge of the life histories of individual baboons is still scant; such information is only available from continuous long-term observations. Once field work on this species is resumed, it will be possible to fill this gap in our knowledge.

This chapter summarizes our present knowledge on the ecology of the extant *Theropithecus*, which will help our attempts to reconstruct habitats and life forms of extinct *Theropithecus* species.

Distribution

Theropithecus gelada is found along the gorges and escarpments of the Ethiopian highlands. The altitude of the habitat ranges between 1500 and 4500 m. Most populations live at altitudes around 2000–3000 m, overlapping with human populations.

Geladas are classified into two subspecies, *T. gelada gelada* and *T. g. obscurus* (Hill, 1970). The latter is characterized by the darker coloured dorsal fur and a flesh-coloured face, and inhabits 'two southern belts of territory, one on the west in Gojjam, the other, a long strip, on the east' (Starck & Frick, 1958). However, as Crook (1966) confirmed that *T. g. gelada* inhabited both Shoa Province and Simen Mountains, the range of *T. g. obscurus* is surrounded by that of *T. g. gelada*.

Quite recently, Mori & Belay (1990) found a new population in the Arussi region. Starck & Frick (1958) referred to the presence of geladas in this region, based on information given by native peoples, but no specimens or observations by scientists had been available to confirm this. However, Mori & Belay (1990) also noted that the animals in this population have brighter and more golden-coloured fur than those in the southern population (around Shoa). It is still not known which subspecies this new population belongs to, and a systematic examination of morphological characteristics on the geladas living in each region is urgently needed.

Habitat

The rainy season on the Amhara plateau extends from June to September, although there are large regional differences. Some southern places also have a small rainy season from March to April. Annual rainfall in all gelada habitats exceeds 1000 mm, mostly being around 1200 mm. Higher regions have more rainfall (e.g. 1465 mm in Gich of Simen Mountains: Iwamoto & Dunbar, 1983).

The average monthly temperature on the plateau (near Addis Ababa) appears to be highest (about 20 °C) from March to May and lowest (about 15 °C) in the rainy season, from July to September. However, on the Simen plateau, it ranges from 10° to 5 °C. Thus, the climate of the gelada habitat is characterized as wet and cool compared to that of lowland inhabiting *Papio* species. This low temperature reduces the rate at which the plants desiccate during the dry season, and thereby extends the availability of the geladas' main food, grass.

The original vegetation on the plateau in the middle altitude (around 2000 m) was *Acacia* woodlands or thick forests of *Juniperus*, *Podocarpus*, and *Olea*, but these now remain only as isolated relict patches. In the Simen Mountains, ridge tops and gorge surfaces are covered by thickets of *Erica*, *Hypericum*, and *Rosa abyssinica*, while the more extensive areas of plateau top are dominated by open grassland composed of grass species such as *Festuca*, *Danthonia*, and *Poa*.

Most of the flatter areas on the Amhara plateau are intensively cultivated for barley, wheat, corn, and tef (the seed of which is used for the Ethiopian staple food, *indjera*). Therefore, gelada populations are usually confined to small areas along escarpment and gorge faces. They sometimes intrude on the cultivated land to damage crops. In more natural habitats like the Simen Mountains National Park, however, geladas use cliffs as sleeping sites, and climb up to the top in the daytime to harvest grass growing on the plateau. The geladas rarely range more than 2 km from the cliff edge, and spend most of their time within 1 km (Crook, 1966; Kawai & Iwamoto, 1979) because they are so sensitive to predators. They tend to move further from the edge when they form bigger herds (see Dunbar, chapter 15). This implies that the big herds frequently formed by geladas are an adaptation to predation pressure and the wider exploitation of their habitat.

The predation pressure on present gelada populations seems minimal, except for humans and domestic dogs (Dunbar, 1977a). Because most of their habitat is adjacent to human settlements, potential predators such as large cats and canids are much reduced. However, considering their sensitivity to potential predators and their persistent preference for cliffs, it may be supposed that the predation pressure on gelada populations in the past was greater.

Dunbar & Dunbar (1974) concluded that three sympatric primates, *T. gelada*, *Papio anubis* and *Cercopithecus aethiops* in the Bole Valley showed little direct interaction and had well-separated niches as a result of differential use of diet and range areas. Dunbar (1978) reported that, in the high altitude habitat in Simen, geladas, Walia ibex, klipspringer, bushbuck, bush duiker, horse, and cattle minimized their niche overlap by using different sectors of the habitat even when food preferences overlapped. He also suggested that, at critical times when food resources were in short supply in the Simen, the species with the most similar diets tended to reduce the overlap between them. This implies that dietary competition becomes more apparent in the more severe season. It should be noted, however, that *T. gelada* was the species with the least dietary overlap with other 'natural' species (i.e. except domestic animals) in the Simen (Dunbar, 1978).

Population density and biomass

The density of gelada baboons is high compared with other sympatric primate species. They maintained a density about three times greater than that of anubis baboons and about 12 per cent higher than that of vervet monkeys in the Bole Valley (Dunbar & Dunbar, 1974). The reproductive success of geladas becomes more conspicuous when their biomass is compared with that of other species. The biomasses of geladas, anubis and vervets were 459, 206, and 201 kg/km², respectively. Geladas show a biomass more than twice that of other primates. Similarly, in the highland habitat, geladas attain the highest density and biomass among all 'natural' herbivores (Dunbar, 1978). In particular, their biomass is about three times greater than that of the klipspringer, the species with the second largest biomass. The density of gelada baboons in the Gich area was 63 animals/km² and the biomass was 352 kg/km², which was a little bit lower than that in the Bole Valley.

Such high densities and biomass might be attributed to the feeding habits of geladas. As explained below, in every habitat they feed mainly on grass whenever there is sufficient biomass of grass for them to harvest. Clutton-Brock & Harvey (1977) noted that population biomass correlates directly with the degree of herbivory in primates in general. This rule is also applicable to gelada. In addition to this, a feeding technique that allows them to select only green grass blades contributes to their ability to maintain a high biomass in grassland habitats. Ungulates are less efficient at selecting green blades because of the wider harvesting edge of the incisor row. Geladas pick grass blades carefully between thumb and forefinger, which enables them to take higher quality food than ungulates. However, in the situation of an extremely low green biomass, this feeding technique becomes inefficient.

A further important factor in the gelada's ability to maintain a high level of productivity is the fact that it is able to utilize all parts of grass plants, including the roots and seeds as well as the leaves (Dunbar, 1978). Adaptations of other feeding behaviours, digestive processes, and also their anti-predator behaviours will be responsible for their success in attaining a high biomass (see also Dunbar, chapter 15).

Feeding behaviour

Feeding technique

When they feed on grass blades, geladas carefully pick only the green ones, using the thumb and forefinger of both hands

Table 16.1 *Diet composition of gelada baboons at Bole, Sankaber, and Gich, Ethiopia.*

Plant type	Part eaten	Bole*	Sankaber**			Gich†		
			Jul.–Aug.	Nov.	Jan.–Feb.	Oct.	Dec.	Feb.
Grasses	Leaves	91.4	93.0	17.3	24.7	80.7	81.0	44.7
	Rhizomes	0.5	–	6.5	66.9	–	8.0	16.0
	Flowers, seeds	5.0	–	69.7	–	14.0	0.4	0.4
Herbs	Leaves, stems	0.3	0.3	1.2	2.8	2.6	8.4	36.0
	Roots, tubers	–	3.3	0.9	–	2.7	2.2	2.5
	Flowers	0.3	0.2	3.0	0.1	–	–	0.4
	Seeds	–	–	1.3	5.2	–	–	–
Bushes	Leaves	0.1	0.2	–	0.2	–	–	–
	Flowers	–	–	–	0.1	–	–	–
	Fruits	2.0	2.9	–	–	–	–	–
Insects		–	0.1	–	0.1	–	–	–
No. of records		764	3051	231	3291	150	224	280

* Dunbar & Dunbar (1974); ** Dunbar (1977b); † Iwamoto (1979).

alternately while sitting. After accumulating blades from 10–20 pickings between the two fingers, they transfer the sheaves to their mouth. One feeding bout consists of several passages of hands to the mouth over a period of one to two minutes. Between two feeding bouts, they shuffle less than one metre or walk a few metres to change the feeding site. They continue this pattern of picking and moving throughout the day during the rainy season.

Geladas largely depend on the rhizomes of grasses and lily bulbs in the dry season. As they dig up foods that lie in the soil at depths of 3–7 cm, they use both hands alternately as hoes to scrape out the ground. They usually dig in the sitting posture.

Sometimes a bipedal stance is adopted in order that foods on the shrubs or trees can be taken and to eat the stem cores of *Lobelia* and the leaves of *Cactus*, although none of these behaviours occur with any frequency.

Diet

Geladas are highly graminivorous. They prefer grass blades to herbs or shrub vegetation, when the former is available (Table 16.1). In both Bole and Sankaber, grasses (blades, rhizomes, flower and seeds) formed more than 90 per cent of diet composition on a 'time spent feeding' basis (Dunbar & Dunbar, 1974; Dunbar, 1977b). In Gich, grasses accounted for almost 90 per cent of the diet in October and December, but decreased to 60 per cent during the middle of the dry season in February. At that time, they ate a large volume of *Trifolium* plants, which suffer less desiccation because they are hidden under the shade of tall grasses. When the grasses set seed, the geladas intensively exploited them. The change was remarkable in Sankaber (about 70 per cent), but rather less so in Gich (14 per cent). Because of exploitation pressure and the desiccation rate of the grass seeds, the period during which the geladas fed on grass seeds lasted about a month. Similarly, geladas at Sankaber were largely dependent on rhizomes of grasses (67 per cent) in the middle of the dry season, but these accounted for a much smaller proportion of the diet (16 per cent) at Gich. Thus, seasonal changes in dietary composition appeared to be more drastic in Sankaber. This might be related to less grass coverage in Sankaber, because of either the low rainfall or higher level of competition from domestic stock. Although the seasonal change of diet in the Bole population was not determined, it is likely that the change is larger there than in Sankaber, because the rainfall is much lower. Low rainfall usually correlates with a more rapid desiccation of vegetation.

Geladas rarely eat insects or other invertebrates. However, if the situation presents itself, they do not ignore opportunities for eating animal materials that arise. I once observed all members of a herd enthusiastically hunting termites when these suddenly swarmed on the gorge top near Debarek. Thus, geladas eat small invertebrates when they do not need to expend much effort doing so, but they usually prefer grasses.

Seasonal changes in the quality of diet In each habitat, geladas seasonally change their main food. This can be attributed to the seasonal change in biomass and nutri-
tional content of the plants. As the rainy and dry seasons are clearly identified in the habitat of gelada baboons, grasses start to desiccate immediately after onset of the dry season, though the speed of desiccation is slower in the highland habitat. The green biomass of plants (*Danthonia*-type grassland) reaches a peak in September (just at the end of the rainy season in the Gich area) and a minimum level in January (Fig. 16.1). However, it should be noted that the green biomass never approaches zero by the middle of the dry season in this area. That is probably one reason why geladas in Gich continued to depend more heavily on grass blades than the geladas inhabiting the drier Sankaber area (Table 16.1).

The crude protein content in green grass blades sampled in the *Danthonia*-type grassland decreased during the dry season (Fig. 16.1). The change was very quick, falling to less than eight per cent in November after the peak content (9.4 per cent) in October. The minimum level of crude protein was observed in January (7.3 per cent). This represented three-quarters of the peak content. On the other hand, the crude fibre content in green blades was 26 per cent at minimum in October and 28 per cent at maximum in February. The latter value was only two per cent higher than the former. This means that the seasonal change in crude fibre content exerted a less important effect on geladas in this area than that of crude protein. In Sankaber, the crude protein content in grasses was 12.5 per cent in the dry season, which is much higher than that in Gich (Iwamoto & Dunbar, 1983). This difference may be due to the

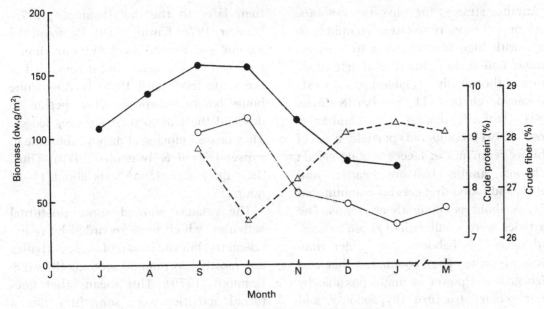

Fig. 16.1. Seasonal change in the green biomass (●), crude protein content (○) and crude fibre content (△) of plants in *Danthonia*-type grassland at Gich, Simen National Park, Ethiopia.

grass species composition in the samples. Some soft grasses show higher protein content than the coarse ones.

As mentioned above, although geladas prefer grasses to herbs they are not exclusively graminivorous. The quality of grass blades quickly declines during the dry season, and at Gich, the geladas shifted from grasses to herbs. *Trifolium* was the main alternative food at this time, probably because of the high protein content (12.6 per cent) compared with that of grasses (about 7 per cent) in the dry season. Also, the crude fibre content of *Trifolium* is much lower (22.4 per cent) than that of grasses (30.3 per cent). Thus, *Trifolium* plants seemed to be an excellent alternative food for geladas in the dry season. However, the fact that they did not feed much on *Trifolium* plants during the

rainy season (in spite of their great availability and fresher condition) implies that *Trifolium* is less than ideal in other respects (see Iwamoto, chapter 17).

Food processing and digestion by geladas
Several primates are known to show very strong folivorous tendencies. Colobus monkeys and langurs possess a forestomach fermentation chamber to decompose plant cell walls by the activity of symbiotic bacteria, and howler monkeys use hindgut fermentation for the same purpose. In spite of their predilection for highly fibrous foods, little is known about the digestive physiology of geladas. Iwamoto (chapter 17) has suggested the possibility of hindgut fermentation by geladas, but the verification by physiological experiments is still lacking.

Another strategy for folivorous animals, with or without fermentation chambers to cope with high fibre diets, is to increase intake and/or to grind the tough foods more thoroughly (Jablonski, 1981; Jablonski, chapter 11; Dunbar & Bose, 1991; Teaford, chapter 12). Dunbar & Bose (1991) compared particle size of fibrous remnants in faeces among gelada baboons, anubis baboons, cattle, and zebra, and found that geladas comminuted grasses about as efficiently as zebras. The particles were smaller than those processed by anubis baboons, but larger than those of cattle. They concluded that this high grinding power is made possible by their molar structure (hypsodonty and deep surface relief) as pointed out by Jolly (1972) and Eck & Jablonski (1987).

Activity pattern

Daily activity rhythm

When geladas feed on grass blades, they usually pick up only the green ones that are shorter than 10 cm in length. Consequently, the feeding rate is limited in itself. Moreover, they have to process a relatively large quantity of food because the quality is low. Consequently, geladas have to feed from early in the morning to late in the afternoon, more or less continuously. Around 07.00 h, very close to the time of sunrise, they start to move up to the cliff top (Kawai & Iwamoto, 1979; Dunbar 1977b), and feeding begins at about 08.00 h. However, part of the morning is taken up with grooming sessions so that the proportion of the feeding individuals in a herd was lower in the morning (30–35 per cent)

than later in the day (Iwamoto, 1975; Dunbar, 1978; Kawai, 1979). By noon, the feeding activity had reached its maximum level (60–80 per cent) and it remained at the same level until 18.00 h. About one hour before sleeping, they began to descend the cliff to their sleeping ledges. They ceased moving at around the time of sunset (Kawai & Iwamoto, 1979). Thus, their daily active time lasts about 11–12 hours.

The geladas showed some nocturnal activities, which were recorded by radio-telemetry, but the length of these activities was mostly less than 30 seconds (Kawai & Iwamoto, 1979). This means that nocturnal activities were something like a small shift of sitting site and body shaking. We did occasionally hear agonistic vocalization during moonlit nights. However, it is clear that geladas are basically diurnal.

Time spent feeding

Although the sampling methods of different investigators varied, several studies give estimates of time spent feeding. Kawai & Iwamoto (1979) reported a value of 78 per cent for adult males and 81.6 per cent for adult females in Gich, which is a little bit longer than that by Iwamoto (1979). He reported the proportion of 65.5 per cent and 67.0 per cent for adult males and females, respectively, in the Gich population. The difference of the values between two reports is due to the sampling season and observation hours. The former estimate derives from the dry season only when geladas feed longer (see chapter 17) and for a shorter day length (from 09.00 to

17.00 h) which did not include the social hours at the beginning and end of the day. Geladas spend 59.0 per cent on feeding in October, but 72.8 per cent in February (Iwamoto, 1979). Thus, during the dry season, geladas tended to increase the time spent feeding.

At Sankaber, geladas spent 40.6 per cent of their time feeding in the rainy season, and 47.1 per cent in the dry season (Dunbar, 1977b). Although the time spent feeding in Sankaber is about 20 per cent lower than that in Gich, the seasonal trends are similar in the two habitats. Dunbar (1978) also reported that in the Bole Valley population, the geladas spent 35.7 per cent of their time feeding, a value that is clearly much lower than those in highland populations. Iwamoto & Dunbar (1983) analyzed this trend and concluded that it is due to a combination of increasing temperature-dependent energy requirements and declining habitat quality.

Length and duration of day journeys

Time spent moving was 14.7, 20.4, and 17.4 per cent at Gich, Sankaber, and Bole, respectively. Combining moving time with feeding time in each area, geladas spend from 85 to 50 per cent of time in foraging (feeding and moving) activities. This enormous amount of time spent feeding and moving (85 per cent at Gich) is the highest so far observed in any primate (Clutton-Brock & Harvey, 1977). This may well impose serious constraints on their social organization (Dunbar, chapter 18).

Geladas travelled less distance (630 m) compared with sympatric anubis baboons

in Bole (1210 m) (Dunbar & Dunbar, 1974), because grasses are distributed uniformly. The day journey length was dependent on the band size. In Sankaber, where the band size was 262, the day journey length was 2160 m, while it was 1008 m in Gich where the band size was 112 (Iwamoto & Dunbar, 1983). Also, when a band joined another band making a large herd, they travelled farther (844 m versus 1606 m in Gich). Thus, although the formation of large herds has fitness benefits for geladas in terms of predator detection and range use (a large herd can travel further from the cliff edge), it has imposed a cost in terms of the need to travel further in order to ensure that all members of the herd can meet their nutritional requirements.

Discussion

The distribution of the gelada baboon is closely related to the escarpments and gorge systems. Since most of the plateau top of the present habitat is heavily cultivated and animals avoid these areas, it is difficult to know whether geladas are able to inhabit the flatter areas without cliffs. However, considering the fact that even at such a place as Gich in the Simen National Park, where a considerable part of the plateau top remains uncultivated and no geladas live completely independent of the cliffs (Ohsawa, 1979), it seems likely that cliffs are essential for their survival. As already mentioned, they can utilize the inland parts of the plateau top only when they form a large herd composed of several hundred animals, while they sleep at and escape to the cliff even in such a time. Therefore, without any effective anti-predator strategies (weapons or behaviour) other than escaping to cliffs, it seems unlikely that they could survive in the savannah plains where refuges such as cliffs is far less available.

In this context, the high predation pressure in the open lands can be considered as one of the factors determining the large body size seen in some *Theropithecus*, particularly *T. oswaldi* (Jolly, 1972) during the Pliocene and Pleistocene. Unlike the ungulates, baboons are not high-speed long-distance runners. The most effective anti-predator strategy for them might be the early detection of predators by group-living to enable the animals to retreat into bushes or trees. A large body size may also make predators less willing to attack (Jarman, 1974). Even yellow baboons sometimes try to chase small predators (Altmann & Altmann, 1970).

There was, however, another *Theropithecus* species, *T. darti*, with body weight and morphological characteristics rather similar to those of *T. gelada*. From his examinations of reduncine bovids found in the same formation as *T. darti*, Eck (chapter 2) suggested that *T. darti* had already abandoned the habitat of moist swamps and woodlands and expanded into the drier woodlands and grasslands. This implies that *T. darti* had been under high predation pressure in the open lands while still of small body size. They could possibly be considered 'beginners' in the open habitat and coped with predators by group living and escaping to trees. The evidence that *T. darti* increased in body size thereafter and that *T. oswaldi*, the descendant of *T. darti*

(Eck, chapter 2), living in open grasslands, had a body size as large as the gorilla seems to suggest an adaptive trend in body size in *Theropithecus* baboons in response to the demands of an open habitat. On the other hand, the ancestor of *T. gelada* could have found its preferred habitat again in the Abyssinian highlands with safe cliffs, far from its putative area of origin. In this habitat, they might have been exposed to fewer selection pressures to become large in body size.

There is another factor that leads to herbivorous animals developing larger body sizes. This is the tendency to exploit lower quality foods (Jarman, 1974; Clutton-Brock & Harvey, 1977). Animals have to possess large (usually long) digestive tracts to extract nutrients more efficiently from a given mass of foods when they want to take foods of lower quality. The increase in body size of late *T. darti* and of *T. oswaldi*, both of which probably mainly fed on gramineous plants in open habitats, may be also explained by this. Some herbivorous mammals, however, solved this problem by also possessing fermentation chambers, which provide them with a more efficient means to live on lower quality foods. Unlike ungulates, the picking behaviours and digestive processes of extinct and extant *Theropithecus* might not be sufficiently efficient to harvest and process the mass of low quality foods (Iwamoto, 1979; chapter 17; Lee & Foley, chapter 19) required to support a large body size. It is not difficult to imagine that only a slight climatic change to a drier habitat adversely influenced the survival of this large *Theropithecus* species.

In contrast, the gelada baboons inhabiting more montane habitats would have been able to survive during these periods, because they could depend on the rich grass vegetation for a longer period during the year (Dunbar, chapter 18). The cliffs in the habitat also provided a safe refuge, where no selection pressure to increase in body size in order to cope with the predation pressures could work. Thus, the small-sized *T. gelada* was the only species that could continue to be a grass-harvesting machine in the evolutionary history of *Theropithecus*.

Acknowledgement

I am greatly indebted to Prof Y. Ono, Kyushu University and Dr M. Kawai, Inuyama Monkey Center for their encouragement in this study. I am grateful to Dr N.J. Jablonski and Dr R.A. Foley for their kindness in offering me the chance to prepare this paper. I also express my sincere thanks to Dr R.I.M. Dunbar for his critical reading of this manuscript and for correcting the English. I am grateful to the Ethiopian Government for permission to study in the Simen Mountains National Park.

References

ALTMANN, S.A. & ALTMANN, J. (1970). *Baboon Ecology: African Field Research*. Chicago: University of Chicago Press.

CLUTTON-BROCK T.H. & HARVEY, P.H. (1977). *Primate Ecology*, ed. T.H. Clutton-Brock, pp. 557–84. London: Academic Press.

CROOK, J.H. (1966). Gelada baboon herd structure and movement. A comparative report. *Symposium of the Zoological Society of London*, 18, 237–58.

CROOK, J.H. & ALDRICH-BLAKE, P. (1968). Ecological and behavioral contrasts between sympatric ground dwelling primates in Ethiopia. *Folia Primatologica*, 8, 192–227.

DUNBAR, R.I.M. (1977a). The gelada baboon: status and conservation. In *Primate Conservation*, pp. 363–83. New York: Academic Press.

DUNBAR, R.I.M. (1977b). Feeding ecology of gelada baboons: a preliminary report. In *Primate Ecology*, ed. T.H. Clutton-Brock, pp. 251–73. London: Academic Press.

DUNBAR, R.I.M. (1978). Competition and niche separation in a high altitude herbivore community in Ethiopia. *East African Wildlife Journal*, 16, 183–99.

DUNBAR, R.I.M. (1984). *Reproductive Decisions: An Economic Analysis of Gelada Baboon Social Strategies*. Princeton: Princeton University Press.

DUNBAR, R.I.M. & BOSE, U. (1991). Adaptation to grass-eating in gelada baboons. *Primates*, 32, 1–7.

DUNBAR, R.I.M. & DUNBAR, E.P. (1974). Ecological and niche separation between sympatric terrestrial primates in Ethiopia. *Folia Primatologica*, 21, 36–60.

ECK, G.G. & JABLONSKI, N.G. (1987). The skull of *Theropithecus brumpti* compared with those of other species of the genus *Theropithecus*. In *Les Faunes Plio-Pléistocènes de la Basse Vallée de l'Omo (Éthiopie)*. Tome 3. Cercopithecidae de la Formation de Shungura, ed. Y. Coppens & F.C. Howell, pp. 11–122. Cahiers de Paléontologie, Travaux de Paléontologie Est-Africaine. Paris: Editions du Centre National de la Recherche Scientifique.

HILL, W.C.O. (1970). *The Primates*, vol. 8 *Cynopithecinae*. Edinburgh: University of Edinburgh Press.

IWAMOTO, T. (1975). Food resource and the feeding activity. *Contemporary Primatology*, ed. S. Kondo, M. Kawai & A. Ehara, pp. 475–80. Basel: Karger.

IWAMOTO, T. (1979). Feeding Ecology. In *Ecological and Sociological Studies of Gelada Baboons*, ed. M. Kawai, pp. 279–330. Tokyo: Kodansha.

IWAMOTO, T. & DUNBAR, R.I.M. (1983). Thermoregulation, habitat quality and the behavioral ecology of gelada baboons. *Journal of Animal Ecology*, 52, 357–66.

JABLONSKI, N.G. (1981). *Functional Analysis of the Masticatory Apparatus of* Theropithecus gelada *(Primates: Cercopithecidae)*. Ph.D. Thesis, University of Washington.

JARMAN, P.J. (1974). The social organisation of antelope in relation to their ecology. *Behaviour*, 48, 215–67.

JOLLY, C.J. (1972). The classification and natural history of *Theropithecus (Simopithecus)* (Andrews, 1916), baboon of the Plio-Pleistocene. *Bulletin of British Museum (Natural History), Geology*, 22, 1–123.

KAWAI, M. (1979). *Ecological and Sociological Studies of Gelada Baboons*. Tokyo: Kodansha.

KAWAI, M. & IWAMOTO, T. (1979). Nomadism

and activities. In *Ecological and Sociological Studies of Gelada Baboons*, ed. M. Kawai, pp. 251–78. Tokyo: Kodansha.

MORI, A. & BELAY, G. (1990). The distribution of baboon species and a new population of gelada baboons along the Wabi-Shebeli, Ethiopia. *Primates*, 31 (4).

OHSAWA, H. (1979). The local population and environment of the Gich Area. In *Ecological and Sociological Studies of Gelada Baboons*, ed. M. Kawai, pp. 3–45. Tokyo: Kodansha.

STARCK, D. & FRICK, H. (1958). Beobachtungen an äthiopischen Primaten. *Zoologische Jahrbuch*, 86, 41–70.

WRANGHAM, R.W. (1976). Aspects of feeding and social behavior in gelada baboons. *Mimeo Report, Science Research Council*, pp. 1–56.

17 Food digestion and energetic conditions in *Theropithecus gelada*

TOSHITAKA IWAMOTO

Summary

1. A hypothesis of hindgut fermentation of gelada (*Theropithecus gelada*) was examined by assessing the ability of animals to digest crude fibre in foods and by evaluating the extent of bulk eating in the dry season.
2. Three caged geladas could digest more than 50 per cent of the crude fibre content of rations; this suggests that gelada have a microbial fauna to decompose fibre in the hindgut (caecum and/or colon).
3. Gelada living in the Gich area of the Simen Mountains significantly increased their food intake going into the dry season. This was brought about by changing their food eating habits from grasses to the herb, *Trifolium*, and by prolonging the time spent feeding.
4. The quality of food eaten, however, did not decrease in the dry season, because of the high nutritional content of *Trifolium*. This indicates that geladas do not fully satisfy the condition stated by Janis (1976) that animals employing hindgut fermentation increase their bulk food intake when faced with lowered food quality.
5. The greater energy intake in the dry season might be explained by factors such as the need for more food for body temperature regulation during the cold dry season, lessened ability to digest *Trifolium*, the toxin content of *Trifolium*, or sampling error.
6. It is suggested that further studies on the digestive physiology of geladas are necessary to understand the evolution of *Theropithecus* to such a highly specialized food habit.

Introduction

Some folivorous primates have developed fermentation chambers in the gut to enable them to extract more nutrients and energy from low quality food than would be possible without such a specialization of the digestive apparatus (Bauchop, 1978). Colobine monkeys have sacculated stomachs that act as fermentation chambers (Hill, 1952; Bauchop & Martucci, 1968; Ohwaki *et al.*, 1974). In contrast, some indriids and New World monkeys have an enlarged caecum and/or colon that appear to be specialized for fermentation, although microbial and nutritional tests are still lacking (Bauchop, 1978; Chivers & Hladik, 1980).

Theropithecus has adapted well to feeding on small and tough food items by evolving highly dexterous fingers and a specialized masticatory apparatus (Jolly, 1972; Jablonski, 1981). More than 80 per cent of the diet of the extant gelada is grass, in most cases (Dunbar & Dunbar, 1974; Dunbar, 1977; Iwamoto, 1979). In spite of the fact that they have such a highly specialized diet, only a few examinations have been carried out on their gut morphology (Osman Hill, 1970) and digestive abilities. Although it is clear that the gelada does not have a sacculated stomach (Iwamoto, unpub. data), the possibility of hindgut fermentation in this species needs to be explored.

This study, therefore, was aimed at examining the question of hindgut fermentation in *Theropithecus gelada*. The direct way to examine this question is to measure the activities of microbial symbionts in the caecum or colon. However, because this was not possible, an indirect method was used in this study. An animal engaging in hindgut fermentation should meet two conditions: (1) they should demonstrate the ability to decompose crude fibre, and (2) they should show an increase in the rate of food intake when faced with deteriorating food quality (Janis, 1976). The former is the necessary and sufficient criterion. The latter may be supplementary, because even animals without fermentation chambers could follow the second strategy when faced with

the situation of low food quality (Sibly, 1981; Dunbar, 1984).

In this study an attempt was also made to calculate the energy budget of geladas that inhabit the Simen Mountains, the most severe habitat occupied by the species today. This was done in order to help us recognize the other possible constraints that *Theropithecus* may have had to overcome in the evolution of highly specialized feeding habits.

Methods

Experiments on digestibility

In order to test whether geladas could decompose dietary fibre, two series of experiments on food digestion were carried out in captive animals. The first was conducted with one adult female gelada (body weight (BW) = 9.4 kg) at the Kyoto University Primate Research Institute in Japan, from 26 July to 11 August, 1981, and the second was conducted with one adult male (BW = 22 kg) and adult female (BW = 10.0 kg) in the Fukuoka Municipal Zoo in Japan, from 5 to 12 August, 1982. After the experiments, the nutritional content of all food and faecal samples was measured with a normal procedure of nutritional analysis (Iwamoto, 1982).

The apparent digestibility (Da per cent) of nutriments was calculated using the following equation:

$$Da = (Qi - Qo) \times 100/Qi$$

where, Qi and Qo are the average daily intake of nutriments and average undigested nutriments, respectively. Crude fibre is not an exact measure of cellulose content in the samples as it also contains hemicellulose and lignin. Hemicellulose is soluble in acid and alkali solutions (Van Soest, 1982). However, the proportion of it in the measures of crude fibre is usually around 20 per cent (Van Soest, 1982). The value was smaller than the level needed for the examination of the gelada's ability to decompose cellulose components in diet.

In the first experiment, cabbage and monkey pellets were served, and the gelada ate 117.8 g dry wt/day (481 kcal/day). In the zoo, the adult male ate 190 g (736 kcal) and the adult female 141 g dry wt/day (550 kcal/day) of food (see Table 17.1) from a provision of sweet potatoes, carrots, oranges, bamboo leaves, and monkey pellets.

Analysis of field data

The amount of daily food intake was estimated by season using the focal animal sampling technique for four geladas living in the Gich area of the Simen Mountains, Ethiopia (see Iwamoto, 1979 for details of sampling methods). This study was undertaken to test the hypothesis that food intake would increase as food quality decreased.

In order to test whether geladas ate significantly more food in the dry season than in the rainy season, the corrected weight of daily food intake was calculated from data in Iwamoto (1979, Table 12.16). The corrected weight was calculated under the assumption that the animals observed for the same duration (630 min.) in each sampling month. The ratios of daily food intake (g dry wt) for December to October,

February to October and February to December were calculated. In order to test whether these ratios were significantly larger than 1.0, a simple two-tailed test of mean was done under the null hypothesis that the true mean was 1.0. If the null hypothesis was rejected, it could be concluded that the amount of daily food intake had really changed between the two months.

Results

Digestibility of food in captive geladas

The food intake per metabolic unit of body size in the captive animals was 89.6, 72.4, and 97.8 kcal/$BW^{0.75}$/day for the Kyoto female, the Fukuoka male, and the Fukuoka female, respectively. Thus, the male had less food compared with the two females. The assimilated energy for these animals was considerably lower than for animals in the field (see below). This low daily energy consumption could have been due to the relatively confined quarters in which the captive animals were held.

The apparent digestibility of dry matter was over 80 per cent, because of the low crude fibre content of the food. The noteworthy finding was that a considerable part of the crude fibre had been digested. The digestibility was 54.2, 72.8 and 56.2 per cent for the three captive geladas, respectively (Table 17.1). This indicated that more than half of the crude fibre content of the food had been digested.

Seasonal change in nutritional intake of geladas under field conditions

The first ratio (DEC/OCT) was significantly larger than 1.0, and the second

Table 17.1 Results of the experiments on food digestibility. The digestibility of dry matter, energy and five nutrients were calculated. All of the weights are shown as daily averages.

	Crude protein (g)	Ether extract (g)	Crude fibre (g)	Nitrogen free extract (g)	Ash (g)	Dry matter (g)	Energy (kcal)
Adult female (BW = 9.4kg)							
Intake	30.5	7.7	2.5	70.0	6.3	117.1	481.2
Assimilated	27.8	6.5	1.4	65.7	1.7	103.1	437.9
Digestibility (%)*	91.0	84.8	54.2	93.8	27.0	88.0	91.0
Content in food (%)	26.1	6.5	2.1	59.8	5.4		
Adult male (BW = 22.0kg)							
Intake	23.1	4.1	12.3	139.3	11.6	190.3	735.6
Assimilated	16.0	2.2	9.0	131.6	7.1	165.9	646.4
Digestibility (%)	69.2	54.8	72.8	94.5	61.2	87.1	87.9
Content in food (%)	12.1	2.1	6.5	73.2	6.1		
Adult female (BW = 10.0kg)							
Intake	18.0	3.4	8.9	102.9	8.1	141.3	549.9
Assimilated	10.9	2.1	5.0	93.0	4.3	115.4	455.1
Digestibility (%)	60.7	63.1	56.2	90.4	52.7	81.6	82.8
Content in food (%)	12.7	2.4	6.3	72.8	5.7		

* Apparent digestibility = (Food intake–faeces) ×100/food intake.

Table 17.2 *Daily food intake in each month for four focal animals and ratios between the months of October, December and February. Oct, Dec, and Feb indicate the values of dry weight (g) of food intake in October, December, and February, respectively.*

Focal animal	Sex	Oct	Dec	Feb	Dec/Oct	Feb/Oct	Feb/Dec
Ayin	Male	747.3	1102.6	1174.7	1.47	1.57	1.07
Nech	Male	877.9	1077.8	903.2	1.23	1.03	0.84
Uso	Female	709.5	936.4	1179.8	1.32	1.66	1.26
Tara	Female	518.4	840.6	767.8	1.62	1.48	0.91
Average					1.41 *	1.44 †	1.02 NS

*: $0.01 < p < 0.05$; †: $0.05 < p < 0.10$; NS: $p > 0.10$ in Two-tailed Test of Mean (null hypothesis: true mean = 1.0).

Table 17.3 *Average ratios of daily intake of dry weight of food, time spent feeding, crude protein, and crude fibre content of food eaten between two months. Oct, Dec, and Feb show the value for October, December, and February, respectively.*

Mass, time or nutrients	Average values in October	Dec/Oct	Feb/Oct	Feb/Dec
Dry weight	713.3 dw.g.	1.41 *	1.44 †	1.02 NS
Time spent feeding	59.0 %	1.14 NS	1.24 *	0.11 NS
Crude protein	9.6 %	0.89 †	1.00 NS	1.12 *
Crude fibre	26.7 %	1.01 NS	0.95 †	0.95 *

See Table 17.1 for symbols *, †, and NS.

(FEB/OCT) just missed the five per cent level but was significantly larger at the ten per cent level (Table 17.2). Thus, when their main food, grass, started to dry up, the geladas increased their food intake rate. The FEB/DEC ratio was not significantly larger than 1.0, probably because the biomass and crude protein content of the grass did not change much between October and February, as noted above.

The increase in the amount of food intake during the dry season was brought about both by the changing diet (see below) and the prolongation of time spent feeding (Table 17.3). The ratio of time spent feeding, FEB/OCT, was significantly greater than 1.0, and the others also showed ratios higher than 1.0 (but these were not significant). The average speed of geladas feeding on *Danthonia* grass blades was 2.6 g dry wt/min, but the speed for *Trifolium*, the herb species mostly preferred in the dry season, was 4.4 g dry wt/min. Thus, the more geladas ate *Trifolium*, the higher the feeding speed they could achieve.

The same tests were carried out for crude protein content and crude fibre content in the foods eaten (Table 17.3). There was a weak significant difference ($p < 0.10$) from 1.0 in the test for the ratio of protein content, DEC/OCT, and a significant one (larger than 1.0) for FEB/DEC. The former is reasonable, considering the decrease in protein content of grass blades going into the dry season (see Fig. 16.1 in Chapter 16), but the latter result requires an explanation. Geladas fed more on *Trifolium* in the dry season, and this food plant had a higher crude protein content than grass. Therefore, the ratio higher than 1.0 in FEB/DEC was probably due to feeding on *Trifolium* in the dry season. The decrease in the crude fibre content of food in the dry season can also be explained by the eating of *Trifolium*.

Thus, geladas increased their food intake during the dry season, which seems to satisfy the second condition set out at the beginning of this study. However, the problem is not a simple one, because geladas change their food eating habits in the dry season, ingesting more protein and less fibre than they did in the rainy season. In other words, the quality of food eaten by geladas did not necessarily deteriorate in the dry season. This is clearly seen in the next section, in which the digestibility of natural foods and the amount of daily food assimilation are discussed.

Fibre content, digestibility and energy budget

The ability of captive geladas to digest foods was high, over 80 per cent, but that of geladas in the wild may be considerably lower than that value because of the low quality of available foods.

Digestibility in herbivorous mammals is strongly correlated with the crude fibre content of foods (Mitchell, 1942). Mitchell (1964) reviewed published data and concluded that the general equation for digestibility (y per cent) is revealed as

$$y = 90 - ax$$

where 'x' is the crude fibre content of food and 'a' is a coefficient based on the animal's ability to decompose fibre. In animals that utilize forestomach fermentation, such as cattle and sheep, the values for 'a' are low, 0.781 and 0.919, respectively. Horses, that utilize hindgut fermentation, show a higher value of 1.196, and swine an even higher value of 1.293. A high value of 'a' means a rapid decline in digestibility as the fibre content in the food increases. Thus, the effect of poor food quality on digestibility is greater in non-ruminant herbivores than it is in ruminants.

Unfortunately, there are very few data on digestibility of non-human primates. Only Iwamoto (1988) has reported an equation, $y = 90.5 - 1.44x$ for Japanese macaques, based on a small sample size. This coefficient, 1.44, is higher than that of swine, which seems reasonable considering that the macaque is primarily a frugivore. In the present study, the digestibility of geladas feeding on low quality foods was not measured, but we may deduce a reasonable value for the coefficient 'a' from the above results. It may be higher than that of horses, specialists in grass eating, but lower than that of Japanese macaques, which are frugivores. Swine have enlarged stomachs and are known to ferment fibre (Janis, 1976). The geladas in this study also digested fibre to some extent, probably in the caecum and/or colon. Therefore, I have adopted an ideal coefficient of 'a' near that of swine, 1.30, for use here (see also Goodall, 1977, for the gorilla). It must be borne in mind, however, that this equation can only provide rough assessment of the energy budget.

Using the equation,

$$y = 90 - 1.30x$$

the digestibility of foods in each month was calculated to be less than 60 per cent, with the value for February being the highest of three month's values, although the difference was not significant (t-test, $p > 0.05$). This means that the food quality improved towards the dry season (until February, at least), as measured by food digestibility.

The digestibilities also gave values for assimilated energy between 1100 to 2600 kcal/animal/day (Table 17.4).

The ratios of assimilated energy in December and February to that in October were also calculated for each baboon (Table 17.4). These ranged from 1.2 to 1.6 ($\bar{x} = 1.38$) in December and from 1.0 to 1.9 ($\bar{x} = 1.5$) in February. Thus, in the dry season, assimilated energy as well as food intake, tended to increase.

Discussion and conclusions

In this study, it was found that geladas can decompose crude fibre in foods, and that they, thus, fulfil the first condition that must be met for an animal with hindgut fermentation. There are three components in so-called cell wall materials: cellulose, hemicellulose, and lignin. The normal method for analyzing crude fibre content gives the proportions for part of hemicellulose, and part of lignin, and most of the cellulose in a sample (Van Soest, 1982). The crude fibre content measured in this study represents these three structural carbohydrates together. Lignin and cellulose cannot be decomposed by normal digestive fluid, but hemicellulose is soluble to acidic and alkaline solutions. Therefore, some of the crude fibre digestion observed in this study might be attributed to the decomposition of hemicellulose by acidic stomach fluid. However, the digestibility of crude fibre was too high (more than 50 per cent) to be explained solely by hemicellulose (see 'Methods').

Perissodactyls ferment fibrous plant materials in the caecum and colon, and they absorb VFA (volatile fatty acids produced by activities of micro-organisms) from the colon wall as an energy source (Janis, 1976). Geladas also have a caecum (but not a very large one) and a fairly long colon (Osman Hill, 1970). Therefore, it is not difficult to consider that geladas can also decompose some part of cellulose by the activities of micro-organisms. This would help geladas to obtain extra energy when the quality of food declines (i.e. when the fibre content increases). The digestibility of fibre under natural food

Table 17.4 *Energy budgets for four gelada. Daily energy intake was calculated from the ideal regression equation between crude fibre content of food and digestibility. Ratios of assimilated energy in December (Dec) and February (Feb) to October (Oct) are also shown.*

Name	Month	Food intake (kcal/day)	Crude fibre content (%)	Digestibility (%)	Assimilated energy (kcal)	Ratio to Oct
Ayin (♂)	Oct	2926	27.1	54.8	1605	1.0
Nech (♂)	Oct	3432	26.0	56.2	1928	1.0
Uso (♀)	Oct	2775	26.9	55.1	1528	1.0
Tara (♀)	Oct	2018	26.2	56.0	1130	1.0
Ayin	Dec	4279	27.8	53.8	2304	1.4
Nech	Dec	4185	28.0	53.6	2244	1.2
Uso	Dec	3618	26.5	55.6	2012	1.3
Tara	Dec	3237	25.5	56.9	1842	1.6
Ayin	Feb	4611	27.1	54.8	2528	1.6
Nech	Feb	3514	25.9	56.3	1978	1.0
Uso	Feb	4574	24.5	58.1	2659	1.9
Tara	Feb	2961	24.5	58.1	1721	1.5

The amount of food intake was based on the corrected dry weights (Table 17.2).

conditions, however, may be less than the value obtained in the experiment reported here, because grass contains many more cell wall components which have to be decomposed than the food provided to captive animals, and also because the mass of food processed in a day by wild geladas is 1.5 to 2.0 times greater than for caged animals.

Geladas in the Gich area increased their daily food intake going into the dry season. The protein content of blades quickly declined after the onset of the dry season. This was the main change in the nutritional condition of the gelada's foods. In order to compensate for this, the gelada increased the amount of food intake by changing their staple food from grass blades to herbs (mainly *Trifolium*) and also by increasing the time spent feeding. As a result, gelada could take foods with higher quality in the dry season as a whole. Therefore, it is concluded that the second condition for hindgut fermentation was not fully satisfied.

It is possible that if *Trifolium* were not available, the second condition might be met as the animals increased their daily intake of low quality foods. If they were forced to choose only grass blades in the dry season, the time spent feeding by gelada should increase from 72.8 per cent (see Iwamoto, chapter 16) to almost 90 per cent to achieve the same energy intake as those measured in this study. With this high proportion of time spent feeding and also additional time spent moving (about 15 per cent or more), gelada might not have any time for their social life (see Dunbar, chapter 15). Thus, *Trifolium* eating appeared to be the final choice in the diet selection of gelada during the dry season in the severe ¬Simen Mountains.

The first condition was, however, satisfied as described above, and so the hypothesis of hindgut fermentation seems very plausible. More data on the rate of food intake in the more severe seasons are needed, however, to resolve this problem. Also, studies on the symbiotic activities of micro-organisms in the hindgut of geladas are essential.

Using the digestibilities derived from the fibre content of natural foods, assimilated energy was calculated for each month, and considerably larger energy intakes were found in the dry season (1.4 to 1.5 times more in December and February to October). Why do geladas need so much energy in the dry season? There are three possible explanations.

First, geladas need considerable energy to maintain body temperature in the dry season. Iwamoto & Dunbar (1983) compared time spent feeding for three separate gelada populations at various altitudes, and concluded that the only possible factor which could explain the large food intake of geladas living in high altitude areas was thermoregulation. This may also explain the seasonal difference in food intake in the Gich area. The average air temperature declined by 4 °C toward the middle of the dry season (Kawai, 1979). According to Kleiber (1961), the metabolic rate of normal endothermal mammals increases by about 20 per cent for this temperature difference.

Secondly, the answer to the following question may offer a solution: why do geladas not choose herbs as food all year round, as they are juicier, more nutritional and are present as a larger biomass during the rainy season? There actually may be two reasons for this. One is that geladas may have adapted to grass eating so completely that they are less efficient in digesting non-gramineacous plants. This may be explained by the idea that the micro-organismal fauna in the gut has not been established for herb digestion, but for grass digestion through a long evolutionary history. This seems unlikely, however, in view of the fact that animals employing hindgut fermentation are usually recognized to be more tolerant of a wide range of quality of vegetation than those employing forestomach fermentation (Janis, 1976). An alternative answer for the question is that plants belonging to genus *Trifolium* sometimes contain toxins including phytooestrogens (Garey, 1984), which are known to inhibit conception in females, and cyanogenic compounds (Conn, 1979; Crawley, 1983) which disturb the animals' physiological activities. If *Trifolium* in Gich contains such toxins which can be detoxified, the more they take *Trifolium*, the more they have to expend energy for detoxification. Note that geladas increased their energy intake when the proportion of *Trifolium* increased in the diet. The detoxification explanation seems plausible, but further studies on the digestive physiology of geladas are needed to confirm this.

Thirdly, as seen in Table 17.4, the ratio

of assimilated energy in February to that in October was extremely high only for one female, Uso (1.9), while the ratios of other baboons (average of 3 geladas = 1.4) showed level similar to those of December (average of 4 geladas = 1.4). As there was no evidence that she had a newborn baby or had been pregnant around the sampling period in February, this abnormally high energy intake could not be due to her reproductive status. Therefore, her higher intake represents a sampling error due to her unusually high intake in the sampling day.

Among these explanations, the first one may be the most highly probable, while the others need further examination to be confirmed. In any case, the result that geladas took more foods and energy in the dry season than in the rainy season is not inconsistent with the hypothesis of bulk eating (Dunbar, 1984), and hence will supply a key to understanding the process of adaptation of geladas to highly graminivorous food habits. The results of this study also suggest the possibility of hindgut fermentation in extant geladas. More attention should be paid to the physiological and morphological features of gelada which have enabled them to become such highly specialized herbivores, unique among living primates.

Acknowledgements

I would like to express my sincere thanks to Prof Y. Ono of Kyushu University and Prof M. Kawai of Kyoto University for their encouragement of my research. I am grateful to the Ethiopian Government for permission to study geladas in the Simen National Park. I would also like to thank Dr N.G. Jablonski and Dr R.A. Foley for their kind invitation to the Cambridge *Theropithecus* symposium and to Dr R.I.M. Dunbar for his valuable suggestions on this study. Thanks also to Dr N.G. Jablonski and Mr G.A. Butterworth of Miyazaki University for their assistance in correcting my English.

References

BAUCHOP, T. (1978). Digestion of leaves in vertebrate arboreal folivores. In *The Ecology of Arboreal Folivores*, ed. G.G. Montgomery, pp. 193–229. Washington, D.C: Smithsonian Institution Press.

BAUCHOP, T. & MARTUCCI, R.W. (1968). Ruminant-like digestion of the langur monkey. *Science*, 161, 698–700.

CHIVERS, D.J. & HLADIK, C.M. (1980). Morphology of the gastrointestinal tract in primates: comparisons with other mammals in relation to diet. *Journal of Morphology*, 166, 337–86.

CONN, E.C. (1979). Cyanide and Cyanogenic Glycosides. In *Herbivores: Their Interaction with Secondary Plant Metabolites*, ed. G.A. Rosenthal & D.H. Janzen, pp. 387–412. New York: Academic Press.

CRAWLEY, M. (1983). *Herbivory: The Dynamics of Animal-Plant Interactions*. Oxford: Blackwell Scientific Publications.

DUNBAR, R.I.M. (1977). Feeding ecology of gelada baboons: a preliminary report. In *Primate Ecology*, ed. T.H. Clutton- Brock, pp. 251–73. London: Academic Press.

DUNBAR, R.I.M. & DUNBAR, P. (1974). Ecological relations and niche separation between sympatric terrestrial primates in Ethiopia. *Folia Primatologica*, 21, 36–60.

DUNBAR, R.I.M. (1984). *Reproductive Decisions: An Economic Analysis of Gelada Baboon Social Strategies*. New Jersey: Princeton University Press.

GAREY, J.D. (1984). A possible role for secondary plant compounds in the regulation of primates breeding cycles. *American Journal of Physical Anthropology*, 63, 160.

GOODALL, A.G. (1977). Feeding and ranging behavior of a mountain gorilla group (*Gorilla gorilla beringei*) in the Tshibinda-Kahuzi region (Zaire). In *Primate Ecology*, ed. T.H. Clutton-Brock, pp. 450–80. London: Academic Press.

HILL, W.C.O. (1952). The external and visceral anatomy of the olive colobus monkey (*Procolobus verus*). *Proceeding of the Zoological Society, London*, 122, 127–›86.

IWAMOTO, T. (1979). Feeding ecology. In *Ecological and Sociological Studies on Gelada Baboons*, ed. M. Kawai, pp. 279–330. Tokyo: Kodansha.

IWAMOTO, T. (1982). Food and nutritional condition of free ranging Japanese monkeys on Koshima Islet during winter. *Primates*, 23, 153–70.

IWAMOTO, T. (1988). Food and energetics of provisioned wild Japanese macaques (*Macaca fuscata*). In *Ecology and Behavior of Food Enhanced Primate Groups*, ed. J.E. Fa & C.H. Southwick, pp. 79–94. New York: Alan R. Liss.

IWAMOTO, T. & DUNBAR, R.I.M. (1983). Thermoregulation, habitat quality and the behavioral ecology of gelada baboons. *Journal Animal Ecology*, 52, 357–66.

JABLONSKI, N.G. (1981). *Functional Analysis of the Masticatory Apparatus of* Theropithecus gelada *(Primates: Cercopithecidae)*. Ph.D. Thesis, University of Washington.

JANIS, C. (1976). Evolutionary strategy of the equids and the origins of rumen and caecal digestion. *Evolution*, 30, 757–74.

JOLLY, C.J. (1972). The classification and natural history of *Theropithecus* (*Simopithecus*) (Andrews, 1916), baboon of the Plio-Pleistocene. *Bulletin of British Museum (Natural History), Geology*, 22, 1–23.

KAWAI, M. (1979). *Ecological and Sociological Studies on Gelada Baboons*. Tokyo: Kodansha.

KLEIBER, M. (1961). *The Fire of Life*. New York: John Wiley & Sons Inc.

MITCHELL, H.H. (1942). The evaluation of feeds on the basis of digestible and metabolizable nutrients. *Journal of Animal Science*, 1, 159–73.

MITCHELL, H.H. (1964). *Comparative Nutrition of Man and Domestic Animals*. London: Academic Press.

OHWAKI, K., HUNGATE, R.E., LOTTER, L., HOFMANN, R.R. & MALOIY, G. (1974). Stomach fermentation in East African colobus monkeys in their natural state. *Applied Microbiology*, 27, 713–23.

OSMAN HILL, W.C. (1970). *Primates: Comparative Anatomy and Taxonomy*, vol. 8, *Cynopithecinae*. Edinburgh: Edinburgh University Press.

SIBLY, R.M. (1981). Strategies of digestion in relation to diet. In *Physiological Ecology*, ed. C.R. Townsend & P. Calow, pp. 109–39. Oxford: Blackwell Scientific Publications.

VAN SOEST, P.J. (1982). *Nutritional Ecology of the Ruminant*. Portland: Durham and Downey, Inc.

18 Socioecology of the extinct theropiths: a modelling approach

R. I. M. DUNBAR

Summary

1. A systems model of the socioecology of extant gelada is used to predict maximum group sizes for populations of extinct theropiths living under various climatic conditions.
2. Under present climatic conditions, poor nutritional quality of the graze at low and high altitudes restricts the extant gelada to habitats lying between 1500–4000 m in altitude, but higher latitudes or lower global temperatures would make the colonization of lower altitudes possible.
3. The theropiths of the Plio-Pleistocene were not large enough to be able to exploit poorer quality vegetation at lower altitudes; consequently, they must have had an even more restricted distribution than the extant gelada.
4. The largest theropiths of the later Pleistocene could only have survived in the localities where they are known to have occurred if ambient temperatures were at least 6 °C cooler than at present or graze quality was at least three times

greater than is the case in modern grassland habitats; this suggests that they were restricted to the immediate vicinity of permanent water and thus that they would have been particularly vulnerable to extinction.
5. The analyses suggest that the largest species could not have lived in groups as large as those of contemporary gelada; nor could they have grown much larger in body size than they did.

Introduction

In this chapter, I use a model of the behavioural ecology of the extant gelada (*Theropithecus gelada*) to explore the likely constraints acting on the ecology the extinct congeners of this species. Part of my purpose in doing so is to comment on the possible reasons for the extinction of this once successful and widespread taxon during the later Pleistocene. This episode of extinction has left us with a single relict population (the extant gelada) occupying a retreat habitat on the high plateaux of northern Ethiopia.

I approach the problem by focusing on

the question of what limits group size. I argue that time budgets are a crucial component of an animal's behavioural ecology and that it is the need to balance its time budget that ultimately imposes a limit on the size of group in which it lives. In effect, I am suggesting that nutrients *per se* do not often limit populations: rather, the problem is that the animals do not have sufficient time to harvest the nutrients they need each day. Given an infinitely long day, an animal could always balance its energy budget even in the poorest quality habitat.

The problem, then, resolves itself into one of being able to determine time budgets for any given population. If we can derive equations that accurately predict the time budgets of living populations, then we should be able to interpolate the appropriate values from extinct populations into these equations and thus derive predictions about their behavioural ecology.

First, a series of equations for the key components of gelada time budgets is presented. These equations are then used to determine the maximum ecologically tolerable group size for populations of extinct theropiths inhabiting different environments. The maximum ecologically tolerable group size represents the limit beyond which any further increase in group size would result in the animals being unable to meet the time budget allocations required by that particular habitat.

The argument here rests on the assumption that the three main components of the time budget (feeding, moving, and social time) are determined by environmental and/or demographic factors, whereas the fourth component (resting time) is made up in part of an environmentally imposed quantity and partly of 'free' time not required for any other activity. Resting time thus acts as a reservoir of uncommitted time that can be used to supplement other activity categories whenever changing conditions require this (see Altmann, 1980; Dunbar & Sharman, 1984; Dunbar & Dunbar, 1988). Since increases in group size also impose demands on feeding, moving and social time, group size can be increased in any given habitat only up to the limit imposed by the quantity of spare capacity in the resting time budget. Hence, we can identify the maximum ecologically tolerable group size as that group size at which all the spare resting time (i.e. other than the habitat-specific minimum) has been allocated to the other activity categories.

It is important to appreciate at the outset that the maximum ecologically tolerable group size is not necessarily equivalent to the actual size of the groups in which animals live. It is simply the limiting size that cannot be exceeded without imposing severe ecological stress on the animals because they would be unable to meet the demands on their time budget. Whether animals actually live in groups as large as this will depend on the strength of the factors favouring large groups relative to those militating against living in groups. The latter consideration is especially important since factors other than the time budget may limit individuals' willingness to live together (see van Schaik, 1983; Dunbar, 1988). These factors mainly concern the consequences of competition for food resources, space, and social partners: as group size increases, so does the amount of aggression. This may, in itself, add further costs by forcing animals to spend more time feeding as a result of disrupted feeding bouts; more importantly, it is known to impose serious costs in the form of reproductive suppression, especially in the large groups typical of the gelada (see Dunbar, 1980). Thus, social factors may constrain group sizes below what is ecologically tolerable.

In addition, whether or not animals can survive in a given habitat will depend not merely on whether they can balance their activity budgets, but also on whether other environmental factors impose a minimum group size. If the minimum group size for survival is greater than the maximum ecologically tolerable group size, then the animals will not be able to survive in that habitat. In this respect, predation risk is likely to be an especially important consideration (see van Schaik, 1983; Dunbar, 1988). In the *Papio* baboons, group size appears to be directly related to the density of predators in the habitat and negatively related to the density of trees (in which to escape from predators). Since tree density is related to rainfall, mean group sizes are negatively related to rainfall (Dunbar, 1988). We cannot, at present, say anything about the precise relationship between predation risk and group size in theropiths. However, since theropiths in general are likely to have been poor tree climbers (Jolly, 1972) as, indeed, are the extant gelada (Dunbar, 1984a), it seems reasonable to assume that they would have relied mainly on group size as a means of reducing predation risk on open grasslands. Judging by

the behaviour of both the gelada and the other baboons, minimum group sizes in open habitats are of the order of 40–50 animals. Since increasing body size also allows animals to reduce their predation risk, the larger species may well have been able to live in very much smaller groups. Even so, minimum group sizes may have been in the order of 10–20 individuals.

In undertaking these analyses, it is important to emphasize that we need to consider particular populations of the extinct species. Conventionally, we have tended to consider the 'typical' population for any given species (i.e. in effect, to treat each species as an homogenous whole). Such a view tells us very little about the constraints acting on a palaeospecies and, thus, even less about the reasons why it became extinct. The biological reality is, of course, that it is populations that die out rather than species that become extinct; consequently, we need to know why it was that individual populations might not have been able to cope with their local environments.

I take the view that congeneric species share a common set of strategic adaptations at the ecological (especially dietary) and reproductive levels (i.e. share the same *Bauplan*). This means that we can consider the differences between them as being mainly due to differences in either body size or environmental conditions. For present purposes, I shall consider just three notional species representing small, medium, and large theropiths. *Theropithecus darti* and *T. oswaldi* will represent the medium and large members of the genus, while *T. gelada* will represent the small members (and hence, presumably, the ancestral condition). *Theropithecus brumpti* would seem to be sufficiently different in its ecology (Eck & Jablonski, 1987) to require separate treatment, and I therefore exclude it from consideration.

It is worth pointing out before we embark on this exercise that the theropiths as a group possess one key feature that makes it possible to undertake such an analysis with more confidence than we might normally do in these circumstances. This is the fact that they are (and were) all grazers (Jolly, 1972; Szalay & Delson, 1979; Lee-Thorp, van der Merwe & Brain, 1989). Being essentially monocultural, grassland is much easier to analyse than woodland or forest and generally produces simpler relationships between environmental variables (such as rainfall or temperature) and primary productivity. This means that the precise nature of plant species can be ignored and the food resources available treated under the single undifferentiated rubric of 'grasses'.

One final caveat is in order. As with all simulation models, the process of refining the model by the inclusion of additional variables as new information becomes available is a continuing one. Hence, the particular results presented below remain subject to change without notice. Nonetheless, the general conclusions do seem to be fairly robust in that a number of alternative formulations tried out during the process of developing the model yielded substantially similar results. I therefore think that we can draw some fairly clearcut conclusions without too much risk of future contradiction.

Methods

The data for the extant gelada derive from three sites in Ethiopia: the Bole Valley in the extreme south of the species' range and two sites (Sankaber and Gich) in the Simen Mountains from the north. Data for the last of these three populations derive from the studies carried out by Kawai and his co-workers (see Kawai, 1979), while data for the first two sites derive from my own work (see Dunbar & Dunbar, 1975; Dunbar, 1984b). Some considerable progress towards understanding the behavioural ecology and population dynamics of the gelada has been made through comparative studies of the data from these three sites (see Iwamoto & Dunbar, 1983; Ohsawa & Dunbar, 1984, Dunbar, 1984b).

In building a model of the gelada socio-ecological system, I concentrated on seven main variables: the four time budget components (feeding, moving, social, and resting time), day journey length and two demographic variables (band and herd sizes). Five environmental parameters were used as potential independent variables: mean annual rainfall, mean ambient temperature, altitude, the percentage of grass cover, and the percentage protein content of grass during the dry season.

I have used a conventional stepwise multiple regression analysis to identify the set of independent variables that best predict each of the dependent variables. However, because the sample sizes are generally rather small (N = 3 in most cases), it was not possible to extract more than two independent variables and, even then, there were no remaining degrees of freedom with which to assess the significance of the regression equation. I have proceeded with the analyses despite this rather serious drawback only because the coefficients of determination are in all cases very close to $r^2 = 1$, suggesting that the regression equations really do predict the dependent variables with some reliability. In addition, I have imposed a fairly stringent test of the equations' validity, namely that they be able to predict very precisely the altitudes above and below which gelada do not live. In general, I have used log-transformed data in order to minimize any problems with data that are not normally distributed, except in those cases where sample sizes are larger and untransformed data yield similar or larger values of r^2.

It is important to differentiate between the two different demographic variables, band size, and herd size. Gelada live in a complex hierarchically organized social system that consists of a number of separate levels of grouping (see chapter 15). Of these, the band is the largest grouping which exhibits consistent membership based on regular intimate interaction. Since it is defined as the group of animals that share a common ranging area, we may expect it to have important ecological and social implications. Thus, the dispersion of the bands essentially determines population density. A band consists of a number of reproductive units; these range in size from 3–28 animals of all ages and both sexes. Reproductive units are the smallest units capable of an independent existence. However, they do not often travel on their own, but prefer to congregate in temporary foraging parties (known as herds). A herd may consist of all or just some of the reproductive units of a given band. Lone units will not normally join herds that only contain units from another band. Since herds constantly change in size over the course of a day as units join and leave, herd size rather than band size may thus be expected to be more relevant to ecological variables such as day journey length. The minimum group size would appear to be that of the reproductive units, which typically average 10 animals in size. Neither a band nor a herd, for example, can be smaller than one reproductive unit in size.

This analysis of gelada behavioural ecology was undertaken in parallel with a comparable analysis of the behavioural ecology of *Papio* baboons. In the baboon case, however, the available sample of populations is very much greater ($12 < n < 23$ for different variables). This much larger sample of data has allowed a more formal approach to regression analysis. Although in most cases, it is inappropriate to apply the baboon equations to the gelada because of the differences between the ecological adaptations of the two genera, nonetheless they provide valuable support for the general form of the gelada equations since many of the same variables turn up in both sets of equations (albeit with different regression coefficients).

In order to simplify the analyses as far as possible, I have endeavoured to reduce to a minimum the number of environmental parameters that we need to consider by expressing some variables as functions of others whenever there are good geophysical or biological reasons for doing so.

Table 18.1. *Regression equations for systems model of gelada socioecology*

Dependent variable	Equation	N	r^2
Feeding time (%)	$\ln(F) = (0.40\ln(W)/1.13) \times (5.94 - 0.60\ln(T) - 0.31\ln(Q))$	3	0.999
Moving time (%)	$\ln(M) = (0.93/0.33\ln(M)) \times (4.75 + 0.26\ln(J) - 0.48\ln(C))$	3	0.999
Social time (%)	$S = 4.53 + 0.08N$	20	0.490*
Resting time (%)	$R = -12.24 + 2.46T$	3	0.972
Day journey (km)	$\ln(J) = 1.25 + 1.08\ln(Nf) - 1.29\ln(C)$	3	0.977
Herd size	$\ln(Nf) = -3.93 + 0.88\ln(N) - 1.29\ln(C)$	3	0.965
Temperature (°C)	$T = 28.36 + U - 0.0048A - 0.176L$	26	0.893*
Grass cover (%)	$C = 55.90 + 1.25T - 0.13T^2$	8	0.891*
Protein content (%)	$\ln(Q) = -26.71 + 23.90\ln(T) - 4.84(\ln T)^2$	9	0.865*

A = Altitude (m); L = absolute latitude (°); N = band size; U = temperature difference from present conditions (°C); * p<0.01.

Thus, I was able to find regression equations that express ambient temperature as a function of altitude and latitude (a well known geophysical relationship) and grass density and quality as functions of ambient temperature. These analyses are more secure since they are based on larger samples of habitats drawn from throughout eastern and southern Africa.

The procedure used in the second part of this chapter is based on a simple algorithm that searches for the group size that uses up all the available resting time, subject to the habitat-specific minimum imposed by environmental parameters. Details are given in Dunbar (1992a).

Results

The basic model

Table 18.1 lists the equations used in the simulations. Full details of their derivation are given in Dunbar (1992a). Note that the equations in Table 18.1 are modified

where appropriate to take account of differences in body size. Though not relevant in the case of the extant gelada, this is an important consideration when we come to assess the impact of environmental parameters on the time budgets of the extinct species.

Body weight enters into the equation for feeding time in two important respects. First, it determines the absolute energy requirement of an animal, this being a function of body weight raised to the 0.75 power (Kleiber's Law: Kleiber, 1960; Peters, 1983). In addition, body size has important implications for an animal's ability to retain ingesta in its gut, and hence for its ability to handle a poor quality diet with a high fibre content (the Jarman-Bell principle: Jarman, 1974). Demment & van Soest (1985) found that the retention time of ingesta in the gut is a function of the 0.346 power of body weight. Hence, although larger animals will require absolutely more food, they will be able to extract nutrients more efficiently from what they eat. Consequently, the net effect on feeding time requirements of increasing adult body weight from a mean of 16.5 for extant gelada to a weight of W kg will be to multiply feeding time by a ratio of:

$$\frac{W^{0.75}}{8.1868} \times \frac{2.6378}{W^{0.346}} = \frac{W^{0.75-0.346}}{3.1036} = \frac{W^{0.404}}{3.1036}$$

Body weight also enters into the equation for time spent moving because larger animals have a longer stride length and can therefore travel further per unit time spent moving (Peters, 1983). Stride length is a simple function of leg length, and linear dimensions of the body are generally a function of the cube root of body weight (Peters, 1983). Hence, we can expect the time spent moving to travel a given distance to decrease in proportion to the cube root of body weight.

Ambient temperature influences feeding time because of the relationship between ambient temperature and the costs of thermoregulation: once temperature falls below a critical threshold, proportionally more energy is required to maintain body temperature at a constant value (Kleiber, 1960; Mount, 1979). Feeding time is assumed to be directly proportional to energy requirement (see Altmann, 1980), but the constant of proportionality will depend on forage quality. As forage quality declines, a larger quantity of food will have to be eaten (and hence proportionately more time spent feeding) in order to ingest the same quantity of nutrients. Estimates of the protein content and digestibility of grass in a number of east African habitats indicated that protein content varies consistently with environmental parameters, whereas digestibility remains more or less constant across habitats. Since proteins provide most of the energy content of grasses, time spent feeding may be expected to vary inversely with grass protein content. The dry season protein content has been used as the relevant variable because the dry season is usually the limiting time of year in tropical environments.

As the length of the day journey has an important impact on the amount of time that has to be devoted to travel, we also need to know what determines day journey length. The data for both the gelada and the *Papio* baboons indicate that group size and food (in this case, grass) density are the two most relevant variables. In the case of the gelada, we would expect the size of the foraging party (herd size) to be the appropriate demographic variable.

The equations for resting and social time require some explanation. An analysis of baboon time budgets indicated that social time (which is required for servicing the social bonds that hold a group of animals together through time) is influenced by a number of factors, including the amount of time spent resting, the ambient temperature and the size of the group (see Dunbar, 1990, 1992b). The first two of these variables appear to be constraints imposed by the environment, and only the third (group size) is a causal determinant, a point emphasized by the finding that time spent grooming correlates with group size in all haplorhine primates (Dunbar, 1991). Since our present concerns are with the minimum amount of time that the animals have to spend in social interaction in order to keep a group of a given size together, the only relationship we need to consider is that between group size and social time. Analyses of both the gelada and baboon data-sets yield regression equations that are similar, but differ in slope (with that for baboons being considerably steeper than that for the gelada). Rather than use either of these equations, however, I have chosen to use a more general equation, relating time spent in social interaction to group size for a large sample of Old World monkeys (see Dunbar, 1991). Since this equation is intermediate in slope between those obtained for the gelada and the baboons, it should act conservatively by minimizing any errors that might be introduced by using the wrong equation. This is an important point because the relationship between social time and group size probably reflects the way in which social relationships are structured: we cannot necessarily assume that the social system of the extinct theropiths is the same as that of living gelada. Because the gelada equation rises less steeply with group size than the baboon equation, its use would have the effect of reducing maximum group sizes in low altitude populations where group sizes will naturally tend to be low.

Evidence for both baboons (Dunbar & Sharman, 1984) and the gelada (Dunbar & Dunbar, 1988) suggest that resting time acts as a reserve of uncommitted time that can be drawn on whenever additional time is required for feeding, travel, or social interaction. More detailed analysis of the time budgets of baboons, however, suggests that there is an additional component that is forced on the animals by the environment. This component is a direct function of ambient temperature and seems to be a consequence of excessive heat loads incurred in more open habitats. In the present analyses, I consider only this component of resting: this gives us the minimum resting time requirement that the animals cannot convert into feeding, moving or social time. The equation given in Table 18.1 derives from the gelada data.

Finally, an equation is given relating herd size to band size and grass density. That these two variables should have been selected by the regression analysis makes sense since band size is likely to place an upper limit on herd size while the number of animals that can be in any one place at any one time is likely to be a function of the density of the grass sward.

These equations suggest that only one environmental variable needs to be determined in most cases, namely ambient temperature. Knowledge of this one parameter gives us surprisingly good estimates of most of the key variables that we need to know about (grass cover, graze quality, etc). Ambient temperature, in turn, is a relatively simple function of altitude and latitude, following known geophysical principles (though the regression equation in this case is based on a sample of 26 sub-Saharan habitats where either baboons or gelada have been studied: see Dunbar, 1992b). I had initially assumed that rainfall would prove to be an important determinant of gelada behavioural ecology, since rainfall is known to be a good predictor of primary productivity in sub-Saharan habitats (Coe, Cummings & Phillipson, 1976; Le

Table 18.2. *Maximum ecologically tolerable group sizes for gelada (mean adult body weight = 16.5 kg) living in different habitats under current climatic conditions*

Altitude (m)	Maximum ecologically tolerable group size at latitude (degrees from equator)						
	0	5	10	15	20	25	30
200	0	0	0	0	0	0	0
600	0	0	0	0	0	3	14
1000	0	0	0	5	17	38	67
1400	1	6	20	43	72	104	137
1800	23	47	77	109	142	174	203
2200	82	115	147	179	207	232	252
2600	153	184	212	236	254	266	267
3000	216	239	257	267	266	251	215
3400	259	268	265	247	206	131	20
3800	263	242	196	114	5	0	0
4200	185	95	0	0	0	0	0
4600	0	0	0	0	0	0	0
5000	0	0	0	0	0	0	0

Houeron & Hoste, 1977; Deshmukh, 1984). Such habitats are, however, often very seasonal in their productivity, with all growth being confined to the relatively short periods when rainfall occurs. Desiccation of grasses during the dry season is then the most significant factor determining plant growth. Consequently, a relationship between temperature and both grass cover and dry season grass quality may not be unreasonable. Equations for these environmental variables are also given in Table 18.1.

Using these equations to calculate the maximum ecologically tolerable group size in any given habitat suggests that, in their current latitudinal range of 9–13° from the equator, gelada would find it impossible to survive at altitudes lower than about 1500 m and higher than about 4000 m

(their current altitudinal limits) (Table 18.2). Thus, the extant gelada are apparently prevented from colonising lowland habitats by the rapidity with which increasing temperatures overload their time budgets.

Note that these results apply only to open grassland habitats and not necessarily to woodland or forested habitats. On the Ethiopian plateau, forest occurs in any habitat lying between 2000 m and 3000 m altitude that receives more than 1000 mm of rainfall per year, while *Acacia* woodland is typical of habitats below 2000 m in altitude (Hurni, 1982). Gelada rarely enter forested areas and appear unable to survive in such habitats (Dunbar & Dunbar, 1974; Dunbar, 1984a), and it is probably the presence of extensive forest that prevents the gelada expanding their current range to the south and west.

These results suggest that the model is sufficiently reliable to warrant extrapolation to other populations. More importantly, the variables that we need to know about turn out to be rather limited (essentially ambient temperature and body weight). This means that the model should be reasonably robust for extrapolation to palaeo-environments where our knowledge of variables such as plant productivity or the protein content of grasses is non-existent.

Ecology of extinct theropiths

In examining the implications of this model for the ecology of the extinct theropiths, I proceed in two steps. First, I determine the maximum tolerable group size at a range of altitudes and latitudes for small, medium, and large species. I then examine the effect that changing climatic conditions would have had on the animals' ability to survive in different habitats.

The crucial parameter whose value we need to determine is the animal's body weight. Inevitably, estimates of body size for extinct species are open to various sources of error. However, I think this less problematic than gaining some idea of how an animal's behavioural ecology responds to changes in body weight and climate. Hence, knowing that the extinct species were all considerably larger than the extant gelada, we can in effect ask how changes in body weight would, in general, influence the animals' ecology. I therefore treat *T. darti* as representative of a medium-sized theropith and the largest *T. oswaldi oswaldi* as representative of the

upper limit on body size. As the smallest of all the theropiths, the gelada can represent the small end of the size spectrum. Since there seems to have been a continuous increase in size over time within the genus (Jolly, 1972), the gelada may also represent the ancestral condition in this respect.

Jolly (1972) gives estimates in the order of 35 kg for *T. darti* males, but was unable to estimate body weights for any females. Since *T. darti* males were of the same body weight as *T. o. mariae* males, I assume that *T. darti* females were about the same weight as the females of this subspecies (namely 20 kg). This gives us an average body weight of around 27.5 kg. Estimates for the largest of the extinct theropith (*T. o. oswaldi*) suggest that males achieved body weights in the order of 65 kg (Jolly, 1972). No estimates are available for female body weights, but if we assume that females weighed around 40 kg, this allows us to take 55 kg for the average adult weight.

We can now examine the impact of increased body weight on maximum tolerable group sizes in different environments. I use the same procedure as with the gelada, namely determining maximum group size for a range of altitudes and latitudes.

Tables 18.2 to 18.4 give the maximum tolerable group size for small, medium and large theropiths, respectively, under current climatic regimes. Three points may be noted. First, the animals are generally restricted to a fairly narrow range of altitudes at any given latitude, with this band of ecologically tolerable habitats occurring at progressively lower altitudes

Table 18.3. *Maximum ecologically tolerable group sizes for medium-sized theropiths (mean adult body weight = 27.5 kg) living in different habitats under current climatic conditions*

Altitude (m)	Maximum ecologically tolerable group size at latitude (degrees from equator)						
	0	5	10	15	20	25	30
200	0	0	0	0	0	0	0
600	0	0	0	0	0	0	1
1000	0	0	0	0	·1	10	29
1400	0	0	2	13	33	61	92
1800	3	15	38	66	97	167	155
2200	42	71	102	132	160	182	199
2600	108	137	164	186	202	209	204
3000	168	189	203	209	202	177	128
3400	205	209	199	171	117	33	0
3800	196	164	104	19	0	0	0
4200	91	7	0	0	0	0	0
4600	0	0	0	0	0	0	0
5000	0	0	0	0	0	0	0

Table 18.4. *Maximum ecologically tolerable group sizes for large theropiths (mean adult body weight = 55 kg) living in different habitats under current climatic conditions*

Altitude (m)	Maximum ecologically tolerable group size at latitude (degrees from equator)						
	0	5	10	15	20	25	30
200	0	0	0	0	0	0	0
600	0	0	0	0	0	0	0
1000	0	0	0	0	0	0	0
1400	0	0	0	0	0	1	12
1800	0	0	0	2	15	35	56
2200	0	4	18	38	60	77	89
2600	21	42	63	80	90	90	76
3000	66	82	91	89	72	38	1
3400	91	87	68	31	0	0	0
3800	63	24	0	0	0	0	0
4200	0	0	0	0	0	0	0
4600	0	0	0	0	0	0	0
5000	0	0	0	0	0	0	0

as latitude increases north or south of the equator. Secondly, the altitudinal range becomes progressively more compressed as body size increases. Finally, the maximum tolerable group size decreases with body size. It seems, then, that the larger theropiths would have had an even more restricted ecological niche than contemporary gelada. The analysis also suggests that the increase in theropith body size during the Pleistocene could not have been a response to deteriorating forage conditions as the climate became progressively warmer. Large species do gain a saving in terms of increased digestive efficiency, but this is more than offset by their increased metabolic costs.

Of particular importance here is the implication that although the small- to medium-sized theropiths could have coped with environmental conditions at the altitudes at which their fossils occur in both the Mediterranean region (latitude 30° N) and southern Africa (latitude 25–30° S), they would have experienced considerable difficulty surviving at those altitudes where they are known to have occurred on the equator in eastern Africa. At 0–5° from the equator, they could not have lived in habitats below 2000 m under current climatic conditions. The situation is even more problematic for the very large species, which are known only from East African sites. The simulation suggests that these species could only have survived at altitudes above 2500 m. Yet, they are known to have been abundant in a number of localities whose present altitudes are as low as 1000 m (assuming that these sites in the Rift Valley have not dropped significantly in altitude since the mid-Pleistocene).

We know that global temperatures, changed considerably during the Plio-Pleistocene. Sub-arctic seabed temperatures, for example, were about 1.5 °C warmer than at present at the start of the Pliocene and declined to around 1.5 °C cooler than at present by the end of the Pliocene, before rising towards their current level during the Pleistocene (Kennett, 1977). To what extent might cooler climatic conditions have allowed the extinct theropiths to expand their putative range beyond those indicated in Tables 18.2 to 18.4?

I reran the simulations for the three species for a range of climatic conditions by allowing the mean global temperature to vary from 2 °C warmer than present to 8 °C cooler than present. The first of these values corresponds to the early Pliocene maximum in global temperature, while the later reflects the most extreme temperatures that might have occurred during the Holocene cool periods. Studies by Hurni (1982) of glacial deposits in the Simen dating from the last cold period 20,000–12,000 BP suggest that local mean temperatures reached a low at about 7 °C below present levels during this period. At the height of these glaciations, the periglacial belt (marking the upper limit for plant growth) was some 700–800 m lower in altitude than at present in the Simen Mountains.

The results are presented in Table 18.5 as the minimum altitude at which each species could have survived under different temperature conditions. These

Table 18.5. *Minimum altitudinal limits below which theropiths of different size could not survive under various temperature regimes*

Temperature difference from present (°)	Minimum altitude for survival* (m) at latitude (degrees from equator)						
	0	5	10	15	20	25	30
Small species (W = 16.5 kg)							
+2	2200	2000	1800	1600	1400	1200	1100
0	1800	1600	1400	1200	1000	900	700
−2	1400	1200	1000	800	600	400	300
−4	900	700	600	400	200	0	0
−6	500	300	100	0	0	0	0
−8	100	0	0	0	0	0	0
Medium species (W = 27.5 kg)							
+2	2500	2300	2100	1900	1700	1500	1300
0	2000	1800	1700	1500	1300	1100	900
−2	1600	1400	1300	1100	900	700	500
−4	1200	1000	800	600	500	300	100
−6	800	600	400	200	0	0	0
−8	400	200	0	0	0	0	0
Large species (W = 55 kg)							
+2	3000	2800	2700	2500	2300	2100	1900
0	2600	2400	2200	2000	1900	1700	1500
−2	2200	2000	1800	1600	1400	1200	1100
−4	1800	1600	1400	1200	1000	800	700
−6	1300	1100	900	800	600	400	200
−8	900	700	500	400	200	0	0

*Defined as the altitude at which the maximum ecologically tolerable group size is at least 20 animals; animals can survive at altitudes above the listed figure (up to a maximum altitude determined by time budget constraints).

results show quite clearly that minimum altitude correlates directly with mean global temperature for all three species.

The first point to note is the effect of climatic changes on the altitudinal distribution of the smallest theropiths, since this is relevant to the later history of the gelada. Jolly (1972) argued that the gelada

probably became isolated in its present mountain retreat at a fairly early stage in the taxon's history since it shares a number of primitive characters with the earliest theropiths. The results suggest that, during the cool periods when high mountains like the Simen became extensively glaciated, the gelada would have

Table 18.6. *Maximum ecologically tolerable group sizes for gelada living in various habitats within their current latitudinal range under conditions where global temperatures are 6–8°C below their current values*

Altitude (m)	Maximum ecologically tolerable group size					
	At 6°C below present latitude (°N)			At 8°C below present latitude (°N)		
	5	10	15	5	10	15
200	10	27	52	60	91	124
600	57	87	120	129	161	192
1000	126	158	188	196	223	245
1400	193	220	242	248	268	262
1800	246	261	268	268	259	232
2200	268	261	237	225	165	65
2600	230	173	78	45	0	0
3000	58	0	0	0	0	0
3400	0	0	0	0	0	0
3800	0	0	0	0	0	0
4200	0	0	0	0	0	0
4600	0	0	0	0	0	0
5000	0	0	0	0	0	0

been forced to move downhill quite a considerable distance. This is clear from Table 18.6 which gives the maximum ecologically tolerable group sizes for the gelada under the most extreme conditions (mean temperatures 8 °C below present). At the latitude of the Simen (13° N), they would have occurred predominantly in the surrounding lowlands at altitudes below 2500 m.

At this time, they ought to have been able to recolonize most of the areas to the south of their present distribution by expanding along the floor of the Rift Valley southwards into Kenya as well as northwards into the Afar region. How far they might have got depends on the rate of migration and the time span available to them. Ohsawa & Dunbar (1984) found that bands in the Simen underwent fission approximately once every 8.4 years (and this in habitats where the natural rate of increase is an astonishing 13 per cent per annum). With an average of 2 km between the centres of adjacent bands' ranges, it would take the gelada 4300 years to cover the 1000 km between their current centre of mass in the Simen and Lake Turkana, and around 7750 years to cover the 1800 km from the Simen to Olduvai. If growth rates were half those observed in the Simen, then the time required to cover these distances would double. Similarly, if temperatures fluctuated over time rather than remaining stable at 8 °C below the present level, then their rate of expansion would have been considerably slowed down. It seems unlikely that the gelada would have had sufficient time (at least during the later cold periods) to migrate very far before conditions began to deteriorate again (i.e. got warmer).

Nonetheless, it seems probable that the gelada have repeatedly abandoned their retreat habitat on the Ethiopian plateau and expanded into the surrounding lowlands during the Holocene and Late Pleistocene. Evidence for this in the form of gelada fossils should exist at lowland sites of these ages. The only factor that might have prevented an expansion of range during these periods would have been a shift to forest rather than grassland at lower altitudes during these cooler periods. This shift of vegetation type would have been especially likely if there was a concomitant change in rainfall patterns. If this was the case, then it is likely that the gelada's geographical range was even more reduced than it is at present.

These results also carry interesting implications for the ancestral theropiths if these were of similar size to the gelada. If temperatures early in the Pliocene were around 2 °C warmer than at present, then these animals would have found it difficult to survive at altitudes below 2000 m in eastern Africa and 1100 m in southern Africa (Table 18.5). Later in the Pliocene, however, temperatures were lower, reaching a low point some 2 °C below current temperatures at the Plio-Pleistocene boundary. In southern Africa, the animals could have comfortably inhabited the low veld grasslands at altitudes below 1000 m as far south as latitude 30° S. Since they would still have been restricted to altitudes above 1400 m in central eastern Africa even under these more benign con-

ditions, these results tend to favour a southern origin for the taxon with an expansion northwards in the wake of the cooling Pliocene climate.

For the two larger species, mean temperatures would have had to be around 6–8 °C cooler than at present for them to have been able to survive in those habitats within 5° latitude of the equator where we know they lived (i.e. Omo, Olorgesailie, Olduvai). This is a considerable temperature difference. Although it is quite possible that global temperatures did fall by this amount for brief periods (for example during the more extreme cool periods), such a temperature differential is far greater than that which occurred in general over this period. Hence, unless the larger species had a rather limited temporal distribution and repeatedly reinvaded eastern Africa from retreat habitats at very high altitudes or from southern Africa, it is difficult to see how these animals could have been as common as they appear to have been during the Pleistocene.

An alternative possibility is that graze density and/or graze quality might have been greater in the habitats occupied by the extinct theropiths than in those east African habitats from which these variables have been calculated. The latter are all open savannah sites at some distance from free-standing water (lakes, large rivers), whereas we know from the palaeontological record that most of the sites at which the later species, at least, occurred were predominantly associated with water (lakeside or riverside habitats) (Jolly, 1972; Eck & Jablonski, 1987). The ready availability of free water undoubt-

Table 18.7. *Minimum altitudinal limits below which large theropiths could not survive when living in habitats where grass density and quality is richer than in contemporary east African grassland habitats*

Relative grassland richness	Minimum altitude for survival* (m) at latitude						
	0	5	10	15	20	25	30
× 1.5	1700	1500	1400	1200	1000	800	600
× 2.0	1500	1300	1100	900	700	500	300
× 3.0	1100	900	700	500	300	200	0
× 4.0	900	700	500	300	100	0	0

*Defined as the lowest altitude at which the maximum ecologically tolerable group size is at least 10 animals.

†Factor by which the grass density and protein content are greater than that for contemporary east African savannah grasslands.

See text for details.

edly has a significant effect on the availability of green grass. I have, therefore, assumed that the presence of water decreases the slope of the regressions for both grass density and grass protein content on temperature (so making the grass both denser and richer in protein) by a specified factor to the right of the hump in the regression equations (i.e. in habitats with temperatures greater than about 10 °C). To do this, I introduced a scaling factor that increased proportionately the predicted values for these two variables. The scaling factor, G, was standardized to a baseline at 20 °C (roughly equivalent to savannah grasslands at moderate altitudes in eastern Africa):

$$G_t = 1 + (G_r - 1)\ (T - 10)/10$$

where G_t is the scaling factor for a habitat with a mean ambient temperature of $T = t$ °C and G_r is the reference scaling factor (defined as the ratio of grass density or quality to contemporary values in a habitat with a mean ambient temperature of $T = 20$ °C). I tried a range of reference scaling ratios between $G_r = 1.5$ and $G_r = 4$. The effect of this is to scale the predicted density and quality of the grass by a factor equivalent to G_r at a temperature of 20 °C, but by a proportionately smaller or larger value at lower or higher temperatures. This takes into account the fact that proximity to water will have less effect on grass when it is already relatively abundant, but more effect when it is less abundant.

The consequences for the lower boundary of the habitable zone for the largest (55 kg) theropiths during the period when temperatures were at their global minimum of 2 °C below present values are shown in Table 18.7. The important point to note is that grass density and quality had to have been at least three times higher at the fossil sites than in the sampled contemporary East African habitats

Table 18.8. *Maximum ecologically tolerable group sizes for the largest theropiths (55 kg) when global temperatures are 2°C below current values and grass productivity is three times that found on east African savannah*

Altitude (m)	Maximum ecologically tolerable group size at latitude (degrees from equator)						
	0	5	10	15	20	25	30
200	0	0	0	0	0	13	41
600	0	0	1	18	47	79	111
1000	2	22	52	85	116	144	168
1400	58	90	121	148	171	189	199
1800	126	153	175	191	200	200	188
2200	178	193	201	199	184	154	102
2600	201	197	181	147	91	34	0
3000	176	139	80	26	0	0	0
3400	67	19	0	0	0	0	0
3800	0	0	0	0	0	0	0
4200	0	0	0	0	0	0	0
4600	0	0	0	0	0	0	0
5000	0	0	0	0	0	0	0

for these animals to have survived there. It is worth noting that the main effect of changing the equations in this way is to progressively increase the range of altitudes at which the animals can survive (compare Tables 18.4 and 18.8).

Changes in the protein content and density of grasses of this magnitude are probably possible only in the vicinity of large bodies of permanent water. We can, I think, infer that the larger theropiths of the later Pleistocene survived only in those habitats whose primary productivity was kept 'artificially' high by local ground water. In other words, these species must have had an ecologically rather restricted (even if geographically widespread) distribution based on lakeside margins and the immediate vicinity of very large rivers

(especially where associated with floodplains).

Extinction of the large theropiths

It is clear that an animal as large *T. o. oswaldi* must have experienced a very hard time ecologically. Demment & van Soest (1985) have pointed out that non-ruminant herbivores of this body size are conspicuous by their absence from contemporary ungulate communities in Africa. They argue that differences in digestive efficiency make ruminants more competitive than non-ruminants within the size range 10–600 kg. The simulations for the extant gelada tend to confirm this by suggesting that they would find it impossible to survive in those lowland

savannah habitats which are now dominated by communities of medium-sized
ruminants (i.e. habitats at 1000–2000 m
which receive less than 1000 mm of rainfall a year).

It seems to me that the distribution of
Theropithecus fossil sites during the
relevant period bears this claim out: those
that have produced the largest specimens
(Omo, East Turkana, Olduvai,
Olorgesailie) are all associated with lake
margins or large river beds. If this is so,
then it seems likely that the larger
theropiths were in fact occupying a rather
fragile ecological niche. At sites like
Olorgesailie, these large animals would
have had to spend around 80 per cent of
their daytime feeding even under the most
benign conditions; even then, group sizes
would have had to be compressed to the
minimum in order to make this possible.
They would have had very little freedom of
movement to respond to a sudden deterioration in climate (as, for example, during
periodic droughts). Drought conditions
occur regularly in the East African grassland habitats (Wood & Lovett, 1974) –
and, indeed, have done so for a very long
time (Nicholson, 1981) – and may be
expected to have had a serious impact on
theropith ecology even in the later
Pleistocene.

In this respect, it is especially interesting to note the Shipman, Bosler & Davis
(1981) finding that juveniles are over-
represented among the *Theropithecus*
remains at Olorgesailie. They interpreted
this as reflecting the greater ease with
which hominids might have been able to
kill younger animals. But carnivores rarely
take juveniles: most often they concentrate on the very young and the old and
sick, which are the easiest to run down. In
fact, the evidence for butchery at
Olorgesailie is compatible with two alternative hypotheses about exactly how the
hominids acquired their prey, namely
active hunting and passive scavenging.
Among contemporary primate populations, juveniles invariably bear the brunt
of mortality when populations crash under
adverse climatic conditions (vervets:
Struhsaker, 1973; macaques: Dittus, 1975;
see Dunbar, 1988). The age structure of
the death assemblage might thus be taken
to reflect the ease with which hominids
were able to pick up carcasses (or, at
worst, kill animals already near to death).
If this interpretation is correct, then it
implies that, at the time the carcasses
were collected, the local theropith population was suffering severely, and may have
done so at intervals over a long period of
time. Bearing in mind the fact that birth
rates are allometrically scaled to body
weight, it seems more than likely that so
large an animal would always have been at
risk of local extinction whenever conditions deteriorated.

This conclusion is reinforced by the fact
that these large species must have been
forced to live in rather small groups. Table
18.8 gives the maximum ecologically tolerable group sizes for the largest theropiths
under the most benign ecological conditions (mean temperatures 2 °C below
present values and grass density and
quality three times greater than in contemporary savannah habitats). These
results suggest that, at sites like
Olorgesailie that are less than 5° latitude
from the equator and lie below 1500 m in

altitude, maximum group sizes would have been in the order of 20–90 animals at most. Bearing in mind that these are maximum values, average group sizes would obviously have been considerably less.

Iwamoto & Dunbar (1983) found that gelada average 147 animals per km² of available grass cover. For a 55 kg theropith, this would scale down to 59.6 animals/km² of grass when scaled for metabolic body weight. In a habitat like Olorgesailie at 1000 m and 2.5° S of the equator with 55.5 per cent grass cover under the optimum conditions given in Table 18.8, this would translate into a true density of 33.1 animals per km². With group sizes of around 20 animals, this means that range sizes would be 0.60 km² in area. Since the influence of standing water on grass productivity falls off quite rapidly, group ranges would probably be distributed linearly along river banks or lake shores, so reducing the apparent density of the animals to around one group for every 1.2 km or more of shoreline. Smaller sized animals would, of course, have been able to maintain proportionately larger groups or smaller ranging areas. Nonetheless, although by no means rare, these animals would not have existed in the large numbers implied by the density of fossil remains at Olorgesailie. Rather, the density of remains probably implies a high rate of attritional mortality, probably at periodic intervals when droughts suddenly resulted in a dramatic deterioration in habitat conditions.

Why were the later theropiths so big?

The one outstanding problem is the question of why the later theropiths evolved such a large body size. It is tempting to view the increase in size through time in this lineage as a response to deteriorating climatic conditions during the Pleistocene. However, despite significant savings in the costs of travel and the ability to handle lower quality forage, it is clear that animals of this size range do not gain a net advantage because the costs of maintaining BMR are still high. This increase in size must, therefore, have evolved for some other reason(s). The most likely options are either an increase in predation risk and/or an increase in the size of the predators themselves (Dunbar, 1988). The Pleistocene was, of course, associated with the evolution of gigantism in many lineages, with a number of very large species of predators (including sabre-toothed cats). It is clear that, for the theropiths, increasing body size could only have been achieved at considerable energetic cost to the animals, so that the selection pressures favouring it must have been considerable. This is most likely to have occurred in relatively open habitats where the fact that the animals do not have access to trees or cliffs as refuges would have exacerbated the effects of an increase in the size of the predators.

This raises the interesting question of whether the theropiths of the Pleistocene could have got any larger. Table 18.9 gives altitudinal limits (i.e. those within which maximum tolerable group sizes are >10) and maximum possible group sizes for theropiths weighing 70 and 85 kg under both the current and the most benign climatic conditions. It is clear from these simulations that theropiths living in the

Table 18.9. *Minimum and maximum altitudes at which very large theropiths would have been able to survive under various climatic conditions*

Climatic conditions	Altitudinal range (m)* at		Maximum tolerable group size**
	0° latitude	5° latitude	
Body Weight = 70 kg			
Current	2800–3900	2600–3700	44 (3300)
Optimum†	1300–3500	1000–3300	154 (2500)
Body Weight = 85 kg			
Current	3300–3400	3100–3200	11 (3300)
Optimum†	1500–3300	1300–3100	113 (2500)

*Defined as the altitude range where the maximum ecologically tolerable group sizes > 10.
**Maximum predicted group size at any altitude and, in parentheses, altitude (in m) at which this occurs.
†Temperatures 2°C below present values and grassland 3 times richer than contemporary savannah habitats.

East African sites (i.e. within 5° latitude of the equator) could not have grown significantly larger than they did. Under the most benign environmental conditions (ambient temperatures 2 °C cooler than at present and habitats close to large bodies of water), they could not have survived at altitudes below 1000 m for animals weighing 70 kg or below 1300 m for animals weighing 85 kg. Under current climatic conditions (i.e. those prevailing towards the end of the Pleistocene), animals weighing 70 kg could only have survived at altitudes of 2600–3900 m (with maximum group sizes at about 44), while animals weighing 85 kg would only have been able to survive in a very narrow band between altitudes of 3100–3400 m providing group sizes never exceeded 11 animals. The largest theropiths of the Pleistocene (with an adult weight averaging about 55 kg) would thus have been close to the ecological ceiling on body weight at the altitudes where they occurred.

Conclusions

In this contribution, my aim has been to explore the extent to which we can use detailed models of the behavioural ecology of a particular taxon to explore the ecology of its extinct congeners. I should stress that this is not the same kind of analogical modelling that has tended to predominate in this area in the past, but rather an attempt to use general biological principles to reconstruct the behaviour of an extinct population. Not only may the predicted behaviour be very different from that observed in any living taxon, but, as with extant species, we can expect behaviour to vary from one population to another. The exercise seems to me to produce results which give us valuable insights into the ecology of the extinct theropithecines, though I would not wish to claim at this stage that we know all we want to know. The point to be emphasized is the fact that we may gain a very much better understanding of why these species evolved and eventually went extinct if we can integrate detailed knowledge of changing climatic conditions with their impact on habitat productivity and the animals' behavioural ecology at particular sites.

We can, I think, extend the analyses very much further than I have done in this case. Our understanding of the forces acting on primate social systems is now very considerable. Given a knowledge of group size and interbirth intervals (the latter being partly a function of body size), we can determine the mating system with a fair degree of precision (see Dunbar, 1988). We would need to know a little more than we do at present about the way in which social systems respond to changes in demography, and particularly about the costs of group-living. However, we understand in principle how these effects work, and I think it will not be too long hence before we can incorporate these factors into our models. Doing so should allow us to say a great deal about the social life of extinct species.

The point that I perhaps want to stress above all is that these analyses are necessarily habitat-specific. It is not enough to be able to talk in generalities, for even contemporary species vary immensely in their ecological and social behaviour across the range of habitats they occupy. We have to know a great deal more about the climatic conditions prevailing in particular areas at specific points in time. Once we know

these, we can say a great deal about how a given taxon with its own particular dietary and reproductive strategies is likely to have responded.

References

ALTMANN, J. (1980). *Baboon Mothers and Infants*. Cambridge: Harvard University Press.

COE, M.J., CUMMINGS, D.H. & PHILLIPSON, J. (1976). Biomass and production of large herbivores in relation to rainfall and primary production. *Oecologia*, 22, 341–54.

DEMMENT, M.W. & VAN SOEST, P.J. (1985). A nutritional explanation for body-size patterns in ruminant and nonruminant herbivores. *American Naturalist*, 125, 641–72.

DESHMUKH, I. (1984). Primary production of a grassland in Nairobi National Park, Kenya. *Journal of African Ecology*, 23, 115–23.

DITTUS, W.P.J. (1975). Population dynamics of toque monkeys, *Macaca sinica*. In *Socioecology and Psychology of Primates*, ed. R. Tuttle, pp. 125–51. The Hague: Mouton.

DUNBAR, R.I.M. (1980). Determinants and evolutionary consequences of dominance among female gelada baboons. *Behavioural Ecology and Sociobiology*, 7, 253–65.

DUNBAR, R.I.M. (1984a). *Reproductive Decisions: An Economic Analysis of Gelada Baboon Social Strategies*. Princeton: Princeton University Press.

DUNBAR, R.I.M. (1984b). Theropithecines and hominids: contrasting solutions to the same ecological problem. *Journal of Human Evolution*, 12, 647–58.

DUNBAR, R.I.M. (1988). *Primate Social Systems*. London: Chapman & Hall.

DUNBAR, R.I.M. (1990). Ecological models in an evolutionary context. *Folia primatologica*, 53, 235–46.

DUNBAR, R.I.M. (1991). Functional significance of social grooming in primates. *Folia primatologica*, 57, 121–31.

DUNBAR, R.I.M. (1992a). A model of the gelada socioecological system. *Primates*. 33, 69–83.

DUNBAR, R.I.M. (1992b). Time: a hidden constraint on the behavioural ecology of baboons. *Behavioural Ecology and Sociobiology*. (In press.)

DUNBAR, R.I.M. & DUNBAR, P. (1974). Ecological relations and niche separation between sympatric terrestrial primates in Ethiopia. *Folia primatologica*, 21, 36–60.

DUNBAR, R.I.M. & DUNBAR, P. (1975). *Social Dynamics of Gelada Baboons*. Basel: Karger.

DUNBAR, R.I.M. & DUNBAR, P. (1988). Maternal time budgets of gelada baboons. *Animal Behaviour*, 36, 970–80.

DUNBAR, R.I.M. & SHARMAN, M. (1984). Is social grooming altruistic? *Zeitschrift für Tierpsychologie*, 64, 163–73.

ECK, G.G. & JABLONSKI, N.G. (1987). The skull of *Theropithecus brumpti* compared with those of other species of the genus *Theropithecus*. In *Les Faunes Plio-Pléistocènes de la Vallée de l'Omo (Éthiopie)*. Tome 3. Cercopithecidae de la Formation de Shungura, ed. Y. Coppens & F.C. Howell, pp. 11–122. Cahiers de Paléontologie, Travaux de Paléontologie Est-Africaine. Paris: Editions du Centre National de la Recherche Scientifique.

LE HOUERON, H.N. & HOSTE, C.H. (1977). Rangeland production and annual rainfall relations in the Mediterranean basin and in the African Sahelo–Sudanian zone. *Journal Range Management*, 30, 181–9.

HURNI, H. (1982). *Simen Mountains – Ethiopia*, vol. 2, *Climate and the Dynamics of Altitudinal Belts from the Last Cold Period to the Present Day*. Bern: University of Bern Geographical Institute.

IWAMOTO, T. & DUNBAR, R.I.M. (1983). Thermoregulation, habitat quality and the behavioural ecology of gelada baboons. *Journal of Animal Ecology*, 52, 357–66.

JARMAN, P.J. (1974). The social organisation of antelope in relation to their ecology. *Behaviour*, 48, 215–67.

JOLLY, C.J. (1972). The classification and natural history of Theropithecus (Simopithecus) (Andrews 1916), baboons of the African Plio-Pleistocene. *Bulletin of the British Museum (Natural History), Geology*, 22, 1–123.

KAWAI, M. (ed) (1979). *Ecological and Sociological Studies of Gelada Baboons*. Basel: Karger.

KENNETT, J.P. (1977). Cenozoic evolution of antarctic glaciation, the circum-antarctic ocean and their implications on global palaeooceanography. *Journal of Geophysical Research*, 82, 3843–60.

KLEIBER, M. (1960). *The Fire of Life*. New York: Methuen.

LEE-THORP, J.A., VAN DER MERWE, N.J. & BRAIN, C.K. (1989). Isotopic evidence for dietary differences between two extinct baboon species from Swartkrans. *Journal of Human Evolution*, 18, 183–90.

MOUNT, L.E. (1979). *Adaptation to Thermal Environment*. London: Edward Arnold.

NICHOLSON, S.E. (1981). The historical climatology of Africa. In *Climate and History*, ed. T. Wigley, M. Ingram & G. Farmer, pp. 249–70. Cambridge: Cambridge University Press.

OHSAWA, H. & DUNBAR, R.I.M. (1984). Variations in the demographic structure and dynamics of gelada baboon populations. *Behavioural Ecology and Sociobiology*, 15, 231–40.

PETERS, R.H. (1983). *The Ecological Relevance of Body Size*. Cambridge: Cambridge University Press.

SHIPMAN, P., BOSLER, W. & DAVIS, K.L. (1981). Butchering of giant gelada at an Acheulian site. *Current Anthropology*, 22, 257–68.

STRUHSAKER, T.T. (1973). A recensus of vervet monkeys in the Masai-Amboseli Game Reserve, Kenya. *Ecology*, 54, 930–2.

SZALAY, F. & DELSON, E. (1979). *Evolutionary History of the Primates*. New York: Academic Press.

VAN SCHAIK, C.P. (1983). Why are diurnal primates living in groups? *Behaviour*, 87, 120–44.

WOOD, C.A. & LOVETT, R.R. (1974). Rainfall, drought and the solar cycle. *Nature*, 251, 594–6.

19 Ecological energetics and extinction of giant gelada baboons

P.C. LEE AND R.A. FOLEY

Summary

1. Extinction is an evolutionary process well represented by the fossil record. The giant gelada, *Theropithecus oswaldi*, is taken as a case study of extinction.
2. Here we show that the effects of climatic change or competitive pressures, the usual proposed mechanisms of extinction, are mediated through behaviour.
3. The energetic requirements of *T. oswaldi* are modelled, along with correlated life history parameters. The extinction of the giant gelada is suggested to result from limited nutritional intake, leading to a reduced reproductive rate as a consequence of size-specific life history parameters.
4. We suggest that individuals are time-limited in terms of energy acquisition when exploiting low quality forage. Inability to meet energetic requirements reduced reproductive rates to the point where populations appeared to become vulnerable to minor, localized environmental perturbations, and hence to extinction within their typical habitats.

Introduction

During the Pliocene and Pleistocene the *Theropithecus* baboons were both far more abundant and considerably larger than their modern representatives (Jolly, 1972). These species became extinct during the Middle Pleistocene or early Late Pleistocene. Predation by early hominids has been implicated in their extinction (Isaac, 1977), along with a generally changing environment (see Leakey, chapter 3) or competitive pressures from other species (see Pickford, chapter 8). However, new insights from evolutionary ecology may throw light on specifically why large terrestrial primates were at risk during this period. The question addressed here is that of extinction: the constraints on animals that increase their vulnerability and the processes and timing of extinction events in the fossil record. Under investigation is the role of behaviour, which mediates between the organism and its environment, in the capacity to respond to ecological change (Lee, 1991).

Large body size has been associated with adaptations for the exploitation of abundant, relatively uniformly dispersed and low-quality food resources (Bell, 1971; Jarman, 1974; Demment & Van Soest, 1985), since an increase in body size results in lower energetic requirements per unit of body mass despite an absolutely greater total requirement (Peters, 1983). These relatively lower requirements can be sustained on low quality abundant foods. Due to constraints on gestation length, the duration of lactation, and interbirth intervals, large body size in mammals has also been associated with stable environments, since long-term fluctuations in food availability have a greater disruptive effect on the reproductive potential of large-bodied species (Boyce, 1979; Kiltie, 1984). The ecological and life history correlates of body size illuminate the conditions under which a mammalian species becomes larger, and, in certain cases, subsequently becomes vulnerable to extinction.

Among extant primates with relatively unspecialized guts, only the gelada baboon (*T. gelada*) exploits a food supply similar to that of the ungulates – high fibre, low protein grasses (Dunbar, 1977; Iwamoto & Dunbar, 1983; and Iwamoto, chapter 17), in contrast to the typical cercopithecine diet of fruit. Grass offers an abundant resource to mammals, but one that poses particular problems. The ratio of fibre to energy is high, which affects passage rate and digestive/absorptive efficiency (Van Soest *et al.*, 1983; Chivers & Hladik, 1980). The problems of nutrient extrac-

tion on a high fibre diet generally are solved through microbial digestion. Among ruminants this occurs in specialized fermenting chambers in the foregut. Among the primates, the leaf-eating colobines are specialized foregut fermenters. Graminivorous taxa such as equids and proboscids, which have some caecal and colonic fermentation, have relatively less specialized guts. Like the equids, the gelada has an enlarged caecum where some minor microbial fermentation occurs (see Iwamoto, chapter 17). For species with unspecialized guts, passage rate, gut volume, and especially acquisition rate, appear to limit nutrient intake on high fibre foods (Van Soest *et al.*, 1983). Gelada are currently confined to the high altitude regions of Ethiopia in habitats with relatively high grass productivity (see Dunbar, chapter 18).

The gelada exhibit various specializations to their predominantly grass diet (Dunbar, 1983a; Jolly, 1985): their hands are particularly dextrous, they employ a 'shuffling' means of locomotion with truncal erectness that minimizes the energetic costs of altering positions when moving between feeding sites, and they have reduced anterior dentition and enlarged posterior dentition as an adaptation to chewing plant material.

These anatomical features are also found in a range of fossil forms recovered from the Pliocene and Pleistocene of eastern and southern Africa (Leakey, 1976; Eck, 1977; Dechow & Singer, 1984). Five fossil species are currently recognized (Eck & Jablonski, 1984), ranging in date from 5.1 to 0.3 Ma, suggesting a rapid radiation in the genus to exploit the emergent grass-dominated savannah woodlands (Jolly, 1972; Dunbar, 1983a; and Leakey, chapter 3, Foley, chapter 9). Some, particularly *T. oswaldi*, are larger than any known cercopithecoid with body mass estimates of up to 66 kg, compared with a maximum of 20.5 kg for living *T. gelada* males (Jolly, 1972). These larger species, however, became extinct by at least 300,000 years ago. Here we model the energy requirements and energy acquisition for fossil *Theropithecus* with large body sizes, focusing on the relationships between activity budgets, nutrient availability, and life history parameters in order to suggest probable conditions leading to extinction.

Methods and models

Energy requirements

Dry matter food intake necessary to sustain basal metabolic requirements (BMR) was calculated using Demment's (1983) equation:

$$I = 0.016 \, D^{-1} \times W^{0.75} \qquad (1)$$

where I = daily intake in kg of dry matter, D = digestibility, and W = body weight in kg. Digestibility (or the proportion not appearing in the faeces) varies with food quality (Van Soest *et al.*, 1983), and three levels were employed in these calculations: high, mid, and low quality (0.70, 0.50, and 0.35 digestible dry matter, respectively). Estimates of digestibility come from the ranges found in tropical forages as grasses mature during the growing season (Coehlho & da Silva, 1976; Coleman *et al.*, 1978).

For an animal of any size, one critical variable for energy acquisition is the time it takes to acquire sufficient food (see Dunbar, chapter 18). For gelada baboons this is dependent upon the pick rate and size – the rate at which grass is harvested. Using actual values derived from field studies (Dunbar, 1977; Iwamoto, 1979) for pick rates (dry matter g/min) and the percentage of time spent feeding per day in the wet (high quality) and dry (low quality) seasons, daily dry matter intake was calculated as 1.37 g/min × 35.7 per cent of 12 h feeding for the wet season, a total of 352 g/day. For the dry season it was calculated as 2.8 g/min × 62.3 per cent of 12 h feeding, totalling 1256 g/day.

Energy intake

In order to model the energy requirements of extinct species, the intakes estimated from pick rate × percentage of time feeding were compared with those derived from Demment's (1983) equation. This equation allows for the use of body weight alone to predict feeding requirements, while the observed intake rates allow these requirements to be adjusted for maintenance, growth, and reproduction. Using a body weight of 15 kg (average of male and female weights; R.I.M. Dunbar pers. comm.; see also Iwamoto, chapter 17), and digestibility of 0.70, the intake necessary to meet basal energy requirements was calculated as 174 g/day, equalling half of the total intake of 352 g/day possible during the wet season. Since average daily metabolic feeding requirements (ADMFR) range from 1.5–3.0 × BMR (Peters, 1983) in order to account for activity, growth, digestion, and thermoregulation, the basal metabolic feeding requirements (BMFR) calculated from the above equation (1) compare extremely well with those observed among extant gelada. Comparisons of real dry season intakes with those calculated when digestibility was 0.35 showed 3.6 × BMR; a result not inconsistent with Iwamoto's (1979) suggestions of very low digestibility, in addition to costs of reproduction and of thermoregulation during cold seasons (see also Dunbar, 1977). The agreement between these two models for deriving estimates of energy intake should allow us to proceed with some confidence in predicting requirements for the extinct gelada.

Results

Energetic intake

Equation (1) was then used to calculate food requirements for extinct animals. The largest known male *T. oswaldi*, estimated to weigh 66 kg, would have needed a wet season dry matter food intake of 988 g/day to meet ADMFR (ADMFR = 2 × BMFR); for the dry season with lower digestibility, 2117 g/day would have been needed. With these estimates of requirements, it is possible to model the length of feeding time and pick rates necessary for male and female *T. oswaldi* to survive on a grass diet (Fig. 19.1). Female weights were estimated to be 39 kg, using a male–female weight ratio of 1.7 (found in extant gelada (Napier & Napier, 1967) and fossil species (Jolly, 1972)). Female ADMFR were calculated

as 3.5 times BMFR, reflecting an additional cost of 1.5 × ADMR for gestation and lactation (Peters, 1983) averaged in with non-reproductive phases.

The pick rates necessary for *T. oswaldi* to meet intake requirements in the poor–low quality seasons would have been difficult to attain (Fig. 19.1). At a maximum observed rate of intake of 6 g/min (found in *Presbytis entellus*, Hladik, 1978), males and females would need to feed for over 50 per cent of time. Maximum pick rates in extant gelada are only about 3 g/min for grass (Iwamoto, chapter 17). Since plucking with thumb and forefinger allows for the removal of the most nutritious portions of the plant, scaling the hand dimensions up with body size would not necessarily improve harvesting rate in a linear fashion. Harvesting rate would continue to be limited by the plant size and structure. If however, *T. oswaldi* used a scything technique for harvesting (Jablonski, 1986), then the volume of intake would scale with body size, but at the expense of selectivity for the higher quality parts of the grasses. The incorporation of higher proportions of plant structural material would also alter gut fill, passage rate, and generally decrease digestibility even further, requiring even more feeding time to meet daily requirements.

Life history parameters

The energetic and time budget constraints described above can be modelled with respect to those life history variables affecting reproductive potential. Three major components of female reproductive

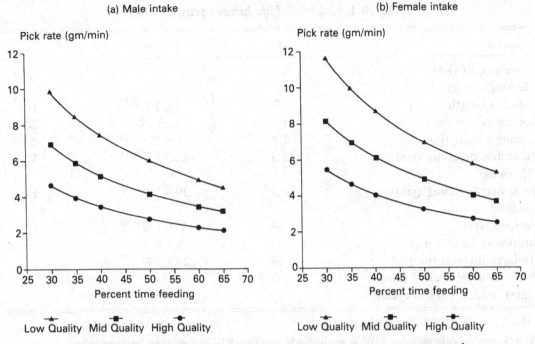

Fig. 19.1. Pick rates (g dry weight/minute) plotted against percentage of time spent feeding for *Theropithecus oswaldi* under three different forage qualities. A: Males at 66 kg with ADMR of 2 × Basal. B: Females at 39 kg with ADMR of 3.5 × Basal. Low forage quality = 35 per cent digestible, Mid = 50 per cent, High = 70 per cent.

success can be considered (Clutton-Brock, 1988): age at first breeding, interbirth intervals, and survivorship of infants and adults. Age at first breeding, neonate weight, and gestation length, scale significantly with body weight (Harvey & Clutton-Brock, 1985; Harvey, Clutton-Brock & Martin, 1987). Interbirth intervals, however, are made up of the gestation length plus the time to reconception; typically the weaning age of infants (Lee, Majluf & Gordon, 1991). Age at weaning bears little relation to adult size, but depends on the nutritional intake of mothers during lactation determining infant growth rates and hence threshold weaning weights (Lee *et al.*, 1991).

Mortality, for juveniles or adults, again bears little relation to adult body size (Promislow & Harvey, 1990), but is dependent on ecological variables (predation, food availability, and group size – Dunbar, 1988). Thus we can make predictions about the basic size-dependent life history parameters, and using the energetic models, simulate patterns of interbirth intervals and mortality.

In Table 19.1, we present the size-dependent life history parameters of *T. oswaldi* using the predictive equations derived from extant primates. In Table 19.2, fecundity and mortality rates are given, drawn from extant primate populations under high, low, and average nutri-

Table 19.1. *Calculated life history parameters*.

Parameter	Value	Equation	Ref.
Female weight (kg)	39		
Male weight (kg)	66		
Gestation length (days)	235	$59.6 \, FW^{0.13}$	1
Neonate weight (kg)	1.5	$0.13 \, FW^{0.66}$	2
Weaning weight (kg)	7.3	$4.6 \, NW^{1.14}$	2
Age at first breeding (yrs) (females)	12.5	$43.5 \, FW^{0.44}$	1
Age at first breeding (yrs) (males)	14.9	$29.6 \, MW^{0.47}$	1
Life span (yrs)	46	$780 \, FW^{0.29}$	1
Duration of lactation (mo)	23	$8.8 \, WW^{0.47}$	2
Interbirth interval (mo)	48	$29.1 \, FW^{0.37}$	1
Min. interbirth interval (mo) (gest + lact + 6 mo cycling)	37		

1. Harvey *et al.*, 1987; 2. Lee *et al.*, 1991.

FW = Female body weight; MW = male body weight; NW = neonate body weight;
WW = weaning weight.

tional regimes, under high predation pressures and with interspecific competition affecting food availability. Using these estimates, we can model some aspects of the population dynamics of *T. oswaldi*. If mortality rates are held constant, Lyle & Dobson (1988) suggest that populations will decline when interbirth intervals double (halving fecundity). Such a doubling is likely to occur under conditions of poor maternal nutrition (Cheney *et al.*, 1988). Alternatively, they show that populations will decline with a decrease in juvenile survival to less than 51 per cent, even if interbirth intervals remain constant (Lyle & Dobson, 1988). The survivorship of juveniles is one of the main factors contributing to individual lifetime reproductive success in some species of

primates (Cheney *et al.*, 1988), and it tends to be influenced primarily by maternal nutrition affecting infant susceptibility to illness and secondarily by predation pressures or infanticide (see Dunbar, 1988).

A model of maternal time budgets in extant gelada (Dunbar & Dunbar, 1988) has shown that females spend between 30 and 70 per cent of time feeding to produce a 3.9 kg weanling over 18 months (growth rate of 5.7 g/day; Lee *et al.*, 1991). For a 39 kg female *T. oswaldi*, weaning an infant at 7.3 kg would require her to sustain growth at 7.6 g/day over the 23 months predicted, or alternatively to increase the duration of lactation to at least 34 months if infants grow at the slower rate of 5.7 g/day. On poorer diets,

Table 19.2 *Population parameters and female reproductive success in different habitat types.*

	Habitat types				
	High predation[1]	Low food	High food	Average food availability	Competition with ungulates
Population parameters					
Juvenile survival	0.57[2]	0.20[3]	0.80[4]	0.56[5]	(0.56)
Adult survival	0.98[2]	0.99	1.00[4]	0.99	(0.99)
Fecundity	(0.39)	0.14[6]	0.67	0.39	0.22[7]
First breeding	(12.5)	13.3[8]	11.7	12.5	(12.5)
Breeding lifespan	(33.5)	32.7	34.3	33.5	(33.5)
Average no. female offspring	3.9	0.46	9.2	3.6	2.1
Gross r_0[9]	0.081	–0.047	0.129	0.077	0.044

[1] High predation is defined as 69 per cent of total mortality due to predation (Cheney *et al.*, 1988).

[2] Total juvenile mortality = 63 per cent; adult mortality = 3 per cent/yr; corrected for mortality due to predation alone (Cheney *et al.*, 1988).

[3] Dittus, 1975.

[4] Loy, 1988.

[5] Average for a range of primate species: Dunbar (1988).

[6] Maximum, minimum and average for *Theropithecus*: Dunbar (1988).

[7] Reduction of 17 per cent of average: Strum & Western (1982); Drucker (1984).

[8] +/– 300 days (*Papio cynocephalus*; Altmann *et al.*, 1988).

[9] Calculated as $\log_e$ no. female offspring/ (breeding lifespan/2).

Values unaffected by the specific habitat type are presented for the 'average' conditions in brackets.

growth rates would slow even further, increasing interbirth intervals. With a growth rate of 3 g/day found in some extant baboons (Sigg *et al.*, 1982; Altmann & Alberts, 1987), interbirth intervals would approach 65 months (average fecundity of 18 per cent) – the doubling necessary for a population decline. As modelled by Fig. 19.1b, female *T. oswaldi* would be spending 55 per cent of their time sustaining average energetic requirements on the highest quality diet. The time budgets of females, limited by energy intake and assimilation, would be further constrained by the amount of time feeding necessary even to sustain these slow (but common) infant growth rates.

A confounding factor is that slowly growing infants are at greater risk of mortality (rhesus; Small & Smith, 1986).

Thus the reduction in growth rates associated with poor quality foods would not only decrease fecundity; it would increase infant mortality and thus the downward population spiral would begin (Table 19.2). Local population extinctions under such severe conditions would be expected, and if such conditions became generalized, then relatively rapid species extinction would follow.

An additional problem may be that of sexual dimorphism. For males attaining a large body size at or around puberty will enforce a continued demand for an increased energy intake to sustain growth beyond the juvenile phase. Males are predicted to begin reproducing later than females (Table 19.1). Meeting the energetic requirements of prolonged growth is known to place an additional mortality burden upon males (Lee & Moss, 1986), potentially skewing sex ratios and affecting female reproductive potential, as well as reducing population numbers.

Discussion

These observations have implications for understanding the evolutionary ecology and ultimate extinction of giant gelada baboons. In pursuing a specialized strategy of bulk grass feeding, T. oswaldi were highly successful under relatively wet and/or less seasonal conditions. The current restricted distribution of members of this genus to the high rainfall and altitude of the Ethiopian plateaux suggests that even the smaller bodied species face energetic constraints when seasonality is more pronounced as at lower elevations (Dunbar, 1983a). Under drier conditions,

such as those prevailing in the later Middle Pleistocene (Eck, chapter 2; Leakey, chapter 3), considerable selective pressure would have been exerted on the upper end of the body size range, as food quality declined with greater seasonality and shorter growing periods. Both the energy available from the food and the time available in a day to harvest that food would place constraints on survival and reproduction for these species. These larger species would be specifically vulnerable to a marked lengthening of interbirth intervals, since mothers must sustain lactation to meet the demands of attaining a threshold weaning weight. When interbirth intervals decline, and juvenile mortality increases under maternal nutritional constraints, the reproductive rate of the population becomes negative. These two factors seem to be critical in affecting the replacement or potential extinction of local populations.

Theropithecus oswaldi could have become more selective in its diet and concentrated on foods that gave a higher energetic return per pick such as fruits. However, both their morphological adaptations to a predominantly grass diet (see also Iwamoto, chapter 17) and the presence of the generalist frugivore *Papio* would limit the foods available to T. oswaldi, especially if the Papio species had greater competitive abilities in terms of locating or defending fruit patches. It is unlikely that T. oswaldi could have fed for longer in a day, in order to take in adequate nutrients. Modern gelada baboons live in complexly structured one-male units embedded in a large social group (Dunbar, 1984). The relationships between the harem holder and his females, which ensure his ability to mate with those females, are maintained through time-consuming activities such as grooming (Dunbar & Sharman, 1984). Reducing the time spent in social interactions tends to lead to harem takeovers and decreases both male and female reproductive potential (Dunbar, 1983b). If the giant gelada also had a social structure similar to that of the modern gelada (see Dunbar, chapter 15), or indeed any of the *Papio* species, then time for additional foraging could only be taken from social time at the expense of the maintenance of those relationships essential to group cohesion, survival, and successful reproduction in a social context (Lee, 1988).

Alternatively, the giant baboons could have become less selective in their diet and maximized intake at the expense of quality as do many grazing species when confronted with declining food biomass (and a correlated decrease in food quality) (Trudell & White, 1983; Short, 1986). However, the larger *Theropithecus* have converged in body size towards the most specialized of the grazers – the ruminants. Ruminant grazers tend to be most efficient at body sizes of greater than 10 kg or less than 650 kg (Demment, 1983; Demment & van Soest, 1985). The adaptive radiation of many medium sized modern bovids during the Early and Middle Pleistocene (Vrba, 1985) may have exerted strong competitive pressure on the relatively less efficient grazing *Theropithecus* and limited the range of foods available for exploitation. Evidence from baboons (Strum & Western, 1982) and Barbary macaques (Drucker, 1984) show that

fertility in primates can be related to stocking levels. As the number of grazers and browsers utilizing the same habitat increases, primate fecundity and biomass decline. The population models presented above suggest, however, that competition in combination with adequate resources will merely reduce the rate of growth, not lead to an overall decline. If competition, affecting fecundity, is compounded by limited food, reducing infant survival, then indeed populations will cease to grow. Predation alone, even at some of the highest rates observed for extant primates (Cheney *et al.*, 1988), would be insufficient to cause the extinction of the giant gelada, since populations could still grow at close to 8 per cent/yr (Table 19.2). This suggests that the role of hominid predation was minimal for the species as a whole, although it could play an important role in localized population extinctions.

It is beyond the scope of a modelling exercise such as this to 'prove' the reasons underlying the extinction of *T. oswaldi*. The approach we have adopted, however, provides two insights that may be of value to understanding the evolutionary history of *Theropithecus*. The first is that models such as these indicate the *conditions* for an animal, of the sort we believe *T. oswaldi* to have been, to come under ecological pressure and hence be vulnerable to extinction. Identification of these conditions in the fossil and palaeoenvironmental record may subsequently form a test of the hypotheses underlying the model. In particular, the model developed here suggests that plant productivity (and associated competitive changes among the herbivorous mammals) was the critical environmental parameter.

Following from this, the second insight is that climatic change does not 'cause' an animal to become extinct. As has been shown, changes in climate and environment alter the conditions under which an animal or population may pursue alternative behavioural and energetic strategies. Whether an animal becomes extinct or not depends ultimately on the range of strategies available to it. In other words, climates and environments set the context for evolution and extinction, but it is behaviour (and the competition underpinning it) that is the true focus for evolutionary events. Individual behaviour furthermore, for the primates in particular, is imbedded in the social context of groups and, for breeding purposes, that of the populations. Changing conditions, provoking altered individual responses, ultimately influence the dynamics of the groups and populations.

Acknowledgements

We thank Robin Dunbar for extensive comments on the manuscript, two referees for their comments, and Iain Gordon and the late Robin Best for helpful discussions of forage quality and feeding rates. Nina Jablonski contributed useful insights into potential behaviour and Meave Leakey graciously allowed us a preliminary look at the collections of *Theropithecus* at the National Museums of Kenya. The research was funded in part by a grant from the Boise Fund.

References

ALTMANN, J. & ALBERTS, S. (1987). Body mass and growth rate in a wild primate population. *Oecologia*, 72, 15–20.

ALTMANN, J., ALTMANN, S.A. & HAUSFATER, G. (1988). Determinants of reproductive success in savannah baboons. In *Reproductive Success*, ed. T.H. Clutton-Brock, pp. 403–18. Chicago: University of Chicago Press.

BELL, R.H.V. (1971). A grazing ecosystem in the Serengeti. *Scientific American*, 224, 86–93.

BOYCE, M.S. (1979). Seasonality and patterns of natural selection for life histories. *American Naturalist*, 114, 569–83.

CHENEY, D.L., SEYFARTH, R.M., ANDELMAN, S.J. & LEE, P.C. (1988). Reproductive success in vervet monkeys. In *Reproductive Success*, ed. T.H. Clutton-Brock, pp. 384–402. Chicago: University of Chicago Press.

CHIVERS, D.J. & HLADIK, C.M. (1980) Morphology of the gastrointestinal tract in primates: comparisons with other mammals in relation to diet. *Journal of Morphology*, 166, 337–86.

CLUTTON-BROCK, T.H. (1988) *Reproductive Success*. Chicago: University of Chicago Press.

COEHLHO DA SILVA, J.F. & DA SILVA, D.J. (1976) Nutritive value of tropical forage in Brazil. In *First International Symposium on Feed Composition, Animal Nutrient Requirements and Computerization of Diets*, ed. V. Fonnesbech, L.E. Harris, & L.C. Kewl, pp. 177–86. Utah: Utah State University Press.

COLEMAN, S.W., NERI-FLORES, O. ALLEN, R.J. & MOORE, J.E. (1978). Effects of pelleting and of forage maturity on quality of two sub-tropical forage grasses. *Journal of Animal Science*, 46, 1130–12.

DECHOW, P.C. & SINGER, R. (1984). Additional fossil *Theropithecus* from Hopefield, South Africa: a comparison with other African sites and a reevaluation of its taxonomic status. *American Journal of Physical Anthropology*, 63, 405–35.

DEMMENT, M.W. (1983). Feeding ecology and the evolution of body size of baboons. *African Journal of Ecology*, 21, 219–33.

DEMMENT, M.W. & VAN SOEST, P.J. (1985). A

nutritional explanation for body size patterns of ruminant and nonruminant herbivores. *American Naturalist*, 125, 641–72.

DITTUS, W.P.J. (1975). Population dynamics of the toque monkey *Macaca sinica*. In *Socioecology and Psychology of Primates*, ed. R.H. Tuttle, pp. 125–52. Mouton, The Hague.

DRUCKER, G.R. (1984). The feeding ecology of the Barbary macaque and cedar forest conservation in the Moroccan Moyen Atlas. In *The Barbary Macaque: A Case Study in Conservation*, ed. J.E. Fa, London: Plenum Press.

DUNBAR, R.I.M. (1977). Feeding ecology of gelada baboons: a preliminary report. In *Primate Ecology*, ed. T.H. Clutton- Brock, pp. 251–73. London: Academic Press.

DUNBAR, R.I.M. (1983a). Theropithecines and hominids: contrasting solutions to the same ecological problem. *Journal of Human Evolution*, 12, 647–58.

DUNBAR, R.I.M. (1983b). Structure of gelada baboon reproductive units. III. The male's relationship with his females. *Animal Behaviour*, 31, 565–75.

DUNBAR, R.I.M. (1984). *Reproductive Decisions*. Princeton: Princeton University Press.

DUNBAR, R.I.M. (1988). *Primate Social Systems*. London: Croom Helm.

DUNBAR, R.I.M. & DUNBAR, P. (1988). Maternal time budgets of gelada baboons. *Animal Behaviour*, 36, 970–80.

DUNBAR, R.I.M. & SHARMAN, M.J. (1984). Is social grooming altruistic? *Zeitschrift fur Tierpsychologie*, 64, 163–73.

ECK, G.G. (1977). Diversity and frequency distribution of Omo group cercopithecoidea. *Journal of Human Evolution*, 6, 55–63.

ECK, G.G. & JABLONSKI, N.G. (1984). A reassessment of the taxonomic status and phyletic relationships of *Papio baringensis* and *Papio quadratirostris* (Primates: Cercopithecidae). *American Journal of Physical Anthropology*, 65, 109–34.

HARVEY, P.H. & CLUTTON-BROCK, T.H. (1985). Life history variation in primates. *Evolution*, 39, 559–81.

HARVEY, P.H., CLUTTON-BROCK, T.H. & MARTIN, R.D. (1987). Life histories in comparative perspective. In *Primate Societies*, ed. B.B.

Smuts, D.L. Cheney, R.M. Seyfarth, R.W. Wrangham and T.T. Struhasker, pp. 181–96. Chicago: University of Chicago Press.

HLADIK, C.M. (1978). Adaptive strategies of primates in relation to leaf eating. In *The Ecology of Arboreal Folivores*, ed. G.G. Montgomery, 373–95. Washington DC: Smithsonian Institution.

ISAAC, G.L. (1977). *Olorgesailie*. Chicago: University of Chicago Press.

IWAMOTO, T. (1979). Feeding ecology. In *Ecological and Sociological Studies of Gelada Baboons*, ed. M. Kawai, pp. 280–330. Basel: Karger.

IWAMOTO, T. & DUNBAR, R.I.M. (1983). Thermoregulation, habitat quality and the behavioural ecology of gelada baboons. *Journal of Animal Ecology*, 52, 357–66.

JABLONSKI, N.G. (1986). The hand of *Theropithecus brumpti*. In *Primate Evolution*, ed. J.G. Else & P.C. Lee, pp. 173–82. Cambridge: Cambridge University Press.

JARMAN, P.J. (1974). The social organization of antelope in relation to their ecology. *Behaviour*, 48, 215–67.

JOLLY, C.J. (1972). The classification and natural history of *Theropithecus* (*Simnopithecus*), (Andrews 1916) baboons of the African Plio-Pleistocene. *Bulletin of the British Museum (Natural History), Geology*, 22, 1–123.

JOLLY, C.J. (1985). The seed-eaters: a new model of hominid differentiation based on a baboon analogy. In *Primate and Human Origins*, ed. R.L. Ciochon & J.G. Fleagle, pp. 323–32. Menlo Park: Benjamin/Cummings.

KILTIE, R.A. (1984). Seasonality, gestation time and large mammal extinctions. In *Quaternary Extinctions: A Prehistoric Revolution*, ed. P.S. Martin & R.G. Klein, pp. 299–314. Tucson: University of Arizona Press.

LEAKEY, M.G. (1976). Cercopithecoidea of the East Rudolf Succession. In *Earliest Man and Environments in the Lake Rudolf Basin*, ed. Y. Coppens, F.D. Howell, G.L. Isaac & R.E.F. Leakey, pp. 345–50. Chicago: University of Chicago Press.

LEE, P.C. (1988). Ecological constraints and opportunities: interactions, relationships and social organization of primates. In *Ecology* and *Behavior of Food- Enhanced Primate Groups*, ed. J.E. Fa & C.H. Southwick, pp. 297–312. New York: Alan R. Liss.

LEE, P.C. (1991). Adaptations to environmental change by primates: an evolutionary perspective. In *Primate Responses to Environmental Change*, ed. H.O. Box, pp. 39–56. London: Chapman & Hall.

LEE, P.C., MAJLUF, P. & GORDON, I. (1991). Growth, weaning and maternal investment from a comparative perspective. *Journal of Zoology*, 225, 99–114.

LEE, P.C. & MOSS, C.J. (1986). Early maternal investment in male and female African elephants. *Behavioural Ecology and Sociobiology*, 18, 353–61.

LOY, J. (1988). Effects of supplementary feeding on maturation and fertility. In *Ecology and Behavior of Food- Enhanced Primate Groups*, ed. J.E. Fa & C.H. Southwick, pp. 153–66. New York: Alan R. Liss.

LYLE, A.M. & DOBSON, A.P. (1988). Dynamics of provisioned and unprovisioned primate populations. In *Ecology and Behavior of Food-Enhanced Primate Groups*, ed. J.E. Fa & C.H. Southwick, pp. 167–98. New York: Alan R. Liss.

NAPIER, J.R. & NAPIER, P.H. (1967). *A Handbook of Living Primates*. New York: Academic Press.

PETERS, R.H. (1983). *The Ecological Implications of Body Size*. Cambridge: Cambridge University Press.

PROMISLOW, D. & HARVEY, P.H. (1990). Living fast and dying young: a comparative analysis of life history variation among mammals. *Journal of Zoology*, 220, 417–38.

SHORT, J. (1986). The effects of pasture availability on food intake, species selection and grazing behaviour of kangaroos. *Journal of Applied Ecology*, 23, 559–71.

SIGG, H., STOLBA, A., ABEGGLEN, J.-J. & DASSER. V. (1982). Life history of hamadryas baboons: physical development, infant mortality, reproductive parameters and family relationships. *Primates*, 23, 473–87.

SMALL, M.F. & SMITH, D.G. (1986). The influence of birth timing upon infant growth and survival in captive rhesus macaques (*Macaca mulatta*). *International Journal of Primatology*, 7, 289–304.

STRUM, S.C. & WESTERN, J.D. (1982). Variations in fecundity with age and environment on olive baboons (*Papio anubis*). *American Journal of Primatology*, 3, 61–76.

TRUDELL, J. & WHITE, R.G. (1983). The effect of forage structure and availability on food intake, biting rate, bite size and daily eating time of reindeer. *Journal of Applied Ecology*, 18, 63–81.

VAN SOEST, P.J., JERACI, J., FOOSE, T., WRICK, K. & EHLE, F. (1983). Comparative fermentation of fibre in man and other animals. *Royal Society of New Zealand Bulletin*, 20, 75–80.

VRBA, E.S. (1985). African bovidae: evolutionary events since the Miocene. *South African Journal of Science*, 81, 263–6.

Appendix I

A partial catalogue of fossil remains of *Theropithecus*

Eric DELSON, Gerald G. ECK, Meave G. LEAKEY & Nina G. JABLONSKI

This appendix has been included to provide readers with basic information on the collections of fossil specimens of *Theropithecus* that exist throughout the world. The lists of fossil specimens included here represent all the sites that have yielded *Theropithecus* fossils. The list of specimens from Ternifine represents the only published complete catalogue for *Theropithecus* at that site, while the remaining lists include only those fossils that could be confidently assigned to a species and that were reasonably complete. This meant the exclusion of small fragments of postcrania and mandibles from several East African sites and the exclusion of a very large collection of isolated teeth and tooth fragments from the Omo Basin. Details on specimens of *Theropithecus* from Kanjera and Olduvai that are housed in the Division of Palaeontology, National Museums of Kenya are provided here, but readers should note that most of the fossils from Kanjera and some from Olduvai are housed in the British Museum (Natural History) and that details of this collection have already been published in Napier (1981).

The lists of fossil specimens are arranged firstly according to site and then by species. At the beginning of each site entry, a short description of the age of the fossiliferous sediments is provided. Information on specimens from the various localities was provided by direct contributions from Eric Delson (ED), Gerald Eck (GE), and Meave Leakey (ML). The specimen lists are arranged within this appendix in the following order:

Locality	Contributor
Ternifine, Algeria	ED
Ain Jourdel, Algeria	ED
Thomas Quarries, Morocco	ED
Melka-Kunturé, Ethiopia	ED
Hadar Formation, Afar Region, Ethiopia	GE
Shungura Formation, Omo Basin, Ethiopia	GE
Usno Formation, Omo Basin, Ethiopia	GE
Kanjera, Kenya	ML
Chemeron Formation, Kenya	ML
Lothagam Hill, Kenya	ML
Kanam East, Kenya	ML

Locality	Contributor
Olorgesailie, Kenya	ML
Kapthurin Formation, Kenya	ML
Koobi Fora Formation, Turkana Basin, Kenya	ML
Nachukui Formation, Turkana Basin, Kenya	ML
Olduvai Gorge, Tanzania	ML
Senga 5A, Semliki, Zaire	ED
Makapansgat Limeworks, South Africa	ED
Hopefield, South Africa	ED
Swartkrans, South Africa	ED
Mirzapur, India	ED

For quick reference, a cross-tabulation showing the time-span of the fossil species of *Theropithecus* at the above sites is given in Table AI.1. Because many of the ages given are based on complex interpretations of stratigraphy and biochronology, the reader is urged to consult the appropriate sections of this Appendix, relevant chapters of this book, and references cited therein for discussions of the ages of the fossiliferous sediments in question.

Table AI.1. *Time spans of* Theropithecus *species by locality.*

Locality	T. darti	T. oswaldi	T. baringensis	T. quadratirostris	T. brumpti	Theropithecus species indeterminate
Ain Jourdel	–	–	–	–	–	2.5–2.0
Chemeron	–	–	3.2	–	–	–
Hadar	3.4–3.0	–	–	–	–	–
Hopefield	–	0.7–0.4	–	–	–	–
Kanam East	–	–	–	–	–	Age uncertain
Kanjera	–	1.85–1.5	–	–	–	–
Kapthurin	–	0.8–0.6	–	–	–	–
Lothagam	–	–	–	–	–	4.5–4.0
Makapansgat	ca.3	–	–	–	–	–
Melka-Kunturé	–	–	–	–	–	1.4–1.0
Mirzapur	–	1.0–0.1	–	–	–	–
Olduvai	–	1.85–0.4	–	–	–	–
Olorgesailie	–	0.99–0.74	–	–	–	–
Omo Basin (Shungura)	–	2.4–2.0	–	–	2.95–2.0	3.3
Omo Basin (Usno)	–	–	–	3.3	–	–
Senga 5A	–	–	–	–	–	2.4–2.0
Swartkrans	–	1.9–1.6	–	–	–	–
Ternifine	–	0.75–0.6	–	–	–	–
Thomas Quarries	–	0.4	–	–	–	–
Turkana Basin (Koobi Fora)	–	2.4–1.0	–	–	3.4–2.9	–
Turkana Basin (Nachukui)	–	2.4–1.0	–	–	3.3–2.5	–

Ternifine (= Palikao = Tighenif), Algeria

As summarized by Delson & Hoffstetter (chapter 6), faunal studies suggest an age near the Early to Middle Pleistocene boundary. Normal palaeomagnetism at the base of the section implies an age of either c. 1 Ma or, as preferred, c. 0.75–0.6 Ma. The specimens are identified as *Theropithecus oswaldi leakeyi*. The specimens are housed in the Muséum national d'Histoire Naturelle, Institut de Paléontologie, Paris, France (MNHN-P).

Specimen number (MNHN–P TER)	Description
1702	male mandibular symphysis with Rt & Lt I_1–P_4
1703	L mandibular corpus (male?) with P_4–M_3
1704	juvenile Rt mandibular corpus fragt. with dP_4–M_1
1706	Rt M_3
1707	Rt M_3
1708	Rt M_3
1709	Rt M_3
1710	Rt $M_{2?}$
1711	Rt $M_{2?}$
1712	Lt M_3
1713	Lt M_3
1714	Lt M_3
1715	Rt $M_{1?}$
1716	Rt $M_{1?}$
1717	Rt $M_{1?}$
1718	Lt $P_{4'}$
1719	Lt $P_{4'}$
1720	Lt $M_{1?}$
1721	Rt P_4
1722	Rt P_4
1723	Rt P_4
1724	Rt $P_{4'}$
1725	Lt C_1 male
1726	Lt C_1 female
1728	Rt $P_{3'}$ fragmentary
1729	Lt P_3

Specimen number (MNHN–P TER)	Description
1730?	Lt P_3
1731?	Rt dP_4
1732	Lt dP_4
1733	Rt dP^4
1734	Lt dP^3, broken fragment
1755	Lt $M_{2?}$
1756	Lt $M_{2?}$
1757	Lt $M_{2?}$
1758	Rt $M^{3?}$
1759	Rt $M^{3?}$
1760	Lt $M^{3?}$
1761	Rt $M^{3?}$
1762	Rt $M^{2?}$
1763	Lt $M^{2?}$
1764	Rt $M^{1?}$
1765	Lt P^4
1767	Rt P^4
1773	Rt $M^{1?}$, associated? with P^{3-4}
1774	Rt P^4
1775	Rt P^3
1778	Lt C^1, male
1779	Rt C^1 male, broken
1780	Rt C^1, male
1781	Rt C^1, male
1815	Rt ramus ascendens

Ain Jourdel, Oran, Algeria

As summarized by Delson (chapter 5), faunal studies suggest an age between 2.5–2.0 Ma, but this is not robust. Only one specimen is known, the holotype of '*Cynocephalus atlanticus*' Thomas; it cannot be allocated definitely to a known species.

Specimen number	Description
MNHN-P (AJO 001)	Isolated unworn R $M_{1?}$

Thomas Quarries, Morocco

Geraads has suggested an age younger than Ternifine on faunal grounds and agreed with an estimate of c. 0.4 Ma.

Specimen designation	Description (all MNHN-P specimens?)
Thomas I	4 isolated lower teeth (left dP_3, dP_4, $M_{2?}$, M_3)
Thomas III	partial male mandibular corpus with right C_1–P_4, left C_1–P_3, M_{2-3}

Melka-Kunturé, Ethiopia

Geraads (1979) has reviewed the fauna of the several fossiliferous sites at Melka, and Cressier (unpub. data) has documented the magnetochronology of the region. Fragmentary *Theropithecus* fossils possibly referable to either *T. oswaldi* or *T. brumpti* are known from two sites: Garba IV (placed later in the inter Olduvai-Jaramillo interval, thus probably 1.4–1.0 Ma) and Garba XII J (placed early in the Jaramillo, thus close to 1.0 Ma). The specimens are housed at the National Museum of Ethiopia (NME), Addis Ababa.

Specimen number	Description
Garba IV, NME 74–7596	distal 2/3 of right M_3
Garba XII J, NME 78–1952	fragment of juvenile right maxilla with erupting P^{3-4} and M^1

Afar Region, Ethiopia – Hadar Formation

This has produced 73 cranial and 24 isolated postcranial specimens of *Theropithecus*. The cranial specimens are listed and described in Appendix 2.1 of Eck (chapter 2). The isolated postcranial specimens are listed below. All the specimens have been referred to *T. darti*.

Specimens of *Theropithecus* have been recovered from all members of the Hadar Formation except the Basal Member and range in age from about 3.4 Ma to about 3.0 Ma (but see Eck, chapter 2, for a summary of the controversy surrounding these ages).

As an addition to the catalogue of the Hadar material, it may be of value to note that fragmentary dental remains of *Theropithecus* cf. *darti* are known from the 'localities' of Leadu (AL 2), Geraru (AL 74), and Ahmado (AL 100). The specimens are housed at the National Museum of Ethiopia (NME), Addis Ababa.

Specimen number	Locality (member)	Description	Specimen number	Locality (member)	Description
AL55–16	DD3–DD3–KH1s	Rt. distal humerus	AL177–1	DD2–DD3	Rt. distal tibia
AL55–17	DD3–DD3–KH1s	Rt.proximal humerus	AL185–18	DD3–DD3	Lt. distal femur
AL113–3	DD1	Rt. distal tibia	AL187–5	DD2–DD3	Lt. proximal ulna
AL120–5	DD3–DD3	Rt. distal tibia	AL206	DD2	Rt. femur
AL126–32	SH1–SH2	Lt. proximal ulna	AL208–1	SH2–SH3	Rt. distal tibia
AL133–1	DD2–DD3	Rt. proximal ulna	AL222–14	SH1–SH2–SH3	Lt. distal humerus
AL133–3	DD2–DD3	Rt. proximal radius	AL259–1	SH2–SH3	Lt. distal humerus
AL143–1	SH2–SH3	Lt. distal femur	AL288–9	KH1–KH1	Rt. proximal femur
AL145–16	SH1–SH2	Rt. distal humerus	AL322–10	DD1–DD2–DD3	Rt. distal humerus
AL145–30	SH1–SH2	Lt. distal femur	AL332–29	SH4–DD1–DD2	Lt. proximal ulna
AL163–10	DD2–DD3	Lt. distal humerus	AL362–3	DD3–DD3	Rt. distal femur
AL175–21	SH1–SH2	Rt. proximal femur	AL368–4	KH2	Rt. distal tibia

Shungura Formation –
Omo Basin, Ethiopia

The Shungura Formation has produced 2379 cranial and 184 postcranial specimens of *Theropithecus*. The majority of these consist of isolated teeth and tooth fragments, but hundreds are more complete specimens. The specimens listed below comprise a short list of the most interesting of these.

All members of the Shungura Formation except the Basal Member have produced specimens of the genus. The oldest of these date about 3.4 Ma and the youngest about 1.0 Ma. The majority of the specimens, however, derive from Member C (36 per cent) and the lower part of Member G (20 per cent); the former range in age between 2.9 Ma and 2.5 Ma, the latter between 2.3 Ma and about 2.1 Ma.

More specific information about the geological provenience of specimens of *Theropithecus brumpti* can be found in Eck & Jablonski (1987) and for *T. oswaldi* in Eck (1987). A brief summary of the geology and dating of the Shungura Formation can be found in Brown & Feibel (1988).

Specimen number	Locality (member)	Description
		Theropithecus brumpti
L1–280	B–11	Lt maxilla fragment with teeth with C, dP4–M2
L1–322	B–11	?Female Lt radius proximal fragment and 1/3 of the shaft
L1–497	B–11	Male Lt ulna proximal fragment
L1–1397	B–10	?Female Lt radius proximal fragment with the radial tuberosity
L17–45a	C–8	Male cranium fragment; muzzle with Rt P4–M3, Lt M1–M3, and other cranial fragments
L17–67	C–8	Female cranium fragments with teeth; muzzle with Lt P4–M1,

Specimen number	Locality (member)	Description
		Rt P3–M2, orbital and vault fragments
L17–72	C–8	Male Lt mandible fragment with teeth; body fragment with M1–M2, lacking the margin
L17–93	C–8	Male Lt femur proximal fragment
L18–17	C–8	Female Rt femur proximal fragment
L18–20	C–8	Male Lt humerus distal fragment
L18–24	C–8	Male Lt radius proximal fragment and the radial tuberosity
L19–10	D–4	Rt mandible fragment with teeth; body with M1–M3, lacking the margin
L23–62	C–6	Male Lt ulna proximal fragment and 1/2 of the shaft
L23–66	C–6	Male Rt ulna proximal fragment and 2/3 of the shaft
L32–154j	C–6	Male cranium fragment; muzzle with Rt P3–M1, LC–M3, zygomatic arch, glenoid region, Rt orbit
L32–155a	C–6	Female cranium fragment; muzzle with M3s only, frontal and zygomatics well preserved
L32–155g	C–6	Male Lt humerus distal fragment and proximal end with 1/4 of shaft
L32–158a	C–6	Male Lt mandible fragment with teeth; hemimandible with symphysis, LM3
L32–159j	C–6	?Female Lt humerus proximal fragment
L32–159k	C–6	Male Lt humerus proximal fragment plus 1/2 of shaft

Table (*cont.*)

Specimen number	Locality (member)	Description
L32–161	C–6	Female Lt cranium fragment with orbital rim, maxilla with M3, Lt temporal, fragments
L32–201	C–6	Female Rt femur proximal fragment
L32–208	C–6	Male Rt femur complete
L32–226	C–6	Lt mandible fragment with teeth; body fragment with M2–M3, lacking the margin
L32–243	C–6	Male Rt femur distal fragment
L37–16	C–8	Rt mandible fragment with teeth; body fragment with M2–M3, lacking the margin
L42–40	C–5	Female Rt ulna proximal fragment and 1/3 of the shaft
L45–1a	C–5	Male cranium fragment with teeth; muzzle Lt M1–M3, Rt P3–M3, frontal and right temporal
L47–41	C–8	Male Lt ulna proximal fragment eroded
L47–44	C–8	Male Lt mandible fragment with teeth with M2, M3, and roots of C, P4, M1
L49–1	E–2	Female mandible fragment with teeth with well preserved body, Lt M1–M3, Rt M1–M3, lacking rami
L54–1	C–7	Female Rt mandible fragment body, edentulous
L55–48	C–6	Male Lt humerus proximal fragment
L58–11	C–6	Male Lt femur proximal fragment
L62–12	C–5	Male Rt ulna distal fragment and 1/3 of the shaft

Specimen number	Locality (member)	Description
L66–17	F–3	Rt mandible fragment with teeth; body fragment with M1–M3, lacking the margin
L69–3	C–8	Male Rt humerus proximal fragment and 1/3 of shaft
L70–4	C–6	Lt mandible fragment with teeth; body fragment with M1–M3
L70–6	C–6	Male Rt radius proximal fragment and 1/4 of the shaft
L70–7	C–6	Male Lt humerus complete
L78–95	C–8	Female Rt ulna proximal fragment
L118–9b	D–5	Rt maxilla fragment with teeth with M2–M3
L119–6	D–3	Male cranium fragment; series of fragments
L122–34	D–3	Female cranium nearly complete but anterior muzzle and vault are damaged
L143–41	C–9	Female Rt femur proximal fragment
L143–43	C–9	Female Rt femur proximal fragment
L143–45	C–9	Female Lt humerus distal fragment
L152–4b	C–8	Female cranium fragments; series of fragments including the Rt maxilla, left temporal, and Lt P3, Rt M2 and M3
L161–24	D–3	Male Lt mandible fragment with teeth; body with M1–M3
L161–25a	D–3	Male Lt maxilla fragment with teeth with C fragment and M3
L161–25b	D–3	Male Rt maxilla fragment with teeth; anterior fragment with C fragment

Table (*cont.*)

Specimen number	Locality (member)	Description
L161–25c	D–3	Male Rt maxilla fragment with teeth; posterior fragment with M3 fragment
L161–43	D–3	Female Lt humerus distal fragment and 1/3 of the shaft
L169–8a	D–4	Male Lt femur distal fragment
L183–32	C–5	Female Lt femur distal fragment
L193–17	C–8	Male Lt maxilla fragment with teeth with M2–M3
L193–28	C–8	Male Rt femur distal fragment
L193–32	C–8	Female Rt mandible fragment with teeth; posterior body with M1–M2
L193–42	C–8	Male Rt ulna proximal fragment and 1/2 of the shaft
L193–43	C–8	Male Lt ulna proximal fragment
L199–2	C–7	Female cranium fragment; small series of fragments including Lt and Rt maxillae and Lt glenoid, Lt P4 and Rt M3
L199–3	C–7	Female mandible fragment with teeth; body with Lt and Rt M2–M3, lacking rami
L199–4	C–7	Female cranium; series of fragments, including right zygomatic and left zygomatic arch
L199–5	C–7	Male mandible fragment with teeth; body with Lt P3–P4, Rt P3–M3, lacking left ramus
L199–6	C–7	Male Lt ulna proximal fragment
L227–5	D–2	Male cranium; series of fragments, including parts of the Rt maxilla and zygomatic, Rt P4, M2 and M3
L227–6	D–2	Male Rt mandible fragment with teeth; hemimandible with M1–M3
L227–7	D–2	Male Lt ulna proximal fragment and 1/3 of the shaft
L236–1b	E–4	Female Rt radius proximal fragment with radial tuberosity
L238–46	F–3	?Male Rt humerus distal fragment and 1/3 of the shaft
L238–50	F–3	Female Lt radius proximal fragment lacking the head but with 2/3 of the shaft
L238–51	F–3	Male Lt radius proximal fragment with radial tuberosity
L238–53	F–3	Female cranium
L238–168a	F–3	Male Rt femur distal fragment
L238–186	F–3	Male Lt femur proximal fragment
L238–244	F–3	Male Lt ulna proximal fragment
L292–8	C–7	Male Lt humerus distal fragment with distal shaft
L292–9	C–7	Male Rt mandible; body, edentulous
L292–28	C–7	Female Lt radius proximal fragment with radial tuberosity
L293–1a	C–4	Male cranium; series of fragments, including the muzzle dorsum, Rt temporal and posterior vault
L300–1	D–1	Male maxilla fragment with teeth; muzzle with Rt P3–M3, Lt M2

Table (*cont.*)

Specimen number	Locality (member)	Description
L303–3	C–7	Lt maxilla fragment with teeth with M1–M3
L303–4	C–7	Male Rt radius proximal fragment with radial tuberosity
L304–9	C–7	?Female Rt humerus proximal fragment
L305–3	C–6	Male Lt mandible fragment with teeth body fragment with M3, lacking the margin
L305–6	C–6	Female Lt mandible fragment with teeth; body with Lt P3–M3, Rt C
L312–4	C–8	Male Lt radius proximal fragment and 2/3 of the shaft
L317–4	C–8	Female Lt mandible fragment with teeth; body with M1–M3
L327–13	C–6	Female Rt mandible fragment with teeth; symphysis and anterior Rt body with P3–M2, Lt P3
L331–4	C–6	Female Rt mandible fragment with teeth; body with M2–M3
L331–9	C–6	Female Lt femur complete
L335–53	C–5	Male Lt femur distal fragment
L338Y–2257	E–3	Male cranium fragment with Lt and Rt C–M3, muzzle well preserved, neurocranium badly damaged
L345–3	C–9	Male cranium fragment; partial muzzle and orbital region with Lt M2–M3, Rt M1–M3
L345–4	C–9	Male Lt mandible fragment with teeth; body with P3–M3, lacking ramus

Specimen number	Locality (member)	Description
L345–8	C–9	Male Lt ulna proximal fragment and 1/4 of the shaft
L345–18	C–9	?Male Rt radius proximal fragment and 2/3 of the shaft
L345–19	C–9	?Male Lt ulna proximal fragment and 1/2 of the shaft
L345–21	C–9	Male Rt femur proximal fragment
L345–25	C–9	Male Rt mandible fragment with teeth; posterior body and ramus with M2–M3
L345–26	C–9	Male Lt mandible fragment with teeth with C–P4
L345–50	C–9	Female Lt femur distal fragment
L345–105	C–9	Rt mandible; posterior body and ramus, edentulous
L345–236	C–9	Male Lt radius proximal fragment and 2/3 of the shaft
L345–287	C–9	Male cranium; nearly complete with Lt C–M3, Rt P3–M3
L351–4	C–7	Male Lt femur proximal fragment
L362–11	C–5	Rt mandible fragment with teeth; body fragment with M2–M3, lacking the margin
L434–8	G–7	Female Lt humerus proximal fragment and 1/4 of the shaft
L440–2	C–8	Male Rt mandible fragment with teeth; body fragment with M1, lacking the margin
L463–7	F–5	Male Lt ulna proximal fragment and 1/2 of the shaft
L467–62	F–1	Rt maxilla fragment with teeth with M1–M2

Table (*cont.*)

Specimen number	Locality (member)	Description
L477–11	G–1	Female Lt femur distal fragment
L489–16	G–4	Female Rt femur proximal fragment
L546–7	C–8	Male Rt ulna proximal fragment
L576–8	C–9	Male mandible fragment with teeth with Lt and Rt M1–M3, lacking Rt ramus
L576–20	C–9	Male Lt femur proximal fragment
L743–3	C–7	Female Rt mandible fragment with teeth; body fragment with P3–M1
L764–1	C–8	Male Rt mandible fragment with teeth; body fragment with M2–M3, lacking the margin
L775–5	C–8	Female Rt radius proximal fragment and 1/3 of the shaft
L790–1b	F–1	Male mandible fragment with teeth; series of fragments with most teeth
L824–4	D–1	Male Rt femur proximal fragment
L830–1	G–13	Male Lt maxilla fragment with teeth with P3–M2
L858–1	C–7	Male Rt ulna proximal fragment
L864–1	G–13	Female Lt mandible fragment with teeth; body with P3–M3
L865–1	E–4	Male partial skeleton
L869–6	C–6	?Male Rt ulna distal fragment and 2/3 of the shaft
L885–1	C–4	Male Rt ulna proximal fragment and 1/4 of the shaft
L885–2	C–4	Female Lt femur proximal fragment

Specimen number	Locality (member)	Description
L886–14	C–8	Male Rt radius proximal fragment with radial tuberosity
L886–15	C–8	Female Lt humerus distal fragment and 1/3 of the shaft
L899–1	C–8	Male Lt humerus proximal fragment and 1/2 of the shaft
F257–3	G–13	Male Lt maxilla fragment; edentulous
P791–5	C–8	Lt maxilla fragment with teeth with M1–M3
OMO 1C–72–2	F	Female cranium fragment – posterior muzzle and anterior neurocranium, edentulous
OMO 4–72–2	E–3	Male Rt maxilla fragment with teeth – with P3–M2
OMO 10B–67–2	E–3	Male Rt mandible fragment with teeth – with M1–M3
OMO 18–67–28	C–6	Male Lt humerus proximal fragment – and 1/2 of the shaft
OMO 18–67–40	C–6	Male Lt humerus proximal fragment – and 1/3 of the shaft
OMO 18–68–368	C–6	Male Rt maxilla fragment with teeth – with P3–M2
OMO 18–68–369	C–6	Male Lt mandible fragment with teeth – with fragmentary M2
OMO 18–68–371	C–6	Lt mandible fragment with teeth – with M2–M3

Table (*cont.*)

Specimen number	Locality (member)	Description
OMO 18–68–2231	C–6	Male Rt mandible fragment with teeth – with M2–M3, eroded
OMO 18–68–2281	C–6	Male Lt humerus proximal fragment – and 2/3 of the shaft
OMO 18–69–495	C–6	Male Rt mandible fragment with teeth – body with fragmentary M3
OMO 18–69–522	C–6	Male Lt femur proximal fragment
OMO 18–70–C1	C–6	Female Lt mandible fragment with teeth – with P4–M1
OMO 18–72–2	C–6	Male Rt radius proximal fragment – with radial tuberosity
OMO 28–67–557	B–10	Rt mandible fragment with teeth – with M2–M3
OMO 28–68–2196	B–10	Male Rt mandible fragment with teeth – with M1–M2
OMO 28–68–2280	B–10	Male Rt femur distal fragment
OMO 33–70–2532	F	Male Rt mandible fragment with teeth – with C–M1
OMO 40–68–1395	C–6	Lt mandible fragment with teeth – with P4–M3
OMO 40–68–1400	C–6	Female Lt maxilla fragment with teeth – with P3–M3
OMO 40–68–1405	C–6	Female Rt mandible fragment with teeth – body with P3 fragment – M1, M3
OMO 40–68–1411	C–6	Male Rt femur distal fragment
OMO 40–69–436	C–6	Male Rt mandible

Specimen number	Locality (member)	Description
		fragment with teeth – with M2–M3
OMO 47–70–1730	G–8	Male Rt mandible fragment with teeth – with M1–M3
OMO 50–73–4449	G–8	Female Lt radius proximal fragment – and 1/2 of the shaft
OMO 52–68–2279	C–7	Male Rt femur complete
OMO 53–68–2282	C–9	Female Rt femur complete
OMO 56–73–5126	C–5	Maxilla fragment with teeth – with Lt I1, dC, dP3–M1; Rt I1, dP3–M1
OMO 57.5–72–121	E–5	Female Rt radius proximal fragment – with the radial tuberosity
OMO 58–68–2225	F	Female Rt humerus distal fragment – and 1/2 of the shaft
OMO 72–69–1899	D	Male Lt mandible fragment with teeth – M2–M3
OMO 75 N–7–C205	G–12	Female Lt humerus proximal fragment
OMO 75–70–C40	G–8	Female mandible fragment with teeth – with Lt & Rt I1–M2
OMO 75–70–888	G–8	Male Rt maxilla fragment with teeth – with C–M2
OMO 75 N–71–C1	G–12	Lt mandible fragment with teeth – body and symphysis with Lt C–M3; Rt P4
OMO 76–72–24	F–2	Male Lt mandible fragment with teeth – with M2–M3
OMO 83–70–C102	E	Female Lt maxilla fragment with teeth – with P3, P4, M2

Table (cont.)

Specimen number	Locality (member)	Description
OMO 92–70–236	E–2	Male Rt mandible fragment with teeth – body with M2–M3
OMO 111–72–11	E–4	Lt maxilla fragment with teeth – with P4–M2
OMO 113–72–41	G–11	Female Lt femur proximal fragment
OMO 130–73–1894	F–3	Male Lt humerus proximal fragment – and 1/3 of the shaft, distal fragment and 1/2 of the shaft
OMO 132–72–5	C–6	Male Rt femur distal fragment
OMO 132–72–103	C–6	Rt maxilla fragment with teeth – with M2–M3
OMO 132–72–145	C–6	Male Lt femur complete
OMO 132–73–724	C–6	Female Lt radius proximal fragment – with the radial tuberosity
OMO 145–72–7	E–1	Female Lt humerus distal fragment
OMO 153–72–1	D–3	Male Lt mandible fragment with teeth – body and symphysis with eroded M2–M3
OMO 154–73–271	C–5	Maxilla fragment with teeth – series of cranial fragments with Lt M1–M2
OMO 158–73–343	C–8	Male Rt mandible fragment with teeth – body with C–M2
OMO 158–73–410	C–8	Male Lt humerus complete
OMO 165–73–608	C–6	Female Lt humerus distal fragment
OMO 169–73–845	E–1	Female Rt mandible fragment

Specimen number	Locality (member)	Description
		with teeth – body with P4–M2
OMO 175.2–73–1573	D–3	Male Rt femur complete
OMO 207–73–1776	E–3	Male Rt femur distal fragment
OMO 207–73–1781	E–3	Lt mandible fragment wih teeth – with M1–M3
OMO 217–73–4390	C–8	Female Rt maxilla fragment with teeth – with M2–M3
OMO 223–73–2864	G–5	Female Lt maxilla fragment with teeth – with M1
OMO 246–73–4748	G–11	Female Rt femur distal fragment
OMO 249–73–4987	G–13	Female Lt radius proximal fragment – with the radial tuberosity
OMO P703–72–2501	C–5	Male Rt mandible fragment with teeth – with P4–M2
OMO P752–70–2495	C–8	Female Rt maxilla fragment with teeth – with P3–M1
OMO SH1–70–C124	G–8	Lt maxilla fragment with teeth – with C, P3, dP4–M2
OMO SH1–70–C125	G–8	Male Rt mandible fragment with teeth – with M2–M3

Theropithecus oswaldi

Specimen number	Locality (member)	Description
L28–130	F–1	Female Lt humerus distal fragment and 2/3 of the shaft
L28–153	F–1	Female Lt ulna proximal fragment and 1/4 of the shaft
L40–15a	E–5	Male Rt mandible fragment with teeth (body with C–M1, M3)

Table (*cont.*)

Specimen number	Locality (member)	Description
L40–15b	E–5	Male Lt mandible fragment with teeth (body fragment with M2–M3)
L67–156	G–8	Male Lt ulna proximal fragment and 1/4 of the shaft
L74–31b	G–4	Female Rt ulna proximal fragment
L74–32a	G–4	Female Rt humerus distal fragment and 2/3 of the shaft
L74–32b	G–4	Female Rt humerus proximal fragment and 1/3 of the shaft
L80–85	G–4	Female Lt mandible fragment with teeth (body with P3–M3)
L178–6	E–5	Male Rt humerus proximal fragment and 1/3 of the shaft
L238–29a	F–3	Female cranium fragment (neurocranium with a damaged Rt side)
L238–29b	F–3	Female Lt maxilla (edentulous)
L238–257	F–3	Female Lt humerus proximal fragment and 1/3 of the shaft
L406–3	G–8	Female Rt femur proximal fragment
L465–82a	F–1	Lt mandible fragment with teeth (body with P4–M3)
L465–82b	F–1	Rt mandible fragment with teeth (body fragment with M1–M2, lacking the margin)
L607–22	G–5	Rt mandible fragment with teeth (body with M1–M3 fragments)
L626–71	G–121	Male Rt ulna proximal fragment
L627–237	G–12	Female Lt ulna proximal fragment
L862–2	F–1	Male Rt femur complete
L867–13	F–2	Male Rt humerus shaft
L867–15	F–2	Male Rt radius proximal fragment (with radial tuberosity)
L884–3	G–15	Male Lt femur distal fragment
F??–1a	G–13 (Locality Number = GS)	Female Rt mandible Symphysis and body with M3
F??–1b	G–13 (Locality Number = GS)	Female Lt mandible fragment with teeth Body fragment with M2
F413–1	L–8	Female Rt femur proximal fragment
F513–1	G–27	Male Lt ulna proximal fragment
OMO–68–1412		Female Rt femur distal fragment
OMO SH1–70–C123	G–8	Lt mandible fragment with teeth – with M2–M3
OMO 1E–67–704	F–2	Male Lt femur distal fragment
OMO 25–67–5	G–4	Rt mandible fragment with teeth – fragment with M2–M3
OMO 33–69–404	F	Female Lt humerus proximal fragment – and 1/4 of the shaft
OMO 33–70–2534	F	Rt mandible fragment with teeth – fragment with M1–M3
OMO 33–73–3092	F	Lt mandible fragment with teeth – fragment with M2–M3
OMO 47–73–1474	G–8	Female Lt humerus distal fragment – and 1/3 of the shaft
OMO 48–73–4729	G–12	Cranium – Series of cranial fragments with M3

Table (*cont.*)

Specimen number	Locality (member)	Description
OMO 57–68–2216	E–5	Female Lt mandible fragment with teeth – body with P3, M2–M3
OMO 75–70–814	G–8	Male Lt radius proximal fragment – and 2//3 of the shaft
OMO 75–70–C50	G–8	Female Lt mandible fragment with teeth – body with P3–M3
OMO 75–70–C51	G–8	Female Lt mandible fragment with teeth – with P3–M3 fragment
OMO 75 I–70–1000	G–5	Female Rt maxilla fragment with teeth – with P3–M3
OMO 75 I–70–1001	G–5	Female Lt mandible fragment with teeth – with M2–M3
OMO 75 I–70–1002	G–5	Female Rt mandible fragment with teeth – with P3–M2
OMO 75 I–70–1003	G–5	Female mandible fragment with teeth (anterior body with Rt I1–C; Lt I1–M1)
OMO 75 I–70–1004	G–5	Female Lt maxilla fragment with teeth – P3–M3
OMO 75 M–71–613	G–12	Female Rt mandible fragment with teeth – with M2
OMO 75 M–71–614	G–12	Female Lt mandible fragment with teeth – body with fragmentary M1 – fragmentary M3
OMO 75 N–7–C201	G–12	Male Rt humerus distal fragment – and 1/2 of the shaft
OMO 75 N–7–C203	G–12	Female Lt humerus distal fragment
OMO 75 N–71–679	G–12	Male Lt femur proximal fragment
OMO 75 N–71–680	G–12	Male Rt femur proximal fragment
OMO 75 N–71–688	G–12	Male Rt humerus proximal fragment – and 1/3 of the shaft
OMO 75 N–71–689	G–12	Male Rt radius proximal fragment – and 1/2 of the shaft
OMO 75 N–71–690	G–12	Female Lt femur distal fragment
OMO 75 N–71–692	G–12	Male Lt humerus distal fragment – and 1/3 of the shaft
OMO 75 N–71–693	G–12	Male Lt humerus proximal fragment – and 1/2 of the shaft
OMO 75 N–71–C24	G–12	Male cranium – nearly complete, in 3 parts, edentulous
OMO 75 S–69–461	G–7	Male Rt humerus proximal fragment – and 1/3 of the shaft
OMO 92–73–984	E–2	Female Lt humerus distal fragment – and 1/3 of the shaft
OMO 93–70–576	G–7	Male Lt radius proximal fragment – and 2/3 of the shaft
OMO 103–72–5	G–4	Lt mandible fragment with teeth – with dP4–M1
OMO 103–73–4401	G–4	Lt mandible fragment with teeth – with P4–M2
OMO 118–72–19	F–1	Male Lt ulna proximal fragment
OMO 135–72–2	E	Male Lt femur proximal fragment
OMO 136–72–5	G–1	Female Rt mandible fragment with teeth – body with P3, P4, M2, M3
OMO 148–72–5	D–1	Female Rt femur proximal fragment
OMO 170–73–834	E–3	Male Rt maxilla fragment with teeth – with P3–M1
OMO 195–73–1355	G–7	Male Rt femur distal fragment

Table (*cont.*)

Specimen number	Locality (member)	Description
OMO 223–73–2870	G–5	Male Rt maxilla fragment with teeth – with C–M2
OMO 233–73–4554	G–1	Female Lt humerus distal fragment – and 1/3 of the shaft

Theropithecus species indeterminate

Specimen number	Locality (member)	Description
L16–59i	G–4	Male Rt mandible fragment with teeth with M1
L16–105	G–4	Female Rt mandible fragment with teeth with LC, P3, dP4, M1, M2
L16–106a	G–4	Female Lt mandible fragment with teeth with P4–M3
L17–41b	C–8	Female mandible fragment with teeth with LdP4, LM1, RM1, RM2; juvenile
L32–157c	C–6	Male Lt mandible fragment with teeth with P3 fragment, P4, M2
L32–283	C–6	?Female Rt radius proximal fragment with radial tuberosity
L55–52	C–6	?Female Rt radius proximal fragment with radial tuberosity
L65–1	G–1	Female Lt mandible fragment with teeth with eroded and worn M1–M3
L67–121a	G–8	Male Lt mandible fragment with teeth
L193–82	C–8	Rt humerus proximal fragment
L238–38	F–3	Female Rt mandible fragment with teeth
L238–261	F–3	Female mandible symphysis
L304–10	C–7	Lt humerus proximal fragment

Specimen number	Locality (member)	Description
L373–1	C–1	Female mandible fragment with teeth
L420–13	F–3	Male Lt mandible fragment with teeth
L785–1	E–1	Female maxilla fragment with teeth
L793–2	B–2	?Female Rt radius proximal fragment with 1/2 of the radial tuberosity
F24–3	G–2	Male mandible
OMO 1B–69–479	G–4	Rt mandible fragment with teeth – with M2–M3
OMO 5.3–67–4	C–9	Male Lt mandible fragment with teeth – with P3–P4, M2
OMO 9–67–601	G–4	Rt mandible fragment with teeth – with M2–M3
OMO 15–67–5	C–5	Rt mandible fragment with teeth – with M2–M3
OMO 29–68–1401	G–3	Lt mandible fragment with teeth – body with M1–M3
OMO 29–69–426	G–3	Lt mandible fragment with teeth – with M2–M3
OMO 29–70–851	G–3	Rt mandible fragment with teeth – with dP4, M1
OMO 29–70–1377	G–3	Rt mandible fragment with teeth – with M2–M3
OMO 33–70–2931	F	Rt mandible fragment with teeth – with P4–M3
OMO 75–70–837	G–8	Rt mandible fragment with teeth – body with M2–M3
OMO 75 N–7–202	G–12	Lt humerus proximal fragment
OMO 153–73–277	D–3	Rt radius proximal fragment
OMO 199–73–1273	F–2	Rt humerus proximal fragment

Table (*cont.*)

Specimen number	Locality (member)	Description
OMO 199–73–1421	F–2	Lt humerus distal fragment
OMO 233–73–4553	G–1	Lt mandible fragment with teeth – body with P4–M3

Usno Formation – Omo Basin, Ethiopia

The Usno Formation has produced 203 cranial specimens of *Theropithecus*, all but one of which are isolated teeth or tooth fragments. The exception is the cranium of *T. quadratirostris* (but see Delson & Dean, chapter 4).

All of these fragments derive from Unit U-12 of the formation and are equivalent in age to those of lower Member B of the Shungura Formation (about 3.3 Ma). The cranium derives from sediments that are also of about this age. Details of geological correlations can be found in de Heinzelin (1983); geological ages in Brown & Feibel (1987).

Chemeron Formation, Kenya

Recent dating of the Chemeron Formation has shown that the site JM90/91 is almost certainly 3.2 Ma (Deino pers. comm.). Many rocks and fossils in the Chemeron Formation are around 5 Ma and others around 2 Ma with apparently no stratigraphic continuity between the two (Hill & Ward, 1988). This has led to problems in dating some sites particularly JM90/91. The specimens *Theropithecus* from Chemeron are housed in the Palaeontology Division of the National Museums of Kenya, Nairobi. Most students would now classify the specimens from Chemeron as *Theropithecus baringensis*, but the reader is referred to Chapters 3, 4, and 7 of the present volume for details of the controversy concerning their classification.

Specimen number (KNM-BC)	Description
3	Skull and mandible
1647	Lt mand (M3) Rt mand (P4-M3) and fragment of femur

Lothagam Hill, Kenya

The single *Theropithecus* molar from Lothagam Hill was recovered from Lothagam-3, a unit of more than 100 m of coarse sandstones, silts, and clays. The Lothagam sill which underlies Lothagam-2 and Lothagam-3 has been dated radiometrically at 3.8 Ma and it shows reversed polarity suggesting a correlation with the terminal third of the Gilbert Reversed Epoch. Faunal correlation suggests a date between 4.0 and 4.5 Ma for Lothagam-3 (Hill & Ward, 1988).

The specimen is housed in the Palaeontology division of the National Museums of Kenya, Nairobi.

Specimen number (KNM-LT)	Description
417	Rt lower M_2

Kanam East, Kenya

The Kanam East sediments have not been accurately dated. The Kanam East specimen is housed in the Palaeontology Division of the National Museums of Kenya, Nairobi.

Specimen number (KNM-KE)	Description
241	M_3 and $M^{1?}$

Olorgesailie, Kenya

The Olorgesailie Formation is divided into 14 members which range in age from 0.99 Ma (Member 1) to 0.49 Ma (Member 14). The upper part of the Formation, Members 10–14, ranges between 0.74 Ma and 0.49 Ma (Bye *et al.*, 1987). The *Theropithecus* fossils, which were all recovered from one site, DE/89B in Member 7 are therefore older than 0.74 Ma (Potts, 1989).

The Olorgesailie *Theropithecus* fossils are classified as *T. oswaldi* and are housed in the Palaeontology Division of the National Museums of Kenya, Nairobi. The most complete specimens are listed below.

Specimen number (KNM-OG)	Description
1	Male Rt maxilla (P3-M2)
2	Female Rt mandible (P3-M3)
4	Rt mandible (P4-M2)
5	Lt mandible (P3-M1)
157	Lt maxilla (P4)
442	Rt distal humerus
454	Rt proximal ulna
542	Rt maxilla (dM1-M1)
565	Lt maxilla (P3-4)
567	Lt maxilla (P3-4)
598	Rt maxilla (P3)
781	Mandible (dM2, M1, erupting C, P3,)
1056	Lt humerus lacking head
1064	Rt distal humerus
1069	Proximal ulna and part shaft
1074	Proximal radius
1076	Lt proximal femur and part shaft
1077	Lt proximal femur and part shaft
1078	Lt proximal femur and part shaft
1088	Rt proximal femur and part shaft
1090	Rt femur lacking distal epiphysis
1109	Distal tibia
1115	Juvenile tibia
1288	Mandible (dM2-M1)
1310	Lt proximal ulna and part shaft
1311	Rt proximal ulna and part shaft

Specimen number (KNM-OG)	Description
1419	Proximal Rt ulna
1420	Rt temporal fragment
1421	Rt temporal fragment
1450	Frontal fragment
1451	Lt frontal fragment
1452	Rt frontal fragment
1454	Scapular glenoid
1455	Lt distal humerus
1459	Rt distal humerus
1460	Rt distal humerus
1461	Rt distal humerus and part shaft
1462	Distal radius
1463	Proximal radius
1467	Proximal Lt ulna
1529	Lt temporal fragment
1536	Occipital fragment
1538	distal ulna
1612	Rt temporal
1662	Rt distal humerus and part shaft

Kapthurin, Kenya

Wood & Van Noten (1986) discuss several dates reported for the Kapthurin Formation and conclude that fauna collected from below the grey tuff appears to be equivalent to post-Bed IV at Olduvai. The single *Theropithecus* specimen from Kapthurin is a mandible, KNM-BK 22638, that preserves a complete dentition.

Kanjera, Kenya

Kanjera is the type locality of *Theropithecus* (='*Simopithecus*') *oswaldi* Andrews, 1916. The Kanjera deposits are not well dated. Plummer & Potts (1989) suggested an age between 1.85 Ma and 1.5 Ma for the *Theropithecus* collection (which is all from one locality), based on similarities with specimens from Bed I, Olduvai Gorge.

Turkana Basin, Kenya

Specimen number (KNM-KJ)	Description
18571	Rt innominate
18575	Rt mandible (M2-3)
18576	Rt. maxilla (P3-4)
18579	Parietal
18580	Fragment mandible ramus (M3)
20517	Prox. Rt radius
20522	Distal Lt humerus
20535	Lt innominate
20598	Partial calvaria

The geology and dating of the Omo Group deposits is given by Feibel, Brown & McDougall (1989), and Leakey (chapter 3) discusses the stratigraphic distribution and age of the *Theropithecus* specimens from the Turkana basin. East of Lake Turkana, *T. brumpti* occurs from the Lokochot to the Tulu Bor Members of the Koobi Fora Formation (about 3.4 Ma to 2.9 Ma), and *T. oswaldi* occurs in the Upper Burgi, KBS and Okote Members (about 2.4 Ma to 1.0 Ma). West of Lake Turkana, *T. brumpti* has been recovered from the Lomekwi Member of the Nachukui Formation (between 3.3 Ma and 2.5 Ma) and *T. oswaldi* between the Kalachoro and Nariokotome Members (about 2.4 Ma to 1.0 Ma).

The *Theropithecus* specimens from the Turkana Basin are housed in the Palaeontology Division of the National Museums of Kenya, Nairobi.

East Turkana, Koobi Fora Formation
Theropithecus brumpti

Specimen number (KNM-ER)	Locality (Member)	Description
124	TB	Rt mandible frag; M2-3
127	TB	Male, isolated teeth; Lt C, P3, M2, M3, Rt M3
1563	TB	Lt mandible; M1-2
1564	TB	Calvaria and Rt & Lt maxilla frags; M2-3
1565	TB	Female Lt maxilla; P3-M1
1566	TB	Male Lt maxilla; broken C-M3
2015	TB	Male mandible; Rt P3-M3, roots Lt C-P4
2017	TB	Rt mandible frag; M3
2018	TB	Rt maxilla; broken M3
2019	TB	Rt temporal
2022	TB	Prox Lt femur, dist Lt femur, prox Lt tibia
2023	TB	Dist Rt radius
2024	TB	Innominate frags
2029	TB	Rt & Lt mandible frags; Rt M2 erupting
3005	TB	Female Rt mandible; M1-3. Lt & Rt maxilla P4-M3
3013	TB	Lt mandible; frag M3, edentulous maxilla frags Postcranial frags including: Rt & Lt prox femora Dist humeri, prox Lt ulna, vertebra and Innominate frags
3018	TB	Male mandible with tooth roots and erupting M3s
3023	TB	Female Lt edentulous mandible
3025	TB	Skull frags including maxilla; Lt & Rt M1-3 Rt mandible; M2-3, supraorbital tori, temporal

Table continued

Specimen number (KNM-ER)	Locality (Member)	Description
3026	TB	Lt mandible; M1-2
3028	TB	Cranial frags and Lt lower M3
3030	LOK	Male mandible, Lt & Rt P4, alveoli Ii-C, roots P4-M2
3035	TB	Rt mandible frag, erupting M
3038	LOK	Mandible; Rt Ii, P3-M3, Lt P4-M3
3045	LOK	Isolated teeth, upper dM2, lower M1
3053	TB	Male isolated upper teeth; Rt 12, C, P3, Lt I1, P3-4
3084	TB	Male mandibluar frags; edentulous, Rt condyle Prox Rt humerus & ulna, glenoid Lt scapula Frags dist Rt humerus & prox Rt femur
3118	?TB	Male mandibular frags; Rt & Lt P3, Lt condyle
3119	TB	Male Lt & Rt mandible; Lt P4, Rt M1-3, Rt femur Shaft Rt humerus
3775	LOK	Rt maxilla; M2-3
3780	TB	Male mandible; Lt P3-M3, Rt P4-M3
3783	LOK	Rt mandible; dM2, germ P4
3792	TB	Lt mandible frag; worn P4
3855	TB	Lt mandible frag; broken P4-M1, root P3
3857	TB	Female Lt mandible frag; 12, P3-M1
4704	TB	Female cranial and postcranial frags
4940	TB	Lt mandible frag; M3
4985	TB	Rt mandible; M2-3
5320	TB	Isolated lower I1-2, upper Lt & I2

TB = Tulu Bor Member
LOK = Lokochot Member

East Turkana, Koobi Fora Formation
Theropithecus oswaldi

Specimen number (KNM-ER)	Locality (Member)	Description
13	U. Burgi	Frag Rt mandible and postcranial elements
28	?U. Burgi	Cranial and postcranial elements
46	?U. Burgi	Cranial and postcranial elements
113	?U. Burgi	Rt mandible; M3
119	?U. Burgi	Female Rt mandible; M2, M3, Lt mandible; P4-M3
122	U. Burgi	Rt mandible, M2-3, Lt mandible M3
123	?U. Burgi	Lt mandible; M3
131	U. Burgi	Male maxilla; Rt C, P3-M2, Lt M1-2
135	?U. Burgi	Juvenile Lt mandible; dM1-2

Table continued

Specimen number (KNM-ER)	Locality (Member)	Description
136	Okote	Cranial fragments; Lt C-P3, Lt & Rt P4-M3
137	?U. Burgi	Lt mandible; M2-3
138	?U. Burgi	Rt mandible; M3
139	?U. Burgi	Lt mandible; M3
147	U. Burgi	Rt mandible; P3-M2, Lt mandible; P4-M1
151	U. Burgi	Broken cranium and mandible
153	U. Burgi	Cranium; Lt & Rt M2 & erupting M3
154	?U. Burgi	Squashed juvenile cranium; Rt I1-M2; Lt I1-2, dm
180	KBS	Female partial cranium; won Lt P4-M3, Rt P3-M3
418	U. Burgi	Female skull; Rt M2, Rt & Lt M3 erupting
547	U. Burgi	Female cranial frags; I1-M2
548	U. Burgi	Juvenile mandible; Lt M1, Rt dM1-M1
550	U. Burgi	Cranial frags
557	KBS	Female mandible; Lt & Rt I-M1, M2-3
558	U. Burgi	Lt mandible; M1-3
560	Okote	Male mandible; Lt C; Lt & Rt P3-M2
566	Okote	Male Skull and mandibular frag
567	Okote	Female cranial and postcranial elements
569	Okote	Cranial and postcranial frags
573	?U. Burgi	Lt mandzble; P4–M3
577	KBS	Male mandible; Lt I1-M3, Rt I1-I2, P3-M3
579	Okote	Female Lt mandible and symphysis; P4-M2
581	Okote	Partial calvaria
582	?Okote	Lt maxilla; M1-M3
585	Okote	Lt mandible; M1-3
587	Okote	Juvenile Lt mandible; dC-dM2
591	Okote	Lower Rt M3, upper P3, P4, female C
597	Okote	Cranial and postcranial frags
601	KBS	Cranial and postcranial frags
602	Okote	Isolated teeth; male upper Rt C, P3, M3
603	Okote	Cranial and postcranial frags
605	KBS	Lt mandible; M3
608	KBS	Mandible; Lt P3-M1
609	Okote	Male cranial and mandibular frags
611	Okote	Rt mandible; M2-3
612	KBS	Juvenile cranium; Lt & Rt M2 and erupting M3
613	U. Burgi	Female Rt mandible; M1-3, Lt mandible; P4-M3
618	Okote	Lt mandible; M3
619	Okote	Male lower Lt C, Rt C, M
620	Okote	Isolated worn upper teeth
621	Okote	Female Lt C, dental frags
622	Okote	Mandibular and dental frags
627	Okote	Female upper Lt C, M, Rt M
632	Okote	Female cranial & mandibular frags
634	KBS	Mandible; Lt P4-M3; Rt M2-3
744	Okote	Rt mandible; M1, dM2, Lt maxilla; M1, dM2

Table continued

Specimen number (KNM-ER)	Locality (Member)	Description
745	Okote	Isolated upper teeth; Lt & Rt C, P4, Lt M, P3, I1 Patella and Rt talus
809	KBS	Lt & Rt unerupted upper P4
828	KBS	Rt upper M3, tooth frags
830	Okote	Male isolated teeth; upper Rt M3, P4, C, Lt C and lower L
832	Okote	Male isolated upper teeth; Lt I1, C, P3, Rt P3, P4
833	Okote	Isolated teeth, Lt & Rt upper M3; lower C, M1, M3
834	Okote	Isolated lower teeth; Rt M2, M3, broken Lt M3
835	KBS	Female isolated lower teeth; Rt C, Lt P3, P4, M1
836	KBS	Isolated teeth; Lt lower 12, Rt upper M
839	Okote	Female isolated lower teeth, Lt P3, P4, M2
840	Okote	Female isolated teeth and bone frags
842	KBS	Lt mandible; M1, M3
849	KBS	Weathered juvenile mandible
850	Okote	Rt mandible; M1-3
851	Okote	Mandibular frags; Rt M3
856	Okote	Female Rt maxilla; C-M1, M3
860	Okote	Cranial and postcranial frags
862	Okote	Rt maxilla; P4-M3
863	Okote	Mandibular symphysis; Rt P3, P4
864	Okote	Male mandible; Lt & Rt P3-M3
865	KBS	Male mandible Lt & Rt C-M3
866	Okote	Postcranial elements
880	KBS	Lt mandible; M1-3
883	KBS	Rt maxilla; M2-3
886	KBS	Lt mandible; M1-2
891	?U. Burgi	Lt mandible frag, distal scapula
967	?U. Burgi	Skull frags
969	Okote	Calvaria
971	U. Burgi	Female cranium
978	U. Burgi	Cranial and postcranial elements
985	KBS	Male isolated teeth; upper Rt C, lower Lt C, M3, M
1194	Okote	Lt mandible; dM2-M1
1522	U. Burgi	Male squashed juvenile cranium; Rt C-M2; Lt C-M1
1525	U. Burgi	Mandible; Rt M1-2, Lt P3-M2
1526	U. Burgi	Cranial and postcranial elements
1527	U. Burgi	Male juvenile cranium
1528	U. Burgi	Female Lt mandible; C-M3
1530	U. Burgi	Lt mandible; M2-3
1531	U. Burgi	Male cranium
1532	U. Burgi	Cranial and mandible frags
1533	U. Burgi	Rt mandible M1-3
1535	U. Burgi	Maxilla; Rt C-M3, Lt M3
1536	U. Burgi	Mandibular symphysis
1537	U. Burgi	Lt mandible; M2-3
1545	U. Burgi	Male mandible, Lt & Rt P4-M3, Rt P3

Table continued

Specimen number (KNM-ER)	Locality (Member)	Description
1547	U. Burgi	Male cranial frags, Lt & Rt M1-2, Lt P3, Cs in crypt
1550	U. Burgi	Rt mandible; M2
1552	U. Burgi	Rt mandible; M2-3
1562	Tulu Bor	Mandible; Rt P4-M3
1567	U. Burgi	Calvaria
1568	U. Burgi	Partial calvaria
1572	U. Burgi	Female cranial and mandibular fragments
1573	KBS	Male mandible and maxilla frags
1574	KBS	Mandible and maxilla frags
1575	KBS	Skull frags
1578	KBS	Cranial and postcranial frags
2001	U. Burgi	Female juvenile mandible; Lt I1-2, M1, Rt I1-2, dC, dM1, M1
2003	U. Burgi	Male mandible; Rt P3, M2-3
2004	U. Burgi	Rt mandible; P4-M1
2005	U. Burgi	Frag Rt mandible, Rt humerus
2006	U. Burgi	Lt mandible; M2-3
2027	KBS	Female mandible; Rt P3-M3
2093	U. Burgi	Rt mandible; M1-3
2112	U. Burgi	Rt maxilla; P3-M2
2116	KBS	Lt juvenile maxilla; dC-dM2
3007	U. Burgi	Female Lt maxilla; C-M3
3066	KS	Cranial frags
3070	U. Burgi	Rt maxilla; M3
3071	KBS	Cranial and postcranial frags
3074	U. Burgi	Lt mandible
3076	U. Burgi	Male mandibular symphysis; Rt & Lt P3-4
3077	Okote	Female Rt mandible; M1-3
3081	U. Burgi	Lt maxilla; M2-3
3082	U. Burgi	Juvenile Lt mandible; dM1-2
3088	U. Burgi	Lt mandible; M2-3
3091	Okote	Male cranial, mandibular and tooth frags
3092	KBS	Lt mandible; P4-M1
3769	U. Burgi	Lt mandible M2-3
3779	Okote	Female mandible; Lt M1, M3, Rt P3-M2
3785	U. Burgi	Juvenile Rt mandible; M1
3807	U. Burgi	Juvenile Rt mandible; M1
3814	U. Burgi	Lt mandible; M3
3819	KBS	Juvenile Rt mandible; dM2, M1 and postcranial frags
3820	U. Burgi	Female Lt mandible; P4-M3
3823	U. Burgi	Male edentulous cranium, erupting M3s, Lt & Rt tibia
3830	U. Burgi	Lt mandible; M in crypt
3837	U. Burgi	Cranial and mandible frags
3839	U. Burgi	Upper Lt M3, frag Rt M3
3840	U. Burgi	Cranial and postcranial frags
3844	KBS	Rt mandible; P4-M3
3845	U. Burgi	Rt mandible; M1-M3

Table continued

Specimen number (KNM-ER)	Locality (Member)	Description
3847	U. Burgi	Male mandible; Lt M2-3
3859	U. Burgi	Isolated teeth (2+ indivs)
3872	U. Burgi	Lt mandible; P4-M3
3874	U. Burgi	Female Lt maxilla; C-M1, M3, Rt M3, Lt mandible; P3-M3
3875	U. Burgi	Juvenile maxilla; Lt dM1-2, M1, Rt M1-2
3876	U. Burgi	Postcranial elements
3877	KBS	Dental & postcranial elements
4412	U. Burgi	Cranial frags; Rt & Lt M1-M3
4415	U. Burgi	Rt edentulous mandible, Lt mandible; M1-3
4416	U. Burgi	Female Rt & Lt edentulous mandible
4417	U. Burgi	Female mandible; Lt M3, Rt P3-4
4418	U. Burgi	Mandibular frags
4425	Okote	Lower molars
4426	Okote	Upper P, lower P4
4444	U. Burgi	Isolated lower teeth; Lt C, Rt M3
4566	KBS	Lt mandible; M2-3
4964	?KBS	Rt mandible; M1-3, Lt mandible; M3
4968	?KBS	Rt mandible, P4-M3
4977	KBS	Rt mandible; M2-3
5308	Okoté	Female lower Lt P3, M3, Rt M3, upper Rt C
5402	KBS	Lt & Rt mandible frags
5403	KBS	Lt mandible; M2-3
5404	Okote	Calvaria
5405	Okote	Juvenile mandibular and dental frags
5410	KBS	Lt mandible; M2, Rt C
5419	KBS	Rt juvenile mandible; erupting P3, M2
5479	?U. Burgi	Lt mandible; P3, M2
5483	U. Burgi	Lt mandible; M3
5490	U. Burgi	Cranial frags
5491	Okote	Cranial and postcranial elements
5564	Okote	Male upper Lt M, male C, Rt I
6001	U. Burgi	Female cranium
6007	Okote	Rt mandible frags; M3
6010	U. Burgi	Mandibular frags
6013	Okote	Rt mandible; M3
6067	U. Burgi	Rt mandible; M1-3
6073	KBS	Unerupted Lt lower M3
7331	KBS	Cranial and postcranial frags
18913	Okote	Male lower Lt male C, P3, M
18917	Okote	Postcranial elements
18918	Okote	Female lower isolated teeth; Lt C, M3, Rt M1
18919	Okote	Female Rt mandible; C-M3
18924	Okote	Female isolated teeth; lower Lt P3, M3, Rt M1, M3, Upper Rt P3, P4, 1t M1, I
18925	U. Burgi	Male cranium and mandible

West Turkana, Nachukui Formation
Theropithecus oswaldi

Specimen number (KNM-WT)	Locality (Member)	Description
14650	KAL	Lt mandible; M3
14659	KAL	Female mandible; Lt & Rt P4, roots Lt & Rt P3, M1
14658		Female mandible; Lt & Rt M2-3
14663	Kaitio	Rt & Lt tibiae frags, prox Lt femur, dist Rt femur
14666	Natoo	Female assoc lower teeth, Lt C, P3, M1, M3, Rt P3
14660	NAR	Rt mandible; M2-3
17435	Kaitio	partial cranium

KAL = Kalochoro
NAR = Narikotome

West Turkana, Nachukui Formation
Theropithecus brumpti

Specimen number (KNM-WT)	Locality (Member)	Description
16746	Lwr LOM	Lt mandible frag; M1-3
16747	Lwr LOM	Female edentulous mandibular symphysis
16749	Lwr LOM	Male mandible with symphysis; Lt P3-M2
16750	Lwr LOM	Lt and Rt mandible frags; Lt M2-3, roots Rt M2
16887	Lwr LOM	Male Rt maxilla; broken P3-M3, root C. Lt mandible; roots I1-M3, Rt mandible; I1, M2-3
16888	Lwr LOM	Female maxilla; Lt P3-M3, Rt P4-M1
16894	Lwr LOM	Rt maxilla; M2-3
16895	Lwr LOM	Rt mandible; P3
17553	Lwr LOM	Male mandible frags with symphysis; Lt broken M3, roots M2-3
17554	Lwr LOM	Female mandible; M3s erupting, roots Lt I1-M2. Lt condyle, Rt talus, prox Lt radius
17555	Lwr LOM	Male mandible; roots Lt I1-M3, Lt & Rt broken M2-3, broken Lt M1, prox Rt ulna
16864	Mid LOM	Female mandible frag, M3
17560	Mid LOM	Male mandible; Lt & Rt C, roots I1-2, Lt P3-4 Prox. Rt ulna
16862	Mid LOM	Maxilla frags; Lt P3, M1-2, Rt P3-4, M2-3
16863	Mid LOM	Mandible frags; Rt P3-4, M3 frag, Lt P4, M1, M3
16753	Upp LOM	Female mandible; Lt I1-P3, Rt P3-4, M2
16801	Upp LOM	Rt mandible frag; M1
16803	Upp LOM	Rt mandible frag; M2-3
16805	Upp LOM	Cranial frags, Lt maxilla; M3, Rt maxilla; M1-3 Temporal frag
16806	Upp LOM	Female mandible; Lt & Rt P3-M3

Table continued

Specimen number (KNM-ER)	Locality (Member)	Description
16808	Upp LOM	Male mandible; Lt I2, C, P4, M1, M3, Rt P3-4, M1, Upper Lt C, prox Rt ulna and shaft, radius shaft,
17571	Upp LOM	Female mandible; Lt P3, Rt C, P4-M2
17569	Upp LOM	Mandible, Lt & Rt M1-3
16828	Upp LOM	Male cranium and postcranial frags; dist Rt & Lt tibia Dist Rt fibula, dist femur frag, prox Lt Mt/I1 Innominate and vertebral frags
16870	Upp LOM	Mandible frag; erupting M2 or 3, prox humerus
16871	Upp LOM	Rt maxilla; M2

upp LOM = upper Lomekwi
mid LOM = middle Lomekwi
lwr LOM = lower Lonekwi

Olduvai Gorge, Tanzania

The geology of Olduvai gorge is described in detail by Hay (1976) who gives radiometric dates for the Olduvai Beds from which *Theropithecus oswaldi* has been recovered as follows:

Bed I; 1.85 Ma to 1.7 Ma. Bed II; 1.7 Ma to 1.15 Ma. Bed III; 1.15 Ma to 0.8 Ma. Masek Beds; 0.6 Ma to 0.4 Ma.

The specimens listed below are the property of the Tanzania Government and are only housed in the Kenya National Museum until such time as the Tanzanians request their return. Some specimens are housed in the British Museum (Natural History) and one of these is listed below.

Specimen number	Locality (bed)	Description
	Bed I	
068/6640		Lower Lt I1, female C, P4, M2, M3, and upper M1
OLD 63/3050		Male Rt mand (P3-M3) Lt mand (P3-M1), Rt max (P3-4), Rt upper C, Lt Max (M2)
OLD 53/117		Lt mand (M1), frag Lt mand (P4-M2)
OLD 61/5170		Max frag (P4-M3)
	Upper Bed II	
OLD 68/ S.9		Male maxilla (Rt P3-M3, broken Lt C-M2)
068/6511		Calvarium
OLD 69/ S.133		Male maxilla (Lt 12, C-M3, Rt I1-2)
OLD 69/ S. 89		Partial calvarium and tooth frags
BK II		Female mandible (Lt & Rt I1-M2)
OLD 77/ S.341		Juvenile skull with damaged teeth
067/5608		Lt max (P4-M3)
SLK II 580		Lt mand (M3)
OLD 60/ S.42		Female Lt mand P3-M1
067/2771		Female Lt mand (M1-3)
OLD 63/3366		Male Rt mand (P3-M3)
OLD 52/ S.694		Lt mand (M2-3)
BM(NH)M/14937		Lt mand (P4-M3)

Table continued

Specimen number	Locality (bed)	Description
067/5603		Male complete lower jaw, upper Bed II
	Bed III	
S.194		Rt max (M1-3), Lt mand (M2-3), Rt mand (M1-3)
S.116, 123, 132		Partial skull, (Lt C-M3, Rt M1, M3)
	Masek Beds	
068/6516		Male complete lower jaw

Senga 5A, Semliki, Zaire

As discussed by Delson (this volume, chapter 5), an estimated faunal age for this site is between 2.4–2.0 Ma, which is confirmed by the size of the two *Theropithecus* teeth recovered, although they cannot be readily allocated to either *T. oswaldi* or *T. brumpti*. The specimens are housed in the Institut des musées nationales de Zaire, Kinshasa.

Specimen number	Description
IMNZ Sn 5A 405	Unworn crown L $M_{2?}$
IMNZ Sn 5A 520	Damaged crown R $M_{1?}$

Makapansgat Limeworks, Transvaal, South Africa

As reviewed by Delson (1984), there are five Members of the Makapan Formation, numbered upwards from 1 to 5. Fossil mammals occur in Members 2 (uppermost, = Basal Red Muds of earlier authors), 3 (Grey Breccia) and 4 (Pink Breccia), although no *Theropithecus* is known from Member 2. Faunal evidence suggests an age in the mid-Pliocene, and paleomagnetic results (which have been interpreted differently by several workers) were evaluated by Delson to indicate an age of just over 3 Ma for Member 3, with the fossiliferous lower part of Member 4 only slightly younger. The holotype of *T. darti* is from Member 4, which also yielded 14 other specimens, while 20 are known from Member 3. The most complete of these are listed below. The specimens are housed in the University of the Witwatersrand Medical School, Department of Anatomy, Johannesburg (UWMA) and in the Bernard Price Institute for Palaeontological Research, University of the Witwatersrand, Johannesburg (BPI).

Specimen number	Locality (Member)	Description
UWMA MP 1	4	Damaged male mandible with P_3–M_3
UWMA MP 44	3	Male mandibular corpus, Rt P_3–M_3 & Lt P_{3-4}, anterior alveoli
UWMA MP 56	3	Female partial Lt mandibular corpus, P_4–M_3
UWMA MP 222	3	Female cranium, well preserved, Rt & Lt P^3–M^3
BPI M 3071	4	Male juvenile mandible, corpora well preserved, Rt & Lt C_1 (erupting) –M_2
BPI M 3073	4	Female subadult cranium with attached mandible, M3s unerupted, most teeth not visible

Hopefield (Saldanha, Elandsfontein), Cape Province, South Africa

The Hopefield deposits represent a long span of time, with elements of several ages represented in the large collection. Klein & Cruz-Uribe (1991) recently suggested that the main fauna was between 0.7–0.4 Myr old, perhaps at the older end of this range. Specimens of *Theropithecus* have been referred to *T. oswaldi leakeyi* by Delson (chapter 5). Among the better-preserved ele-

ments are the following, which are all housed in the South African Museum, Capetown (SAM).

Specimen number	Description
8400	Partial calvaria, reconstructed
16647	Subadult female mandible lacking rami, M_3s erupting
16648	Male mandible lacking rami, Lt C_1, Rt & Lt P_3 (damaged) $-M_3$
16649	Female Lt mandibular corpus, damaged P_3-M_3
16680	Fragments of male mandibular corpus, Rt M_3, Lt P_4-M_3 and loose P_3 and I^2

Swartkrans, Transvaal, South Africa

As reviewed by Delson (1984; and see chapter 5), the fossils from the Swartkrans Formations derive from five Members, numbered from 1 to 5 upwards. The faunally estimated age of Member 1 is 1.9–1.6 Ma, and the great similarity between that large assemblage and the material from Members 2 and 3 suggests that the total time span represented by these three units is less than 0.25 Ma. *Theropithecus* fossils are known from all

three units, and the holotype of '*Simopithecus danieli*' (TMP 563) derives from Member 1 (the 'Hanging Remnant'); all specimens are referred by Delson to *T. oswaldi oswaldi*. Twenty-five specimens derive from the 'Hanging Remnant' and two from the 'Lower Bank' or Orange Breccia (all Member 1); one has been identified from Member 2 and five from Member 3. Only the most complete ones are listed here. The specimens are housed in the Department of Palaeontology, Transvaal Museum, Pretoria (TMP).

Specimen number	Locality (Member)	Description
TMP SK 402/405/563	'Hanging Remnant'	Female partial maxilla, Rt P^4-M^3, and probably associated mandibular corpora and part ramus, C_1-M_3
TMP SK 403/426/567		Juvenile male R mandibular corpus, C_1 (erupting) $-M_2$, Lt corpus M_{1-3}, crushed palate R & L P^3-M^2
TMP SK 411		Female mandibular corpora, all teeth represented
TMP SK 561		Female reconstructed skull, all teeth present
TMP SK 564		Female crushed maxilla, Rt C^1-M^2, Lt P^4-M^1
TMP SK 575		Associated crushed mandibular corpora, Rt & Lt M_{1-3}, and maxillae, Rt P^4-M^2, Lt P^3-M^2
TMP Skx 9579	1	Partial Rt mandibular corpus, P_4-M_3
TMP Skx 38376	1	Isolated Rt P_4
TMP Skx 2996	2	Isolated M_1 (or possibly M_2) in corpus fragment
TMP Skx 27586/87	3	Lt M_3 crown
TMP Skx 28490	3	Lt M_3 crown
TMP Skx 37323	3	Rt M_3
TMP Skx 28812	3	Rt M_2 (or M_1)
TMP Skx 32148	3	Rt $M^{2?}$

Mirzapur, Punjab, India

As reviewed by Delson (chapter 5), the age would appear to lie in the later Early Pleistocene or Middle Pleistocene, between 1.0 and 0.1 Ma. A single specimen is known, identified as *Theropithecus oswaldi delsoni* Gupta & Sahni (holotype). The specimen is housed in the Geology Museum, Panjab University, Chandigarh (PUC-GM).

Specimen number	Description
PUC-GM A/643	Partial Rt maxilla with M^{2-3}

References

BROWN, F., & FEIBEL, C. (1987). 'Robust' hominids and Plio-Pleistocene paleogeography of the Turkana Basin, Kenya and Ethiopia. In *Evolutionary History of the 'Robust' Australopithecines*, ed. F.E. Grine, pp. 325–41. New York: Aldine de Gruyter.

BYE, B.A., BROWN, F., CERLING, T. & McDOUGALL, I. (1987). Increased age estimate for the Lower Paleolithic hominid site at Olorgesailie, Kenya. *Nature*, 329, 237–9.

DELSON, E. (1984). Cercopithecid biochronology of the African Plio-Pleistocene: correlation among eastern and southern hominid-bearing localities. *Courier Forschungs-Institut Senckenberg*, 69, 199–218.

ECK, G.G. (1987). *Theropithecus oswaldi* from the Shungura Formation, Lower Omo Basin, southwestern Ethiopia. In *Les Faunes Plio-Pléistocènes de la Basse Vallée de l'Omo (Éthiopie)*. Tome 3. Cercopithecidae de la Formation de Shungura, ed. Y. Coppens & F.C. Howell, pp. 124–39. Cahiers de Paléontologie, Travaux de Paléontologie Est-Africaine. Paris: Editions du Centre National de la Recherche Scientifique.

ECK, G.G. & JABLONSKI, N.G. (1987). The skull of *Theropithecus brumpti* compared with those of other species of the genus *Theropithecus*. In *Les Faunes Plio-Pléistocènes de la Basse Vallée de l'Omo (Éthiopie)*. Tome 3. Cercopithecidae de la Formation de Shungura, ed. Y. Coppens & F.C. Howell, pp. 11–122. Cahiers de Paléontologie, Travaux de Paléontologie Est-Africaine. Paris: Editions du Centre National de la Recherche Scientifique.

FEIBEL, C.S., BROWN, F.H. & McDOUGALL, I. (1989). Stratigraphic context of fossil hominids from the Omo Group deposits: northern Turkana Basin, Kenya and Ethiopia. *American Journal of Physical Anthropology*, 78, 595–622.

GERAADS, D. (1979). La faune des gisements de Melka-Kunturé (Ethiopie): Artiodactyles, Primates. *Abbay*, 10, 21–49.

GERAADS, D. (1987). Dating the northern African cercopithecid fossil record. *Human Evolution*, 2, 19–27.

HAY, R. (1976). *Geology of the Olduvai Gorge: a study of sedimentation in a semiarid basin*. Berkeley: University of California Press.

DE HEINZELIN, M. (1983). The Omo Group: Archives of the International Omo Research Expedition. *Annales*, S.8, Sciences Géologiques, (85), Musée Royal de l'Afrique Centrale, Tervuren.

HILL, A. & WARD, S. (1988). Origin of the Hominidae: the record of African large hominoid evolution between 14 and 4 My. *Yearbook of Physical Anthropology*, 31, 49–83.

KLEIN, R. & CRUZE-URIBE, K. (1991). The bovids from Elandsfontein, South Africa, and their implications for the age, paleoenvironment and origins of the site. *African Archaeological Review*, 9, 21–79.

NAPIER, P.H. (1981). *Catalogue of Primates in the British Museum (Natural History) and elsewhere in the British Isles. Part II: Family Cercopithecidae, Subfamily Cercopithecinae*. London: British Museum (Natural History).

PLUMMER, T.W. & POTTS, R. (1989). Excavations and new findings at Kanjera, Kenya. *Journal of Human Evolution*, 18, 269–76.

POTTS, R. (1989). Olorgesailie: new excavations and findings in Early and Middle Pleistocene contexts, southern Kenya rift valley. *Journal of Human Evolution*, 18, 477–84.

WOOD, B.A. & VAN NOTEN, F.L. (1986). Preliminary observations on the BK 8518 mandible from Baringo, Kenya. *American Journal of Physical Anthropology*, 69, 117–27.

Appendix II

Conservation status of the gelada

R. I. M. Dunbar

Summary

1. Gelada are confined to an area of highlands lying above 1700 m in the central Ethiopian plateau.
2. Estimates suggest that in the mid-1970s, the total population was in the order of half a million animals. Although in no immediate danger of extinction, the species has recently been added to the IUCN *Red Data Book* because of its uniqueness and limited distribution.
3. Although protected under both Ethiopian law and the international CITES convention, gelada are found in only one conservation area (Simen Mountains National Park).

Introduction

Although only officially collected and described in the 1830s by the explorer-naturalist Ernst Rüppell, the gelada has probably been familiar to the inhabitants of the eastern Mediterranean for well over 2000 years. Jolly & Ucko (1969) have documented its occurrence in travelling menageries in Medieval times in Europe, where it was generally known by its early Greek name *sphingion* ('sphinx monkey'). While these authors dismiss the possibility that the gelada provided the original model for the better known Great Sphinx of Giza (the Egyptian Sphinx), it is likely that its presence in the nearby highlands of Ethiopia was known to the ancient Egyptians.

Indeed, the Greek geographer Agatharcides who travelled extensively in Egypt and the Indian Ocean littoral during the early part of the second century BC gives a very clear description of the gelada (or sphinx, as he calls it) and notes that animals were regularly sent from Ethiopia to the Egyptian port of Alexandria. One and a half centuries later, the Roman historian Pliny gives several accounts of the sphinx-monkeys of the Ethiopian highlands. Jolly & Ucko (1969) summarize the many subsequent references to what can only be the gelada in the geographical accounts and bestiaries of the Roman and Medieval periods.

Distribution

The present-day distribution of the gelada is limited to the steep escarpments and gorges that border the eastern side of the central Ethiopian Amhara highlands between the Tacazze River in the north and the Awash River in the south (Fig. AII.1). There are no records of gelada to the west of the Blue Nile, despite the suitability of the habitat on the river's right-hand bank. The Amhara plateau slopes upwards from southwest to northeast. In contrast to the northern and eastern sectors which terminate in abrupt escarpments that rise up to 1.5 km above the surrounding lowlands, the western and southern sectors give way imperceptibly to lowland. It seems likely that the northern and eastern extension of the species' range is limited by the aridity of the surrounding lowlands. The western sector lacks suitably precipitous habitat, while the southern sector is bordered by the extensive forests of the Guraghe highlands and Kaffa province. Since contemporary gelada rarely enter forested areas even when they have ready access to them, it seems likely that they do not find forested environments especially attractive.

Throughout their distribution on the Ethiopian plateau, the gelada occur only in the immediate vicinity of the extensive cliff lines that border the plateau and the deep gorges that dissect it. Gelada retreat onto the cliffs both to roost at night and for safety from predators and humans during the day. They are rarely found more than half a kilometre from the escarpment rim. Gelada only occur at altitudes above 1700 m, where they range almost to the peak of Ras Dedjen (4500 m). Ethiopia's highest mountain.

The gelada's distribution on the Amhara plateau appears to be discontinuous (Fig. AII.1). Yalden, Largen & Kock (1977) collated information from both published sources and museum records and found that there was a clear disjunction between the distributions of the two subspecies (*Theropithecus gelada gelada* in the north and *T.g. obscurus* in the south) to the east of Lake Tana. There are several references in the early literature (e.g. Neumann, 1902; Stark & Frick, 1959) to reports of gelada populations in the Arussi Mountains to the south of the Amhara plateau, but later authorities have mostly dismissed these reports as unlikely (Osman Hill, 1970; Yalden *et al.*, 1977; Dunbar, 1977). Nonetheless, a specimen of uncertain provenance which differed significantly from both the other two gelada subspecies was described and named as *T.g. senex* by Pucheran (1857).

The possibilty that there might be a third gelada population in the southern mountain ranges has recently been confirmed by Mori & Belay (1990) who observed gelada rather similar in appearance to the description of *T.g. senex* on the northern bank of the Webi Shebelli River near the village of Indeltu (Fig. AII.1). Clearly, further surveys are required to confirm both the subspecific status and the detailed distribution of this new population, but in principle there is no reason why the species should not occur in this area. The habitat lies at a similar altitude to the Amhara plateau and is characterized by steep-sided gorges. The vegetation is similar to that on the Amhara plateau and other species otherwise only known to occur on the Amhara plateau (e.g. the Simen jackal, *Canis simensis*, and the hare, *Lepus starcki*) also occur in the vicinity.

The most plausible explanation for the occurrence of these species in the Arussi and Bale highlands clearly lies in the lower temperatures that occurred during the last glaciation 18–12,000 years BP (see Hurni, 1982). During this period, global temperatures would have been low enough to allow gelada (and presumably the other high altitude species) to colonize the lower-lying Rift Valley (see chapter 18). As temperatures warmed up again in the wake of the retreating ice caps, the populations in these low lying areas would have been forced to move uphill. The Guraghe highlands to the west of the Rift Valley are lower in altitude and more heavily forested than either the Amhara highlands to the north or the Arussi/Bale highlands to the east, and would have been less attractive to gelada than either of these alternatives. The result would have been populations isolated from each other on mountain-tops separated by substantial areas of low-lying habitat that were too warm and arid for the animals to cross.

Conservation status

The gelada is listed in Class A of the 1969 African Convention, which permits it to be hunted, killed, or collected only on government authority, but only providing it is in the national interest or for the purposes of science. It is also listed in Appendix 2 of the 1973 Convention on International Trade in Endangered Species (the so-called CITES or Washington Convention); this permits monitored trade between acceding nations subject to the national regulations of the habitat country.

Within Ethiopia, the gelada has been protected from hunting within conservation areas since the 1960s. However, the Simen Mountains National Park is still the only conservation area where gelada occur. Although they do not often compete with the human population nor raid their crops as other baboons do, the gelada have been subject to

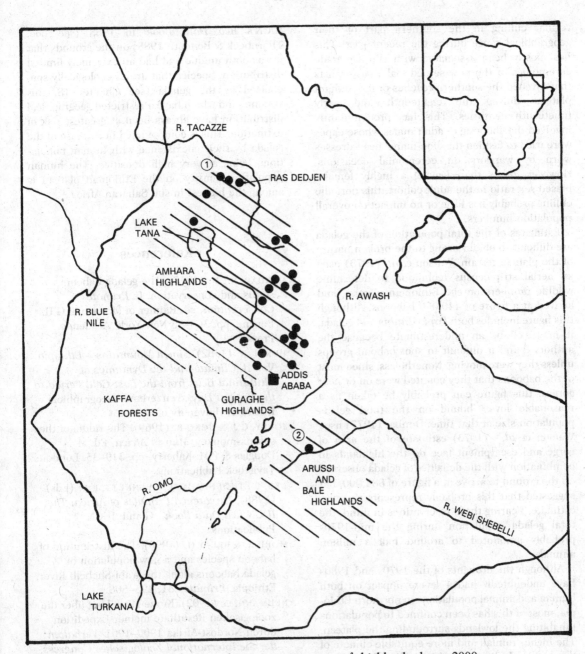

Fig. AII.1. Ethiopia, showing main areas of highland above 2000 m (shaded) and the localities where *gelada* are known to occur (solid circles). 1. Simen Mountains; 2. The new population recently discovered by Mori & Belay (1990) at Indeltu, Arussi highlands.

regular culling in the southern part of their geographical range during the recent past. This has mostly been associated with the age-grade ceremonies of the now settled Galla pastoralists that invaded the southern reaches of the Amhara plateau during the eighteenth and early nineteenth centuries. This has predominantly involved the shooting of adult males whose capes were used to fashion the 'lion-mane' head-dresses worn by warriors on ceremonial occasions. However, other than leaving a highly female-biased sex ratio in the adult cohort, this periodic culling probably has little or no impact on overall population numbers.

Estimates of the total population of the gelada are difficult to obtain owing to the broken nature of the plateau terrain. Watson *et al.* (1973) used an aerial strip-census technique to determine wildlife numbers on the Ethiopian plateau, and arrived at a figure of 440,000 baboons. Although this figure includes both *Papio anubis* and gelada, it may well be an underestimate because the authors found it difficult to spot baboon groups unless they were moving. Nonetheless, since most of the baboons that they counted were on or near gorges, this figure can probably be taken as a reasonable lower bound on the total gelada population size at that time. Dunbar (1977) used Watson *et al.*'s (1973) estimate of the area of gorge and escarpment face on the highlands in combination with the densities of gelada observed on the ground to arrive at a figure of 884,000, but suggested that this probably represents an over-estimate. Bearing these reservations in mind, the total gelada population during the mid-1970s probably amounted to around half a million animals.

Although the droughts of the 1970s and 1980s have undoubtedly had a severe impact on both human and animal populations in northern Ethiopia, most of this has been confined to populations inhabiting the lowlands surrounding the plateau. The higher rainfall and more equitable climate of the highlands has tended to buffer these areas against the worst depredations of the droughts. Thus, it is unlikely that the gelada population has been markedly reduced by the severe droughts of this period.

Despite the relatively large numbers of gelada in Ethiopia, the species was included in the IUCN's *Red Data Book* in 1988 (see Lee, Thornbeck & Bennett, 1988) on the grounds that it was both unique and had an extremely limited distribution. Species that are as ecologically specialized as the gelada (see Chapter 18, this volume) and which have a restricted geographical distribution have always been at greatest risk of extinction. This is compounded in the case of the gelada by their coexistence with human populations living at very high densities. The human population density on the Ethiopian plateau is among the highest in sub-Saharan Africa.

References

DUNBAR, R.I.M. (1977). The gelada baboon: status and conservation. In *Primate Conservation*, ed. Rainier of Monaco & G.H. Bourne, pp. 363–83 New York: Academic Press.

HURNI, H. (1982). *Simen Mountains – Ethiopia.* Vol. II. *Climate and the Dynamics of Altitudinal Belts from the Last Cold Period to the Present Day.* Switzerland: Geographical Institute, University of Berne.

JOLLY, C.J. & UCKO P. (1969). The riddle of the sphinx-monkey. *Man in Africa*, Ed. M. Douglas & P.M. Kaberry, pp. 319–35. London: Tavistock Publications.

LEE, P., THORNBACK, J. & BENNETT, E.L. (Eds) (1988). *Threatened Primates of Africa: The IUCN Red Data Book.* Gland: IUCN Publications.

MORI, A. & BELAY G. (1990). The distribution of baboon species and a new population of gelada baboons along the Wabi-Shebelli River, Ethiopia. *Primates* 31, 495–508.

NEUMANN, O. (1902). Kurze Mitteilung über die zoologischen Resutltate meiner Expedition durch Nordost-Afrika 1900–1901. *Verhalten der 5te International Zoologisches Congress, Berlin 1901*, pp. 201–8.

OSMAN HILL W.C. (1970). *The Primates.* Vol. VIII. *Cynopithecinae.* Edinburgh: Edinburgh University Press.

PUCHERAN, J. (1857). Notices mammalogiques. *Rev. Mag. Zool.* 9, 243.

STARK, D. & FRICK H. (1959). Beobachtungen an

aethiopischen Primaten. *Zoologisches Jahrbuch (Systematisches)*, 86, 41–70.

WATSON, R.M., TIPPETT, C.I., TIPPETT, M.J. & MARRIAN, S.J. (1973). Aerial livestock, land-use and land potential surveys for the central highlands of Ethiopia. Mimeo report, Ethiopian Government Livestock and Meat Board, Addis Ababa.

YALDEN, D.W.M., LARGEN, M.J. & KOCK D. (1977). Catalogue of the mammals of Ethiopia. 3. Primates. *Monitor Zoologico Italiano* N.S. (Suppl. IX), 1, 1–52.

Index